Lecture Notes in Artificial Intelligence 3403

Edited by J. G. Carbonell and J. Sie~~

Subseries of Lecture Notes

Bernhard Ganter   Robert Godin (Eds.)

Formal Concept Analysis

Third International Conference, ICFCA 2005
Lens, France, February 14-18, 2005
Proceedings

 Springer

Series Editors

Jaime G. Carbonell, Carnegie Mellon University, Pittsburgh, PA, USA
Jörg Siekmann, University of Saarland, Saarbrücken, Germany

Volume Editors

Bernhard Ganter
Technische Universität Dresden
Institut für Algebra
01062 Dresden, Germany
E-mail: ganter@math.tu-dresden.de

Robert Godin
UQAM: Université du Québec à Montréal
Département d'informatique
Case postale 8888, succursale Centre-ville, Montreal (Qc), H3C 3P8, Canada
E-mail: godin.robert@uqam.ca

Library of Congress Control Number: 2004118381

CR Subject Classification (1998): I.2, G.2.1-2, F.4.1-2, D.2.4, H.3

ISSN 0302-9743
ISBN 3-540-24525-1 Springer Berlin Heidelberg New York

Springer is a part of Springer Science+Business Media

springeronline.com

© Springer-Verlag Berlin Heidelberg 2005
Printed in Germany

Typesetting: Camera-ready by author, data conversion by Scientific Publishing Services, Chennai, India
Printed on acid-free paper   SPIN: 11384816   06/3142   5 4 3 2 1 0

Preface

This volume contains the Proceedings of ICFCA 2005, the 3rd International Conference on Formal Concept Analysis. The ICFCA conference series aims to be the premier forum for the publication of advances in applied lattice and order theory, and in particular scientific advances related to formal concept analysis.

Formal concept analysis is a field of applied mathematics with its mathematical root in order theory, in particular in the theory of complete lattices. Researchers had long been aware of the fact that these fields have many potential applications. Formal concept analysis emerged in the 1980s from efforts to restructure lattice theory to promote better communication between lattice theorists and potential users of lattice theory. The key theme was the mathematization of *concept* and *conceptual hierarchy*. Since then, the field has developed into a growing research area in its own right with a thriving theoretical community and an increasing number of applications in data and knowledge processing, including data visualization, information retrieval, machine learning, data analysis and knowledge management.

ICFCA 2005 reflected both practical benefits and progress in the foundational theory of formal concept analysis. Algorithmic aspects were discussed as well as efforts to broaden the field. All regular papers appearing in this volume were refereed by at least two, in most cases three independent reviewers. The final decision to accept the papers was arbitrated by the Program Chairs based on the referee reports. It was the involvement of the Program Committee and the Editorial Board that ensured the scientific quality of these proceedings.

The Organizing Chair of the ICFCA 2005 conference, held at the Université d'Artois, Lens, France, was Engelbert Mephu Nguifo. The success of the conference was the result of many hours of tireless planning and work by many volunteers, including the Conference Organization Committee, the Editorial Board and the Program Committee, to whom we convey our sincerest gratitude.

February 2005

Bernhard Ganter
Robert Godin

Organization

Executive Committee

Organizing Conference Chair

Engelbert Mephu Nguifo   IUT de Lens, Université d'Artois, France

Conference Organization Committee

Amedeo Napoli   LORIA, CNRS INRIA, Nancy, France
Huaiguo Fu   CRIL, Université d'Artois, Lens, France
Hugues Delalin   CRIL, Université d'Artois, Lens, France
Olivier Couturier   CRIL, Université d'Artois, Lens, France

Program and Conference Proceedings

Program Chairs

Bernhard Ganter   Technische Universität Dresden, Germany
Robert Godin   Université du Québec à Montréal, Canada

Editorial Board

Peter Eklund   University of Wollongong, Australia
Sergei Kuznetsov   VINITI and RSUH Moscow, Russia
Uta Priss   Napier University, Edinburgh, United Kingdom
Gregor Snelting   University of Passau, Germany
Gerd Stumme   University of Kassel, Germany
Rudolf Wille   Technische Universität Darmstadt, Germany
Karl Erich Wolff   University of Applied Sciences, Darmstadt, Germany

Program Committee

Claudio Carpineto   Fondazione Ugo Bordoni, Rome, Italy
Richard J. Cole   University of Queensland, Brisbane, Australia
Paul Compton   University of New South Wales, Sydney, Australia
Frithjof Dau   Technische Universität Darmstadt, Germany
Brian Davey   La Trobe University, Melbourne, Australia
Vincent Duquenne   Université Pierre et Marie Curie, Paris, France
Ralph Freese   University of Hawaii, Honolulu, USA

Wolfgang Hesse	Universität Marburg, Germany
Léonard Kwuida	Technische Universität Dresden, Germany
Wilfried Lex	Universität Clausthal, Germany
Rokia Missaoui	Université du Québec en Outaouais (UQO), Gatineau, Canada
Lhouari Nourine	LIMOS, Université de Clermont Ferrand, France
Sergei Obiedkov	RSUH Moscow, Russia
Alex Pogel	New Mexico State University, Las Cruces, USA
Sandor Radeleczki	University of Miskolc, Hungary
Stefan Schmidt	New Mexico State University, Las Cruces, USA
Tamas Schmidt	Budapest University of Technology and Economics, Hungary
Bernd Schröder	Louisiana Tech University, Ruston, USA
Selma Strahringer	University of Applied Sciences, Cologne, Germany
Petko Valtchev	DIRO, Université de Montréal, Canada
GQ Zhang	Case Western University, Cleveland, Ohio, USA

Sponsoring Institutions

Le Centre de Recherche en Informatique de Lens
L'Institut Universitaire de Technologie de Lens
L'Université d'Artois
La Ville de Lens
La CommunAupole de Lens-Liévin (CALL)
Le Conseil Général du Pas-de-Calais
La Chambre de Commerce et d'Industrie de Lens
La Caisse d'Epargne du Pas-de-Calais
La Banque Populaire

Table of Contents

Towards Generic Pattern Mining*

Mohammed J. Zaki**, Nagender Parimi, Nilanjana De, Feng Gao,
Benjarath Phoophakdee, Joe Urban, Vineet Chaoji,
Mohammad Al Hasan, and Saeed Salem

Computer Science Department,
Rensselaer Polytechnic Institute, Troy NY 12180

Abstract. Frequent Pattern Mining (FPM) is a very powerful paradigm
for mining informative and useful patterns in massive, complex datasets.
In this paper we propose the Data Mining Template Library, a collection
of generic containers and algorithms for FPM, as well as persistency and
database management classes. DMTL provides a systematic solution to
a whole class of common FPM tasks like itemset, sequence, tree and
graph mining. DMTL is extensible, scalable, and high-performance for
rapid response on massive datasets. Our experiments show that DMTL
is competitive with special purpose algorithms designed for a particular
pattern type, especially as database sizes increase.

1   Introduction

Frequent Pattern Mining (FPM) is a very powerful paradigm which encompasses an entire class of data mining tasks. The specific tasks encompassed
by FPM include the mining of increasingly complex and informative patterns,
in complex structured and unstructured relational datasets, such as: Itemsets
or co-occurrences [1] (transactional, unordered data), Sequences [2, 29] (temporal or positional data, as in text mining, bioinformatics), Tree patterns [30, 3]
(XML/semistructured data), and Graph patterns [12, 16, 26, 27] (complex relational data, bioinformatics). Figure 1 shows examples of these different types of
patterns; in a generic sense a pattern denotes links/relationships between several objects of interest. The objects are denoted as nodes, and the links as edges.
Patterns can have multiple labels, denoting various attributes, on both the nodes
and edges.

The current practice in frequent pattern mining basically falls into the
paradigm of incremental algorithm improvement and solutions to very specific
problems. While there exist tools like MLC++ [15], which provides a collection of algorithms for classification, and Weka [25], which is a general purpose

---

* This work was supported by NSF Grant EIA-0103708 under the KD-D program,
  NSF CAREER Award IIS-0092978, and DOE Early Career PI Award DE-FG02-
  02ER25538.
** We thank Paolo Palmerini and Jeevan Pathuri for their work on an early version of
  DMTL.

B. Ganter and R. Godin (Eds.): ICFCA 2005, LNAI 3403, pp. 1–20, 2005.

Java library of different data mining algorithms including itemset mining, these systems do not have an unifying theme or framework, there is little database support, and scalability to massive datasets is questionable. Moreover, these tools are not designed for handling complex pattern types like trees and graphs.

Our work seeks to address all of the above limitations. In this paper we describe Data Mining Template Library (DMTL), a generic collection of algorithms and persistent data structures, which follow a generic programming paradigm[4]. DMTL provides a systematic solution for the whole class of pattern mining tasks in massive, relational datasets. The main contributions of DMTL are as follows:

- Isolation of generic containers which hold various pattern types from the actual mining algorithms which operate upon them. We define generic data structures to handle various pattern types like itemsets, sequences, trees and graphs, and outline the design and implementation of generic data mining algorithms for FPM, such as depth-first and breadth-first search.
- Persistent data structures for supporting efficient pattern frequency computations using a tightly coupled database (DBMS) approach.
- Native support for both vertical and horizontal database formats for highly efficient mining.
- Developing the motivation to look for unifying themes such as right-most pattern extension and depth-first search in FPM algorithms. We believe this shall facilitate the design of a single generic algorithm applicable across a wide spectrum of patterns.

One of the main attractions of a generic paradigm is that the generic algorithms for mining are guaranteed to work for **any** pattern type. Each pattern is characterized by inherent properties that it satisfies, and the generic algorithm exploits these properties to perform the mining task efficiently. We conduct several experiments to show the scalability and efficiency of DMTL for different pattern types like itemsets, sequences, trees and graphs. Our results indicate that DMTL is competitive with the special purpose algorithms designed for a particular pattern type, especially with increasing database sizes.

2 Preliminaries

The problem of mining frequent patterns can be stated as follows: Let $\mathcal{N} = \{x_1, x_2, \ldots, x_{n_v}\}$ be a set of n_v distinct nodes or vertices. A pair of nodes (x_i, x_j) is called en edge. Let $\mathcal{L} = \{l_1, l_2, \ldots, l_{n_l}\}$, be a set of n_l distinct labels. Let $L_n : \mathcal{N} \to \mathcal{L}$, be a node labeling function that maps a node to its label $L_n(x_i) = l_i$, and let $L_e : \mathcal{N} \times \mathcal{N} \to \mathcal{L}$ be an edge labeling function, that maps an edge to its label $L_e(x_i, x_j) = l_k$.

It is intuitive to represent a *pattern* P as a graph (P_V, P_E), with labeled vertex set $P_V \subseteq \mathcal{N}$ and labeled edge set $P_E = \{(x_i, x_j) \mid x_i, x_j \in P_V\}$. The number of nodes in a pattern P is called its *size*. A pattern of size k is called a k-pattern, and the class of frequent k-patterns is referred to as F_k. In some applications P is a symmetric relation, i.e., $(x_i, x_j) \equiv (x_j, x_i)$ (undirected edges), while in other applications P is anti-symmetric, i.e., $(x_i, x_j) \not\equiv (x_j, x_i)$ (directed edges). A path in P is a set of distinct nodes $\{x_{i_0}, x_{i_1}, x_{i_n}\}$, such that $(x_{i_j}, x_{i_{j+1}})$ in an edge in P_E for all $j = 0 \cdots n - 1$. The number of edges gives the length of the path. If x_i and x_j are connected by a path of length n we denote it as $x_i <_n x_j$. Thus the edge (x_i, x_j) can also be written as $x_i <_0 x_j$.

Given two patterns P and Q, we say that P is a *subpattern* of Q (or Q is a *super-pattern* of P), denoted $P \preceq Q$ if and only if there exists a 1-1 mapping f from nodes in P to nodes in Q, such that for all $x_i, x_j \in P_V$: i) $L_n(x_i) = L_n(f(x_i))$, ii) $L_e(x_i, x_j) = L_e(f(x_i), f(x_j))$, and iii) $(x_i, x_j) \in P_V$ iff (if and only if) $(f(x_i), f(x_j)) \in Q_V$. In some cases we are interested in embedded subpatterns. P is an *embedded subpattern* of Q if: i) $L_n(x_i) = L_n(f(x_i))$, iii) $L_e(x_i, x_j) = L_e(f(x_i), f(X_j))$, and iii) $(x_i, x_j) \in P_E$ iff $f(x_i) <_l f(x_j)$ for some $l \geq 0$, i.e., $f(x_i)$ is connected to $f(x_j)$ on some path. If $P \preceq Q$ we say that P is contained in Q or Q contains P.

A database $\mathcal{D}$ is just a collection (a multi-set) of patterns. A database pattern is also called an *object*. Let $\mathcal{O} = \{o_1, o_2, \ldots, o_{n_o}\}$, be a set of n_o distinct *object identifiers (oid)*. An object has a unique identifier, given by the function $O(d_i) = o_j$, where $d_i \in \mathcal{D}$ and $o_j \in \mathcal{O}$. The number of objects in $\mathcal{D}$ is given as $|\mathcal{D}|$.

The *absolute support* of a pattern P in a database $\mathcal{D}$ is defined as the number of objects in $\mathcal{D}$ that contain P, given as $\pi^a(P, \mathcal{D}) = |\{P \preceq d \mid d \in \mathcal{D}\}|$. The *(relative) support* of P is given as $\pi(P, \mathcal{D}) = \frac{\pi^a(P, \mathcal{D})}{|D|}$. A pattern is *frequent* if its support is more than some user-specified minimum threshold, i.e., if $\pi(P, \mathcal{D}) \geq \pi^{\min}$. A frequent pattern is *maximal* if it is not a subpattern of any other frequent pattern. A frequent pattern is *closed* if it has no super-pattern with the same support. The frequent pattern mining problem is to enumerate all the patterns that satisfy the user-specified $\pi^{\min}$ frequency requirement (and any other user-specified conditions).

The main observation in FPM is that the sub-pattern relation $\preceq$ defines a partial order on the set of patterns. If $P \preceq Q$, we say that P is more general than Q, or Q is more specific than P. The second observation used is that if Q is a frequent pattern, then all sub-patterns $P \preceq Q$ are also frequent. More important is the converse, i.e. if P is infrequent and $P \preceq Q$ then Q shall also be infrequent (follows from the anti-monotonicity of frequency). The different FPM algorithms differ in the manner in with they search the pattern space.

2.1 FPM Instances

Some common types of patterns include itemsets, sequences, trees, and graphs, as shown in Figure 1. In fact, every pattern can be modeled as a graph; the nodes (x_i) are shown under each circle and the node labels ($L_n(x_i)$) are shown inside the circle, whereas edge labels have been omitted.

In an itemset [1] no two nodes have the same label. Let $V = \{x_1, x_2, \cdots x_k\}$ be a node set such that $L_n(x_i) \neq L_n(x_j)$ for all $x_i, x_j \in V$, and $L_n(x_i) < L_n(x_{i+1})$ for all $1 \leq i \leq k - 1$. There are several possible formulation of the itemset pattern: i) *vertex-only*: An itemset pattern P is just a of vertices, i.e., $P_V = V$ and $P_E = \emptyset$, this is shown in Figure 1, ii) *linear*: in another formulation the

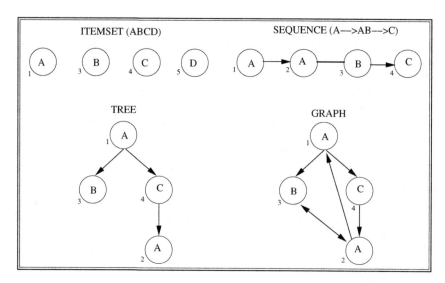

Fig. 1. FPM Instances

itemset is defined as $P_V = V$, and $P_E = \{(x_i, x_{i+1})|x_i, x_{i+1} \in P_V\}$, iii) *clique*: A third alternative is to represent itemset P as a clique, i.e., $P_V = V$ and $P_E = \{(x_i, x_j) \mid i < j \text{ and } x_i, x_j \in P_V\}$.

In sequence mining [2], a sequence is modeled as an ordered list of itemsets, and thus the different nodes in a sequence can have the same label. We can model a sequence pattern P as being made up of a sequence of n itemsets P^i, $i = 1, \cdots n$, using the linear formulation (as shown in Figure 1); note that using the vertex-only formulation is problematic, since it results in a disconnected pattern. Thus P has a vertex set made up of n disjoint subsets $P_V = \bigcup_{i=1}^{n} P_V^i$. The edge set P_E contains all the edges within P^i (consecutive and undirected), and it also contains a directed edge for every pair of consecutive itemsets, i.e., from the last node of P^i to the first node of P^{i+1}.

In tree mining [30, 3], typically rooted, ordered and labeled trees are considered. Thus a tree pattern P consists of the vertex set $P_V = \{r, x_1, x_2, \cdots\}$, where r is a special node called root. A tree pattern must satisfy all tree properties, namely i) the root has no parent, i.e., $(x_i, r) \notin P_E$ for any $x_i \in P_V$, ii) the edges are directed, i.e., if $(x_i, x_j) \in P_E$, then $(x_j, x_i) \notin P_E$), iii) a node has only one parent, i.e., if $(x_i, x_j) \in P_E$, then $(x_k, x_j) \notin P_E$ for any $x_k \neq x_i$, iv) the tree is connected, i.e., for all $x_i \in P_V$, there exists a path from the root r to x_i, and v) tree has no cycles. Furthermore for ordered trees the order of a nodes' children matters. This means that there is an ordering of edges in P_E, such that (x_i, x_j) comes before (x_i, x_k) in P_E only if x_j is before x_k in the ordering of x_i's children. Embedded trees can be defined by following the definition of embedded patterns introduced earlier.

Finally, by definition a pattern can model any general graph, as well as any special constraints that might appear in graph mining [12, 16, 26], such as connected graphs, or induced subgraphs. It is also possible to model other patterns such as DAGs (directed acyclic graphs). DMTL currently supports pattern mining of i) itemsets, ii) sequences, iii) embedded, rooted trees with ordered edges and iv) induced, undirected graphs with no single loops or multiple edges. As we shall soon see, the toolkit can be extended to incorporate mining of other user defined patterns as well.

2.2   Database Format

In a typical FPM task, the database is in the *horizontal* format i.e. a set of transactions, where each transaction is an object of the pattern type being mined [1]. Recently, *vertical* database formats have been proposed for mining itemsets, sequences and trees [28, 29, 30]. The vertical format is the more attractive alternative since it enables fast computation of supports by avoiding repeated database accesses. It does so by associating an entity called *Vertical Attribute Table*, VAT with each pattern. For an itemset, the VAT is the list of tids in which it is contained; VATs for sequences and trees are more complex and are described later. There currently does not exist a vertical scheme for graphs; the introduction of a new and efficient VAT scheme for graphs is one of our main contributions. DMTL introduces two modes of persistency: i) the collection of frequent patterns itself may be too large to fit in main memory, and hence persistent containers are provided to hold them, and ii) persistent storage and access to VATs. Both these modes of persistency are entirely transparent to the user.

3   DMTL: Data Structures and Algorithms

The C++ Standard Template Library (STL) provides efficient, generic implementations of widely used algorithms and data structures, which tremendously aid effective programming. Like STL, DMTL is a collection of generic data mining algorithms and data structures. In addition, DMTL provides persistent data and index structures for efficiently mining any type of pattern or model of interest. The user can mine custom pattern types, by simply defining the new pattern types, but the user need not implement a new algorithm - the generic DMTL algorithms can be used to mine them. Since the mined models and patterns are persistent and indexed, this means the mining can be done efficiently over massive databases, and mined results can be retrieved later from the persistent store.

Following the ideology of generic programming, DMTL provides a standardized, general, and efficient implementation of frequent pattern mining tasks by isolating the concept of data structures or containers, as they are called in generic programming, from algorithms. DMTL provides container classes for representing different patterns (such as itemsets and sequences) and collection of patterns, containers for database objects (horizontal and vertical), and containers for temporary mining results. These container classes support persistency when required.

Generic algorithms, on the other hand are independent of the container and can be applied on any valid container. These include algorithms for performing intersections of the vertical lists [28, 29, 30] for itemsets, sequences or other patterns. Generic algorithms are also provided for mining itemsets, sequences and trees [1, 20, 28, 29], as well as for finding the maximal or closed patterns [11, 31]. Finally DMTL provides support for the database management functionality, pre-processing support for mapping data in different formats to DMTL's native formats, as well as for data transformation (such as discretization of continuous values). It should be noted that some of the algorithms designed for the C++ STL were inherently generic i.e. independent of the underlying datatype or container (e.g. sort). However devising a generic algorithm for FPM was a significant design challenge; we present it in Figure 3.

In this section we focus on the containers and algorithms for mining. In later sections we discuss the database support in DMTL as well as support for pre-processing and post-processing.

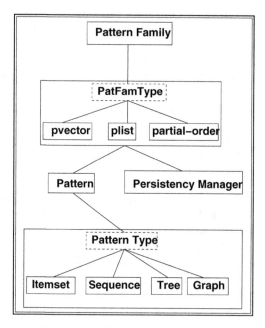

Fig. 2. DMTL Container Hierarchy

3.1   Containers

Figure 2 shows the different DMTL container classes for PMT (the Pattern Mining Toolkit) and the relationship among them. At the lowest level are the different kinds of pattern-types one might be interested in mining. A pattern is a generic container instantiated for one of the pattern-types. There are several pattern family types (such as pvector, plist, etc.) which together with a persistency manager class make up different pattern family classes. More details on each class appears below.

Pattern. In DMTL a pattern is a generic container, which can be instantiated as an itemset, sequence, tree or a graph, specified as `Pattern<class P>` by means of a template argument called Pattern-Type (P). A generic pattern is simply a Pattern-Type whose frequency we need to determine in a larger collection or database of patterns of the same type.

Pattern-Type. A pattern type is the specific pattern to be mined, e.g. itemset, and in that sense is not a generic container. DMTL has the itemset, sequence, tree and graph pattern-types defined internally; however the users are free to define their own pattern types, so long as the user defined class provides implementations for the methods required by the generic containers and algorithms. We shall later show how a new pattern type may be added to the library.

Pattern Family. In addition to the basic pattern classes, most pattern mining algorithms operate on a collection of patterns. The pattern family is a generic container `PatternFamily <class PatFamType>` to store groups of patterns, specified by the template parameter `PatFamType`. `PatFamType` represents a persistent class provided by DMTL, that provides seamless access to the members, whether they be in memory or on disk.

Pattern Family Type. This class provides the required persistency in storage and retrieval of patterns. DMTL provides several pattern family types to store groups of patterns. Each such class is templatized on the pattern-type (P) and a persistency manager class `PM`. An example is `pvector <class P, class PM>`, a persistent vector class. It has the same semantics as a STL vector with added memory management and persistency. Another class is `plist<P,PM>`. Instead of organizing the patterns in a linear structure like a vector or list, another persistent family type DMTL class, `partial-order <P,PM>`, organizes the patterns according to the sub-pattern/super-pattern relationship. While pvector and partial-order provide the same interface, certain operations will be more efficient in one class than the other. For example, inserts and deletions are cheaper for `plists`, while the maximality and closed testing functions will be cheaper for partial-orders, since the patterns are already organized according to sub/super-pattern relation.

3.2   Persistency Manager for Patterns

An important aspect of DMTL is to provide a user-specified level of persistency for all DMTL classes. To support large-scale data mining, DMTL provides automatic support for out-of-core computations, i.e., memory buffer management, via the persistency manager class PM. The PatternFamilyType class uses the persistency manager (PM) to support the buffer management for patterns. The details of implementation are hidden from PatternFamily; all generic algorithms continue to work regardless of whether the family is (partially) in memory or on disk. The interface of a persistent container (like pvector) is similar to that of a volatile container (like STL vector), hence encapsulating the implementation behind the common interface. More details on the persistency manager will be given later.

3.3   Generic Mining Algorithms

The pattern mining task can be viewed as a search over the pattern space looking for those patterns that match the minimum support constraint. For instance in itemset mining, the search space is the set of all possible subsets of items. Within DMTL we attempt to provide a unifying framework for the wide range of mining algorithms that exist today. Figure 3 shows the pseudo-code for the generic mining algorithm, which was devised by combining the unifying aspects of mining itemsets, sequences, trees and graphs [28, 29, 30, 26]. Note that mining F_2 (i.e., level-2) often creates performance and memory bottleneck in FPM tasks, hence we employ a preemptive horizontal scan to accumulate estimated supports of level-2 patterns (line 3). This is an optimization intended for level-2 only, and we use the vertical approach thereon. The *extend* routine outlines the important tasks for mining any pattern: i) systematic candidate generation (line 8), ii) isomorphism checking (line 9) and iii) support counting which we accomplish through the vertical approach (lines 10-11). Partitioning frequent patterns into equivalence classes leads to a $F_k \times F_k$ candidate generation i.e. an F_{k+1} candidate is generated by *joining* two F_k sized patterns. It should also be noted that for graphs $g \in F_k$ implies g has k edges (not k nodes). Some of the salient features of our algorithm's design are:

Search Strategy. Several variants exist, depth-first search (DFS) and breadth-first search (BFS) being the primary ones. BFS has the advantage of providing better pruning of candidates but suffers from the cost of storing all of a given level's frequent patterns in memory. Recent algorithms for mining complex patterns like trees and graphs have focused on the DFS approach, hence it is the preferred choice for our toolkit as well. Nevertheless, support for BFS mining of itemsets and sequences is provided.

Vertical Mining. It has been shown that efficient vertical mining typically outperforms the horizontal approaches [28, 29, 30]. The vertical approach accomplishes fast support counting by intersection of VATs, thereby avoiding repeated database accesses. Section 4 gives details of the support we provide for vertical as well as horizontal mining.

Right-Most Extension. Recent algorithms towards solving tree and graph mining [30, 26] have focused on an approach of *right-most extension* i.e. a new node is added to the pattern only on the right most path from the root. This method has been shown to exhaustively enumerate all candidates for trees and graphs, and we believe that it can be augmented to work for itemsets and sequences as well. Though in the current framework the extension strategy is an internal component of each pattern's specialized routine, part of the proposed future work is devising a completely generic pattern mining algorithm, leveraging aspects such as right most extension and depth-first search which are common across a wide range of patterns. We believe that developing the motivation to look for such unifying themes in pattern mining is one of the key contributions of this toolkit.

DMTL provides generic algorithms encapsulating these search strategies; by their definition these algorithms can work on any type of pattern: Itemset, Sequence, Tree or Graph. An example is the generic algorithm DFS-Mine<class PatFamType> (PatternFamily<PatFamType> &pf, DB &db, ...), which mines the frequent patterns using a depth-first search (DFS) [28, 29]. The DFS algorithm in turn relies on other generic subroutines for creating equivalence classes, for generating candidates, and for support counting. There is also a generic BFS-Mine that performs Breadth-First Search [1, 20] over the pattern space.

---

dfs_mine (DB,result_pats):
1. $F_1 = \{$level-1 frequent patterns$\}$
2. result_pats = result_pats $\cup F_1$
3. $F_2 = \{$optimized mining of level-2 patterns$\}$
4. result_pats = result_pats $\cup F_2$
5. $F_2 = \{$partition F_2 into equivalence classes$\}$
6. **for** each equivalence class $[P]_1$ in F_2 **do**
7.   extend(DB, result_pats, $[P]_1$)

extend (DB, result_pats, $[P]$):
  //DFS, equivalence class-based extension
6. $F_{k+1} = \emptyset$
7. $\forall$ patterns $P_i, P_j \in [P]$ such that $i \neq j$
8.   new_pat = $P_i \odot P_j$ //generate new candidate
9.   **if** new_pat.canonical_code is minimal **then**
      //candidate has passed isomorphism test
10.   new_pat.vat = $P_i.vat \otimes P_j.vat$ //vat intersection
11.   **if** |new_pat.vat| $\geq$ minsup **then** //new_pat is frequent
12.   result_pats = results_pat $\cup$ new_pat
13. $F_{k+1} = F_{k+1} \cup$ new_pat
14. $F_{k+1} = \{$partition F_{k+1} into equivalence classes$\}$
15.   **for** each equivalence class $[P]_k$ in F_{k+1} **do**
16.   extend(DB, result_pats, $[P]_k$)

Fig. 3. Generic DFS Pattern Mining

Figure 3 seeks to illustrate the major steps of **DFS-Mine**, our equivalence class-based vertical mining algorithm. The toolkit employs templates to provide for efficient compile time polymorphism based on the pattern type: the underlying algorithm stays the same but each distinct pattern has its specialized implementation of the key steps. For instance, the isomorphism check in line 9 is necessary only for graphs, and is omitted for other simpler patterns. Isomorphism checking is achieved through the *canonical_code* member of each pattern. Each graph has a canonical code representation, and an ordering is defined on the code such that among all isomorphic graphs only one has the the least canonical code; all other graphs shall be discarded at line 9. DMTL applies the *DFS minimal code* of gSpan [26] but is not constrained by the choice of the canonical code. It is also to be noted that the equivalence class partitioning is omitted for graphs since $F_k \times F_1$ candidate generation does not lend itself easily to equivalence partitions.

3.4   Candidate Generation

We now provide a brief review of our extension routine ($\odot$) for the four primary pattern types, details of the VAT intersection follow later.

Itemset: Itemset join is the simplest and DMTL employs a vertical mining approach based on [28]. The join operation is defined on two itemsets Px and Py, belonging to the same equivalence class, $[P]$, which yields $Pxy \in [P_x]$.

Sequence: An equivalence class of sequences can comprise members which are sequence atoms $(P \rightarrow X)$ or event atoms (PY). As described in [29], a join of two sequences within the same equivalence class $[P]$ can lead to one of three possibilities – i) joining PB with PD yields PBD (join of two event atoms); ii) join of PB with $P \rightarrow A$ results in $PB \rightarrow A$ (join of event atom with sequence atom) and iii) join of two sequence atoms, $P \rightarrow A$ with $P \rightarrow F$ leads to three outcomes: an event atom $P \rightarrow AF$ and two sequence atoms, $P \rightarrow A \rightarrow F$ and $P \rightarrow F \rightarrow A$.

Tree: An equivalence class of trees comprises members which share the common prefix, but differing in the last node of the tree and the position where it is attached to the prefix. Hence members of the same equivalence class $[P]$ may be denoted as pairs of $(last\_node, position)$. A join of (x, i) with (y, j) leads to the following possibilities: i) if $i = j$ add (y, j) and $(y, n_i))$ to $[P_x]$, where n_i is the depth-first number of node x; ii) if $i > j$ the new candidate is (y, j) in class $[P_x]$; and iii) no candidates are possible when $i < j$. We refer the reader to [30] for elaboration on the prefix based representation scheme used for trees.

Graph: To assist in systematic candidate generation and isomorphism testing, DMTL uses the ordering of vertex and edge labels to generate graphs from a core tree structure [26]. An $F_k \times F_1$ join on graphs is a complex operation; at each such extension a new edge is added to the given graph. Two types of edge extensions are defined: a back edge which introduces a cycle, and a forward edge which adds a new node to the graph. See [26] for more details.

3.5   Isomorphism Checking

Since a graph encompasses other simpler patterns (itemset, sequence, tree) we define the isomorphism problem for graphs: a graph p is isomorphic to q if there exists a mapping $M : p_v \rightarrow q_v$ such that for all $x_i, x_j \in p$, $L_p(x_i) = L_q(M(x_i))$

and $(x_i, x_j) \in p_e$ iff $(M(x_i), M(x_j)) \in q_e$. It has been shown that for itemsets, sequences and ordered trees the isomorphism checking may be averted by intelligent candidate generation, e.g., for the case of itemsets, AB and BA are isomorphic, but the algorithm can avoid generating BA by joining an itemset P_i only with a lexicographically greater itemset P_j (where both belong to the equivalence class $[P]$). Such schemes exist for sequences and ordered trees as well, but more complex patterns like unordered trees, free trees, directed acyclic graphs (DAGs) and generic graphs shall require some form of isomorphism testing.

Isomorphism Checking in Graphs: We follow the scheme outlined in [26] to achieve isomorphism checking for graphs. Based on a linear order on vertex and edge labels, a unique depth-first traversal is defined for any given graph. Each vertex in the graph is assigned a depth-first id, which is its order in the depth-first traversal. Each edge is represented by a 5-tuple (i, j, l_i, l_{ij}, l_j) where i is the DFS id of the first vertex of the edge and j of the second one, and l_i, l_{ij} and l_j are labels of the first vertex, the edge and second vertex respectively. Isomorphism checking is accomplished by defining an order on such 5-tuples.

4   DMTL: Persistency and Database Support

DMTL employs a back-end storage manager that provides the persistency and indexing support for both the patterns and the database. It supports DMTL by seamlessly providing support for memory management, data layout, high-performance I/O, as well as tight integration with database management systems (DBMS). It supports multiple back-end storage schemes including flat files, embedded databases, and relational or object-relational DBMS. DMTL also provides persistent pattern management facilities, i.e., mined patterns can themselves be stored in a pattern database for retrieval and interactive exploration.

DMTL provides native database support for both the horizontal [1] and vertical [28, 29, 30] data formats. It is also worth noting that since in many cases the database contains the same kind of objects as the patterns to be extracted (i.e., the database can be viewed as a pattern family), the same database functionality used for horizontal format can be used for providing persistency for pattern families. It is relatively straightforward to store a horizontal format object, and by extension, a family of such patterns, in any object-relational database. Thus the persistency manager for pattern families can handle both the original database and the patterns that are generated while mining. DMTL provides the required buffer management so that the algorithms continue to work regardless of whether the database/patterns are in memory or on disk.

4.1   Vertical Attribute Tables

To provide native database support for objects in the vertical format, DMTL adopts a fine grained data model, where records are stored as *Vertical Attribute Tables* (VATs). Given a database of objects, where each object is characterized by a set of properties or attributes, a VAT is essentially the collection of objects that share the same values for the attributes. For example, for a relational table, `cars`, with the two attributes, `color` and `brand`, a VAT for the property `color=red` stores all the transaction identifiers of cars whose color is red. The main advantage of VATs is that they allow for optimizations of query intensive applications like data mining where only a subset of the attributes need to be processed during each query. As was mentioned earlier these kinds of vertical representations have proved to be useful in many data mining tasks [28, 29, 30].

In DMTL there is one VAT per pattern-type. Depending on the pattern type being mined the vat-type class may be different. Accordingly, their intersection (line 10, Figure 3) shall vary as well:

Itemset. For an itemset the VAT is simply a `vector <tid>`, where each tid may be stored as an `int`. VAT intersection in this case is straight forward, $new\_pat.vat = \{t | t \in P_i.vat$ and $t \in P_j.vat\}$, where $new\_pat = P_i \odot P_j$.

Sequence. The VAT for a sequence is defined as a `vector<pair<tid, vector <time-stamp>>>`. In this case the intersection has to take into account the type of extension under consideration (refer to the section on sequence extension). The intersection operation is a simple intersection of tid-lists for a join of two event atoms, but requires comparison of the timestamps when doing sequence joins. For instance, when computing the VAT intersection for $P \rightarrow A \rightarrow F$ from its subsequences $P \rightarrow A$ and $P \rightarrow F$, one needs to match the tid *and* ensure that the time-stamp of A in that tid is less than that of F.

Tree. Define `triple` to be (`tid`, `scope`, `match-label`), then the VAT for a tree pattern is a `vector<triple>`. The `tid` identifies a tree in the input database; `scope` is an interval [`l,u`] which denotes the range of DFS ids which lie embedded under the last depth-first node of the tree, and `match-label` is a list of DFS positions at which the current tree is embedded in that tree of the database. Intersection of tree VATs is an involved operation, comprising in-scope and out-scope tests corresponding to the two types of tree extensions described earlier [30].

Graph. The VAT for a graph is defined as a `vector<edge_vat>` where an `edge_vat` is defined as `vector<tid, vids>` where `vids` is a `vector <pair<int, int>>`. A graph may be viewed as a collection of edges; following this approach an `edge_vat` is in essence the VAT for an edge of a graph. It stores the tid of the graph in which the edge is present, and a collection of pair of vertex ids – each pair denoting an occurrence of the edge in that graph. Intersection of graph VATs is complicated due to isomorphism checking, and the details are beyond the scope of this paper.

DMTL provides support for creating VATs during the mining process, i.e., during algorithms execution, as well as support for updating VATs (add and delete operations). In DMTL VATs can be either persistent or non-persistent. Finally DMTL uses indexes for a collection of VATs for efficient retrieval based on a given attribute-value, or a given pattern.

4.2   Storage and Persistency Manager

The database support for VATs and for the horizontal family of patterns is provided by DMTL in terms of the following classes, which are illustrated in Figure 4. Vat-type is a class describing the vat-type that composes the body of a VAT, for instance int for itemsets and pair<int,time> for sequences. VAT<class V> is the class that represents VATs. This class is composed of a collection of records of vat-type V. Storage<class PM> is the generic persistency-manager class that implements the physical persistency for VATs and other classes. The class PM provides the actual implementations of the generic operations required by Storage. For example, PM_metakit and PM_gigabase are two actual implementations of the Storage class in terms of different DBMS like Metakit [24], a persistent C++ library that natively supports the vertical format, and Gigabase [14], an object-relational database. Other implementations can easily be added as long as they provide the required functionality. MetaTable<class V, class PM> represents a collection of VATs. It stores a list of VAT pointers and the adequate data structures to handle efficient search for a specific VAT in the collection. It also provides physical storage for VATs. It is templatized on the vat-type V and on the Storage implementation PM. In the figure the H refers to a pattern and B its corresponding VAT. The Storage class provides for efficient lookup of a particular VAT object given the header. DB<class V, class PM> is the database class which holds a collection of Metatables. This is the main user interface to VATs and constitutes the database class DB referred to in previous sections. It supports VAT operations such as intersection, as well as

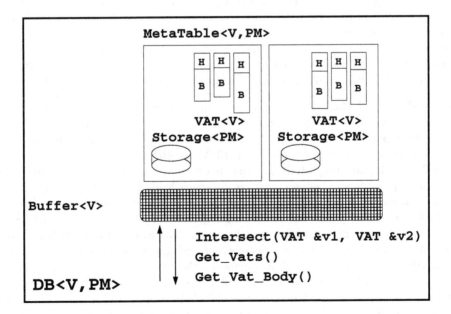

Fig. 4. DMTL: High level overview of the different classes used for Persistency

the operations for data import and export. The DB class is a doubly templated class where both the vat-type and the storage implementation need to be specified. An example instantiation of a DB class for itemset patterns would therefore be DB<int,PM_metakit> or DB<int, PM_gigabase>. DB has as data members an object of type Buffer<V> and a collection of MetaTables<V,PM>. Buffer<class V> provides a fixed-size main-memory buffer to which VATs are written and from which VATs are accessed, used for buffer management to provide seamless support for main-memory and out-of-core VATs (of type V). When a VAT body is requested from the DB class, the buffer is searched first. If the body is not already present there, it is retrieved from disk, by accessing the Metatable containing the requested VAT. If there is not enough space to store the new VAT in the buffer, the buffer manager will (transparently) replace an existing VAT with the new one. A similar interface is used to provide access to patterns in a persistent family or the horizontal database.

5   Extensibility of DMTL

DMTL provides a highly extensible yet potent framework for frequent pattern mining. We provide this flexibility via two central distinctions built into the library by design.

Containers and Algorithms. DMTL makes a clear delineation between patterns and the containers used to store them, and the underlying mining algorithm. This enables us to introduce the concept of a generic pattern mining algorithm, e.g., dfs_mine. The algorithms presented are the first step towards that end, and in our conclusions we outline the future challenges. We believe the benefits of a generic framework are at least two-fold: firstly, it provides a single platform for the field of frequent pattern mining and facilitates re-use of mining techniques and methodologies among various patterns, and secondly it may yield insight into discovering algorithms for newer patterns, e.g. DAGs.

Front-End and Back-End. We provide an explicit demarcation between the roles played by the containers and methods used by the actual mining algorithms (called the *front-end* operations) and those employed by the database to provide its functionality (*back-end* operations). FPM algorithms so far have mainly focused on a highly integrated approach between the front-end operations and back-end procedures. Though such an approach leads to efficient mining algorithms, it compromises on their extensibility and scalability. For instance, there is little support for persistency, buffer management, or even adding new DBMSs. DMTL addresses this issue by demonstrating a clean way of seamlessly integrating new pattern types, buffer management techniques or even support for a new DBMS. Furthermore, such a framework also enables us to define distinctly the roles played by its various components, especially in the vertical mining approach, e.g., a pattern need not be aware of its VAT representation at all, and this appeals intuitively too. A pattern is characterized completely by its definition only, and its VAT is an entity defined by us in order to achieve vertical mining.

This concept is again depicted cleanly in our toolkit - the pattern is aware of only the high-level methods add_vat() and get_vat(); it is not restricted by the specific VAT representation used.

This design enables DMTL to provide extensibility in three key ways.

Adding a New Pattern-Type. Due to the inherent distinction between containers and algorithms, a new pattern type can be added to DMTL in a clean fashion. We demonstrate how it may be extended for unordered, rooted trees [17]. The order of a node's children is relevant in ordered trees, while it is not so in unordered trees. We observe that DMTL already provides tree mining, hence much of the infrastructure may be re-used. The only significant modification required is isomorphism checking. Hence the user can define an unordered tree class, utree, similar to the in-built tree class. utree should provide an implementation for its canonical_code member, which our algorithm shall use to determine isomorphism. In addition, a vertical representation (VAT) needs to be provided for utree. Since utree is essentially a tree itself, it may utilize tree's vat_body but needs to provide its distinct implementation of VAT intersection. In this instance, due to its similarity to tree, utree could utilize many of the common algorithms and routines. We acknowledge that this may not always be the case; nevertheless for a new pattern-type, the user needs to define specialized implementations of the main containers, viz., utree and utree_vat and their methods, but can reuse the toolkit's infrastructure for vertical/horizontal and DFS/BFS mining, as well as buffering and persistency. This way all algorithms are guaranteed to work with **any** pattern as long as certain basic operations are defined.

Buffering Scheme. The Buffer class provides memory management of patterns and VATs. A new buffer manager may be put in place simply by defining an appropriate new class, say NFU_Buffer employing a *not frequently used* strategy. NFU_Buffer should define methods such as add_vat(vat_body&) which shall implement the appropriate buffering of VATs. No other modification to the toolkit is necessary.

DBMS Support. The back-end DBMS and buffer manager are interleaved to provide seamless retrieval and storage of patterns and VATs. The buffer manager fetches data as required from the DBMS, and writes out excess patterns/VATs to the DBMS as the buffering strategy may dictate. In order to provide support for a new DBMS, appropriate methods shall have to be defined, which the toolkit would invoke through templatization of the DB class. Again, the design ensures that this new DBMS can be cleanly integrated into the toolkit.

6   Experiments

Templates provide a clean means of implementing our concepts of genericity of containers and algorithms; hence DMTL is implemented using the C++ Standard Template Library [4]. We present some experimental results on the time taken by DMTL to perform different types of pattern mining. We used the IBM synthetic database generator [1] for itemset and sequence mining, the tree generator from [30] for tree mining and the graph generator by [16], with sizes ranging from $10K$ to $500K$ (or 0.5 million) objects. The experiment were run on a Pentium4 2.8Ghz Processor with 6GB of memory, running Linux.

Figure 5 shows the DMTL mining time versus the specialized algorithms for itemset mining (Eclat [28]), sequences (Spade [29]), trees (TreeMiner [30]) and graphs (gSpan [26]). For the DMTL algorithms, we show the time with different persistency managers/databases: flat-file (Flat), metakit backend (Metakit) and the gigabase backend (Gigabase). The left hand column shows the effect of minimum support on the mining time for the various patterns, and the column on the right hand size shows the effect of increasing database sizes on these algorithms. Figures 5(a) and 5(b) contrast performance of DMTL with Eclat over varying supports and database sizes, respectively. As can be seen in, Figure 5(b), DMTL(Metakit) is as fast as the specialized algorithm for larger database sizes. Tree mining in DMTL (figures 5(e) and 5(f)) substantially outperforms TreeMiner; we attribute this to the initial overhead that TreeMiner incurs by reading the database in horizontal format, and then converting it into the vertical one. We have accomplished high optimization of the mining algorithm for itemsets and trees; proposed future work is to utilize similar enhancements for sequences and graphs. For graph and sequence patterns, we find that DMTL is at most, within a factor of 10 as compared to specialized algorithms and often much closer (Figure 5(d)). Overall, the timings demonstrate that the performance and scalability benefits of DMTL are clearly evident with large databases. For itemsets, another experiments (not shown here) reported that Eclat breaks for a database with 5 million records, while DMTL terminated in 23.5s with complete results.

7   Future Work: Generic Closed Patterns

Our current DMTL prototype allows the mining of all frequent patterns. However, in the future we also plan to implement generic mining of other pattern spaces such as maximal patterns, and closed patterns. Informally, a maximal frequent pattern is a pattern which is not contained in another longer frequent pattern, whereas a closed frequent patterns is not contained in a longer frequent pattern which has the same frequency. We are especially interested in closed patterns since they form a lossless representation for the set of all frequent patterns.

Mining closed patterns has a direct connection with the elegant mathematical framework of formal concept analysis (FCA) [9], especially in the context of closed itemset mining. Using notions from FCA one can define a closure operator [9] between the item ($\mathcal{N}$) and transaction ($\mathcal{O}$) subset spaces, which allows one to define a closed itemset lattice. This in turn provides significant insight into the structure of the closed itemset space, and has lead to the development of efficient algorithms. Initial use of closed itemsets for association rules was studied in [32, 18]. Since then many algorithms for mining all the closed sets have been proposed, such as Charm [31], Closet [19], Closet+ [22] Closure [8], Mafia [6] and Pascal [5]. More recent algorithms have been studied in [10].

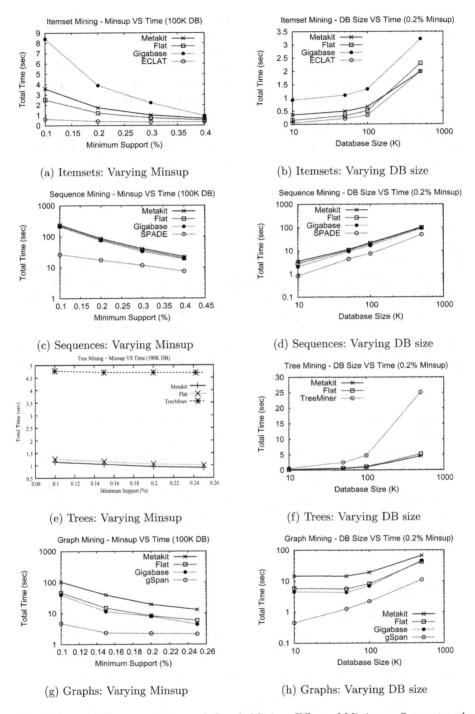

Fig. 5. Itemset, Sequence, Tree and Graph Mining: Effect of Minimum Support and Database Size

Recently, there has also been a surge of interest in mining other kinds of closed patterns such as closed sequences [23], closed trees [21, 7] and closed graphs [27]. For trees and graphs patterns there is currently no good understanding on how to construct the closure operator, and to leverage that to develop more efficient algorithms. The methods cited above use the intuitive notion of closed patterns (i.e., having no super-pattern with the same support) for mining. Recently, for ordered data or sequences, a closure operator has been proposed [13]. In our future work, we would like to develop the theory of a generic closure operator for any pattern and we will also develop generic data structures (e.g., `partial-order` pattern family) and algorithms to efficiently mine the set of all closed patterns.

8   Conclusions

In this paper we describe the design and implementation of the DMTL prototype for important FPM tasks, namely mining frequent itemsets, sequences, trees, and graphs. Following the ideology of generic programming, DMTL provides a standardized, general, and efficient implementation of frequent pattern mining tasks by isolating the concept of data structures or containers, from algorithms. DMTL provides container classes for representing different patterns, collection of patterns, and containers for database objects (horizontal and vertical). Generic algorithms, on the other hand are independent of the container and can be applied on any valid pattern. These include algorithms for candidate generation, isomorphism testing, VAT intersections, etc.

The generic paradigm of DMTL is a first-of-its-kind in data mining, and we plan to use insights gained to extend DMTL to other common mining tasks like classification, clustering, deviation detection, and so on. Eventually, DMTL will house the tightly-integrated and optimized primitive, generic operations, which serve as the building blocks of more complex mining algorithms. The primitive operations will serve all steps of the mining process, i.e., pre-processing of data, mining algorithms, and post-processing of patterns/models. Finally, we plan to release DMTL as part of open-source, and the feedback we receive will help drive more useful enhancements. We also hope that DMTL will provide a common platform for developing new algorithms, and that it will foster comparison among the multitude of existing algorithms.

References

1. R. Agrawal, H. Mannila, R. Srikant, H. Toivonen, and A. Inkeri Verkamo. Fast discovery of association rules. In U. Fayyad and et al, editors, *Advances in Knowledge Discovery and Data Mining*, pages 307–328. AAAI Press, Menlo Park, CA, 1996.
2. R. Agrawal and R. Srikant. Mining sequential patterns. In *11th Intl. Conf. on Data Engg.*, 1995.
3. T. Asai, K. Abe, S. Kawasoe, H. Arimura, H. Satamoto, and S. Arikawa. Efficient substructure discovery from large semi-structured data. In *2nd SIAM Int'l Conference on Data Mining*, April 2002.

4. M. H. Austern. *Generic Programming and the STL*. Addison Wesley Longman, Inc., 1999.

5. Y. Bastide, R. Taouil, N. Pasquier, G. Stumme, and L. Lakhal. Mining frequent patterns with counting inference. *SIGKDD Explorations*, 2(2), December 2000.

6. D. Burdick, M. Calimlim, and J. Gehrke. MAFIA: a maximal frequent itemset algorithm for transactional databases. In *Intl. Conf. on Data Engineering*, April 2001.

7. Yun Chi, Yirong Yang, Yi Xia, and Richard R. Muntz. Cmtreeminer: Mining both closed and maximal frequent subtrees. In *8th Pacific-Asia Conference on Knowledge Discovery and Data Mining*, 2004.

8. D. Cristofor, L. Cristofor, and D. Simovici. Galois connection and data mining. *Journal of Universal Computer Science*, 6(1):60–73, 2000.

9. B. Ganter and R. Wille. *Formal Concept Analysis: Mathematical Foundations*. Springer-Verlag, 1999.

10. B. Goethals and M.J. Zaki. Advances in frequent itemset mining implementations: report on FIMI'03. *SIGKDD Explorations*, 6(1), June 2003.

11. K. Gouda and M. J. Zaki. Efficiently mining maximal frequent itemsets. In *1st IEEE Int'l Conf. on Data Mining*, November 2001.

12. A. Inokuchi, T. Washio, and H. Motoda. An apriori-based algorithm for mining frequent substructures from graph data. In *4th European Conference on Principles of Knowledge Discovery and Data Mining*, September 2000.

13. J.L.Balcazar and G.Casas-Garriga. On horn axiomatizations for sequential data. In *10th International Conference on Database Theory*, 2005.

14. Konstantin Knizhnik. Gigabase, object-relational database management system. http://sourceforge.net/projects/gigabase.

15. R. Kohavi, D. Sommerfield, and J. Dougherty. Data mining using mlc++, a machine learning library in c++. *International Journal of Artificial Intelligence Tools*, 6(4):537–566, 1997.

16. M. Kuramochi and G. Karypis. Frequent subgraph discovery. In *1st IEEE Int'l Conf. on Data Mining*, November 2001.

17. Siegfried Nijssen and Joost N. Kok. Efficient discovery of frequent unordered trees. In *1st Int'l Workshop on Mining Graphs, Trees and Sequences*, 2003.

18. N. Pasquier, Y. Bastide, R. Taouil, and L. Lakhal. Discovering frequent closed itemsets for association rules. In *7th Intl. Conf. on Database Theory*, January 1999.

19. J. Pei, J. Han, and R. Mao. Closet: An efficient algorithm for mining frequent closed itemsets. In *SIGMOD Int'l Workshop on Data Mining and Knowledge Discovery*, May 2000.

20. R. Srikant and R. Agrawal. Mining sequential patterns: Generalizations and performance improvements. In *5th Intl. Conf. Extending Database Technology*, March 1996.

21. A. Termier, M-C. Rousset, and M. Sebag. Dryade: a new approach for discovering closed frequent trees in heterogeneous tree databases. In *IEEE Int'l Conf. on Data Mining*, 2004.

22. J. Wang, J. Han, and J. Pei. Closet+: Searching for the best strategies for mining frequent closed itemsets. In *ACM SIGKDD Int'l Conf. on Knowledge Discovery and Data Mining*, August 2003.

23. Jianyong Wang and Jiawei Han. Bide: Efficient mining of frequent closed sequences. In *IEEE Int'l Conf. on Data Engineering*, 2004.

24. Jean-Claude Wippler. Metakit. http://www.equi4.com/metakit/.

25. I. H. Witten and E. Frank. *Data Mining: Practical Machine Learning Tools and Techniques with Java Implementations*. Morgan Kaufmann Publishers, 1999.

26. X. Yan and J. Han. gspan: Graph-based substructure pattern mining. In *IEEE Int'l Conf. on Data Mining*, 2002.

27. X. Yan and J. Han. Closegraph: Mining closed frequent graph patterns. In *ACM SIGKDD Int. Conf. on Knowledge Discovery and Data Mining*, August 2003.

28. M. J. Zaki. Scalable algorithms for association mining. *IEEE Transactions on Knowledge and Data Engineering*, 12(3):372-390, May-June 2000.

29. M. J. Zaki. SPADE: An efficient algorithm for mining frequent sequences. *Machine Learning Journal*, 42(1/2):31–60, Jan/Feb 2001.

30. M. J. Zaki. Efficiently mining frequent trees in a forest. In *8th ACM SIGKDD Int'l Conf. Knowledge Discovery and Data Mining*, July 2002.

31. M. J. Zaki and C.-J. Hsiao. CHARM: An efficient algorithm for closed itemset mining. In *2nd SIAM International Conference on Data Mining*, April 2002.

32. M. J. Zaki and M. Ogihara. Theoretical foundations of association rules. In *3rd ACM SIGMOD Workshop on Research Issues in Data Mining and Knowledge Discovery*, June 1998.

Conceptual Exploration of Semantic Mirrors

Uta Priss and L. John Old

Napier University, School of Computing
{u.priss, j.old}@napier.ac.uk

Abstract. The "Semantic Mirrors Method" (Dyvik, 1998) is a means for automatic derivation of thesaurus entries from a word-aligned parallel corpus. The method is based on the construction of lattices of linguistic features. This paper models the Semantic Mirrors Method with Formal Concept Analysis. It is argued that the method becomes simpler to understand with the help of FCA. This paper then investigates to what extent the Semantic Mirrors Method is applicable if the linguistic resource is not a high quality parallel corpus but, instead, a medium quality bilingual dictionary. This is a relevant question because medium quality bilingual dictionaries are freely available whereas high quality parallel corpora are expensive and difficult to obtain. The analysis shows that by themselves, bilingual dictionaries are not as suitable for the Semantic Mirrors Method but that this can be improved by applying conceptual exploration. The combined method of conceptual exploration and Semantic Mirrors provides a useful toolkit specifically for smaller size bilingual resources, such as ontologies and classification systems. The last section of this paper suggests that such applications are of interest in the area of ontology engineering.

1   Introduction

Dyvik (1998, 2003, 2004) invented the "Semantic Mirrors Method" as a means for automatic derivation of thesaurus entries from a word-aligned parallel corpus. His on-line interface[1] uses a parallel corpus of Norwegian and English texts, from which users can interactively derive thesaurus entries in either language. A feature set is derived for each sense of each word. The senses then form a semi-lattice based on inclusion and overlap among feature sets. Priss & Old (2004) note (without providing any details) that Dyvik's method is similar to certain concept lattices derived from monolingual lexical databases. The Semantic Mirrors Method is briefly described in section 2 of this paper. Section 3 explains how the Semantic Mirrors Method can be represented with respect to Formal Concept Analysis (FCA). We believe that the Semantic Mirrors Method is of general interest to the FCA community because there may be other similar applications in this area.

In section 4, the FCA version of the Semantic Mirrors Method from section 3 is applied to an English-German dictionary. An advantage of using bilingual dictionaries instead of parallel corpora is that bilingual dictionaries are freely available on the Web whereas word-aligned parallel corpora are expensive. A disadvantage of using bilingual

---

[1] http://ling.uib.no/~helge/mirrwebguide.html

B. Ganter and R. Godin (Eds.): ICFCA 2005, LNAI 3403, pp. 21–32, 2005.

dictionaries is that the semantic information which can be extracted from them is less complete, at least with respect to the creation of Semantic Mirrors. Therefore, in section 5 of this paper we analyse how conceptual exploration (cf. Stumme (1996)) can be used to improve the incomplete information extracted from bilingual dictionaries. Even though conceptual exploration is a semi-automated process, we believe that in combination with the Semantic Mirrors Method, this approach has potential applications with respect to ontology merging as described in section 6.

This paper attempts to provide sufficient details of the Semantic Mirrors Method to be understandable for non-linguists, but it is assumed that readers are familiar with the basics of FCA, which can be found in Ganter & Wille (1999).

2   The Semantic Mirrors Method

The Semantic Mirrors Method intends to extract semantic information from bilingual corpora, which are large collections of texts existing in two languages and which are aligned according to their translations. The assumption is that if the same sentence is expressed in two different languages, then it should be possible to align words or phrases (or "lemmata") in one language with the corresponding words or phrases in the other language. This word alignment is not trivial because languages can differ significantly with respect to grammar and syntactic ordering. Computational linguists have developed a variety of statistical algorithms for such word-alignment tasks. These algorithms perform with different degrees of accuracy. One of Dyvik's interfaces allows for users to vary the parameters used in these algorithms to explore their impact on the extracted Semantic Mirrors. For comparison, Dyvik has also experimented with manually aligned corpora[2]. For the purposes of this paper, only the resulting lists of aligned translations are of interest. The quality or accuracy of the word alignment algorithms are not discussed in this paper.

2.1   Step 1

Once a bilingual corpus is word-aligned, one can select a word in either language and list all translations of that word occurring in the corpus. These lists of words and their respective lists of translations form the basis of the Semantic Mirrors Method. Dyvik (2003) calls the set of translations of a word a from language A its "(first) t-image" in language B. One can then form the t-images (in language A) of the t-image (in language B) of word a from language A. This set of sets is called the "inverse t-image of a". This algorithm of collecting the translations of the translations of a word has been mentioned by other authors (for example, Wunderlich (1980)) and is called the "plus operator" by Priss & Old (2004). This algorithm presents the first step of Dyvik's Mirrors Method. In contrast to this first step which has independently been discovered by different authors, to our knowledge, the next steps of the Semantic Mirrors Method are unique to this method.

---

[2] http://ling.uib.no/~helge/mirrwebguide.html\#bases

2.2   Step 2

The second step is to partition the t-image of a word into distinct senses. As an example, a t-image of English "wood" in German could be {Wald, Holz, Gehölz}. Intuitively, these three words belong to two senses: the sense of "wood" as a collection of trees ("Wald" and "Gehölz") and the sense of "wood" as a building material ("Holz"). These senses can be derived automatically by analysing the inverse t-image, i.e., the set of sets of t-images of the initial t-image. In this example, it is assumed that the t-image of "Holz" is {timber, wood}, the t-image of "Gehölz" is {grove, wood}, and the t-image of "Wald" is {grove, forest, wood}. Because the t-images of "Wald" and "Gehölz" overlap in more than one word, they are considered one sense of "wood" denoted by "wood1". Because the t-image of "Holz" overlaps with the other two t-images only in the original word "wood", "Holz" is considered a second sense of "wood" denoted by "wood2".

Once each sense of a word is individuated, it can be associated with its own t-image. Thus the t-image of "wood1" is {Wald, Gehölz}; the t-image of "wood2" is {Holz}. These two senses belong to different "semantic fields". According to Dyvik (2003): "traditionally, a semantic field is a set of senses that are directly or indirectly related to each other by a relation of semantic closeness. In our translational approach, the semantic fields are isolated on the basis of overlapping t-images: two senses belong to the same semantic field if at least one sense in the other language corresponds translationally with both of them." This means that "grove" and "forest" belong to the same semantic field as "wood1"; "timber" belongs to the same semantic field as "wood2". Of course, before assigning "grove", "forest", and "timber" to semantic fields, one would need to determine their own inverse t-images to see whether or not they have more than one sense themselves.

Dyvik (2003) explains that because the translational relation is considered symmetric, i.e. independent of the direction of the translation, one obtains corresponding semantic fields in two languages. These fields are not usually exactly structurally identical because the t-images in each language can be of different sizes and their sub-relationships can be different. But each semantic field imposes a subset structure on the corresponding semantic field in the other language. Thus each semantic field is structured by its own subset relationships and receives further structures from the corresponding field in the other language. We defer a more detailed description of these relationships to the next section because they are easier to explain with the help of FCA. The Semantic Mirrors Method receives its name from the fact that the semantic structures from one language can be treated as a "Semantic Mirror" of structures in the other language.

2.3   Step 3

In the third step of the Semantic Mirrors Method a feature hierarchy is formed based on the set-structures. Again this is more easily explained with FCA in the next section. The idea of expressing semantic information in feature hierarchies (or lattices) is common in the field of componential semantics. But in contrast to componential semantics where features often represent abstract ontological properties (such as "material", "immaterial"), in the Semantic Mirrors Method features are automatically derived as pairs of senses from the two languages, such as "[wood1, Holz]". Thus there is no attempt to manually de-construct features into any form of primitives or universals.

2.4   Step 4

As a last step of the Semantic Mirrors Method, thesaurus entries are generated. This is achieved by operations which extract synonyms, hypernyms, hyponyms and related words for a given word. In contrast to the feature structures which are graphically displayed as lattices, the thesaurus entries are displayed in a textual format. Users of the on-line interface can vary the parameters of "SynsetLimit" and "OverlapThreshold" which influence how wide or narrow the notion of "synonymy" is cast and thus how the thesaurus entries are constructed. The FCA description in the next section does not include an analysis of Step 4 because we have not yet determined whether there is any advantage of using FCA at this stage. That will be left for future research.

3   An FCA Description of Semantic Mirrors

Dyvik (1998) uses Venn diagrams as means of visualising and explaining the different steps of the Semantic Mirrors Method. A disadvantage of Venn diagrams is that they are difficult (or even impossible) to draw for more complex examples. This section demonstrates how concept lattices can be used to visualise the first three steps of the Mirrors Method. As Priss & Old (2004) observe, the first step of the Semantic Mirrors Method is similar to what Priss & Old call "neighbourhood lattices" with respect to lexical databases. By modelling the Semantic Mirrors Method with FCA, the techniques developed for neighbourhood lattices can now also be applied to the Semantic Mirrors Method and vice versa.

3.1   Step 1: Forming a Neighbourhood Lattice

The first step of the Semantic Mirrors Method consists of constructing a formal context, which has a union of t-images as a set of objects and a union of corresponding inverse t-images as a set of attributes. Figure 1 shows an example for English "good", "clever", "cute" and "pretty". The data for this example comes from one of Dyvik's "toy" examples[3]. The t-images (or translations) of "good", "clever", "cute" and "pretty" are the objects in figure 1. The inverse t-images (i.e., the translations of the translations) are the attributes. This kind of lattice is a "neighbourhood lattice" in the sense of Priss & Old (2004).

Instead of t-images and inverse t-images, one can also use inverse t-images and inverse t-images of inverse t-images, and so on. In many cases the continuous search for translations may not converge until large sets of words from both languages are included. For example, Dyvik and Thunes started with the Norwegian words "god", "tak" and "selskap"[4]. After translating back and forth between Norwegian and English four times, they collected a set of 2796 Norwegian words and 724 English words! Therefore, it may be sensible for some applications to terminate the search for t-images after a few iterations. The resulting neighbourhood lattice is complete for the initial set but incomplete either with respect to the translations of the objects or the attributes. If

---

[3] http://ling.uib.no/~helge/mirrwebguide.html
[4] http://ling.uib.no/~helge/mirrwebguide.html\#bases

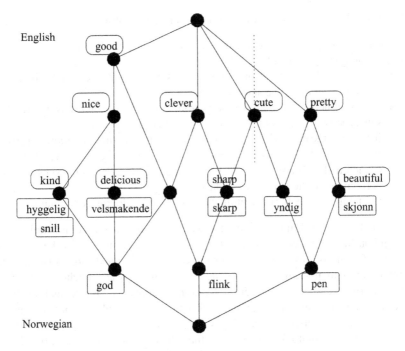

Fig. 1. A neighbourhood lattice for "good/god" (Step 1)

the translations are stopped after adding attributes, then some of the attributes which were added last may not have their complete set of translations among the objects. If the translations are stopped after adding objects, then some of these may be missing some of their translations.

3.2   Step 2: The Sense Distinction Algorithm

The second step of the Semantic Mirrors Method consists of identifying which different senses each word has. The different senses are then used to form different semantic fields. Modelled with respect to FCA, we call this algorithm the "Sense Distinction Algorithm". This algorithm can be applied to any finite formal context, but we do not know whether the algorithm produces any interesting results for other formal contexts than those describing neighbourhood lattices. It should be noted that this algorithm focuses on attributes attached to co-atoms (the lower neighbours of the top concept) and on objects attached to atoms (the upper neighbours of the bottom concept). Thus in figure 1 only the attributes "good", "clever", "cute" and "pretty" and the objects "god", "flink", and "pen" are of interest. A "contingent" of a concept is defined as the set of attributes and objects, which are in the extension of the concept but not in the extension of any subconcept and in the intension of the concept but not in the intension of any superconcept. Thus these attributes and objects belong directly to the concept and are not inherited from sub- or superconcepts. In line diagrams, such as figure 1, the objects and attributes attached to each node form the contingent of that concept.

The Sense Distinction Algorithm can be described as follows:

- For each co-atom c which has attribute a in its contingent collect the set S of concepts immediately below and adjacent to c.
    - i) If c also has at least one object in its contingent, then each object in the contingent defines one sense of a.
    - ii) If the meet of S is above the bottom concept $\perp$, then a has one remaining sense. Skip iii) and continue with the next co-atom.
    - iii) Else, if the meet of S is the bottom concept, construct a relation R as follows: for $c_1, c_2 \in S : c_1 R c_2 :\Longleftrightarrow c_1 \wedge c_2 > \perp$. Form the transitive closure of R (which makes R an equivalence relation on S). The remaining senses of a now correspond to the equivalence classes of R on S.
- Determine the senses for each atomic object in an analogous, dual manner.

Step i) of the algorithm relates to what was said above about the incompleteness of the neighbourhood lattices. If an object is attached to a co-atom then the chances are that some of its translations are missing from the formal context. This is because many words have more than one translation, which means that they are attached to the meet (or dually join) of several co-atomic (or dually atomic) concepts. Objects that are attached to a co-atom because their translational information is incomplete, would move further down in the lattice if their translations were added to the set of attributes. Therefore objects attached to co-atoms can indicate that the word which is the attribute of that co-atom has several senses. Step 1 provides information about whether the set of objects or whether the set of attributes may have incomplete translations. Therefore step i) can be rewritten to incorporate this information as "If c has an object in its contingent and the translations of this object may be incomplete in the formal context, then each object attached to c corresponds to one sense of a". But this rewritten version of step i) is different from Dyvik's (1998) Mirrors Method.

In figure 1, only the attribute "cute" has two senses. The dotted line in figure 1 indicates that the lattice contains two separate semantic fields: one for each sense of "cute". Figure 2 shows these two semantic fields. The algorithm which leads from figure 1 to figure 2 can be described as deleting the top and bottom concept and all atomic or co-atomic concepts which were identified as having words with more than one sense in the Sense Distinction Algorithm. The different senses are numerically labelled and move to adjacent concepts. For example in figure 2, "cute2" is now attached to the same concept as "sharp", and "cute1" is now attached to the same concept as "yndig". The left diagram in figure 2 is not a lattice anymore but it can be thought of as a lattice whose top and bottom concepts are omitted in the graphical representation.

3.3   Step 3: Creating Mirror Images

Figure 1 shows that the neighbourhood lattice in this example is almost symmetric with respect to a horizontal line in the middle. This line can be thought of as the "Semantic Mirror" between the two languages. In this example, the two languages are very similar. Except for "nice", all other words have a corresponding translation in the other language. For "nice" there are two possibilities to find translations as explained further below. Figures 3 to 5 show the resulting semantic fields.

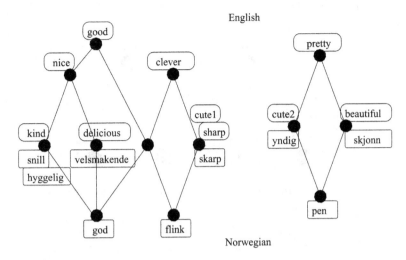

Fig. 2. Two semantic fields (Step 2)

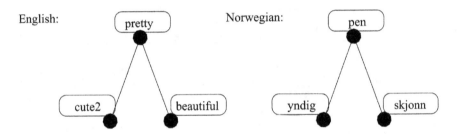

Fig. 3. The mirror images for "pretty" in English and Norwegian (Step 3)

In the example in figure 2, all attributes belong to concepts above or equal to the mirror line and all objects to concepts below or equal to the mirror line. This indicates that the lattices in this example can be cut apart along the mirror line so that each half represents the structures in one language. In general, this may not always be possible. Step 2 ensures only that co-atoms have no objects and atoms have no attributes in their contingents. The concept for "nice" could still have an object attached. It may be useful to apply the Sense Distinction Algorithm to any concept above or below the mirror line if the concept has both objects and attributes in its contingent. In any case, for each lattice resulting from Step 2, a "mirror" M can be defined as the set of all concepts which have both objects and attributes in their contingents plus those concepts which have no objects and attributes in their contingents but which are "equi-distant" from the top and bottom concept. This notion is somewhat fuzzy because there are different possibilities for defining "equi-distance" in a lattice.

In figure 2, the left lattice has a mirror M containing 4 concepts (the anti-chain in the middle), the right lattice has a mirror M containing two concepts. Each lattice is now split into two halves as follows: a formal context C_1 is formed which has an object for each concept in the mirror M and which has the original set of attributes; a second

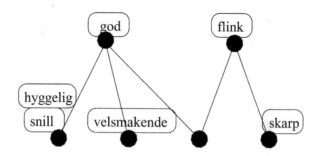

Fig. 4. The mirror image of English "good" in Norwegian (Step 3)

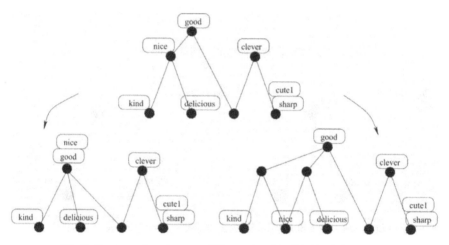

Fig. 5. The two mirror images of Norwegian "god" in English (Step 3)

formal context C_2 is formed which also has an object for each concept in the mirror M but has the original set of objects as a set of attributes. The crosses for each of these formal contexts are inserted according to the relation between the mirror elements and the objects (or attributes, respectively) in the original formal context.

If C_1 and C_2 are structurally identical, then their lattices are "exact mirror images" of each other (such as in figure 3). Otherwise, Step 3 of Dyvik's method attempts to relate the concepts from C_1 and C_2 in a top-down manner, which corresponds to context isomorphisms in FCA. The resulting lattices are "distorted mirror images". If C_1 is a subrelation of C_2, then C_1 is left unchanged (such as the Norwegian "god" in figure 4). For any column in the relation of C_2 that is not contained in the relation of C_1 there are several possibilities (cf. figure 5), some of which require addition of structural attributes, which do not correspond to words in the language (such as the two missing attributes in figure 5).

4   Semantic Mirrors in Bilingual Dictionaries

As mentioned in the introduction, a disadvantage of the use of parallel corpora is that they are expensive to obtain or construct. On the other hand, bilingual dictionaries for many languages are available on-line for free[5]. These bilingual dictionaries can be of questionable quality, but methods such as the Semantic Mirrors Method should be applicable even to slightly faulty data because errors should be detectable in the end result. The lists of possible translations which Dyvik (1998) utilises for his method are also not without errors if they are based on statistical automatic word alignment. The Semantic Mirrors Method is designed to cope with such data.

A more significant problem relating to bilingual dictionaries is not their quality but the fact that they contain fewer translations than parallel corpora because, in a corpus, words are not only translated into their direct counterparts but can also be translated into their hypernyms or hyponyms. This is because in natural language it is in general possible to use hypernyms and hyponyms for the same reference. For example, in a conversation a single person could be referred to as "the man", "that guy", "he", "Paul", and so on. Therefore in a parallel corpus of sufficient size one can expect these kinds of relationships to occur across languages. The separation into semantic fields in the Semantic Mirrors Method depends on these relationships. In a bilingual dictionary, however, it is usually attempted to translate words into exact counterparts if possible and to provide only as few translations as necessary. Therefore, one can expect that different translations of a word in a bilingual dictionary will more often refer to different senses than to synonyms within the same semantic field. For a single sense fewer translations can be expected than would be found in a parallel corpus.

The following example is constructed using a freely available German-English dictionary[6]. The dictionary has more than 400,000 entries and is thus of reasonable size. A manual comparison of the translations of a few words with other dictionaries shows that the dictionary is of reasonable quality. Figure 6 shows a neighbourhood lattice generated from this dictionary for the starting word "wood". This lattice is very "shallow" in that it has only two levels of concepts between the top and bottom concept. If the Sense Distinction Algorithm from the last section was applied to this lattice, every single translational pair would be a separate sense. For example, "wood" would have three senses "Wald", "Gehölz", and "Holz". The resulting semantic fields would all be lattices consisting of a single concept. On the other hand, manual inspection of the lattice indicates that there are two larger fields contained among the words: one for the "set of trees"-sense of "wood" and one for the building material sense of "wood". Several words indicate other semantic fields, which are incomplete in the context, such as the "beam" sense of "timber" and the other senses of "lumber". Clearly, the Sense Distinction Algorithm is insufficient in this case because it does not result in such fields. The reason for this insufficiency, however, is not a shortcoming of the algorithm but instead the differences in the nature of the data derived from bilingual dictionaries opposed to parallel corpora.

---

[5] For example at www.fdicts.com
[6] http://www.dict.cc

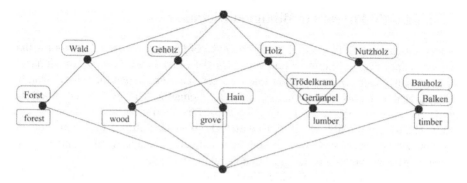

Fig. 6. A neighbourhood lattice generated from a bilingual dictionary

5   Conceptual Exploration of Semantic Mirrors

Since the Sense Distinction Algorithm is insufficient with respect to bilingual dictionaries, the question arises as to how it can be improved. A common FCA technique for improving incomplete data sets is "conceptual exploration". Stumme (1996) lists four cases of conceptual exploration: attribute exploration, which refers to the process of interactive addition of attributes to a formal context, object exploration, which refers to the process of interactive addition of objects to a formal context, concept exploration, which refers to the process of interactive addition of objects and attributes to a formal context, and an un-named fourth type of conceptual exploration, which refers to the process of interactive addition of crosses to a formal context. Stumme notes that so far there exists no exploration software for this last type of exploration.

The main shortcoming of the lattice in figure 6 is the fact that it is too shallow. From a linguistic view this means that hypernyms are missing. For example, "wood" in its first sense could be considered a hypernym of "grove" and "forest", but that is not depicted in the lattice. From an FCA view, for a word to be a hypernym of another word, there must be a subset-superset relation between the intensions or extensions. If "wood" is to become a hypernym of "grove" it must also be a translation of "Hain" and a cross for "Hain/wood" must be added in the formal context. Thus hypernyms can be established by adding certain crosses to the relation of a formal context. But these crosses cannot be randomly chosen because they must result in subset-superset relations.

Stumme's fourth type of conceptual exploration is relevant for this situation. The other types of conceptual exploration can also be relevant, because in some cases a hypernym may exist in a language but may not yet be included among the objects or attributes of the formal context in question. The conceptual exploration algorithm can work as follows: the inverse t-image is formed for each co-atomic object. In each case, a user is asked whether a hypernym in the other language can be found for the set. Since there are four co-atoms in figure 6, a user will be asked four questions. The inverse t-image of "Wald" is "Wald, Gehölz, Forst, Holz". A user might decide that "wood" is in fact a hypernym in English for this set. The inverse t-image for "Gehölz" is the set "Wald, Gehölz, Hain, Holz" and also has the English hypernym "wood". For the inverse t-images of "Holz" and "Nutzholz" no English hypernyms can be found. The first two

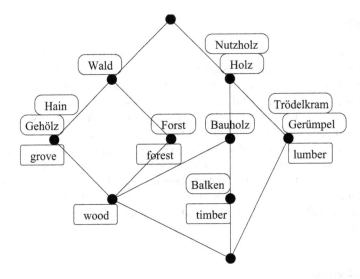

Fig. 7. The lattice after conceptual exploration

questions result in adding "Forst/wood" and "Hain/wood" to the relation. Next, any hypernym that was found is tested against any words of the other language that have not been considered as translations. Therefore a user is asked whether "wood" can also be a translation of any of the words "Trödelkram, Gerümpel, Nutzholz, Bauholz, Balken". A user might decide that "Nutzholz/wood" and "Bauholz/wood" also need to be added.

After finishing with the objects, the lattice is recalculated. It now has only three atoms: "wood", "timber" and "lumber". Each of the three inverse t-images is checked: a German hypernym of "wood, grove, forest" is identified as "Wald"; "wood, grove, lumber" has no hypernym; a German hypernym of "wood, lumber, timber" is determined to be "Holz". The crosses "Wald/grove" and "Holz/timber" are added. As a last step, the newly found hypernyms "Holz" and "Wald" are checked against the remaining English words, which does not lead to any further added crosses. The recalculated lattice is depicted in figure 7. The Sense Distinction Algorithm applied to the lattice in figure 7 identifies two senses of "wood", two senses of "Holz", two senses of "lumber" and two senses of "timber". All other words have only a single sense. This division of senses and semantic fields is much more compatible with an intuitive notion than the initial results obtained from figure 6.

This algorithm does not identify every possibly missing cross in the original context. It only checks for hypernyms because these are essential for the division into semantic fields. We have tested the algorithm with a few other examples. It works better in areas where hypernyms are easily identifiable, such as for concrete nouns. With respect to other types of words, more research is needed. It is hoped that a heuristic rule set can be developed that adjusts the conceptual exploration to the specific requirements of the types of semantic fields that are involved.

6   Conclusion

This paper argues that a combination of the Semantic Mirrors Method with conceptual exploration may yield promising results in certain areas. Because conceptual exploration is an interactive method, it can be labour-intensive. In the example in the last section, the relation contained 30 possibilities for added crosses, but a user was asked only 10 questions. Thus the conceptual exploration was more efficient than asking a user to manually complete a given formal context without any further tools.

We believe that these kinds of methods can be suitably applied in the area of ontology engineering and ontology merging. Ontologies are much smaller than bilingual dictionaries. While bilingual dictionaries of natural languages are too large to be processed with semi-automated methods, ontologies might still be in a range where semi-automated methods are feasible. In fact, many current methods of ontology merging in AI are semi-automated.

Acknowledgements

We wish to thank Helge Dyvik and three anonymous reviewers for helpful comments and suggestions.

References

1. Dyvik, Helge (1998). *A Translational Basis for Semantics.* In: Johansson and Oksefjell (eds.): Corpora and Crosslinguistic Research: Theory, Method and Case Studies, Rodopi, p. 51-86.
2. Dyvik, Helge (2003). *Translations as a Semantic Knowledge Source.* Available on-line at www.hf.uib.no/i/LiLi/SLF/ans/Dyvik/transknow.pdf
3. Dyvik, Helge (2004). *Translations as semantic mirrors: from parallel corpus to wordnet.* Language and Computers, Vol. 49, iss. 1, Rodopi, p. 311-326.
4. Ganter, B.; Wille, R. (1999). *Formal Concept Analysis.* Mathematical Foundations. Berlin-Heidelberg-New York: Springer, Berlin-Heidelberg.
5. Priss, U.; Old, L. J. (2004). *Modelling Lexical Databases with Formal Concept Analysis.* Journal of Universal Computer Science, Vol 10, 8, p. 967-984.
6. Stumme, Gerd (1996). *Exploration Tools in Formal Concept Analysis.* In: E. Diday, Y. Lechevallier, O. Opitz (Eds.): Ordinal and Symbolic Data Analysis. Proc. OSDA'95. Studies in Classification, Data Analysis, and Knowledge Organization 8, Springer, p. 31-44.
7. Wunderlich, Dieter (1980). *Arbeitsbuch Semantik.* Athenaeum, Königstein im Taunus.

Towards a Formal Concept Analysis Approach to Exploring Communities on the World Wide Web

Jayson E. Rome and Robert M. Haralick

Department of Computer Science,
The City University of New York, New York NY 10016, USA

Abstract. An interesting problem associated with the World Wide Web (Web) is the definition and delineation of so called Web communities. The Web can be characterized as a directed graph whose nodes represent Web pages and whose edges represent hyperlinks. An authority is a page that is linked to by high quality hubs, while a hub is a page that links to high quality authorities. A Web community is a highly interconnected aggregate of hubs and authorities. We define a community core to be a maximally connected bipartite subgraph of the Web graph.

We observe that the web subgraph can be viewed as a formal context and that web communities can be modeled by formal concepts. Additionally, the notions of hub and authority are captured by the extent and intent, respectively, of a concept. Though Formal Concept Analysis (FCA) has previously been applied to the Web, none of the FCA based approaches that we are aware of consider the link structure of the Web pages. We utilize notions from FCA to explore the community structure of the Web graph. We discuss the problem of utilizing this structure to locate and organize communities in the form of a knowledge base built from the resulting concept lattice and discuss methods to reduce the complexity of the knowledge base by coalescing similar Web communities. We present preliminary experimental results obtained from real Web data that demonstrate the usefulness of FCA for improving Web search.

1 Introduction

Traditional techniques for information retrieval involve text based search and various indexing methods. The presence of hyperlinks between documents presents challenges and opportunities that traditional information retrieval techniques have not had to deal with. By viewing the set of n pages on the World Wide Web as nodes V and links (similarity, association) between pages as directed edges E of a directed graph $\Gamma = (V, E)$, the graph of n nodes can be stored in an $n \times n$ matrix. A nonzero entry in the $(i, j)^{th}$ position of the matrix indicates an edge (possibly weighted or labelled) from node i to node j. A hyperlink implies some form of endorsement, or conferral of authority, by citing document to the cited document. A large portion of the current research in improving web search

B. Ganter and R. Godin (Eds.): ICFCA 2005, LNAI 3403, pp. 33–48, 2005.

is concerned with utilizing the hyperlinked nature of the web. Kleinberg's HITS algorithm [1], and various extensions [2], [3], [4], [5], and the Google PageRank algorithm [6], [7] demonstrate the success of link based ranking in refining Web search. Henziger's recent survey [8] enumerates the following open algorithmic challenges for Web search engines:

- Finding techniques to generate random samples of the Web in order to determine statistical properties of the Web,
- Modeling the web to explain observed properties,
- Detecting duplicates and near duplicates to improve search efficiency,
- Analyzing temporal trends in data streams that result from user access logs,
- Finding and analyzing dense bipartite subgraphs, or Web communities,
- Finding eigenvector-induced partitionings of directed graphs in order to cluster the Web graph.

1.1   Hubs and Authorities

Consider the problem of finding "definitive" or "authoritative" sources in the mass of information available on the web. The user should be provided with relevant pages of the highest quality. The hyperlink structure of the web contains a tremendous amount of latent information in that the creation of a link from page a to page b in some way represents a's endorsement of b. Purely text based search methods fail to find authoritative sources. For example if one uses the query "operating systems," there is no guarantee that Windows, Linux, Apple or any other operating system vendor will be among the pages returned because these pages may not explicitly contain the query terms. These pages are, however, relevant and of high quality and should in fact be returned. We can define these pages to be *authorities* because they are linked to by a large number of other pages. We can define a *hub* to be a page with a large collection of links to related pages. A good hub should point to many good authorities and a good authority should be pointed to by good hubs [1].

1.2   Web Communities

An interesting problem associated with the Web is the definition and delineation of so called Web communities [9], [10], [11], [12], [13], [14]. A *web community* is loosely defined to be a collection of content creators that share a common interest or topic and manifests itself as a highly interconnected aggregate or subgraph [9]. Kumar et al define a web community as being "characterized by dense directed bipartite subgraphs [10]." Figure 1 illustrates a simple community centered around a densely interconnected set of hubs and authorities. The World Wide Web contains many thousand explicitly defined communities and many more that are implicitly defined or are emerging [10]. The systematic extraction of emerging communities is useful for many reasons, including communities provide high quality information to interested users, they represent the sociology of the web and they can be used for target advertising [9]. In addition, community linkage can be used to find association between seemingly unconnected topics.

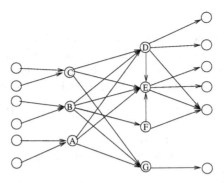

Fig. 1. A web community is characterized by a maximally complete bipartite subgraph, or diclique. In this example, nodes $\{A, B, C\}$ are hubs and nodes $\{D, E, G\}$ are authorities for a Web community

1.3   Previous Work

Previous work on defining and delineating structures of the web graph can be roughly broken down into graph theoretic, spectral graph theoretic, distance based and probabilistic approaches.

Botafogo and Schneiderman [15] proposed a method for determining aggregates from the hyperlink structure of a small hypertext system by using the graph theoretic notions of biconnected components and strongly connected components. Flake, et al [12] describe a method for identifying Web communities based on solving for the maximal flow/minimal cut through the network. Kumar, et al [10] present a method for enumerating bipartite cores of a snapshot of the web graph.

Spectral graph theoretic methods [16] form the foundation for many link based approaches, including the Google Page Rank algorithm [7], [6] and Klienberg's HITS algorithm [1]. Klienberg uses non principle eigenvectors of matrices derived from the graph matrix to partition the graph into communities. Pirolli et al, [17] propose a procedure based on the paradigm of information foraging that "spreads activation" through the network in order to utilize both link and text information for the purposes of locating useful structures and aggregates. The method can be viewed in terms of a random walk on a weighted graph in which the nodes with the greatest activation are selected from the steady state distribution of the random process. He et al [18] describe a spectral graph partitioning method based on the Normalized Cut criterion [19].

Modha and Spangler [20] describe a method of clustering results returned by a search engine using a geometric technique based on a similarity measure that incorporates word and link information.

Almeida and Almeida [21] propose a Bayesian network approach to combine content information with user behavior to model interest based communities.

1.4   Knowledge Bases

Once we have found the communities we require a method for organizing and analyzing them. A *knowledge base* is a system for enumerating, indexing and annotating all occurrences of a specific subgraph on the web and organizing the information into a useful structure [9]. Knowledge bases are constructed by:

- Identifying a signature subgraph that is likely to characterize a specific phenomena.
- Devising a method for enumerating all occurrences of the signature subgraph.
- From each enumerated subgraph, reconstructing the associated element of the knowledge base.
- Annotating and indexing the elements of the knowledge base [9].

Reasons for building such knowledge bases include the fact that they can provide a better starting point than raw data for analysis and mining and can aid in navigation and searching. In addition, fine-grained structures can be used for targeted market segmentation and the time evolution of such structures can provide information regarding the sociological evolution of the web [10].

Though Formal Concept Analysis (FCA) has been applied to the Web, [22], [23], [24], [25], [26], [27], these approaches focus on the terms found in Web documents rather than links between documents. Kalfoglou et al [28] report using FCA to analyze program committee membership, evolution of research themes and research areas attributed to published papers. The techniques that they describe are also used to identify communities of practice [29] which are clusters of individuals defined on a weighted association network. Tilley et al [30] report results of applying FCA to the transitive closure of the citation graph within a set of survey papers. None of the approaches that we are aware of consider the context defined by the link structure of the Web graph.

In the remainder of the paper we show an example of how we use notions from Formal Concept Analysis to explore the algebraic structure of Web communities, describe a prototype system for building web community knowledge bases using FCA and present preliminary experimental results obtained from real Web data that demonstrate the usefulness of FCA for improving Web search.

1.5   FCA for the Web

We formally define a community to be a set of hub pages that link to a set of authority pages. We model a community as a maximally complete directed bipartite subgraph, or diclique [31]. Our model is similar to the bipartite cores used in [10] except that we require the cores to be maximal. This definition suffers from being too strong in that it allows no exceptions, and at the same time being too weak in that it allows communities with few members as well as communities that are defined by a single page. Our current approach to addressing these problems is to apply a post processing step in which similar concepts are coalesced together.

In most applications of FCA the sets G and M are disjoint. However, if we take both the set of objects and the set of attributes to be a set of web pages so that $G = M$, and the link matrix of the web pages to be the incidence relation I, then concepts of the context (G, M, I) correspond directly to communities of the subgraph $\Gamma = (V, E)$. For a given concept $C = (A, B)$, the extent A corresponds to the set of hub pages and the intent B corresponds to the set of authorities.

1.6   Concept Coalescing

Given a set of concepts, the concept lattice provides a convenient hierarchical description. Contexts constructed from the Web can be very large in terms of the number of nodes and dense in terms of the number of edges. It is well known that the size of the lattice grows with the size of the relation [27].

For the purposes of our investigations we are interested in reducing the complexity of the lattice by merging concepts that are in some sense similar. One approach is to look for an algebraic decomposition of the lattice [32]. Funk, et al [33] present algorithms for horizontal, subdirect and subtensorial decompositions.

In addition, homomorphisms that results from a congruence relation can be used to reduce the complexity of the lattice by coalescing of concepts, while preserving much of the underlying structure. A complete congruence relation of a concept lattice $\mathfrak{B}(G, M, I)$ is an equivalence relation θ on $\mathfrak{B}(G, M, I)$ such that $x_t \theta y_t$ for $t \in T$ implies $\left(\bigwedge_{t \in T} x_t\right) \theta \left(\bigwedge_{t \in T} y_t\right)$ and $\left(\bigvee_{t \in T} x_t\right) \theta \left(\bigvee_{t \in T} y_t\right)$ [34]. Define $[x]\theta = \{y \in \mathfrak{B}(G, M, I) | x \theta y\}$ to be the equivalence class of θ that contains element x and $\mathfrak{B}(G, M, I)/\theta := \{[x]\theta | x \in \mathfrak{B}(G, M, I)\}$ to be the factor lattice for $\mathfrak{B}(G, M, I)$ and congruence θ [34]. The set of all complete congruences forms a complete lattice. Ganter and Wille describe a method to construct the congruence lattice for a formal context [34].

Another approach is to utilize notions from association rule mining in order to relax the restrictions on what constitutes a concept. The confidence is a measure of the strength of a rule , while support is a measure of the statistical significance of a rule [35].

Needed Links. Often the definition of the binary relation I is prone to error, and this is especially true when dealing with the web. These errors could be an error of omission, in which a pair (g, m) that should be in the relation is left out, or an error of commission, in which the pair (g, m) has erroneously been included in the relation. Omission tends to have more dramatic effect than commission [31] in that it is often easier to detect a spurious association from a given set of associations than to find a previously unknown association.

In order to collapse the lattice by the coalescing procedure we used in our experiments, we need to introduce edges to the original relation. The ability to isolate needed links automatically can be of great benefit in the case of the Web due to the fact that the Web is a volatile and dynamic environment. Additionally many communities that are forming or emerging may not know that they are, or that they should be, part of another community.

Table 1. The incidence relation I

x	$I(x)$	$I^{-1}(x)$
1	2,6,9	10,11,12
2	10,11	1,3,5,7
3	2,6,9	10,11,12
4	6,9	0
5	2,6	0
6	10,12	1,3,4,5,7
7	2,6,9	0
8	10,12	0
9	10,11,12	1,3,4,7
10	1,3	2,6,8,9
11	1,3	2,9
12	1,3	6,9

2   A Simple Example

We shall take an example first presented by Haralick [31] that is sufficiently complex to illustrate the manipulation of communities. We are given the incidence relation I, shown in tabular form in table 1 and graphical form in figure 2, which we can take to be the link structure of some subgraph of the Web. The communities of this relation are graphically enumerated in figure 3. Even with only 12 nodes and 28 links, finding these communities by inspection is nontrivial. The concept lattice, shown in figure 4, reveals the underlying structure in a relatively straightforward manner. Viewing the lattice structure is only useful for a small number of communities and for larger lattices we require a method of collapsing, or coalescing similar communities.

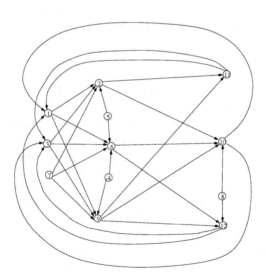

Fig. 2. The graph of binary relation I

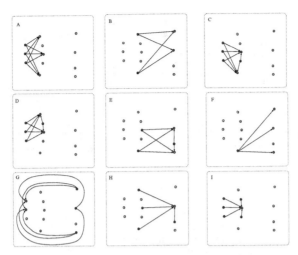

Fig. 3. A graphical enumeration of the concepts (community cores) of relation I

Coalescing finds "needed links," i.e. those links that need to be added to the relation I such that the concepts of the new coalesced relation $\hat{I}$, shown in figure 6 are themselves concepts. For example, let us again consider relation I. In order to coalesce concepts A, C, D and the concept labelled I, we need to introduce edges $(4, 2)$ and $(5, 9)$ to relation I in order to make the bipartite graph defined by $\{1, 3, 4, 5, 7\}, \{2, 6, 9\}$ complete. To coalesce concepts $B, E, F,$ and H, we need to add $(2, 12), (6, 11)$ and $(8, 11)$ to the original relation to make the bipartite graph defined by $\{2, 6, 8, 9\}, \{10, 11, 12\}$ complete.

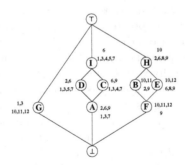

Fig. 4. The concept lattice for incidence relation I with intent (top) and extent (bottom) for each concept written out in full

3   Experimental Setup

We built a prototype system, whose architecture is shown in figure 7, to empirically verify the effectiveness of our approach. We chose to use live data from the web, rather than a test collection like the TREC WT10g dataset or the

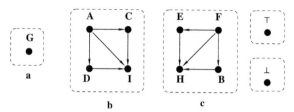

Fig. 5. A graphical representation of the relationship between concepts of I. The coalesced concepts (equivalence classes) are shown with dashed boxes

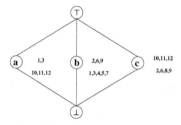

Fig. 6. The concept lattice for the coalesced binary relation $\hat{I}$ with intent and extent for each concept written out in full

crawls available from the Internet Archive, because it is free, easily available and supported by a large set of software tools and search services. A simple Web crawler is used to query a search engine and retrieve a set of urls that make up the nodes of the subgraph. For our experiments we used the text based search engine AltaVista. The top 20 results of the query are used to form a root set. A base set is constructed by adding those pages that point to any page in the root set and those pages which are pointed to by a page in the root set. Inbound links to page URL were obtained from the search engine using the link:URL command. This procedure can be repeated recursively up to a depth k, though the number of pages increases drastically with k. For each crawled page an index is created, the html source is stored and an entry is added into a graph that stores the link information. Text processing is used to create a descriptive feature vector for each page. These vectors are used to create summaries for each community in the knowledge base. The object set G and the attribute set M are the pages in the base set. The graph of the base set is constructed and used as the incidence relation I. The concepts and concept lattice $\mathfrak{B}(G, M, I)$ of the context (G, M, I) are computed and the concepts are coalesced produce a new relation $\hat{I}$ and a new concept lattice $\mathfrak{B}(G, M, \hat{I})$. Finally, a knowledge base is created by combining the concept lattice with the feature vectors.

A variety of algorithms are available to compute concepts and concept lattices [34], [36], [27], [37]. Given the lattice structure of the concepts a separate procedure is used to coalesce the concepts and form a new concept lattice. Additional research needs to be done to define criterion by which an "optimal" coalescing procedure can be determined.

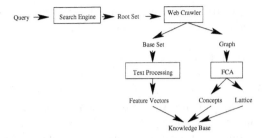

Fig. 7. System flow diagram for the prototype

Experimental Coalescing Procedure. For the experiments presented here, we utilized a coalescing procedure similar to the horizontal decomposition described in [33]. This procedure was chosen because it is straightforward to compute and is easily understood intuitively. Recall that a lattice L is said to be *horizontally decomposable* if it can be expressed as a horizontal sum $L = \{\top, \bot\} \bigcup \sum_{i=1}^{N} L_i \setminus \{\top_i, \bot_i\}$, where the summands $L_i \cap L_j = \emptyset$ for $i \neq j$ are lattices.

We define a relation R on $\mathfrak{B}(G, M, I) \setminus \{\top_i, \bot_i\}$ as follows $R = \{(\alpha, \beta) | \exists \gamma, \delta \ni \gamma \wedge \delta = \alpha, \text{ and } \gamma \vee \delta = \beta\}$ for $\alpha, \beta, \gamma, \delta \in \mathfrak{B}(G, M, I) \setminus \{\top_i, \bot_i\}$ and a relation S on $\mathfrak{B}(G, M, I)$ as $S = (R \cup R^{-1})^T \cup \underline{\mathfrak{B}}(G, M, I)$, where T signifies the transitive closure operation. The relation S is reflexive, symmetric and transitive and is therefore an equivalence relation on $\mathfrak{B}(G, M, I)$. Thus S forms a partition of the set of concepts. All concepts within a given equivalence class are merged to form a new concept.

We apply this approach by constructing a directed graph whose nodes are the concepts of $\mathfrak{B}(G, M, I) \setminus \{\top_i, \bot_i\}$. For every pair of concepts we connect the supremum and the infimum by a directed edge. For the relation I the graph is shown in figure 5. The connected components of this graph, shown by the dashed boxes, correspond to the equivalence classes of the equivalence relation S, and therefore represent the coalesced concepts. The coalesced concept that contains element x is formed by $C_{[x]S} = \bigcup_{i \in [x]S} C_i = (\bigcup_{i \in [x]S} A_i, \bigcup_{i \in [x]S} B_i) = (A_{[x]S}, B_{[x]S})$. Finally, the relation I is updated by forming the new relation $\hat{I}$ defined by $a \hat{I} b \; \forall a \in A_{[x]S}, \; b \in B_{[x]S}$.

3.1   The Knowledge Base

The hierarchy that results from the lattice structure forms a knowledge base that contains information about the relationships between various communities. The knowledge base is created by:

- Identifying a maximally complete bipartite subgraph (concept) to model a community core.
- Using FCA to enumerate all community cores (concepts).
- From each core, computing a representative description in the form of a feature vector.

– Using the concept lattice and the community representatives for annotating and indexing the elements of the knowledge base.

4   Experimental Results

To verify the effectiveness of the approach we performed several experiments on real Web data. Experiments were performed by posting a query to a text based search engine (in this case Alta Vista) and the base set was grown to depth $k = 1$. A summary of the results is given in table 2. The number of pages returned for a given query can be larger than what is realistically computable. Due to computational complexity, some preprocessing may be required to reduce the context to a reasonable size. The simplest method is to remove those nodes whose edge degree is below some threshold τ_I. For each query the resulting concept lattice is shown. There is clearly structure in the lattice as indicated by the number of concepts found and the density of the concept lattice.

Table 2. Summary of experimental results

Query	$|G| = |M|$	Number of Communities	Number Coalesced
formal concept analysis	382	43	10
support vector machine	442	51	7
ronald rivest	631	31	14
sustainable energy resources	2256	72	10
jaguar	253	41	10

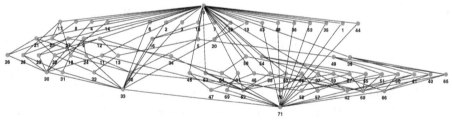

Fig. 8. The community knowledge base in the form of the concept lattice for the subgraph resulting from query *sustainable energy resources*. The base set was constructed to depth 1

4.1   Community Evaluation

Community representatives are needed for the evaluation of the quality and content of a given community and can be used for annotating and indexing the knowledge base. We expect that the communities found will have some cohesiveness in terms of content. Standard approaches to clustering of text documents involve expressing each document in terms of a feature vector and then grouping the documents into clusters based on some measure of similarity. Hotho and

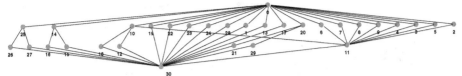

Fig. 9. The community knowledge base in the form of the concept lattice for the subgraph resulting from the query *ronald rivest*. The base set was constructed to depth 1

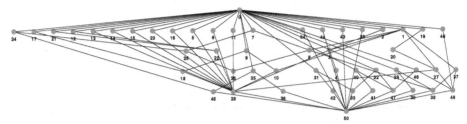

Fig. 10. The community knowledge base in the form of the concept lattice for subgraph resulting from the query *support vector machine*. The base set was constructed to depth 1

Stumme describe a system that uses text based clustering as a preprocessing step prior to conceptual clustering using FCA [38]. To evaluate the quality of the communities found we use the text features in a post-processing step to create descriptions of the concepts found by FCA.

Documents were processed by first extracting the text from the html documents and discarding all html markup commands. The text was broken up into a list of single words, or tokens. Stop words were removed from the token list, using a standard list of stop words. Each token was then stemmed to its root word using the Porter Stemming algorithm [39]. For each document, terms were constructed from the token list by considering all sets of words up to size s, for the purposes of the experiments described here $s = 3$. Thus, for example, the sentence "The `quick brown fox jumps over the lazy dog`" becomes `quick brown fox jump over lazi dog`" after stemming and stop word removal and gives the features: {`quick`, `quick-brown`, `quick-brown-fox`, `brown`, `brown-fox`, `brown-fox-jump`, ...`"`} This procedure captures a great deal of the word interaction and semantic content of the documents. For each term t in each document d the *term frequency - inverse document frequency* $tfidf(d, t)$ is computed so that:

$$tfidf(d, t) = tf(d, t) \times log\left(\frac{|V|}{|V_t|}\right)$$

where $tf(d, t)$ is the frequency of term t in document d, V is the set of Web documents and V_t is the number of documents in which term t occurs [38].

In the prototype system community representatives are determined by computing a set of mean vectors over members of the given community. A mean vector for the object set and a mean vector for the attribute set as well as a

Fig. 11. The community knowledge base in the form of the concept lattice for subgraph resulting from the query *jaguar*. The base set was constructed to depth 1

Fig. 12. The factor lattice for concept lattice resulting from the query *jaguar*

combined mean that considers pages in both the object and attribute sets are all computed. For each community the top n features, ranked in terms of tf-idf score, from the community representatives are used to create a description of the community.

Table 3. The urls returned from the search engine for the query *jaguar*

URL
http://www.jaguar.com
http://www.jaguarcars.com
http://www.jaguar.co.uk
http://www.apple.com/macosx
http://www.jag-lovers.org
http://www.jagweb.com
http://ds.dial.pipex.com/agarman/jaguar.htm
http://www.jaguar.com.au
http://www.jaguar-racing.com/uk/flash
http://www.psgvb.com/Products/jaguar.html
http://www.jaguarmodels.com
http://www.jaguar.ca
http://www.digiserve.com/eescape/showpage.phtml?page=a2
http://www.jaguar.is
http://hem.passagen.se/isvar/jaguar_server/jserver.html
http://www.primenet.com/ brendel/jaguar.html
http://www.jec.org.uk
http://www.bluelion.org/jaguar.htm
http://www.oneworldjourneys.com/jaguar

4.2   Query : jaguar

In this section we will look at results for the query "jaguar," which has become a frequently used query for evaluating web search [40], [1], [27]. There are many reasonable answers for the query "jaguar," including Jaguar Automobiles, Animals, the Jaguar Operating System and the Atari Jaguar Game system. The urls returned from the search engine are listed in table 3. This query is an illustrative example because we expect to find many disjoint communities with interconnections between communities, documents with varied quality and with text drawn from a large vocabulary with wide variation. The concept lattice for the query is shown in figure 11 and the factor lattice in figure 12. Examining

Table 4. The top 10 features for *jaguar* Community 28 ($|O| = 2$ and $|A| = 2$) and Community 34 ($|O| = 1$ and $|A| = 34$)

O	OA	A	O	OA	A
jag-lov	triumph	triumph	atari	atari	action
th-image	british	british	tagid	action	atari
xj	mg	mg	lynx	game	bit
find	tr	tr	sid	bit	game
cmpage	car	car	tag	processor	processor
jag-lovers	usa	usa	game	telegam	telegam
ord	british-car	british-car	title	padport	padport
brochures	restor	restor	crs	hz	hz
archives	club	club	akamai	jaguar	jaguar
forums	Mini	Mini	referrer	arcade	arcade

the top ranked features for the representatives of each community gives us an indication of the semantic content of the community. For tables 4 through 5 O indicates the mean for the object set, A indicates the mean for the attribute set and OA indicates the mean for the combined sets. For example, the top ranked features for concept 28, shown in table 4, indicate that the community is focused on automobiles, concept 34 in table 4 is concerned with the Atari game system, while concept 38 in table 5 deals with the Macintosh operating system. After applying the coalescing procedure we observe that many communities re-

Table 5. The top 10 features for *jaguar* Community 38 ($|O| = 1$ and $|A| = 29$) and CoConcept 2 ($|O| = 158$ and $|A| = 51$)

O	OA	A	O	OA	A
mac	mac	mac	species	maya	maya
tagid	tagid	panther	bluelion	cat	coat
tag	tag	os	geovisit	coat	rosett
os	os	tagid	previou	bluelion	leopard
sid	sid	tag	lion	leopard	cat
apple	apple	apple	image	rosett	captiv
mac-os	mac-os	sid	skip	species	civil
crs	panther	ll	suitabl	geovisit	jaguar
akamai	crs	crs	cat	captiv	differ
contentgroup-wtl	akamai	akamai	wild	speci	speci

main unchanged while many communities are merged together. In the factor lattice, concept 34 gets mapped uniquely to coconcept 5 while concepts 28 and 38 get merged with many other concepts to form coconcept 2. Looking at the top ranked features for coconcept 2 shown in table 5 we see that information about the Macintosh operating system and the Jaguar automobile have been diluted by the information about the animal. So while coalescing can be a useful tool for reducing the complexity of the concept lattice, it should not be done blindly. The equivalence relation that we used in the experiments described here is very coarse. What we require is a finer coalescing procedure that considers some measure of goodness of a given congruence to select the "best" equivalence relation based on the specified criterion.

5   Conclusions

We have demonstrated the utility of Formal Concept Analysis in the problem domain of defining and delineating communities on the Web. A formal concept can be used to model a Web community, the extent of the concept corresponding to the set of hubs and the intent of the concept corresponding to the set of authorities. The lattice of communities can be used to investigate the relationships between various communities as well as provide a method of coalescing communities that are similar. The size of the contexts involved when dealing with the Web require some preprocessing. Coalescing is a powerful tool that can be used to greatly simplify the concept lattice as well as isolate needed links, though it needs to be done carefully. Additional research needs to be done to determine a criterion that can be used to determine an optimal coalescing procedure.

Special Thanks. The authors would like to thank the anonymous reviewers for their insightful comments and suggestions.

References

1. Kleinberg, J.M.: Authoritative sources in a hyperlinked environment. Journal of the ACM **46** (1999) 604–632
2. Dean, J., Henzinger, M.R.: Finding related pages in world wide web. In: Proceedings of the Eighth International World Wide Web Conference. (1999)
3. Bharat, K., Henzinger, M.R.: Improved Algorithms for Topic Distillation in a Hyperlinked Environment. In: Proceedings of the 21st annual international ACM SIGIR conference on Research and development in information retrieval, Melbourne Australia (1998) 104–111
4. Chakrabarti, S., Dom, B., Raghavan, P., Rajagopalan, S., Gibson, D., Kleinberg, J.M.: Automatic resource compilation by analyzing hyperlink structure and associated text. Computer Networks and ISDN Systems **30** (1998) 65–74
5. Chakrabarti, S., Dom, B.E., Gibson, D., Kumar, S.R., Raghavan, P., Rajagopalan, S., Tomkins, A.: Experiments in topic distillation. In: ACM SIGIR Workshop on Hypertext Information Retrieval on the Web, Melbourne, Australia (1998)
6. Brin, S., Page, L.: The anatomy of a large-scale hypertextual Web search engine. Computer Networks and ISDN Systems **30** (1998) 107–117
7. Page, L., Brin, S., Motwani, R., Winograd, T.: The PageRank Citation Ranking: Bringing Order to the Web. Technical Report Stanford Digital Libraries Working Paper SIDL-WP-1999-0120, Stanford University (1999)
8. Henzinger, M.R.: Algorithmic Challenges in Web Search Engines. Internet Mathematics **1** (2004) 115–126
9. Kumar, S.R., Raghavan, P., Rajagopalan, S., Tomkins, A.: Extracting Large-Scale Knowledge Bases from the Web. In: The VLDB Journal. (1999) 639–650
10. Kumar, S.R., Raghavan, P., Rajagopalan, S., Tomkins, A.: Trawling the Web for Emerging Cyber-Communities. WWW8 / Computer Networks **31** (1999) 1481–1493
11. Gibson, D., Kleinberg, J.M., Raghavan, P.: Inferring Web Communities from Link Topology. In: Proceedings of the Ninth ACM Conference on Hypertext and Hypermedia. (1998) 225–234

12. Flake, G.W., Lawrence, S.R., Giles, C.L.: Efficient Identification of Web Communities. In: Sixth ACM SIGKDD International Conference on Knowledge Discovery and Data Mining, Boston, MA (2000) 150–160
13. Flake, G.W., Lawrence, S.R., Giles, C.L., Coetzee, F.M.: Self-Organization and Identification of Web Communities. IEEE Computer **33** (2002)
14. Deshpande, A., Huang, R., Raman, Riggs, T., Song, D., Subramanian, L.: A study of the structure of the web. Technical Report 284, EECS Computer Science Division, University of California, Berkeley, Berkely, CA (1999)
15. Botafogo, R.A., Shneiderman, B.: Identifying Aggregates in Hypertext Structures. In: Third ACM Conference on Hypertext, San Antonio, TX (1991) 63–74
16. Chung, F.R.K.: Spectral Graph Theory, vol. 92 of CBMS. American Mathematical Society, Providence, RI (1997)
17. Pirolli, P., Pitkow, J.E., Rao, R.: Silk from a Sow's Ear: Extracting Usable Structures from the Web. In: Proceedings of the ACM Conference on Human Factors in Computing Systems, CHI, ACM Press (1996)
18. He, X., Ding, C.H.Q., Zha, H., Simon, H.D.: Automatic Topic Identification Using Webpage Clustering. In: ICDM. (2001) 195–202
19. Shi, J., Malik, J.: Normalized cuts and image segmentation. IEEE Transactions on Pattern Analysis and Machine Intelligence **22** (2000)
20. Modha, D.S., Spangler, W.S.: Clustering Hypertext with Applications to Web Searching. In: Proceedings of the eleventh ACM Conference on Hypertext and Hypermedia, San Antonio, TX USA, ACM Press, New York, US (2000) 143–152
21. Almeida, R.B., Almeida, V.A.: A community-aware search engine. In: WWW2004. (2004)
22. Cole, R.J., Eklund, P.W.: Browsing Semi-structured Web Texts Using Formal Concept Analysis. Lecture Notes in Computer Science **2120** (2001)
23. Cole, R.J., Eklund, P.W.: Analyzing an Email Collection Using Formal Concept Analysis. In: PKDD. (1999)
24. Kim, M., Compton, P.: A Web-based Browsing Mechanism based on Conceptual Structure. In Mineau, G.W., ed.: The 9th International Conference on Conceptual Structures (ICCS 2001). (2001)
25. Kim, M., Compton, P.: Formal Concept Analysis for Domain-Specific Document Retrieval Analysis. In: AI 2001: Advances in Artificial Intelligence: Australian Joint Conference on Artificial Intelligence, Adelaide, Australia (2001)
26. Berendt, B., Hotho, A., Stumme, G.: Towards semantic web mining. In: First International Semantic Web Conference, Sardinia, Italy (2002)
27. Carpineto, C., Romano, G.: Concept Data Analysis : Theory and Applications. Wiley (2004)
28. Kalfoglou, Y., Dasmahaptra, S., Chen-Burger, J.: Fca in knowledge technologies: experiences and opportunities. In: 2nd International Conference on Formal Concept Analysis (ICFCA'04). (2004)
29. Alani, H., Dasmahapatra, S., O'Hara, K., Shadbolt, N.: Identifying communities of practice through ontology network analysis. IEEE Intelligent Systems **18** (2003)
30. Tilley, T.A., Cole, R.J., Becker, P., Eklund, P.: A survey of formal concept analysis support for software engineering activities. In: 1st International Conference on Formal Concept Analysis (ICFCA'03). (2003)
31. Haralick, R.M.: The diclique representation and decomposition of binary relations. Journal of the ACM **21** (1974) 356–366
32. Snelting, G., Tip, F.: Reengineering Class Hierarchies Using Concept Analysis. In: Proceedings of the ACM SIGSOFT Symposium on the Foundations of Software Engineering. (1998)

33. Funk, P., Lewien, A., Snelting, G.: Algorithms for Concept Lattice Decomposition and their Application. Technical report, TU Braunschweig, FB Informatik (1998)
34. Ganter, B., Wille, R.: Formal Concept Analysis : Mathematical Foundations. Springer Verlag, Berlin – Heidelberg – New York (1999)
35. Agrawal, R., Imielinski, T., Swami, A.N.: Mining Association Rules between Sets of Items in Large Databases. In: Proceedings of the 1993 ACM SIGMOD International Conference on Management of Data, Washington, D.C. (1993)
36. Lindig, C.: Fast Concept Analysis. (In: Working with Conceptual Structures - Contributions to ICCS 2000)
37. Kuznetsov, S., Obedkov, S.A.: Comparing Performance of Algorithms for Generating Concept Lattices. (In: ICCS'01 International Workshop on Concept Lattices-based KDD,)
38. Hotho, A., Stumme, G.: Conceptual Clustering of Text Clusters. In: Proceedings FGML Workshop, Hannover (2002)
39. Porter, M.F.: An algorithm for suffix stripping. Program **14** (1980)
40. Chekuri, C., Goldwasser, M., Raghavan, P., Upfal, E.: Web Search Using Automatic Classification. In: Proceedings of WWW-96, 6th International Conference on the World Wide Web, San Jose, US (1996)

Automatic Selection of Noun Phrases as Document Descriptors in an FCA-Based Information Retrieval System

Juan M. Cigarrán, Anselmo Peñas, Julio Gonzalo, and Felisa Verdejo

Dept. Lenguajes y Sistemas Informáticos**
E.T.S.I. Informática, UNED, Madrid Spain
{juanci, anselmo, julio, felisa}@lsi.uned.es
http://nlp.uned.es

Abstract. Automatic attribute selection is a critical step when using Formal Concept Analysis (FCA) in a free text document retrieval framework. Optimal attributes as document descriptors should produce smaller, clearer and more browsable concept lattices with better clustering features. In this paper we focus on the automatic selection of noun phrases as document descriptors to build an FCA-based IR framework. We present three different phrase selection strategies which are evaluated using the *Lattice Distillation Factor* and the *Minimal Browsing Area* evaluation measures. Noun phrases are shown to produce lattices with good clustering properties, with the advantage (over simple terms) of being better intensional descriptors from the user's point of view.

1 Introduction

The main goal of an Information Retrieval (IR) system is to ease information access tasks over large document collections. Starting from a user's query, usually made in natural language, a classic IR system retrieves the set of documents relevant to the user needs and shows them using ranked lists (e.g. Google, Yahoo or Altavista).

The use of ranked lists, however, does not always satisfy the user's information needs. Ranked lists are best suitable when users know exactly what they are looking for and how to express it using the right words (e.g. the last driver for a specific graphics card or the papers published by any author). More generally, ranked lists can be useful when the task is to retrieve a very small number of relevant items. However, when there is a need to retrieve relevant information from many sources, or when the query involves fuzzy or polysemous terms, the use of a ranked list implies to read almost the whole list to find the maximum

---

** This work has been partially supported by the Spanish Ministry of Science and Technology within the following projects: TIC-2003-07158-C04 Answer Retrieval from Digital Documents, R2D2; and TIC-2003-07158-C04-02 Multilingual Answer Retrieval Systems and Evaluation, SyEMBRA.

B. Ganter and R. Godin (Eds.): ICFCA 2005, LNAI 3403, pp. 49–63, 2005.

number of relevant documents. For instance, if we ask *Google* (*www.google.com*) with the query *'jaguar'* looking for documents related with the jaguar as animal, we obtain 7.420.000 of web pages as a result. Of course, not all the retrieved pages are relevant to our needs and, based on the ranking algorithm of Google [1], pages containing the term *'jaguar'* but with different senses (i.e. jaguar as a car brand or jaguar as a Mac operating system) are mixed up in the resulting ranking, making the information access task tedious and time consuming.

As an alternative, clustering techniques organize search results allowing a quick focus on specific document groups and improving, as a consequence, the final precision of the system from a user's perspective. In this way, some commercial search engines (i.e. www.vivisimo.com) apply clustering to a small set of documents obtained as a result of a query or a filtering profile. The use of clustering as a post-search process applied only to a subset of the whole document collection makes clustering an enabling search technology halfway between browsing (i.e. as in web directories) and pure querying (i.e. as in Google or Yahoo).

We propose the use of Formal Concept Analysis (FCA) as an alternative to classic document clustering, not only considered as an information organization mechanism but also as a tool to drive the user's query refinement process. Advantages of FCA over standard document clustering algorithms are: a) FCA provides an intensional description of each document cluster that can be used for query modification or refinement, making groups more interpretable; and b) the clustering organization is a lattice, rather then a hierarchy, which is more natural when multiple classification is possible, and facilitates recovering from bad decisions while browsing the lattice to find relevant information.

The main drawbacks of FCA disappear when dealing with small contexts (i.e. with a small set of documents obtained as the result of a search process): a) FCA is computationally more costly than standard clustering, but when it is applied to small sets of documents (i.e. in the range of 50 to 500 documents) is efficient enough for online applications; and b) lattices generated by FCA usually are big, complex and hence difficult to use for practical browsing purposes. Again, this should not be a critical problem when the set of documents and descriptors are restricted in size by a previous search over the full document collection.

But the use of FCA for clustering the results of a free text search is not a straightforward application of FCA. Most Information Retrieval applications of FCA are domain-specific, and rely on thesauruses or (usually hierarchical) sets of keywords which cover the domain and are manually assigned as document descriptors [12, 6, 7, 5, 8, 16, 9]. The viability of using FCA in this scenario implies to solve some challenges related with: a) the automatic selection of suitable descriptors for context building, b) the rendering of node descriptions, c) the visualization of concept lattices obtained, and; d) the definition of suitable query refinement tasks. Most importantly, the (non-trivial) issue of how to evaluate and compare different approaches has barely been discussed in the past.

This paper is presented as a continuation of the research presented in [4], where the problem of automatic selection of descriptors was first addressed. In that previous research we focused on the automatic selection of single terms as document descriptors. The problem with single terms is that, even if the lattice has good clustering/browsing properties, the intensional descriptions are not descriptive enough from a user's point of view. Noun phrases, however, tend to be excellent descriptors of the main concepts in a document, and are easily interpretable to users. Therefore, in this paper we will refine our proposal by using noun phrases as document descriptors. We will propose and compare three algorithms to select noun phrase document descriptors, using a shallow, efficient phrase extraction technique and the evaluation framework introduced in [4].

The paper is organized as follows. First of all we will present the information retrieval and organization system in which our evaluation framework is based; we will introduce the information organization model used and the system architecture. Then, we will describe the phrase extraction methodology and the set of phrase selection strategies presented for evaluation. Finally, we will present the evaluation experiments and discuss the results.

2   Information Retrieval and Organization System

2.1   The Information Organization Model

Using the ranked list of documents retrieved by a search engine, we generate a concept lattice to organize these search results. Lattices generated are based on a formal context $K := (G, M, I)$, where $G = \{doc_1, doc_2, \ldots, doc_n\}$ represents a subset of the retrieved documents, $M = \{desc_1, desc_2, \ldots, desc_k\}$ is a subset of document descriptors and I is the incidence relationship.

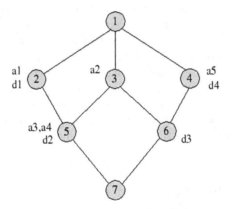

Fig. 1. Example concept lattice

This model relies on the set of concepts generated and its corresponding concept lattice, while introducing some assumptions about what concept information is going to be considered for showing, browsing or evaluation purposes.

Table 1. Formal concepts and their corresponding information nodes of the example lattice of Figure 1

Concept Node	Information Node
$c_1 = (\{d_1, d_2, d_3, d_4\}, \{\emptyset\})$	$n_1 = (\{\emptyset\}, \{\emptyset\})$
$c_2 = (\{d_1, d_2\}, \{a_1\})$	$n_2 = (\{d_1\}, \{a_1\})$
$c_3 = (\{d_2, d_3\}, \{a_2\})$	$n_3 = (\{\emptyset\}, \{a_2\})$
$c_4 = (\{d_4, d_3\}, \{a_5\})$	$n_4 = (\{d_4\}, \{a_5\})$
$c_5 = (\{d_2\}, \{a_1, a_2, a_3, a_4\})$	$n_5 = (\{d_2\}, \{a_1, a_2, a_3, a_4\})$
$c_6 = (\{d_3\}, \{a_2, a_5\})$	$n_6 = (\{d_3\}, \{a_2, a_5\})$
$c_7 = (\{\emptyset\}, \{a_1, a_2, a_3, a_4, a_5\})$	$n_7 = (\{\emptyset\}, \{a_1, a_2, a_3, a_4, a_5\})$

In our context, the remade formal concepts will be called *information nodes* and are defined as follows. Being A_i and B_i the extent and the intent of a generic formal concept c_i, we define its corresponding information node n_i as:

$$n_i = (AI_i, BI_i) \equiv \begin{cases} AI_i \subseteq A_i, \text{where } \forall \alpha \in AI_i \cdot \gamma(\alpha) = c_i \\ BI_i = B_i, \end{cases} \quad (1)$$

We also define a *connection node* as a information node where $AI_i = \emptyset$.

Information nodes are based on the assumption that a concept node should not display all its extent information. Working with the whole extent implies no differences between those documents which are object concepts (i.e. they are not going to appear as extent components of lower nodes) and those documents that can be specialized.

Figure 1 shows, as an example, a concept lattice where concept nodes are represented in Table 1 followed by its corresponding information nodes. Showing concept extent implies, for instance, that a user located at the top node of the lattice would be seeing the whole list of the documents retrieved at once. This situation would make our system essentially identical to a ranked list for browsing purposes. The use of information nodes overcome this problem, granting the document access only when no more specialization is possible. This model agrees with the access model used by most web directories (e.g. Open Directory Project ODP or Yahoo! Directory), where it is possible to find categories with no documents (i.e. categories that being very general do not completely describe any web page).

2.2   System Architecture

Our proposal to integrate FCA in a information retrieval and organization system is based on the architecture presented in Figure 2. It is divided in four main subsystems that solve the indexing, retrieval, lattice building and lattice representation and visualization tasks. Interactions between the subsystems are represented in the same figure and can be summarized as follows:

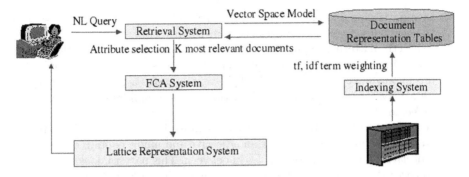

Fig. 2. System Architecture

1. The user makes a query using natural language as query language. The retrieval subsystem processes the query, removing stopwords and lemmatising meaningful query terms. The output of this step is a query representation which is used in the next step to make the retrieval process.
2. Relevant documents are retrieved by the retrieval subsystem using the query representation and the vector space model cosine measure.
3. The FCA subsystem builds a formal context using the first n most relevant documents retrieved and the k most suitable descriptors, which generates a concept lattice.
4. The Lattice representation subsystem applies the information organization model to the lattice generated, displaying a suitable visualization for user browsing purposes.

Currently, two prototypes have been developed based on this architecture, JBraindead and JBraindead2, which are used as our framework for evaluation purposes.

3   Phrase Selection

Dealing with noun phrases involves three main steps [14]: phrase extraction, phrase weighting and phrase selection.

3.1   Phrase Extraction

Phrase extraction is aimed at detecting all the possible candidate phrases that could be relevant and descriptive for the document in which they appear, and for the domains related to the document. These phrases are sequences of words that match the pattern of terminological phrases [13].

The phrase extraction procedure is divided in four main steps:

1. Document tokenisation, which identifies all possible tokens in the document collection.
2. Text segmentation according to punctuation marks.

3. Lemmatisation and part-of-speech tagging, in order to associate a base form to each word, together with its grammatical category.
4. Syntactic pattern recognition to detect the sequences of words that match a terminological phrase structure. The pattern is defined in Formula 2 as morpho-syntactic tag sequences. If the text contains a word sequence whose gramatical tags match the pattern, then a new candidate phrase has been detected.

   According to the pattern, a candidate phrase is a sequence of words that starts and ends with a noun or an adjective and might contain other nouns, adjectives, prepositions or articles in between. This pattern does not attempt to cover the possible constructions of a noun phrase, but the form of termi-nological expressions. The pattern is general enough to be applied to several languages, including English and Spanish [13, 15].

The result of the term detection phase is a list of terminological phrases with their collection term and document frequencies. For practical purposes, in our experiments we only consider those phrases of two or three terms (longer phrases usually have very low frequency values, which is of little help for concept clustering purposes).

$$[Noun|Adjective]\,[Noun|Adjective|Preposition|Article]*[Noun|Adjective]$$

$$(2)$$

3.2   Selection Strategies

Once suitable phrases are extracted, the next step is to select the subset of phrases that best characterizes the retrieved document set. Results of this phase are critical to a) reach a reasonable cardinality for the concept set; and b) reach an optimal distribution of the documents in the lattice. A good balance between the cardinality of the descriptors set and the coverage of each descriptor should provide meaningful and easy to browse concept lattices.

With this objective in mind, we introduce three different phrase selection strategies to select candidate phrases as document descriptors: a) selection of generic phrases that occur with highest document frequencies (*Generic Balanced* strategy), b) selection of those phrases that, containing at least one query term as phrase component, have the highest document frequency values *(Query Specific Balanced* strategy) , and c) selection of phrases which are terminologically relevant to describe the retrieved document set (*Terminological* strategy).

Generic Balanced Strategy. This strategy selects the k phrases with the highest document frequency covering the maximum number of documents re-trieved. Noun phrases occur much less frequently than terms, so the main idea is to describe, at least with one descriptor, the maximum number of documents. This is the algorithm to select the k set of descriptors according to this principle:

- Being $D = \{doc_1, doc_2, \ldots, doc_n\}$, the set of n most relevant documents selected from the retrieved set, and $P = \{phr_1, phr_2, \ldots, phr_m\}$, the set of

m phrases that appear in the n documents. We define a set $G = \{\emptyset\}$ that will store the covered documents, and a set $S = \{\emptyset\}$, that will store the final selected phrases.

- Repeat until $|S| = k$ or $|D| = \emptyset$, where k is the number of phrases to select for the document descriptors set.
  1. From P extract the phr_i with the highest document frequency in current D. If two or more phrases should have the same document frequency, then select the phrase that appear in the most relevant document of D (i.e. documents are ranked by the search engine).
  2. Store in an empty auxiliary set (AUX) those documents, belonging to current D, where phr_i appears.
  3. Delete the selected phrase from the candidate phrases set. $P = P \backslash \{phr_i\}$.
  4. Delete the selected documents from the documents set. $D = D \setminus AUX$
  5. Add the selected phrase to the final descriptors set. $S = S \cup \{phr_i\}$
  6. Add the selected documents to the used documents set. $G = G \cup AUX$
- The S set will contain the k highest document frequency phrases with maximal document coverage.

Query Specific Balanced Strategy. This is the same strategy, but restricting the set of candidate phrases P to those phrases containing one or more query terms as phrase components. First, we directly add to the S set the k' phrases with more than one query term and with a document frequency greater than one. Then we apply the above algorithm to calculate the best $k - k'$ phrases containing one query term. The main idea of this approach is to extract query related phrases that, due to its lower document frequencies, are not selected as document descriptors by the Generic Balanced selection strategy. In addition, phrases containing query terms should be better suggestions for users.

Terminological Strategy. Here we apply the terminological formula introduced in [4], but computed on phrases instead of terms. The main motivation of this formula is to weight with higher values those phrases that appear more frequently in the retrieved document set than in the whole collection. Formula 3 reflects this behavior, where w_i is the terminological weight of phrase i, $tf_{i,ret}$ is the relative frequency of phrase i in the retrieved document set, $f_{i,ret}$ is the retrieved set document frequency of phrase i, and $tf_{i,col}$ is the relative frequency of phrase i in the whole collection minus the retrieved set.

$$w_i = 1 - \frac{1}{log_2 \left(2 + \dfrac{tf_{i,ret} \cdot f_{i,ret} - 1}{tf_{i,col} + 1} \right)} \tag{3}$$

4   Evaluation

4.1   Information Retrieval Testbed and Evaluation Measures

The Information Retrieval testbed for our experiments has been the same as in [4]. The new JBraindead2 prototype, which was tested with a set of 47 TREC-like topics coming from the CLEF 2001 and 2002 campaigns, and having extensive manual relevance assessments in the CLEF EFE 1994 text collection.

The main evaluation measures used were the *Lattice Distillation Factor* (LDF) and the *Minimal Browsing Area* (MBA) defined and motivated in [4]. LDF is the precision gain between the original ranked list and the minimal set of documents which should be inspected in the lattice in order to find the same amount of relevant information. The biggest the LDF, the better the lattice. The MBA is the percentage of nodes in the lattice which have to be considered with an optimal browsing strategy. The smaller the MBA, the better the lattice.

Previous research on FCA applied to IR has barely focused on evaluation issues. Two exceptions are [11, 3], where empirical tests with users were conducted. In both cases, documents were manually indexed and the lattices were built using that information. Therefore, the problem of choosing optimal indexes was not an issue. In free-text retrieval, however, selecting the indexes is one of the main research challenges. Our LDF and MBA measures estimate the quality of the lattices for browsing purposes on different index sets, permitting an initial optimization of the attribute selection process prior to experimenting with users.

4.2   Experiments

We made three main experiments to test which selection strategy had best performance values in our information organization framework. Experiments are described in the following subsections.

Table 2. Experimental results for Experiment 1 with $k = 10$ and $k = 15$ with a number of documents $n = 100$. The averaged precision of the baseline ranked list was 0.15

	GB	QSB	T
LDF(%) (k=10)	255.7	84.17	16.7
LDF(%) (k=15)	562.72	127.01	25.17
MBA(%) (k=10)	46.97	59.25	69.14
MBA(%) (k=15)	42.48	60.64	73.29
Nodes (k=10)	114.27	35.89	13.31
Nodes (k=15)	161.24	48.07	19.87
Obj. Concepts (k=10)	52.8	27.56	12.09
Obj. Concepts (k=15)	65.73	36.98	17.93
Connect. Nodes (k=10)	61.47	8.33	1.22
Connect. Nodes (k=15)	95.51	11.09	1.93

Experiment 1. We evaluated the system using only noun phrases as document descriptors. We applied the three selection strategies (i.e. generic balanced, query-specific balanced and terminological) presented to extract the k most relevant phrases. The strategies were tested with the first 100 most relevant documents retrieved and a set of descriptors with $k = 10$ and $k = 15$. Results are summarized in Table 2 and in Figures 3 and 4, where GB stands for the generic

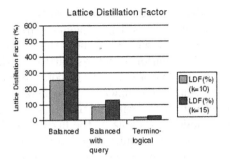

Fig. 3. Lattice Distillation Factor in Experiment 1

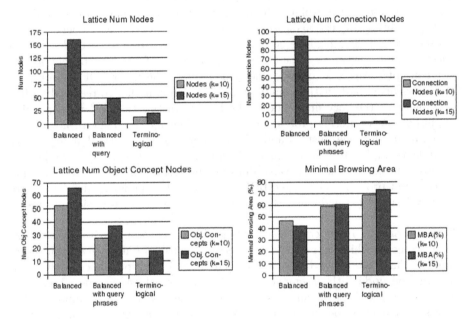

Fig. 4. Number of nodes, object concepts, connection nodes and Minimal browsing Area for Experiment 1

balanced phrase selection strategy, QSB is the query specific balanced strategy and T the terminological selection strategy.

Although the best lattice distillation factor (LDF) values are obtained using the Generic Balanced strategy (with improvements of 256% and 563% in the retrieval precision of the system), the size of the concept lattices generated makes them too complex for practical browsing purposes. In addition, minimal browsing area (MBA) values (i.e. which imply to explore a 47% and a 42% of the lattices) imply that a substantial proportion of a big lattice has to be inspected to find all relevant information. We estimate that the size (i.e. number of nodes) of a lattice should in any case be below 50% of the size of the document set; otherwise, the lattice makes document inspect even more complex than the original ranked list.

The main reason for the high LDF values in the Generic Balanced approach is the nature of the selected phrases. High document frequency phrases selected are shared by many documents, which enhances the possibility of combinations between documents. This situation generates lattices with a great number of connection nodes (i.e. nodes that are not object concepts and, as a consequence, do not have any document to be shown), which makes very easy to find the optimal ways to the relevant nodes without traversing any non-relevant node. A large set of connection nodes also explains the high values for the MBA.

Query Specific Balanced and Terminological strategies obtain lower LDF values but, in contrast, the number of nodes generated and the MBA values are more appropriate to our requirements. These two selection strategies choose more specific phrases (i.e. with lower document frequencies) than the Generic Balanced recipe, which explains the smaller size of the concept lattices obtained. At this point, two questions arise: due to the lower document frequency of the phrases selected, is the whole document space covered by the set of descriptors selected? and how many documents are generators of the top concept as object concept?. Answers to these questions explain the lower LDF values: a) a top information node with a small set of documents implies to read only a few documents at the very first node of the lattice. In this case, a low LDF value implies a poor clustering process with the relevant and non relevant documents mixed up in the same clusters, and; b) a top information node with a large set of documents implies to read too many documents at the first node of the lattice. In this situation, a low LDF value does not necessarily imply a bad clustering process, but probably a damaged LDF where lots of non relevant documents are clustered in the top node, and therefore always counted for precision purposes.

Experiment 2. In order to solve this problem, we proposed to characterize the documents which do not receive descriptors with a dummy descriptor '*other topics*'. This new context description generates a concept lattice with a top information node with an empty set of documents, which ensures a first node of the lattice with no documents to show. If the new node containing "other topics" documents does not contain relevant documents, then it will not affect negatively to the LDF measure.

Using this "other topics" strategy, we re-evaluated the three selection strategies with the first 100 most relevant documents retrieved and a set of descriptors with $k = 10$. Results are shown in Table 3 and in Figure 5, where GB represents the generic balanced phrase selection strategy, QSB the query specific balanced strategy and T the terminological selection strategy.

The results show much better LDF values for the Specific Query Balanced and the Terminological selection strategies than in the previous experiment (with improvements of 182% and 1300% respectively). The LDF value for the Generic Balanced selection strategy is also improved (237.84%). The values obtained in the previous experiment indicate that the bad clustering performance of the proposed selection strategies were due to the generation of top information nodes with too many non relevant documents (which the user is forced to read) which damage the final LDF values.

Table 3. Experimental results for Experiment 2 with $k = 10$ and with a number of documents selected of $n = 100$. The averaged precision of the baseline ranked list was 0.15

	GB	QSB	T
LDF(%) (k=10)	863.86	237.64	233.81
MBA(%) (k=10)	48.4	65.55	84.62
Nodes (k=10)	115.27	36.87	14.31
Obj. Concepts (k=10)	52.8	27.56	12.09
Connect. Nodes (k=10)	62.47	9.31	2.22

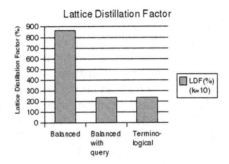

Fig. 5. Lattice Distillation Factor in Experiment 2

LDF values for the Specific Query Balanced and the Terminological selection strategies are very similar; a new question has to be asked to differentiate both: having similar LDF values, which of these strategies groups documents best?. Both strategies generate an acceptable number of nodes, but looking at the number of object concepts generated, we can see that the Specific Query Balanced strategy doubles the number of object concepts in the Terminological one. The number of object concepts is directly related with the number and size (i.e. in average) of the clusters generated and, as a consequence, the Specific Query Balanced strategy generates more clusters with a smaller size than those generated by the Terminological one. In this situation, a small set of large clusters gives a vast, poorly related view of the clustered document space where the user is not able to specialize the contents of the large relevant clusters selected. We think that this scenario is not desirable for browsing purposes and, provided that acceptable lattice sizes and similar LDF values are obtained, the selection strategy which gives the maximum number of clusters should be preferred.

Finally, there is a practical reason to select Specific Query Balanced as the preferred selection strategy: it is very simple to compute, and does not need collection statistics to be calculated. This is relevant, e.g., if the goal is to cluster the results of a web search without having collection statistics from the full web.

Experiment 3. As an additional experiment to avoid the overload of the top information nodes, we tested the effect of adding the query terms as document

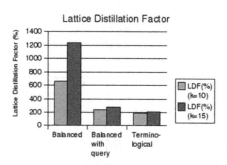

Fig. 6. Lattice Distillation Factor in Experiment 3

descriptors, in addition to noun phrases. The idea is based in the fact that query terms, that should appear as attribute concepts in the first levels of the lattice, are not only a good aid to drive the initial user navigation, but also they make a natural partition of the document space clarifying basic document relations and generating top information nodes with an empty document set.

We evaluated the three selection strategies with the first 100 most relevant documents retrieved and a set of descriptors with $k = 10$ and $k = 15$ built using the qt query terms and the $k - qt$ phrases selected. Results are summarized in Table 4 and in Figures 6 and 7, where GB represents the generic balanced phrase selection strategy, QSB the query specific balanced strategy and T the terminological selection strategy.

Although the results show that the Generic Balanced selection strategy obtains the best LDF values, the large number of nodes generated and, as a consequence, the small size of the clusters generated and the large number of connection nodes lead us to reject this selection strategy as optimal for our information organization purposes.

Query Specific Balanced and Terminological selection strategies perform lower LDF values but generate better-sized lattices. Query Specific obtains better LDF

Table 4. Experimental results for Experiment 3 with $k = 10$ and $k = 15$, with a number of documents $n = 100$. The averaged precision of the baseline ranked list was 0.15

	GB	QSB	T
LDF(%) (k=10)	664.34	235.97	192.04
LDF(%) (k=15)	1239.75	277.99	208.14
MBA(%) (k=10)	43.57	59.74	66.27
MBA(%) (k=15)	38.36	55.43	63.55
Nodes (k=10)	92.53	38.96	16.33
Nodes (k=15)	191.31	53	23.49
Obj. Concepts (k=10)	49.22	27.91	13.84
Obj. Concepts (k=15)	66.36	38.36	20.16
Connect. Nodes (k=10)	43.31	11.04	2.49
Connect. Nodes (k=15)	124.96	14.64	3.33

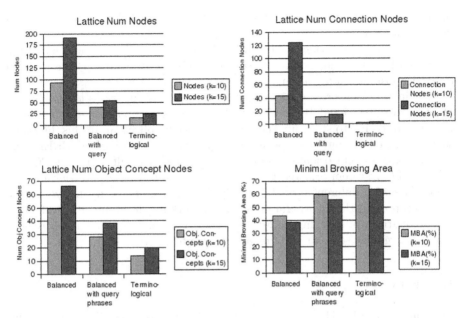

Fig. 7. Number of nodes, object concepts, connection nodes and Minimal browsing Area in Experiment 3

values than Terminological, which, in addition to our previous experiment discussion, makes this strategy better than the others in our information organization framework. Furthermore, although the growth of k does not significantly improve the LDF values, a better improvement is detected with the Query Specific Balanced selection strategy (i.e. Query Specific selection strategy improves in a 18% LDF values while, in contrast, Terminological selection strategy improves LDF only in a 8%).

5   Conclusions

Starting from the organization framework presented in [4], this paper has explored the possibility of using noun phrases as document descriptors to generate lattices for browsing search results. We have focused our research on the development of different phrase selection strategies, which have been tested using the LDF and MBA evaluation measures specifically designed for this task and introduced in [4].

The experiment results reveal a high clustering power for lattices built using all three selection strategies studied, being the Query Specific Balanced selection strategy the most suitable for user browsing purposes. The use of noun phrases, in contrast with the use of terms as document descriptors, deals with low document frequency values so the use mixed approach based on the use of query terms and phrases as document descriptors gives us the best LDF values. In addition, noun phrases are more adequate as intensional descriptions of lattice nodes from a user's point of view.

Our current work is focused on two main objectives: a) the evaluation of the generated lattices in an interactive setting involving users, and b) the research on lattice visualization and query refinement aspects.

The information organization framework proposed illustrates the scalability of FCA to unrestricted IR settings if it is applied to organize search results, rather than trying to structure the whole document collection. In this direction, some recent efforts have also been made by other systems such as CREDO [2] (i.e. an FCA system oriented to organize web search results) or DOCCO [10] (i.e. and FCA system oriented to manage the organization and retrieval of PC stored files with different formats) with promising results. A distinctive feature of our proposal is the incorporation of a framework for the systematic evaluation and comparison of indexing strategies within this general paradigm of FCA as a tool to organize search results in generic free-text retrieval processes.

References

1. Brin S., Page L.: The anatomy of a large-scale hypertextual Web search engine. Computer Networks and ISDN Systems. 30 **1–7** 107–117 1998.
2. Carpineto C., Romano G. Concept Data Analysis. Theory and Applications. Ed. Willey ISBN: 0-470-85055-8. 2004.
3. Carpineto, C. and Romano, G. A Lattice Conceptual Clustering System and its Application to Browsing Retrieval. Machine Learning (1996) 24, 95-122.
4. Cigarrán J.M., Gonzalo J., Peñas A., Verdejo F.: Browsing Search Results via Formal Concept Analysis: Automatic Selection of Attributes. Concept Lattices, Second International Conference on Formal Concept Analysis, ICFCA 2004. Sydney, Australia. Ed. Eklund P.W. LNCS **2961** 74–87 Springer-Verlag, Berlin 2004.
5. Cole, R. J. The management and visualization of document collections using Formal Concept Analysis. Ph. D. Thesis, Griffith University. 2000.
6. Cole, R. J. and Eklund, P. W. Application of Formal Concept Analysis to Information Retrieval using a Hierarchically structured thesaurus.
7. Cole, R. J. and Eklund, P. W. A Knowledge Representation for Information Filtering Using Formal Concept Analysis. Linkoping Electronic Articles in Computer and Information Science (2000), 5 (5).
8. Cole, R. J. and Eklund, P. W. Scalability in Formal Concept Analysis. Computational Intelligence (1999), 15 (1), pp. 11-27
9. Cole, R., Eklund, P. and Amardeilh, F. Browsing Semi-structured Texts on the web using Formal Concept Analysis. Web Intelligence (2003).
10. Docco Project home page. http://tockit.sourceforge.net/docco/
11. Godin, R., Missaoui, R. and April, A. Experimental Comparision of Navigation in a Galois Lattice with Conventional Information Retrieval Methods. Int. J. Man-Machine Studies (1993) 38,747-767.
12. Godin, R., Gecsel, J. and Pichet, C. Design of a Browsing Interface for Information Retrieval. In 12th Annual International ACM SIGIR Conference on Research and Development in Information Retrieval, (Cambridge, MA, 1989), ACM SIGIR Forum, 32-39.

13. Peñas, A., Verdejo, F., Gonzalo, J. Terminology Retrieval: towards a synergy between thesaurus and free text searching. Advances in Artificial Intelligence - IBERAMIA 2002. Ed. F.J. Garijo, J.C. Riquelme, M. Toro. LNAI **2527**. Springer-Verlag 2002.

14. Peñas, A., Verdejo, F., Gonzalo, J. Corpus-based terminology extraction applied to information access. Proceedings of the Corpus Linguistics 2001, Technical Papers, Special Issue. University Centre for Computer Corpus Research on Language, Lancaster University. **13**, 458–465, 2001.

15. Peñas, A., Gonzalo, J., Verdejo, F., Cross-Language Information Access through Phrase Browsing. Applications of Natural Language to Information Systems. Proceedings of 6th International Workshop NLDB 2001, Madrid, **P-3**, 121–130. Lecture Notes in Informatics (LNI), Series of the German Informatics (GI-Edition). 2001.

16. Priss, U. Lattice-based Information Retrieval. Knowledge Organization (2000), 27 (3), p. 132-142.

Combining Spatial and Lattice-Based Information Landscapes

Jon Ducrou and Peter Eklund

School of Information Technology and Computer Science,
The University of Wollongong,
Northfields Avenue, Wollongong, NSW 2522, Australia
{jrd990, peklund}@uow.edu.au

Abstract. In this paper we report on practical information visualization aspects of Conceptual Knowledge Processing (CKP), realizing and illustrating Wille's "conceptual landscapes" in the context of developing a conceptual information system to determine surfing conditions on the South Coast of New South Wales, Australia. This novel application illustrates some (if not all) of Wille's CKP tasks: exploring, searching, recognizing, identifying, analyzing, investigating, deciding, restructuring and memorizing (all but improving). It does this by concentrating on combining an information landscape with maps of the physical world.

1   Introduction

Conceptual Information Systems (CIS) conform to the 10 tasks of conceptual knowledge processing (CKP) defined in Wille's Landscape paradigm [1]. These include exploring, searching, recognizing, identifying, analyzing, investigating, deciding, restructuring and memorizing. Experimenting with Wille's 10 methods motivates this work and provides a design framework for the development of practical problem solving tools in CIS.

A survey of existing Web-based surfing portals reveals their reliance on Webcams and the absence of any analytical features[1][2][3][4]. These sites rely on low-quality streamed video inputs that are unreliable – they are often off-line, do not work in poor lighting conditions or at night and often "point" in the wrong direction to give any clear indication of the prevailing conditions (see Fig. 1).

Our objective is to improve on these portals by providing a more principled analysis of surfing breaks based on geographic information & weather inputs and showing prototypical images of the breaks based on a variety of weather conditions. By tying this information with maps and a concept lattice we have engineered the first Web-based Spatial CIS.

---

[1] http://www.coastalwatch.com.au
[2] http://www.surfit.com.au
[3] http://www.realsurf.com
[4] http://www.wannasurf.com

B. Ganter and R. Godin (Eds.): ICFCA 2005, LNAI 3403, pp. 64–78, 2005.

Fig. 1. http://www.coastalwatch.com is a popular surfing portal that features many Web-cams and weather forecasting tools for locations around Australia and a limited number of International locations including New Zealand

While the dimensionality of the input data to the system we describe is low, reflecting the detail (or lack thereof) contained within the input primary data sources, the SURFMACHINE system is a prototypical example of a Web-based CIS that integrates (in a natural way) with spatial data and improves the predominant Web-cam paradigm for portals aimed at surfers. The "landscape of knowledge" that results therefore closely reflects the practical knowledge that surfers apply when deciding where to surf.

This paper is structured as follows. Section 2 gives an overview of how beaches are affected by their shape and orientation. Section 3 describes the input source data. Section 4 describes the SURFMACHINE Conceptual Information System (CIS). Section 5 describes elements of the CIS as they relate to Wille's CKP tasks and Section 6 describes extensions and limitations of this work.

2   The Mechanics of Beach Selection in Surfing

One of the most important aspects of surfing is locating the best waves. Different
surf locations – or breaks – on a coastline have different physical properties
setting them apart. The shape of the coastline and underwater terrain determines
how and where waves will form. Ideally, a wave should break along the beach,
rather than all at once, thus headlands and curved beaches usually provide better
surfing than long straight stretches of sand (See Fig. 2). The exception is the
occurance of reefs[5] which cause the waves to break on both sides of the reef.
With knowledge of the structure of the local breaks, the determining factor on
any given day is based on the current weather conditions.

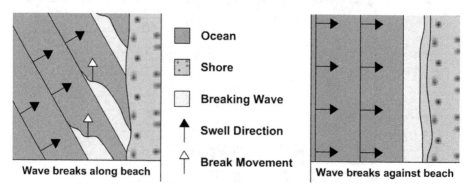

Fig. 2. The angle between the swell and the shore should be sufficient that the wave
breaks along the beach (left), rather than all at once against the beach (right)

The primary weather condition which determines if a break will be good is
the wind direction and strength. Wind should optimally be offshore, i.e. blowing
against the oncoming wave. This holds the wave up, slowing the break, and
giving the surf-rider time to navigate the face of the wave. Wind is affected by
the land structures surrounding or adjacent to a break; for example, wind can
be diverted around a headland, becoming an offshore breeze when it reaches the
sea. The strength of the wind should never be too strong (in any direction), but
some breaks are protected from strong offshore breezes by their geography.

The secondary condition which affects the quality of a break is the swell
direction and size. Swell waves are generated by wind and storm conditions far
from the point where they reach the coastline. As swell travels from its point of
origin, it becomes organised into groups with similar height and period[6]. These
groups can travel thousands of kilometers without significant changes to their
height and period. An example of this are Australian "cyclone swells", which

---

[5] Underwater structures, either natural (rocks) or man-made (pipes and purpose-built
artificial reefs).

[6] Distance between consectutive waves.

may cause a swell to start in northern Queensland and are felt as far south as the east coast of Victoria (some 3,000kms away)[7].

When swell reaches a coast, the way it interacts with the shape and orientation of each break dictates how the wave will break. As mentioned earlier, the wave should break along the beach. When a break is curved it will be better in a wider range of swell directions. Point breaks[8] have an interesting property that causes swell to wrap around the headland and so also allows for better surfing. Swell size, for the most part, relates to the skills and fitness of the surfer, but in situations like reef breaks, it can directly influence the break's quality (i.e. the wave has to be large to form over a reef).

Tidal conditions are the third factor of importance. In all but the smallest swell an incoming tide will have an effect on the swell size, increasing it. This is called tidal surge. The shape of the sand banks on a beach at any given time will determine the surf quality for a given tidal condition, either incoming or outgoing tide, but this is not usually something that can be codified in a surfing guide and is largely dependent on shifting sand. Ideal tide conditions can however be a predicting factor for more permanent undersea terrains, such as rock or point breaks. For the novice surfer, point and rock breaks are to be avoided because they require much higher skill levels to be ridden safely.

In general the combination of wind, swell direction and tidal condition mean that each break has its own characteristics and nuances which are usually learnt through experience and local knowledge.

3   Sources of Data

For this study, the primary data source was, *"The Surf Report - Journal of World-wide Surfing Destinations"*. Despite being published in the US by Surf Publications (Volume 17 No. 2, 1996)[9], this is the most authoritative guide on Australian surfing we have seen. It is a detailed (often obsessively so) list of all breaks – listing some 120 different breaks over almost 500 kilometers of coastline on the South New South Wales coast. Only 44 of the breaks were used in our study (considered "local" breaks), mainly due to data entry constraints. The 44 selected breaks are those around the Illawarra area from the Royal National Park (near Sydney) down to Minnamurra (north of Kiama) covering 94 kms of coastline.

An excerpt from "The Surf Report" is shown in Fig. 3. Data inconsistencies were checked with a secondary source "Mark Warren's Australian Surfing Atlas" [2] and with local surfers, mostly to normalize the local name of the break rather than use the official map location name[10].

---

[7] A similar European phenomenon is the Icelanic trough ("Island-Tief" in German).

[8] A wave that breaks along a headland or promontory.

[9] First Published in 1984.

[10] Surfers often use colloquial names for breaks, for example, Kilalea Beach is known as "The Farm" because for many years access to that beach required surfers to pay a toll to the local farmer in order to cross his land.

THE SURF REPORT VOL. 17#2   PAGE 3

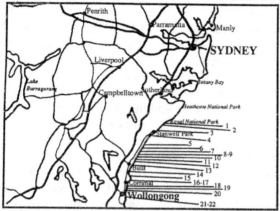

best or NW-W winds. **32. The Shallows.** Excellent reef breaks, ridable from 3-15'. All three reefs hook up as the swell gets gigantic. One of the few spots surfable when the wind is hard from the S and the swell is huge. It is the proposed site of a boat harbor. **32A. Redsands.** A hollow peak that breaks both left and right. For experienced surfers only. The rights are longer. Breaks to 10-12'. At 12' can have tubes to drive a truck through. Best on SE/E swell over 6' and SW/SE winds. E/NE swells actually break even bigger, but quality is not as good. **33. The Farm.** One of the most famous beach breaks in Australia. One of only two spots where the NE winds of summer are off shore, and one of the few beach breaks in the area with consistent sand banks. S swells. Beach-break surfing at its best. Ridable to 10' in the alley. **34. Minnamurra Mystics.** When the swells are from the NE, this sandspit of the Minnamurra River forms perfect lefts, with perfect rights on S swells. W winds required. Sharks.

(#'s 35-40 Kiama Area.) **35. Jones Beach.** Beach break with several rock outcroppings that hold sand and provide fairly stable banks. Any swell with NW-SW winds. **36. Boneyard.** Very powerful right reef break with heavy bowl sections, experienced surfers only. Never closes out. Ridable from 4-12', best at 6-10' on any swell with S wind. Similar to Sunset Beach, Hawaii. **37. Bombo.** Long sand beach with a variety of banks and wind directions. There's a left reef break off the headland called Pump Rock that can peel all the way through the beach break when the sand is right. **38. The Bay.** Lefthander in front of pool, breaking into the boat harbor. Ridable on NE swells or giant S swells. Good spot for grommets. NW-SW winds. **39. Rock Pool Reef.** Right reef break on the Blowhole Headland. Novelty value only. Dry reef takeoff, square barrel, ridable to 6' if you're crazy enough. NE swell, S winds. **40. Surf Beach.** Mediocre beach break in small cove, breaks on any swell with NE-SW winds. Can get good but usually isn't. **41. Werri Beach (Ourie Beach).** North end is a left reef/beach break and takes a NE swell with NE-NW winds. The middle of the beach takes any swell with NW-SW winds, ridable to 8'. The south end is a right reef/point that gets hot, ridable to 10' on NE-S swell with W-S winds. **42. Boat Harbour.** Breaks right or left in the Gerringong Boat Harbour on very large and especially clean swells. Located below a graveyard. **43. Ava-Go-Bernie.** Super hollow, grinding left reef break on north side of Black Point. Low tide only: NE swell and NW wind. The right is also hollow but short, S swell with S-SW winds to 8'. **44. Gerroa Bommie.** Very hollow and hard-breaking reef, similar to Velzyland. Right is more hollow, currents off the left. Works only a few times a year on super clean NE swells with NE winds, 4-8'. **45. Chips.** Left point break running along a rock ledge on the headland (Black Point).

sand bottom and reef over the years. Best as reef bottom. Small waves to 5-6' from NE best. Winds SE/W. **24. Flshos Reef.** A righthand reef break that breaks several times a decade. Really needs a large cyclone swell to get going, as Rabbit Island blocks most swell. The takeoff, when good, is a dry-rock suck-out in the 6-8' range. The wave then runs into deeper water down the point and becomes a slow, fullish wall. When this wave works, the Shellharbour area will be going off, so why bother? **24A. Stoneys.** A mysto spot that breaks off the lee of Rabbit Island, a full quarter-mile paddle off shore. Works a couple of times a year. Strong currents and sharks should deter all but the desperate, but it does work well on an E wind (rare in all of NSW). **25. Port Reef.** Lefthand reef break. Best around 8'. Can hold to 10' if tides and the banks of Port Kembla Beach do not interfere. Tubing takeoff is right in front of a board and body-eating rock called "Big Charley." Can get really good when the swell is clean. Best on S/SE swell and a NE/NW wind. **25A. Port Beach.** A very consistent beach break with generally good banks in northern corner. Can hold to 8-10' if the banks are right. From here to Windang the beach stretches about five miles. Various parts are known as Primbee,

Fig. 3. Excerpt from "The Surf Report"

The context derived from this data uses the breaks as the object set G. This is because the breaks form part of the desired result set.

$$G := \{Garie\ Beach, ..., Minnamurra\}$$

The attribute set M is comprised of possible wind and swell conditions, which form a constraint over the objects. These conditions are broken into eight possible compass directions. (Note: W_N is representitive of Northerly Wind. Also, this does not include swells that cannot occur, i.e., swells do not orginate from inland.)

$$M := \{W_N, W_{NE}, W_E, W_{SE}, W_S, W_{SW}, W_W, W_{NW}, S_N, S_{NE}, S_E, S_{SE}, S_S\}$$

The incidence relation I between the elements of G and the elements of M can be thought of as "works well in" and is represented by,

$$gIm \Leftrightarrow break\ g\ works\ well\ in\ condition\ m,\ where\ g \in G\ and\ m \in M.$$

The formal context is shown in Table 1.

	NW Wind	W Wind	SW Wind	N Swell	NE Swell	E Swell	SE Swell	S Swell	S Wind	N Wind	NE Wind	SE Wind	E Wind
Garie Beach	×	×	×	×	×	×	×	×					
Burning Palms	×	×	×	×	×	×	×	×					
Stanwell Park	×	×	×	×	×	×	×	×					
Coalcliff	×	×	×		×								
Scarborough	×	×	×	×	×	×	×	×					
Wombarra	×	×	×	×	×	×	×						
Coledale	×	×	×	×	×	×	×	×					
Sharkies													
Headlands			×	×	×	×	×	×	×				
Austinmer	×	×	×	×	×	×	×	×					
Thirroul	×	×	×	×	×	×	×	×					
Sandon Point			×	×	×			×					
Peggy's Point			×		×								
Woonona	×				×				×	×			
Bellambi			×	×	×	×	×	×					
Corrimal	×	×	×	×	×	×	×	×					
East Corrimal	×	×	×	×	×	×	×	×					
Towradgi	×	×	×	×	×	×	×	×					
Pucky's			×					×					
North Wollongong			×					×					
City Beach Wollongong	×	×		×	×	×	×	×					
Bastian's	×	×		×	×								
Shitties	×	×		×									
MM Reef					×	×							
MM Beach			×	×		×	×		×			×	
MM Bay			×		×			×				×	
Fisho's Reef													
Stoney's													×
Port Reef	×						×	×	×	×			
Port Beach	×			×	×	×	×	×	×	×			
Lake Illawarra Entrance													
Sharkies (Windang)			×		×				×			×	
Windang Island	×	×			×	×			×	×			
Warilla Beach	×	×	×										
Barrack Point		×	×	×				×					
Madman's			×		×								
Suck Hole	×	×			×	×			×	×			
Cowries			×	×			×	×	×				
Pools / The Bombie			×	×	×								
South Shellharbour	×				×				×				
The Shallows								×					
Redsands							×	×	×			×	
The Farm								×			×		
Minnamurra			×		×			×					

Table 1. The formal context for the SURF-MACHINE. Note Sharkies, Fisho's and Lakes Entrance. The description for Fisho's in the guide says "a right-hand reef break that breaks several times a decade" which tells us something but nothing about the Wind or Swell conditions. Likewise Sharkies, "short shallow left reef, break named after shark attack 30 years ago". Lakes Entrance also contains no extractable information. This explains why there are no attributes for these objects and helps underline the difficulty of the information extraction problem in texts of this sort

Fig. 4. The above photos show two breaks in the Illawarra region in good conditions. (Top: Sandon Point, Bottom: Port Kembla)

4   The SurfMachine

The SURFMACHINE system is a simple Web-based CIS (in the style of RENTAL-FCA [3]) and comprises an area to capture the user's query, a lattice display, a geographical map of the areas covered and a database of individual photos of indicative conditions.

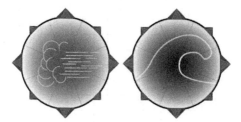

Fig. 5. Wind and Swell Queries are captured with a Pair of Compasses

The user enters a query with 2 facets, wind and swell direction via a graphical representation of 2 compasses. This provides an input control so that only valid compass points can be entered (see Fig. 5). Both wind and swell directions are

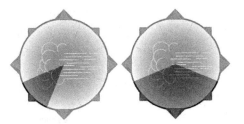

Fig. 6. Queries are restricted to ranges achieved by clicking on the start compass point followed by the end compass point. Above left, South-West is clicked, then, above right, South-East is clicked. This would be representative of the query 'Wind equals South-West, South or South-East'

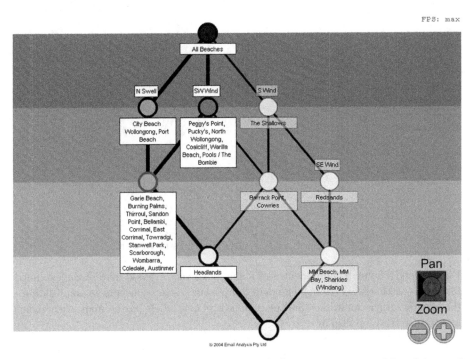

Fig. 7. The lattice component displays the break data as a conceptual breakdown of the current weather conditions

split into 8 discrete values (*North, North-East, East, South-East, South, South-West, West, North-West*) and queries are limited to a range of directions not exceeding a range over 4 compass points, e.g. North through South-East, which is 4 values (*N, NE, E, SE*). Another example is shown in Fig. 6. Each of the values are used as a scale representing *current weather* and the corresponding concept lattice is generated (see Fig. 7). The extent of each concept of the lattice translates to a set of locations on the map, and clicking the concept realises its extent as locations on the map (see Fig. 8).

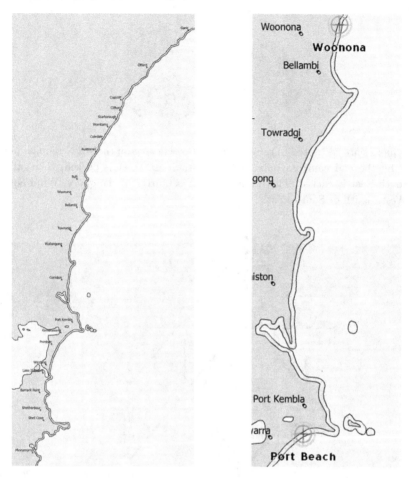

Fig. 8. (left) The map component uses a simplified vector-based map of the extended Illawarra coastline; and (right) The map component realises concepts as point data for the given weather conditions, and places markers and labels on the map, then zooms to optimize the map display

In this way, the resulting formal concepts map to sets of geographic point data. A given formal concept may therefore contain multiple surf breaks geographically dispersed over the total area of the study. Visualizing the information space as a concept lattice allows the user to view multiple results by representing the objects in the extent as points on a map.

An important feature of the SURFMACHINE is that the surfer can see breaks which are "similar to one another" from the concept lattice. In so doing, the concept lattice reaffirms local knowledge about which breaks work in which conditions and suggests alternate locations to those the surfers may know and like. For instance, from the concept lattice in Fig. 7 we can see that *Garie Beach, Thirroul, Sandon Point* and *Austimer* etc all cluster to the same formal concept

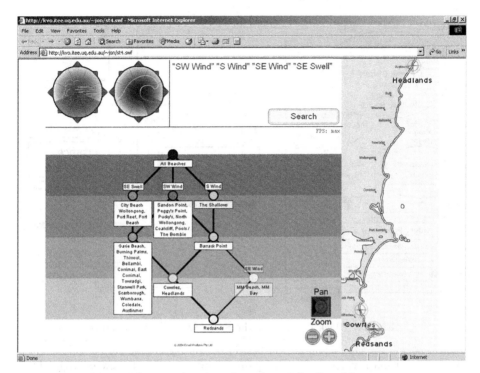

Fig. 9. The complete interface view of the SURFMACHINE

under the input weather conditions. They will therefore all be similar in a SW Wind and a N Swell, so the surfer would feel inclined to visit the closest on the day or if he/she is feeling adventurous try a different break with knowledge that that break will be similar to one already surfed.

The SURFMACHINE uses the Model–View–Controller design pattern, with the server holding all data and performing processing of the conceptual data (the *model*). The client controls all transactions and data flow (the *controller*) and uses an embedded plug-in for lattice drawing and a map view component to render the spatial data and images (the *view*). The architecture is shown in Fig. 10.

The lattice visualization component has been based on code from the MAIL-SLEUTH[11][4] program. It has been extensively re-worked to be used as a general flash plug-in and is now part of the Tockit project[12]. It uses the TOSCANAJ [5] look-and-feel (also used in HIERMAIL[13]), and uses a more advanced user interaction method than HIERMAIL/MAIL-SLEUTH[14]. The map component is a

---

[11] http://www.mail-sleuth.com

[12] http://tockit.sourceforge.net

[13] The academic version of MAIL-SLEUTH at http://www.hiermail.com

[14] SURFMACHINE uses a simple attribute additive interaction, rather than using the ideal.

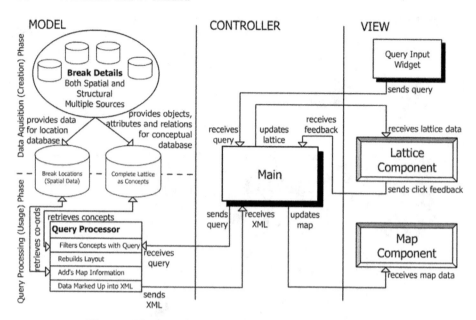

Fig. 10. Block Diagram of the SURFMACHINE architecture

vector-based map which scales and pans to optimise the display of highlighted locations.

The *model* goes through two phases, a *creation* phase and a *usage* phase. The creation phase makes use of the various data sources to create a concept lattice that encompasses the entire context and a database of coordinates. This phase was used initially to set up the databases and then subsequently to restructure the data when faults or deficiencies were found. The usage phase centers around a Query Processor which accepts the query from a client and returns XML representing the lattice to be displayed. The Query Processor is divided into four steps; filtering, rebuilding, adding map information and mark-up into XML. The term 'filtering' is used in the lattice theory sense, whereby the attribute set of a lattice is restrict to produce a sub-lattice. The next step rebuilds the layout so that it is optimised and the filtered lattice will graphically fit better in the lattice viewer component. Then the point data is added to the concepts by querying the spatial database and extracting the breaks for each concept. Finally, the various data is brought together, and formatted into XML for transport to the browser client.

5   SURFMACHINE and CKP Tasks

Having given the details of the domain knowledge and implementation architecture, we return in this last section to the landscapes paradigm and identify the conceptual CKP tasks that SURFMACHINE demonstrates.

The surfer is looking for a place to surf so he knows that he is looking for a break, but which one? He can constrain his choice by giving the input weather conditions and this corresponds to **exploring** the conceptual landscape of knowledge by specifying constraints on the attribute space. Depending on the strength of the specified constraints, the user is alternatively **searching** or **recognizing** knowledge. For instance, he can search for the best break in current conditions or he can see which day (given a several day forecast) would be best for a specific break. The user therefore shifts from a search focus based on the attribute space to recognition by focusing on individuals from the object space.

Recognizing in the context of the SURFMACHINE is being able to clearly see which breaks have similar properties, and those which are anomolies in the data. For example, the "*Garie Beach, Burning Palms, Thirroul, Corrimal, East Corrimal, Towradgi, Stanwell Park, Scarborough, Wombarra, Coledale, Austinmer*" concept is the *only* concept that has a contingent size greater than 3 and is also the upper cover of the base concept. This group shares 8 of the 13 attributes used[15] attributes "*NW Wind, W Wind, SW Wind, N Swell, NE Swell, E Swell, SE Swell, S Swell*" and is not broken down any further. This means the group represented by this formal concept is always seen together and because they share a large portion of the possible attributes these beaches can be considered "safe" beaches, namely because they are recommended in the majority of conditions.

At the other end of the spectrum is a break by the name of '*Stoney's*', which is the only break in the data which has '*E Wind*' as an attribute. Initially this was assumed to be an error, but after some research it was found that '*Stoney's*' is in fact the only break on the entire east coast of Australia considered surfable in an onshore breeze because it breaks on the western edge of an off-shore island called Big Island (aka Rabbit Island) near Port Kembla.

The information space in SURFMACHINE also suggests a topology of surf breaks that corresponds to **identifying** in the landscapes paradigm. The concept lattice represents a conceptual landscape where it is possible to see a hierarchy of breaks in the current weather conditions; from top to bottom, and as an order ranking, worst to best.

Analyzing is what is revealed by the structure of the conceptual language and the way it supports the collected knowledge. For instance, two of the most reliable beaches in the Illawarra during summer are *Port Beach* and *Woonona*, everyone who surfs locally "understands" this. It is therefore reassuring that SURFMACHINE shows these two beaches as a single formal concept when the prevailing summer conditions are used as the query (*North East* wind & *North, North East* swell).

When looking for a break, if the 'best' break(s) (contained in the bottom most concept of the lattice, i.e., those that have all query conditions as attributes) are undesirable – because it is unsafe to park your car for instance – relaxing the

---

[15] The swell never comes from inland, so NW Swell, W Swell and SW Swell are attributes that are never used.

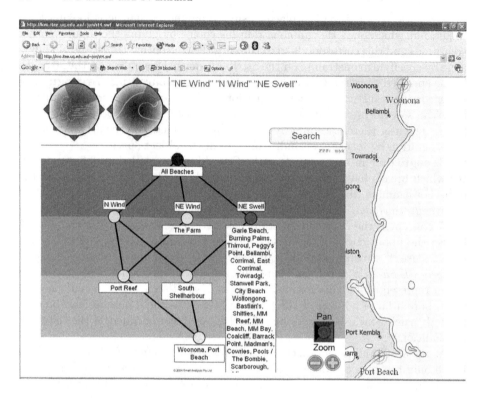

Fig. 11. The two best beaches in summer on the Illawarra are Woonona and Port Beach because they 'work' on the prevailing summer winds and swells

concept to its upper neighbors requires a methodology for **investigating** its direct upper cover. Generalization of a best concept point should therefore proceed by choosing concepts above that have both swell and wind attributes. This is illustrated in Fig. 11: if we can't surf *Port Beach* or *Woonona* (for whatever reason) the next best generalization is *South Shellharbour* combining a *Northerly Wind* and *North-East Swell*. The formal concept with the extension *South Shellharbour* is preferred to the formal concept with the extension *Port Reef* (also in the direct cover) which combines *Northerly* and *North-East* Wind conditions in its intent (and therefore ignores all swell information).

Restructuring the presentation of the conceptual landscape from its original paper-based form **improves** the presentation of the source data and introduces support for inter-subjective argumentation – namely **decision** support based on mixed initiative. This enhances the user's capacity to **investigate** the domain. Further **restructuring** occurs at fault points in the data; these were found when domain experts (local surfers) could *see* breaks that didn't belong with others, or breaks that were missing from natural groupings in formal concepts of the concept lattice.

With the aid of the SURFMACHINE, the process of gaining experience with a surfer's local coastline is accelerated, allowing an entire lifetime of local experience (often based on trial and error) to be committed to the concept lattice for **memorizing** the quality of various local breaks under various conditions. The resulting conceptual landscape is intuitive and anecdotally correct.

6   Limitations

We mentioned in the introduction that the dimensionality of the input data is low, reflecting the detail (or lack thereof) contained within the input primary data sources. This statement needs some clarification.

Wind and swell directions are uniformly presented in *"The Surf Report - Journal of World-wide Surfing Destinations"* by Surf Publications (1996) making data acquisition in these dimensions straightforward. The same cannot be said for other attributes such as *tide conditions, wind strength* and *swell size*. These attributes are somewhat obscured in the text, for instance the use of fuzzy phrases like "works well on large swells", and surf idioms like "southerly buster", "double over-head.." and "ebb-tide danger-spot..". These give character to the narrative as well as useful information about swell size, wind and tidal conditions (if you can understand the idioms). These features are harder to extract and quantify but represent ways to extend the data.

In the same way that the dimensionality of the data can be extended to include tide conditions, wind strength and swell size, so too can the analysis be improved by these attributes being filtered using the same kind of control input widgets as swell and wind direction. Likewise, filtering is complemented by object zooming, allowing the user to explore the attribute space for a select group of objects, e.g. analysis of all reef breaks, beach breaks only, good high-tide point breaks, double overhead spots etc.

Finally, the current local weather conditions can be extracted directly from the Bureau of Meteorology[16] and presented as the default view. This eliminates the need for user data inputs but it also limits the exploration aspect of SURF-MACHINE.

7   Conclusion

Our claim is that SURFMACHINE is the first Web-based Spatial Conceptual Information System. It demonstrates a number of Wille's Conceptual Knowledge Processing tasks: namely exploring, searching, recognizing, identifying, analyzing, investigating, deciding, restructuring and memorizing – all but improving. It does this by concentrating on combining a conceptual information landscape with maps of the physical world.

---

[16] http://www.bom.gov.au

References

1. Wille, R.: Conceptual landscapes of knowledge: A pragmatic paradigm for knowledge processing. In Gaul, W., Locarek-Junge, H., eds.: Classification in the Information Age, Springer (1999) 344–356
2. Warren, M.: Atlas of Australian Surfing. HarperCollins (1998)
3. Cole, R., Eklund, P.: Browsing semi-structured web texts using formal concept analysis. In: Proceedings 9th International Conference on Conceptual Structures. LNAI 2120, Berlin, Springer (2001) 319–332
4. Eklund, P., Ducrou, J., Brawn, P.: Concept lattices for information visualization: Can novices read line diagrams. In Eklund, P., ed.: Proceedings of the 2nd International Conference on Formal Concept Analysis - ICFCA'04, Springer (2004)
5. Becker, P., Hereth, J., Stumme, G.: ToscanaJ - an open source tool for qualitative data analysis. In: Advances in Formal Concept Analysis for Knowledge Discovery in Databases. Proc. Workshop FCAKDD of the 15th European Conference on Artificial Intelligence (ECAI 2002). Lyon, France. (2002)
6. Carpineto, C., Romano, G.: Concept Data Analysis: Theory and Applications. John Wiley & Sons (2004)
7. Colas, A.: The Stormrider Guide: The World. Low Pressure Publishing (2001)

Explaining the Structure of FrameNet with Concept Lattices

Francisco J. Valverde-Albacete*

Dpto. de Teoría de la Señal y de las Comunicaciones,
Universidad Carlos III de Madrid
Avda. de la Universidad, 30. Leganés 28911. Spain
and
International Computer Science Institute,
1947 Center St. Berkeley, CA, USA
fva@tsc.uc3m.es

Abstract. This paper reports on ongoing work to use Formal Concept Analysisas an auxiliary tool in understanding and visualising the wealth of data produced by lexical-resource building as embodied in the construction of FrameNet, a database to capture the syntax and semantics of language use in Frame Linguistics. We present proof of the abundance of concept lattices both in the theory of frames and in its present day incarnation, the FrameNet resource, with contributions that range from data-visualisation to the fine-tuning of some lexico-theoretical concepts better understood in terms of Formal Concept Analysis.

1 Introduction

FrameNet[1, 2, 3] is a lexical resource and database being developed at the International Computer Science Institute, ICSI. The name "FrameNet" reflects the twin facts that the project exploits the theory of Frame Semantics and that it is concerned with the semantics of networks through which word meanings are connected with each other. In this paper, we will be concerned with the modelling of Frame Semantics in terms of concept lattices and we will use the data in FrameNet to build concrete examples of such lattices.

The modelling of linguistic phenomena figures among the first applications of Formal Concept Analysis. Early references first applied the technique to the modelling of linguistic paradigms [4, 5, 6]. Lately the spectrum of applications has widened [7] for instance to the formalization of the sign relation [8]. In the lexical enterprise proper, a formalization of WordNet in terms of concept lattices has already been attempted [9, 10]; non-standard techniques like neighbourhood concept lattices have even been used to add structure to such a widely-used

* This work has been partially supported by the Local Government of Madrid, Spain. grant CAM-07T/0014/2003-1 and by a grant for "Estancias de Tecnólogos Españoles en el International Computer Science Institute" of the Spanish Ministry of Industry.

B. Ganter and R. Godin (Eds.): ICFCA 2005, LNAI 3403, pp. 79–94, 2005.
© Springer-Verlag Berlin Heidelberg 2005

resource as Roget's thesaurus [11, 12] and even machine learning techniques have been used to induce the concept lattice of some lexical relations [13, 14]. We believe, however, that ours is the first attempt to model Frame Semantics with the help of Formal Concept Analysistechniques.

The unit of description in FrameNet is not the word of traditional dictionaries, or the Synsets of WordNet, but rather *frames*, the "schematic representations of the conceptual structures and patterns of beliefs, practices, institutions, images, etc. that provide a foundation for the meaningful interaction of a given speech community"[3]. The central idea of the whole frame semantic enterprise is that *word meanings must be described relative to semantic frames*. FrameNet also considers frames as being partially ordered by several special semantic relations, among which the "inherits" relation stands out as the most informative and restrictive.

Our contention is that Formal Concept Analysismethods and constructions can help greatly in formalizing frames, frame elements and other frame-theoretic concepts. Because of the constructive approach of concept lattices, such formalization would automatically entail visualisation, data-mining and (lattice-oriented) application-building capabilities around the FrameNet data. Our purpose with this paper is to show the abundance of concept lattices in FrameNet and how they can help not only in the above-mentioned roles, but also in bringing formalization to strictly frame-theoretic notions.

In the rest of the paper we first review the general enterprise of FrameNet with special attention to the concept of "coreness" so that the reader may later understand lattices built from a lexicographical perspective. In section 3 we introduce the types of lattices we will be considering and how to build them from FrameNet data. Section 4 is dedicated to the analysis of the *coreness* of frame elements and invokes Formal Concept Analysisto obtain concept lattices that help define the concept of "coresets", crucial to Frame Semantics, producing some analyses of coreness and coresets as obtained from the FrameNet database. In this section we also pose the question whether frame structure is inherited through the frame hierarchy. We conclude with some suggestions of research into Frame Semantics using concept lattices.

2   FrameNet: An Embodiment of Frame Semantics

Regardless of their theoretical interest, lexical resources capable of serving natural language processing applications should at least include the following types of information:

- representations of the meaning of each lexical unit (LU);
- various types of relations between lexical units;
- information about a word's capacity to combine with other linguistic units to form phrases and sentences;
- semantic information associated with individual words that allows us to interpret the phrases that contain them.

However, FrameNet does not cover all these issues, but rather concentrates on the first, third and fourth points, in the hope of complementing previously existing lexical resources more concerned with the second point, such as Word-Net. On the other hand, FrameNet is arguably the paramount example of a frame-based resource for English (and has spawned a number of versions for other languages) hence its perspective on the task of defining the meaning of linguistic items is unique.

2.1 Introducing Frame Semantics

The unit of description in FrameNet is not the word of traditional dictionaries, or the synsets of WordNet, but rather *frames*, "(the) schematic representations of the conceptual structures and patterns of beliefs, practices, institutions, images, etc. that provide a foundation for the meaningful interaction of a given speech community"[3].

Specifically, FrameNet endeavours [3]:

1. "to describe the conceptual structures, or frames, a given lexical unit belongs to;
2. to extract sentences from the corpus that contain the word and to select sentences that exemplify the intended sense of each of the lexical units;
3. to annotate selected sentences by assigning frame-relevant labels to the phrases found in the sentences containing the lexical unit; and
4. to prepare reports that summarize the resulting annotations, showing succinctly the combinatorial possibilities of each lexical unit; these are called *valence descriptions*."

An Example: The Revenge frame. In order to better understand these concepts we now review a frame definition example. We have somehow[1] detected the need for:

- a frame we will call the REVENGE frame;
- to accomodate lexical units like *avenge, avenger, revenge(noun), revenge (verb), get back (at), get even (with),* etc.
- where *salient* participants in the idea would be an AVENGER, who inflicts a PUNISHMENT on an OFFENDER as a consequence of the INJURY, an earlier action perpetrated by the offender. (Note that the AVENGER may or may not be the same individual as the INJURED PARTY suffering the INJURY.)
- And a description of this REVENGE idea has to do with the idea of inflicting punishment in return for a wrong suffered, where the OFFENDER has been judged independent of the law (so as to distinguish REVENGE from legally sanctioned Punishment).

---

[1] In whatever manner lexicographers detect the need for a new entry in their inventory of items.

All the items above involve Frame Semantics-distinguished concepts:

- "REVENGE" is the *frame name*, an univocal identifier;
- We say that the set of lexical units ascribed to a frame *evoke* it in the discourse.
- The events, participants and props (tools), involved in instances of the frame scenario are its *frame elements*;
- The description itself constituting a *definition* of the frame in question.

The immediate consequence of having achieved such a definition is that we may properly interpret the following *annotated sentences* with respect to the REVENGE frame:

(1)   [AVENGER Ethel] eventually **got even** [OFFENDER with Mildred] [INJURY for the insult to Ethel's family]

(2)   Why hadn't [AVENGER he] sought to **avenge** [INJURED PARTY his child]?

2.2   Frame to Frame Relations

FrameNet considers the set of frames as partially ordered by several special semantic relations, notably:

- the **inherits** relation, which holds between a more general and a more specific frame when all of the properties assigned to the parent correspond to some properties of the child frame. For instance, the MOTION_DIRECTIONAL frame inherits all of its properties from the MOTION frame and further specifies some of the frame elements.
- in contrast, the **using** relation is posited to hold between two frames when one makes reference to the structure of a more abstract schematic frame, but it may fail to acquire all the defining properties of the parent, for instance the set of core frame elements.
- a different kind of ordering is introduced by the **subframe** relation, which holds between a frame that is temporally complex, like a process, and the frames describing possible states of affairs in the process and transitions between them. For instance, in the *Employment* frame describing the employing process from the viewpoint of the employer, we could have frames *hiring*, *employing* and *firing* describing the initial event, mid-state and final event of the employment scenario.

2.3   Frame Structures

For each frame f, the set of frame elements which may appear as salient in a sentence evoking the frame will be referred to as the *frame structure of f, $\mathcal{E}(f)$*. A frame element for a frame f can be classified as [1] :
core if it "instantiates a conceptually necessary participant or prop of a frame, while making the frame unique and different from other frames". We take this to

imply that all uses of f-evoking lexical units should express it[2]. Hence, the core elements for f must appear in each sentence in the annotated subcorpus of f. For instance, AVENGER is a core frame element for frame REVENGE. The importance of *coreness* resides in the fact that all *core* frame elements are inherited by every child frame.

non-core if it does not "introduce additional, independent or distinct events from the main reported event (...) They do not uniquely characterize a frame, and can be instantiated in any event frame". What we take to mean that their relation with the frame is rather contingent, hence they may or may not be expressed in any sentence annotated in the subcorpus of f. For instance, TIME is extrathematic in REVENGE, because not all uses of lexical units evoking it require a salient time coordinate.

Coreness of frame elements is crucial for a theory of frame-semantics because it describes the syntagmatic behaviour of lexical units evoking that frame: all core elements have to appear in every use of a frame somehow or other. As such, it has received much theoretical attention in recent FrameNet work. It is defined in [1] under the heading "Recent innovations and Future plans" (hence we may consider it a topic undergoing consideration and testing) and further discussed in the subsection "Coreness sets", where the following facts about frame element behaviour are recorded:

- "some *groups* of frame elements seem to act like sets, in that the presence of any member of the set is sufficient to satisfy a semantic valence of the predicator." [3]
- "...it is not necessary for all the frame elements (in such sets) to occur ..."
- "...if only one of them (i.e. frame elements) occurs in a sentence annotated in a ... frame, we consider it to be sufficient to fulfill the valence requirement of the target word..."
- "...In some cases, the occurrence of one frame element in a coreness set requires that another occurs as well".
- "...In some cases, if one of the frame elements in a CoreSet shows up, no other frame element in that set can".

We prefer "core sets" to "coreness sets" for reasons of terminology.

Missing Elements or *Null Instantiation*. One more phenomenon complicates questions of coreness: sometimes annotation is put on a sentence to indicate *null instantiation*, i.e. the fact that some salient frame elements are omitted in the sentence. There are three types of null instantiation:

- *Constructional Null Instantiation (CNI, structural omission)*, when the "omission (is) licensed by a grammatical construction in which the target word

---

[2] Allowing here for the consideration of phenomena related to null-instantiation, see below.

[3] Italics added.

appears and are therefore more or less independent of the lexical unit", for instance, "the omitted subject of imperatives, the omitted object of passive sentences, the omitted subjects of independent gerunds and infinitive", "missing objects in instructional imperatives", etc;

— *Definite Null Instantiation (DNI, anaphoric null instantiation)*, a lexically specific type of null instantiation, "in which the missing element must be something that is already understood in the linguistic discourse context";
— *Indefinite Null Instantiation (INI, existential null instantiation)*, as "illustrated by the missing objects of verbs like *eat*, *sew*, *bake*, *drink*, etc., that is, cases in which these ordinarily transitive verbs can be spoken of as used intransitively." We also learn from [1] that:
  • "There are often special interpretations of the existentially understood missing objects."
  • "However, the essential difference between indefinite/existential ad definite/anaphoric omissions is that with existential cases the nature (or at least the semantic type) of the missing element can be understood given the conventions of interpretation, but there is no need to retrieve or construct a specific discourse referent."
  • "...usually verbs in a frame differ in this respect. For instance, while *eat* allows its object to be omitted, *devour* does not," although both belong in the same frame.

Null instantiation is important for the synthesis of concept lattices because crosses have to be supplied differently for different formal contexts, depending on the type of null instantiation in each sentence.

3   Concept Lattices for FrameNet

Given the above state of affairs from the perspective of FrameNet as a lexicographic enterprise and the present-day development of Formal Concept Analysisas outlined in [15], the question is whether the methods of Formal Concept Analysiscan help alleviate the titanic effort of developing a resource such as FrameNet. From the lexicographic point of view, we hope to glean some understanding from the alternate view on FrameNet data offered by concept lattices.

From the lattice-application point of view we try to find rich, real-world data demonstrate the adequacy of Formal Concept Analysisfor data analysis, as well as making these techniques available to a broader community.

3.1   Prominent Contexts and Lattices in FrameNet

To apply Formal Concept Analysistechniques to FrameNet data the following must hold: objects should include theoretically meaningful concepts of Frame Semantics; attributes should depict theoretically meaningful properties or property-bearing items of Frame Semantics; and the incidences should capture linguistically meaningful relations between the above-mentioned items and properties or property-bearing items. For the above-mentioned fragment of Frame Semantics only the following theoretical objects are necessary[4]:

- The set $\mathcal{F}$ of frames (amounting to some 450 annotated frames as of Fall 2004).
- A class of partial order relations in this set, included in its partial orders, $\mathcal{O}(\mathcal{F})$:

$$\{inherits, subframe\_of, uses, \cdots\} \subseteq \mathcal{O}(\mathcal{F}) \subseteq 2^{\mathcal{F} \times \mathcal{F}}$$

- The set $\mathcal{E}$ of frame elements, and in particular the *frame structure of* $f \in \mathcal{F}$, i.e. the set of frame elements salient in the frame $\mathcal{E}(f) \subseteq \mathcal{E}$.
- The set of sentences in the corpus, $\mathcal{S}$, and in particular the *subcorpus*[5] *for* $f \in \mathcal{F}$, i.e. the set of sentences with targets evoking a particular frame $\mathcal{S}(f) \subseteq \mathcal{S}$.

With those structures in mind, there are at least the following types of concept lattices that prominently come to mind when considering the lexicographic data:

- The lattices derived from any ordering relation between frames, i.e. lattices of contexts $[H](r) = (\mathcal{F}, \mathcal{F}, r)$, where $r \subseteq \mathcal{F} \times \mathcal{F}$ may range over all possible orders induced by the covering relations between frames. These are the Dedekind-MacNeille completions of such orders and allow, for instance, to observe the inheritance relation between frames in FrameNet. We may obtain a lattice for each of these relations between frames, hence we call these the *frame lattices*.
- The lattice of frames and their associated frame elements, or *overall frame structure*, i.e. $[S] = (\mathcal{F}, \mathcal{E}, I)$. The description of $I \subseteq \mathcal{F} \times \mathcal{E}$ is one of the main endeavours of FrameNet and may be paraphrased as "frame f sports frame element e". However, because frame elements are defined as frame-dependent, essentially no structure would be observed in this lattice, as each frame element (as attribute) ranges exactly one object: its frame. Without further constraints, this lattice would look like $\mathbb{M}_{|\mathcal{F}|}$, the Dedekind-MacNeille completion of the antichain of $|\mathcal{F}|$ elements: the flattest complete bounded lattice of cardinal $|\mathcal{F}| + 2$.

Although some further structure can be gained from considering frame element to frame element relations, we now concentrate on a different device to see *a part of* the overall frame structure lattice. Specifically, for a particular frame f we consider the lattice of sentences evoking it $\mathcal{S}(f)$ and its frame elements $\mathcal{E}(f)$, i.e. those uses of a particular frame $[U](f) = (\mathcal{S}(f), \mathcal{E}(f), I(f))$, where the incidence $I(f) \subseteq \mathcal{S}(f) \times \mathcal{E}(f)$ describes uses of f *attested in some corpus*. We call this a *frame structure lattice*.

---

[4] Note that we choose to ignore at present the underlying set $\mathcal{L}$ of lexical units evoking frames already in the database, although its elements are somehow implied in the definition of the set of sentences bearing frame-evoking lexical units.

[5] Note also that the use of "subcorpus" here differs from that of FrameNet practice.

3.2   Frame Lattices

These lattices can easily be obtained as the concept lattice of up-sets and down-sets of points in the ordering relation achieved with the transitive and reflexive closure of covering relation recorded in FrameNet, $[H](r) = (\mathcal{F}, \mathcal{F}, r)$. For instance, let r be the *inherits* relation in the following. The concepts of this lattice are proven to be of the kind

$$(\downarrow f, \uparrow f)$$

([16], p. 73), i.e. the concept for a particular frame $f \in \mathcal{F}$ has for its extent all those frames inheriting from it and for its intent all those frames it inherits from. Figure 1 shows the part of the frame lattice where the MOTION frame resides.

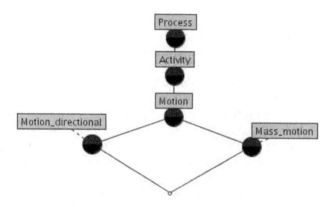

Fig. 1. A partial view of the frame lattice in an environment of frame MOTION

3.3   Frame Structure Lattices

Next, consider the annotation subcorpus for a frame f: it is composed of those sentences with a target element evoking the frame $\mathcal{S}(f)$ for which all incident frame elements $\mathcal{E}(f)$ have been annotated. Hence, we may view the sentences in the subcorpus as the objects, and the frame elements for the particular frame in the subcorpus as the attributes for a formal context $\mathbb{U}(f) = (\mathcal{S}(f), \mathcal{E}(f), I(f))$ whose incidence relation $sI(f)e$ quite intuitively reads: "for frame f sentence s expresses frame element e." Note that there is one such context for each frame f in the set of frames, i.e. around 450 as noted previously.

Lattice Building Procedure. The data for frame structure contexts can be obtained straightforwardly from the FrameNet database through the use of MySQL statements. For each frame, if we choose to represent the sentences in its annotation subcorpus by their identifiers and its frame elements by their names in the database, the data extraction procedure can proceed as follows:

sentence × FE	THEME	PATH	GOAL	MANNER	AREA	...
948874	×	×	×			
948886	×	×				
948888	×		×			
948891	×	×				
948893	×			×	×	
...						

Fig. 2. Part of the sentence-by-FE formal context for MOTION_DIRECTIONAL

1. For each frame, the FrameNet database is consulted, by selecting all data for lexical units in the appropriate annotation status[6]. For instance, the context for frame MOTION_DIRECTIONAL looks something like that shown in Fig. 2.
2. The data returned by the DBMS is filtered by a program to write the frame structure context in a suitable interchange format. Since we are interested in quantifying the support for certain concepts in the frame structure, no rows or columns are eliminated from the observed data, i.e. no clarification

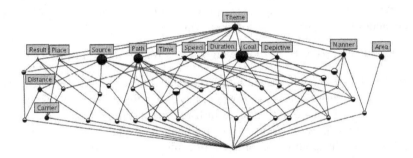

Fig. 3. The frame structure lattice for MOTION_DIRECTIONAL

3. Finally, the data are uploaded in the CONEXP tool[7] [17], the resulting graph of the selected core frame element rearranged to improve its readability and its picture written to secondary storage for later transformation to printable format (PostScript), as exemplified in Fig. 3.

4   Lattice Analysis of Core Frame Structures

Because core sets are so important in the theory of frame elements, our purpose in this paper is to contribute to their understanding by providing a model in

---

[6] In the following, we only used annotation in the *FN1_Sent* or *Finished_Initial* status of annotation, and only considered the annotation of lexical units which are in fact verbs.

[7] http://sourceforge.net/projects/conexp

which both qualitative (competence-inspired) and quantitative (data-inspired) hypotheses can be tested. We will start by reducing the question of describing all aspects of "coreness" related to a frame to the description of an inventory and typology of the core sets for the frame, and then proceed by providing a model based on concept lattices that fulfills the claims made above.

4.1   Core Frame Structure Lattices

For such purposes, we take the linguistic implications of the microtheory of frame elements to be:

- Core elements for $f \in \mathcal{F}$ are necessary in each sentence in the subcorpus for f .
- Non-core elements for f are contingent in sentences evoking f .
- The existence of one frame element in one sentence and a *requires* relation between that element and another implies the occurrence of the other frame element in the annotation for that sentence.
- The existence of one frame element in one sentence and an *excludes* relation between that element and others prevents the occurrence of the other frame elements in the annotation for that sentence.

Consequently, we take the following actions regarding null instantiation:

- CNI and DNI being licensed implies that they appear under some syntactically-induced disguise, hence we consider them examples of actual instantiation, and substitute them with crosses in frame structure contexts.
- INI and incorporated frame elements being idiosyncratic of each verb (entailing a slight modulation of meaning) might be treated differently, but the fact remains that the frame element is present in the use (it is just brought in the scene by the lexical unit instead of being a different phrase).

In practice, we carry out the following scaling of the frame element values (acting as attributes):

- for each frame element e with value INI we create an attribute $e.INI$
- for each sentence in which a frame value e appeared with value INI, we blank attribute e and put a cross in $e.INI$, to signal the special status.

Finally, the lattices obtained in the previous section are neither simple nor elucidating of the behaviour of the core frame elements. For instance in the lattice for MOTION_DIRECTIONAL the incidence of contingent, non-core frame elements obscures the structure due to core frame elements namely THEME, SOURCE, PATH, GOAL and AREA, as shown in Fig. 3. Keeping the two preceding paragraphs in mind, we *select* only the core attributes to be visualised and adjust the visualisation to obtain Fig. 4, an instance of a *core frame structure*, the sublattice of a frame structure lattice obtained by projecting only core elements in the frame structure context.

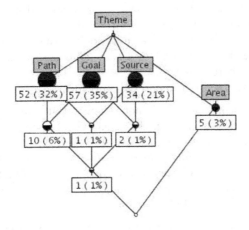

Fig. 4. The core frame structure lattice for MOTION_DIRECTIONAL

4.2   Core Set Analysis

In core frame structure lattices we will consider reduced labelling on the attributes and a particular labelling on the objects: instead of showing the rather meaningless set of sentence labels at the lowest concept in which they appear, we show the number of sentences that support that concept, and their overall percentage in the lattice. With these provisos, each node in Fig. 4 is read as follows:

- Its *extent* is composed of all the sentences in which a lexical unit has evoked the frame, i.e. for the node below frame element PATH, the sentences it contains are those found in itself ("52(32%)"), those of the node marked with "10(6%)" directly below it, those of the node marked with "1(1%)" down to the right and finally those of the node marked "1(1%)" below the latter two[8]. This means the concept has a total of 64 sentences in its extent.
- More interestingly, its *intent* is composed of all the frame elements collected by navigating the lattice towards the top, i.e. PATH (for itself) and THEME (for the top concept).

The next step is to try to profit from this formal apparatus to help in the task of understanding and formalizing coreness. We claim that a number of issues regarding the coreness of frame elements are better exposed with core frame structure lattices.

For that purpose, consider the top concept: its extent is formed by all sentences in nodes below it (i.e. all of them in this example), while its intent is formed by the single attribute THEME. This means that THEME is expressed in

---

[8] Different settings in CONEXP allow to view this quantities aggregated in a different way.

all sentences for MOTION_DIRECTIONAL, i.e. it is indeed a *core frame element* after the definition above. The reason not to reduce the context is now clear: to test the coreness of frame elements against their actual support. We call frame elements appearing in the top concept, *singleton core sets*, because each one works as a separate core element[9].

Next, consider the concept annotated with AREA:

- Clearly such a frame element is not salient on all frame instances as there are sentences which do not express it (e.g., all the sentences in the concepts below those marked with SOURCE, PATH or GOAL).
- Furthermore, there isn't any sentence marked with AREA **and** either SOURCE, PATH or GOAL. Hence we may safely conclude that AREA *excludes the set* {SOURCE, PATH, GOAL} *as a whole*. Since `excludes` is a symmetric constraint, any of the latter also excludes the former.

These results agree completely with standard FrameNet practice.

However, consider the almost perfect cube of concepts representing all the combinations in which SOURCE, PATH and GOAL may appear:

- None of these combinations is exhaustive, i.e. all leave out all the sentences annotated with AREA.
- Further, the presence of any of these frame elements does not require or prevent the presence or absence of any of the others except in a purely stochastic way: out of the 71 times GOAL appears, in 63 it is on its own, 4 it appears associated with PATH, in 3 to SOURCE, and only once with both of them. We may infer a conditional distribution from this fact but we cannot rule out or assert the presence of PATH or SOURCE given GOAL.

This is only *implicitly assumed* in FrameNet description and not enforced in any way: it is a sort of "default" rule.

The preceding paragraphs rather belabour the following points we are trying to clarify:

- *Neither* {SOURCE, PATH, GOAL} *together, nor* AREA *on its own are properly speaking core frame elements.* Rather they are a *core set*, i.e. a set of frame elements related by co-occurrence constraints, at least one of which must appear to license the use of the frame-evoking target.
- Furthermore, the structure of the core set is not merely that of a set in which any possible combination of elements in the powerset may appear. For example, AREA *excludes* the rest.
- *This structure cannot be made clear by the mere use of requires and excludes constraints, because there is no way the co-occurence patterns of* SOURCE, PATH *and* GOAL *may be described with these constraints alone.*
- *There is no formal reason why* {SOURCE, PATH, GOAL, AREA}, *a (proper) core set, should be considered with different ontological status from* {THEME}, *a (singleton) core set.*

---

[9] This is not standard FrameNet practice, but has been suggested by this research, as noted further ahead.

In the next section we will be visiting more examples of the use of lattice methods to visualise sentence-by-FE lattices and core sets, and better grounding a new intuition of how core sets behave with respect to a number of phenomena.

4.3 Discussion: The Inheritance of Core Frame Structure in the Lattices of Motion and Daughter Frames

We now turn to exploring the effect of the `inherit` relation on the structure of the core frame elements of a frame, in order to contrast the order imposed on the frame lattice by the relation with the order suggested by the structure of the different core frame structure lattices for each frame in the relation.

For that purpose, recall that MOTION_DIRECTIONAL and MASS_MOTION both `inherit` from MOTION. The single difference is a name change from the THEME frame element in MOTION to MASS_THEME in MASS_MOTION.

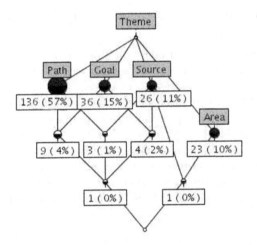

Fig. 5. The core frame structure sublattice for MOTION

Figures 5 and 6 together with Fig. 4 above describe this evolution through the hierarchy and show remarkable consistency: the only discordant note is a seemingly spurious case of MOTION in which AREA and SOURCE cooccur. In this case, we think the data would have to be consulted to see if the original annotation is correct and unproblematic or to adopt an interpretation more coherent with the rest of the annotation. Examples such as these have been used to detect patterns of annotation in the database requiring further attention.

As it turned out, in this particular example the annotation was right and the spurious link demonstrated a peculiar, but perfectly valid instance of the use in which some odour was coming from a source and spreading through some area. Because we can no longer say that AREA excludes SOURCE, this example

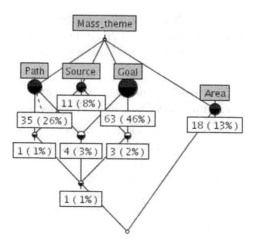

Fig. 6. The core frame structure sublattice for MASS_MOTION

demonstrates the adequacy of frame structure lattices for capturing core set and the inadequacy of mere constraint statements to do so.

As to the larger, more interesting question, whether not only the frame elements are inherited, as dictated by the manually encoded `inherit` relation, but also the lattice structure percolates from parents to children, the behaviour of the MOTIONXX lattices suggests rather that structure is *refined* through inheritance. This abduction may easily be proven wrong by subsequent frame subhierachies, however, and further work needs to be carried out to test it. The point here is that visualising the structure as a lattice has eased the scientific process of hypothesis building and testing in the lexicographic enterprise.

5   Conclusion

In this paper we have introduced the use of Formal Concept Analysisto explore FrameNet, a database of lexical items embodying Frame Semantics principles, with the intention of clarifying its concepts and bringing the power (and rigor) of formalization to its theory. Among the abstractions in FrameNet we have found the set of *frames*, their *frame elements*, and their instances, in particular sentences, the most amenable to analysis using the techniques of Formal Concept Analysis.

We have first described the use of standard Formal Concept Analysistools and concepts to visualize the orders with which the set of frames is endowed: *inherits_from*, *uses* and *subframe_of*. In this instance the use of Formal Concept Analysisis more of a data structuring tool than a real aid in developing the theory of Frame Semantics, which already had made provisions to develop such structuring.

Second and more interestingly, we have shown that the linguistically-conceived concept of a *core set* can be appropriately captured by the intents of an appropriate formal context. These contexts are built around the sentences attested in a corpus evoking each frame and annotated with the salient frame elements for that frame. We have argued that the description of core sets with `requires` and `excludes` constraints could be profitably replaced with the core frame structure. We now further hint at the fact that each core set can by itself be considered the attribute set of one of the core frame structure factor lattices whose decomposition process would considerably ease the lexicographic enterprise of defining core frame elements

We are also considering how to better model the overall frame structure lattice by taking into consideration equivalence and ordering relations between not only frames, but also frame elements.

Finally, the consideration of several types of decomposition, and the inclusion of lexical units as new objects with which to build contexts encourages us to look forward to the use of lattice techniques in Frame Semantics with barely contained glee.

Acknowledgements. We would like to thank C.Baker, M. Ellsworth, M. Petruck and the anonymous reviewers for their comments and suggestions on earlier versions of this paper.

References

1. Johnson, C.R., Petruck, M.R., Baker, C.F., Ellsworth, M., Ruppenhofer, J., Fillmore, C.J.: FrameNet: Theory and Practice. 1.1 edn. International Computer Science Institute, 1947, CenterBerkeley, USA (2003) Available as: http.//framenet.icsi.berkeley.edu/framenet/book/book.html.
2. Petruck, M.R.L.: Frame Semantics. In Verschueren, J., Östman, J.O., Blommaert, J., Bulcaen, C., eds.: Handbook of Pragmatics. John Benjamins, Philadelphia (1996)
3. Fillmore, C.J., Ruppenhoffer, J., Baker, C.F.: FrameNet and Representing the Link between Semantic and Syntactic Relations. In: Frontiers in Linguistics I. Language and Linguistics Monograph Series B.Academia Sinica (2004) 19-62
4. Kipke, U., Wille, R.: Formale Begriffsanalyse erläutert an einem Wordfelt. LDV Forum (1987)
5. Grosskopf, A.: Formal Concept Analysis of verb paradigms in Linguistics.In E. Diday, Y.L., Opitz, O., eds.: Ordinal and Symbolic Data Analysis. Springer, Berlin (1996) 70-79
6. Grosskopf, A., Harras, G.: Begriffiche Erkundung semantischer Strukturen von Sprechaktverben. In Stumme, G., Wille, R., eds.: Begriffiche Wissensverarbeitung: Methoden und Anwendungen, Berlin, Springer (1999) 273-295
7. Priss, U.E.: Linguistic applications of Formal Concept analysis. In: Concept Lattices. First International Conference on Formal Concept Analysis, ICFCA2003, Springer (2003) (To appear).
8. Priss, U.E.: Signs and formal concepts. [8] 28–38

9. Priss, U.E.: Relational Concept Analysis: Semantic structures in dictionaries and lexical databases. PhD thesis, Technical University of Darmstadt, Aachen, Germany (1999)

10. Priss, U.E.: The formalization of WordNet by methods of Relational Cencept Analysis. In Fellbaum, C., ed.: WordNet: An Electronic Lexical Database and some of its Applications. MIT press (1998) 179-196

11. Old, L.J.: Unlocking thesemantics of Roget's Thesaurus using Formal Concept-Analysis. [18] 244-251

12. Priss, U., Old, L.J.: Modellinglex lexical databases with Formal Concept Analysis. Journal of Computer Science 10 (2004)

13. Sporleder, A Galois Lattice approach to lexical inheritance hierarchy learning. In: ECAI 2002 Workshop on Machine Learning and Natural Language Processing for Ontology Engineering (OLT 2002), Lyon, France (2002)

14. Petersen, W.: A set-theoretical approach for the induction of inheritance hierarchies. Language and Computation 1 (2001) 1-14

15. Ganter, B., Wille, R.: Formal cencept Analysis: Mathematical Foundations. Springer, Berlin, Heidelberg (1999)

16. Davey, B., Priestley, H.: Introduction to lattices and order. 2nd edn. Cambridge University Press, Cambridge, UK (2002)

17. Yevtushenko, S.A.: System of data analysis "concept explorer". (in Russian). In: Proceedings of the 7th national conference on Artificial Intelligence KII-2000, Russia, ACM (2000) 127-134 http://sourceforge. net/projects/conexp.

18. Eklund, P., ed.: ICFA2004-Proceedings of the 2nd International Conferenon Formal Concept Analysis. Number 2961 in LNAI, Sydney, Australia, Springer (2004

Lessons Learned in Applying Formal Concept Analysis to Reverse Engineering

Gabriela Arévalo, Stéphane Ducasse, and Oscar Nierstrasz

Software Composition Group,
University of Bern, Switzerland
www.iam.unibe.ch/~scg

Abstract. A key difficulty in the maintenance and evolution of complex software systems is to recognize and understand the implicit dependencies that define contracts that must be respected by changes to the software. Formal Concept Analysis is a well-established technique for identifying groups of elements with common sets of properties. We have successfully applied FCA to complex software systems in order to automatically discover a variety of different kinds of implicit, recurring sets of dependencies amongst design artifacts. In this paper we describe our approach, outline three case studies, and draw various lessons from our experiences. In particular, we discuss how our approach is applied iteratively in order to draw the maximum benefit offered by FCA.

1   Introduction

One of the key difficulties faced by developers who must maintain and extend complex software systems, is to identify the *implicit dependencies* in the system. This problem is particularly onerous in object-oriented systems, where mechanisms such as dynamic binding, inheritance and polymorphism may obscure the presence of dependencies [DRW00, Dek03]. In many cases, these dependencies arise due to the application of well-known programming idioms, coding conventions, architectural constraints and design patterns [SG95], though sometimes they may be a sign of weak programming practices.

On the one hand, it is difficult to identify these dependencies in non-trivial applications because system documentation tends to be inadequate or out-of-date and because the information we seek is not explicit in the code [DDN02, SLMD96, LRP95]. On the other hand, these dependencies play a part in implicit contracts between the various software artifacts of the system. A developer making changes or extensions to an object-oriented system must therefore understand the dependencies among the classes or risk that seemingly innocuous changes break the implicit contracts they play a part in [SLMD96]. In short, implicit, undocumented dependencies lead to *fragile systems* that are difficult to extend or modify correctly.

Due to the complexity and size of present-day software systems, it is clear that a software engineer would benefit from a (semi-)automatic tool to help cope with these problems.

B. Ganter and R. Godin (Eds.): ICFCA 2005, LNAI 3403, pp. 95–112, 2005.
© Springer-Verlag Berlin Heidelberg 2005

Formal Concept Analysis provides a formal framework for recognizing groups of elements that exhibit common properties. It is natural to ask whether FCA can be applied to the problem of recognizing implicit, recurring dependencies and other design artifacts in complex software systems.

From our viewpoint, FCA is a metatool that we use as tool builders to build *new* software engineering tools to analyze the software. The *software engineer* is considered to be the end user for our approaches, but his knowledge is needed to evaluate whether the results provided by the approaches are meaningful or not.

Over the past four years, we have developed an approach based on FCA to detect undocumented dependencies by modeling them as recurring sets of properties over various kinds of software entities. We have successfully applied this approach to a variety of different reverse engineering problems at different levels of abstraction. At the level of classes, FCA helps us to characterize how the methods are accessing state and how the methods commonly collaborate inside the class [ADN03]. At the level of the class hierarchy, FCA helps us to identify typical calling relationships between classes and subclasses in the presence of late binding and overriding and superclass reuse [Aré03]. Finally, at the application level, we have used FCA to detect recurring collaboration patterns and programming idioms [ABN04]. We obtain, as a consequence, views of the software at a higher level of abstraction than the code. These high level views support *opportunistic* understanding [LPLS96] in which a software engineer gains insight into a piece of software by iteratively exploring the views and reading code.

In this paper we summarize our approach and the issues that must be taken into consideration to apply FCA to software artifacts, we briefly outline our three case studies, and we conclude by evaluating the advantages and drawbacks of using FCA as a metatool for our reverse engineering approaches.

2   Overview of the Approach

In this section we describe a general approach to use FCA to build tools that identify recurring sets of dependencies in the context of object-oriented software reengineering. Our approach conforms to a pipeline architecture in which the analysis is carried out by a sequence of processing steps. The output of each step provides the input to the next step. We have implemented the approach as an extension of the *Moose* reengineering environment [DLT00].

The processing steps are illustrated in Figure 1. We can briefly summarize the goal of each step as follows:

- *Model Import:* A model of the software is constructed from the source code.
- *FCA Mapping:* A FCA Context (Elements, Properties, Incidence Table) is built, mapping from metamodel entities to FCA elements (referred as *objects* in FCA literature) and properties (referred as *attributes* in FCA literature)[1].

---

[1] We prefer to use the terms *element* and *property* instead of the terms *object* and *attribute* in this paper because the terms *object* and *attribute* have a very specific meaning in the object oriented programming paradigm.

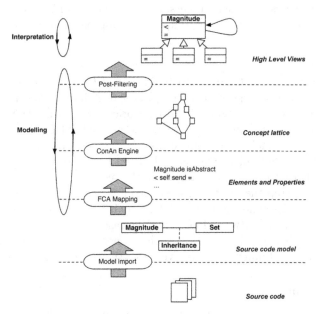

Fig. 1. The overall approach

- *ConAn Engine:* The concepts and the lattice are generated by the ConAn tool.
- *Post-Filtering:* Concepts that are not useful for the analysis are filtered out.
- *Analysis:* The concepts are used to build the high level views.

A key aspect of our approach is that one must iterate over the modeling and the interpretation phases (see Figure 1). The *modeling* phase entails a process of experimentation with smaller case studies to find a suitable mapping from the source code model to FCA elements and properties. A particular challenge is to find a mapping that is efficient in terms of identifying meaningful concepts while minimizing the quantity of data that must be processed.

The *interpretation* phase is the other iterative process in which the output of the modeling phase is analyzed in order to interpret the resulting concepts in the context of the application domain. The useful concepts can then be flagged so that future occurrences can be automatically detected. As more case studies are analyzed, the set of identifiably useful concepts typically increases up to a certain point, and then stabilizes.

We must remark that, from our experiences, there are two main participants in the approach: the *tool builder* and *the software engineer*. The *tool builder* builds the FCA-based tool to generate the high level views, and the *software engineer* uses the results provided by the tool to analyze a piece of software. Both of them work together in the *modeling* and *interpretation* phase of the approach, because *software engineer* has the knowledge of analyzing a system and the *tool builder* can represent this knowledge in the tool.

We will now describe each processing step in detail and outline the key issues that must be addressed in order to apply the approach.

Model Import

Description: Our first step is to build a *model* of the application from the source code. For this purpose we use the *Moose* reengineering platform, a research vehicle for reverse and reengineering object-oriented software [DDN02]. Software models in *Moose* conform to the FAMIX metamodel [TDDN00], a language-independent metamodel for reengineering. *Moose* functions both as a software repository for these models, and as a platform for implementing language-independent tools that query and manipulate the models.

Issues: In this step the most important issue is how to map the source code to metamodel entities. The main goal of this step is to have a language-independent representation of the software. In our specific case, we use the FAMIX metamodel, which includes critical information such as method invocations and attribute accesses. The tool builder can, however, choose any suitable metamodel.

FCA Mapping

Description: In the second step, we need to map the model entities to *elements* and *properties*, and we need to produce an *incidence table* that records which elements fullfil which property. The choice of elements and properties depends on the view we want to obtain.

Issues: This is a critical step because several issues must be considered. Each of these issues is part of the iterative *modeling* process.

- *Choice of Elements:* First we must decide which metamodel entities are mapped to FCA elements. This is normally straightforward. In most cases there are some particular metamodel entities that are directly adopted as FCA elements (*e.g.*, classes, methods). Alternatively, a FCA element may correspond to a set of entities (*e.g.*, a set of collaborating classes).
- *Compact Representation of Data:* In some cases, a naïve mapping from metamodel entities to FCA elements may result in the generation of massive amounts of data. In such cases, it may well be that many elements are in fact redundant. For example, if method invocations are chosen as FCA elements, it may be that multiple invocations of the same method do not add any new information for the purpose of applying FCA. By taking care in how FCA elements are generated from the metamodel, we can not only reduce noise, but we can also reduce the cost of computing the concepts.
- *Choice of Properties:* Once the issue of modeling FCA elements is decided, the next step is to choose suitable properties. Well-chosen properties achieve the goal of distinguishing groups of similar elements. This means that they should neither be too general (so that most elements fulfill them) nor be too specific (so only few elements fulfill them).
- *Use of Negative Properties:* Nevertheless, in some cases the developer needs still more properties to distinguish the elements. But simply adding more

properties may only increase the complexity of the approach. The use of "negative" information (built by negating existing properties) may help.

– *Single or Multiple FCA Contexts:* In some cases, multiple FCA contexts may be required. For example, in the XRay views case study, one context was used to analyze state and another to analyze behavior.

– *Computation of Properties or Elements:* When building the FCA context of a system to analyze, there are two alternatives for the FCA mapping. In an *one-to-one* mapping, the developer directly adopts metamodel entities and metamodel relationships as FCA elements and properties respectively. In a *many-to-one* the developer builds more complex FCA elements and properties by computing them from the metamodel entities and relationships, meaning for example that an FCA element can be composed of several metamodel entities, or an FCA property must be calculated based on metamodel relationships between different entities. This issue is one of the bottlenecks in the total computation time of the approach, because the incidence table must be computed in this step and if the FCA property must be calculated, this time can also compromise the total computation time.

ConAn Engine

Description: Once the elements and properties are defined, we run the ConAn engine. The ConAn engine is a tool implemented in VisualWorks 7 which runs the FCA algorithms to build the concepts and the lattice. ConAn applies the Ganter algorithm [GW99] to build the concepts and our own algorithm to build the lattice [Aré05].

Issues: In this step, there are two main issues to consider

– *Performance of Ganter Algorithm:* Given an FCA Context $C = (E, P, I)$, the Ganter algorithm has a time complexity of $O(|E|^2|P|)$. This is the second bottleneck of the approach because in our case studies the number of FCA elements is large due to the size of the applications. We consider that $|P|$ is not a critical factor because in our cases studies the maximum number of properties is 15.

– *Performance of Lattice Building Algorithm:* Our algorithm is the simplest algorithm to build the lattice but the time complexity is $O(n^2)$ where n is the number of concepts calculated by Ganter algorithm. This is the last bottleneck of the approach.

– *Unnecessary Properties:* It may happen that certain properties are not held by any element. Such properties just increase noise, since they will percolate to the bottom concept of the lattice, where we have no elements and all properties of the context.

Post-filtering

Description: Once the concepts and the lattice are built, each concept constitutes a potential *candidate* for analysis. But not all the concepts are relevant. Thus we have a *post-filtering* process, which is the last step performed by the tool. In this way we filter out meaningless concepts.

Issues: In this step, there are two main issues to consider:

- *Removal of Top and Bottom Concepts:* The first step in *post-filtering* is to remove the *top* and *bottom* concepts. Neither provides useful information for our analysis because each one usually contains an empty set. (The intent is empty in the top concept and the extent is empty in the bottom concept).
- *Removal of Meaningless Concepts:* This step depends on the interpretation we give to the concepts. Usually concepts with only a single element or property are candidates for removal because the interesting characteristic of the approach is to find *groups* of elements sharing *common* characteristics. Concepts with only a single element occur typically in nodes next to the bottom of the lattice, whereas concepts with only one property are usually next to the top of the lattice.

Analysis

Description: In this step, the software engineer examines the candidate concepts resulting from the previous steps and uses them to explore the different *implicit* dependencies between the software entities and how they determine or affect the behavior of the system.

Issues: In this step, there are several issues to consider. All of them are related to how the software engineer interprets the concepts to get meaningful or useful results.

- *Concept Interpretation Based on Elements or Properties:* Once the lattice is calculated, we can interpret each concept $C = (\{E_1 \ldots E_n\}, \{P_1 \ldots P_m\})$ using either its elements or its properties. If we use the properties, we try to associate a meaning to the conjunction of the properties. On the other hand, if we focus on the elements, we essentially discard the properties and instead search for a domain specific association between the elements (for example, classes being related by inheritance).
- *Equivalent Concepts:* When we interpret the concepts based on their properties, we can find that the meaning of several concepts can be the same. This means that for our analysis, the same meaning can be associated to different sets of properties.
- *Automated Concept Interpretation:* The interpretation of concepts can be transformed into an automatic process. Once the software engineer establishes a meaning for a given concept, this correspondence can be stored in a database. The next time the analysis is performed on another case study, the established interpretations can be retrieved and automatically applied to the concepts identified. Once this process is finished, the software engineer must still check those concepts whose meaning has not been identified automatically.
- *Using Partial Order in the Lattice:* The concepts in the lattice are related by a partial order. During analysis, the software engineer should evaluate if it is possible to interpret the partial order of the lattice in terms of software relationships. This means that the software engineer should only not interpret the concepts but also the relationships between them.

– *Limit of Using FCA as a Grouping Technique:* When additional case studies fail to reveal new meaningful concepts, then the application of FCA has reached its limit. At this point, the set of recognized concepts and their interpretations can be encoded in a fixed way, for example, as logical predicates over the model entities, thus fully automating the recognition process and bypassing the use of FCA.

3   High Level Views in Reverse Engineering

Using the approach we have introduced in the previous section, we have generated different views at different abstraction levels for object oriented applications. We present each analysis according to a simple pattern: *Reengineering Goal, FCA Mapping, High Level Views, Validation, Concrete Examples* and *Issues. Reengineering Goal* introduces the analysis goal of the corresponding high level view. *FCA Mapping* explains which are the chosen elements and properties of the context for the analysis. *High level view* presents how the concepts are read to interpret which of them are meaningful or not for our analysis and how we use them to build the corresponding high level views. *Validation* summarizes some results of the case studies. *Concrete Examples* presents some examples found in the specific case study. *Issues* summarizes briefly which were the main issues taken into account to build the high level view.

3.1   Understanding a Class: XRay Views

Reengineering Goal: The goal is to understand the internal workings of a class by capturing how methods call each other and how they collaborate in accessing the state. We focus on analyzing a class as an isolated development unit [ADN03].

FCA Mapping: The elements are the *methods* (labelled m or n) and the *attributes* (labelled a) of the class, and the properties, are:

– m reads or writes the value of a
– m calls via-self n in its body
– m is abstract in its class
– m is concrete in its class
– m is an "interface" (not called in any method defined in the same class)
– m is "stateless" (doesn't read or write any attribute defined in the class)

High Level Views: A XRay view is a combination of concepts that exposes specific aspects of a class. We have defined three XRay views: State Usage, External/Internal Calls and Behavioural Skeleton. These three views address different but logically related aspects of the behavior of a class.

State Usage focuses on how the behavior accesses the state of the class, and what dependencies exist between groups of methods and attributes. This view helps us to measure the class cohesion [BDW98] revealing whether there are methods using the state partially or totally and whether there are attributes working together to provide different functionalities of the class.

BEHAVIOURAL SKELETON focuses on methods according to whether or not they work together with other methods defined in the class or whether or not they access the state of the class. The way methods form groups of methods that work together also indicates how cohesive the class is [BDW98].

EXTERNAL/INTERNAL CALLS focuses on methods according to their participation in internal or external invocations. Thus, this view reveals the overall shape of the class in terms of its internal reuse of functionality. This is important for understanding framework classes that subclasses will extend. Interface methods, for example, are often generic template methods, and internal methods are often hook methods that should be overridden or extended by subclasses.

Validation: We have applied the XRay views to three Smalltalk classes: *Ordered-Collection* (2 attributes, 54 methods), *Scanner*(10 attributes, 24 methods) and *UIBuilder*(18 attributes, 122 methods). We chose these particular three classes because they are different enough in terms of size and functionality and they address well-known domains. In general terms, we discovered that in *Ordered-Collection* most of the methods access all the state of the class, that there is little local behavior to be reused or extended by subclasses because the class works with inherited methods, and there are few collaborations among the methods. *UIBuilder* is a class where most of the methods are not invoked in the class itself, meaning the internal behavior is minimal, and we have a large *interface*. This class offers a lot of functionality to build complex user interface and also several ways to query its internal state. In *Scanner*, the collaboration between methods occurs in pairs and there are no groups of methods collaborating with other groups.

Concrete Examples: We illustrate this case study with the XRay STATE USAGE found in the Smalltalk class *OrderedCollection*. This view is composed of several concepts, and it clusters attributes and methods according to the way methods access the attributes. The motivation for this view is that, in order to understand the design of a class, it is important to gain insight into how the behaviour accesses the state, and what dependencies exist between groups of methods and attributes. This view helps us to measure the cohesion of the class [BDW98], thereby revealing any methods that use the state partially or totally and any attributes that work together to provide different functionalities of the class.

Some of the concepts occurring in STATE USAGE are the following:

- {before, removeAtIndex:, add:beforeIndex:, first, removeFirst, removeFirst:, addFirst } *reads or writes* {firstIndex } represents the Exclusive Direct Accessors of firstIndex.
- {after, last, removeIndex:, addLastNoCheck:, removeLast, addLast:, removeLast: } *reads or writes* {lastIndex } represents the Exclusive Direct Accessors of lastIndex
- {makeRoomAtFirst, changeSizeTo:, removeAllSuchThat:, makeRoomAtLast, do:, notEmpty:, keysAndValuesDo:, detect:ifNone:, changeCapacityTo:, isEmpty, size, remove:ifAbsent:, includes:, reverseDo:, find:, setIndices, insert: before:, at:, at:put:,

includes: } *reads or writes* {firstIndex, lastIndex } represents the Collaborating Attributes
- Stateful Core Methods = the same set as Collaborating Attributes

Before analysing the concepts identified by this view, we posed the hypothesis that the two attributes maintain an invariant representing a memory zone in the third anonymous attribute. From the analysis we obtain the following points:

- By browsing Exclusive Direct Accessors methods, we confirm that the naming conventions used help the maintainer to understand how the methods work with the instance variables, because we see that the method removeFirst accesses firstIndex and removeLast: accesses lastIndex.
- The numbers of methods that exclusively access each attribute are very similar, however, we discover (by inspecting the code) that firstIndex is mostly accessed by readers, whereas lastIndex is mostly accessed by writers.
- It is worth noting that Collaborating Attributes are accessed by the same methods that are identified as Stateful Core Methods. This situation is not common even for classes with a small number of attributes, and reveals a cohesive collaboration between the attributes when the class is well-designed and gives a specific functionality, in this specific case, dealing with collections.
- We identified 20 out of 56 methods in total that systematically access *all* the state of the class. By further inspection, we learned that most of the accessors are readers. There are only five methods, makeRoomAtFirst, makeRoomAtLast, setIndices, insert:before:, and setIndicesFrom:, that read and write the state at the same time. More than half of the methods (33 over 56) directly and indirectly access both attributes. This confirms the hypothesis that the class maintains a strong correlation between the two attributes and the anonymous attribute of the class.

All these facts confirm the hypothesis that the class maintains a strong correlation between the two attributes and the anonymous attribute of the class.
Issues: We mention some important issues about this approach.

- *Choice of Elements and Properties.* Elements and properties are mapped directly from the metamodel: the elements are attributes and methods, and the properties are accesses to attributes and calls to methods.
- *Compact Representation of Data.* Supposing you have two methods m and n and one attribute a, if we have several calls to the method n or accesses to the attribute a in the method body of m, we just keep one representative of n and a related to the method m.
- *Multiple FCA Contexts.* In this case, we have used two lattices. The first one is used to analyze the state of the class and the second one is used to analyze the invocations of the class. We did not combine this information in a single lattice because we consider them to be completely different aspects of the class.
- *Unnecessary Properties.* In some classes, the following properties *isAbstract, isStateless, isInterface* are discarded. This is normal because in any given

class, it commonly occurs that all methods are concrete, or that all the methods access the state, or that most methods are called inside the class.

- *Meaningless Concepts.* We discarded concepts with a single method in the set of elements, because we were more focused on groups of methods (represented in the elements) collaborating with another group of methods (represented in the properties).

3.2   Analyzing Class Hierarchies: Dependency Schemas

Reengineering Goal: Using the state and behavior of a class we analyze the different recurring dependencies between the classes of a hierarchy. They help us to understand which are the common and irregular design decisions taken when the hierarchy was built, and possible refactorings that were carried out [Aré03].

FCA Mapping: The elements are the accesses to any attribute defined in any classes of the hierarchy, and the called methods in any class of the hierarchy. If i is an invoked method or accessed attribute, and C, C_1, C_2 are classes, the properties are grouped as follows:

- Kind of calls: C invokes i *via self* and C invokes i *via super*
- Location of accessed state: i accesses { *local state, state in Ancestor C_1, state in Descendant C_1* }
- Kind and Location of invoked method: { *is abstract, is concrete, is cancelled* } × { *locally, in ancestor C_1 of C , in descendant C_1 of C* }

High Level Views: A *dependency schema* is a recurring set of dependencies (expressed with the properties of the concepts) over methods and attributes in a class hierarchy. We have identified 16 dependency schemas that are classified as:

Classical schemas representing common idioms/styles that are used to build and extend a class hierarchy.

Bad Smell schemas representing doubtful designs decisions used to build the hierarchy. They are frequently a sign that some parts should be completely changed or even rewritten from scratch.

Irregularity schemas representing irregular situations used to build the hierarchy. Often the implementation can be improved using minimal changes. They are less serious than *bad smell* schemas.

Thus we see that the *dependency schemas* can be a good basis for identifying which parts of a system are in need of repair. These schemas can be used in two different ways: Either we obtain the global view of the system and which kinds of dependencies and practices occur when analyzing a complete hierarchy, or we get detailed information about how specific classes are related to others in their hierarchy by restricting the analysis to just those classes.

Validation: We have validated *dependency schemas* in three Smalltalk class hierarchies: *Collection* (104 classes distributed in 8 inheritance levels, 2162 methods, 3117 invocations, 1146 accesses to the state of the classes), *Magnitude* and

Model. Collection is an essential part of the Smalltalk system and it makes use of subclassing for different purposes. In this class hierarchy, the most used *classical* schemas are (1) the reuse of superclass behavior, meaning concrete methods that invokes superclass methods by *self* or *super*, (2) local behavior, meaning methods defined and used in the class that are not overridden in the subclasses and (3) local direct access, meaning methods that directly access the class state. In general terms, this means that the classes define their own state and behavior but they exhibit heavy superclass reuse. Within *bad smell* schemas, the most common is the ancestor direct state access meaning methods that directly access the state of an ancestor, bypassing any accessors. This is not a good coding practice since it violates class encapsulation. Within *irregularity* schemas the most common case is that of inherited and local invocations where methods are invoked by both *self* and *super* sends within the same class. This may be a problem if the super sends are invoked from a method with a different name. This is an *irregular* case because the class is overriding the superclass behavior but is indirectly using the superclass behavior.

Concrete Examples: We illustrate this case study with the *schema* named *Broken super send Chain*, which is categorized as a *Bad Smell* schema. It was found in the analysis of the Smalltalk class OrderedCollection.

Within the *"Bad Smell"* category, we have the schema *Broken* super *send Chain* (shown in Figure 2). It is composed of the following elements and properties:

- *C invokes i via super*: {representBinaryOn:, =} are *super-called* in SortedCollection
- *i is concrete locally*: {representBinaryOn:, =} has concrete behavior in SortedCollection.
- *i is concrete in ancestor C_1 of C* : {representBinaryOn:, =} has concrete behavior in ancestor SequenceableCollection of SortedCollection.
- *i is concrete in descendant C_1 of C* : {representBinaryOn:, =} has concrete behavior in descendant SortedCollectionWithPolicy of SortedCollection.

This schema identifies methods that are extended (*i.e.*, performing a *super* send) in a class but redefined in their subclasses without calling the overridden behavior, thus giving the impression of breaking the original extension logic. In SortedCollection the methods = and representBinaryOn: invoke hidden superclass methods. But the definitions of these methods in the subclass SortedCollectionWithPolicy do not invoke the super methods defined in SortedCollection. Such a behavior can lead to unexpected results when the classes are extended.

Issues: We mention some important issues about this approach.

- *Choice of Elements and Properties.* We map the attribute accesses and method calls directly from the metamodel. The choice of properties requires some analysis, because we need to cover the different possible inheritance relationships of the elements. The properties in the category *Kinds of calls* are mapped directly from the metamodel the rest of the properties are calculated based on the relationships expressed in the metamodel.

– *Single Context.* In this case, we just use only one lattice, because we analyze only one aspect of classes: inheritance relationships.

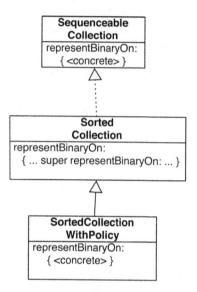

Sequenceable Collection
representBinaryOn:
{ <concrete> }

Sorted Collection
representBinaryOn:
{ ... super representBinaryOn: ... }

SortedCollection WithPolicy
representBinaryOn:
{ <concrete> }

Fig. 2. Broken *super* send Chain

– *Compact Representation of Data.* Supposing you have one method m and one attribute a in a class C. If we have several calls to the method m or accesses to the attribute a in several methods of the class C, we just keep one representative of the call to n and of the access to a related to the class C.

– *Use Negative Properties.* We define three negative properties because they help us to complement the information of the elements considering the three inheritance relationships used in the approach: *local*, *ancestor* and *descendant* definitions.

– *Meaningless Concepts.* All the meaningful concepts must show either a positive or negative information about the 3 relationships: *local*, *ancestor* and *descendant*, and have at least one property of category *kinds of calls* (invokes i *via self* or *via super*). The rest of the concepts are discarded.

3.3   Collaborations Patterns on Applications

Reengineering Goal: We analyze the different patterns used in a system using structural relationships between classes in a system. We call them *Collaboration Patterns* and they show us not only classical design patterns [GHJV95] but any kind of repeated patterns of *hidden contracts* between the classes, which may represent design patterns, architectural constraints, or simply idioms and conventions adopted for the project. With these patterns, we analyze how the system was built and which are the main constraints it respects [ABN04]. This approach refines and extends that which was proposed by Tonella and Antoniol [TA99] for detecting classical design patterns.

FCA Mapping: The elements are tuples composed of classes from the analyzed application. The order refers to length of the tuples. In our cases, we have experimented with order 2, 3 and 4. The properties are relations inside one class tuple, and are classified as *binary* and *unary* properties. The *binary* property characterizes a relationship between two classes inside the tuple, and the *unary* property gives a characteristic of a single class in the tuple. Given a tuple composed of classes $(C_1, C_2, ...C_n)$, the properties are:

– Binary Property: For any i, j, $1 \le i, j \le n$, C_i is subclass of C_j, C_i accesses C_j, C_i has as attribute C_j, C_i invokes C_j, C_i uses locally C_j

- Unary Property: For any i, $1 \leq n$, C_i is abstract, C_i is root, C_i is singleton, C_i has local defined method.

High Level Views: The concepts are composed of groups of tuples of classes that share common properties. The conjunction of these properties character- izes the pattern that all the tuples in the concept represent. Each *meaningful* concept represents a candidate for a pattern. Based on our case studies using tuples of classes of length 3 and 4, we have identified 8 collaboration patterns. 4 of 8 patterns exhibit the structural relationships of known design patterns (*Facade, Composite, Adapter* and *Bridge*). The rest of the identified patterns called *Subclass Star, Subclass Chain, Attribute Chain* and *Attribute Star* show the relationships among classes based on inheritance or on invocations, but not combined at the same time. The frequency of the patterns helps the developer to identify coding style applied in the application. The most interesting contri- bution of this high level view is the possibility of establishing a called *pattern neighbourhoods* over detected patterns. With the *neighbours* of the patterns, we can detect either missing relationships between classes needed to complete a pat- tern, or excess relationships between classes that extend a *pattern*. We can also analyze the connections of the identified patterns with the classes implemented in the analyzed application.

Validation: We have investigated the *collaboration patterns* in three Smalltalk applications: (1) *ADvance* [2](167 classes, 2719 methods, 14466 lines of code) is a multidimensional OOAD-tool for supporting object-oriented analysis and de- sign, reverse engineering and documentation, (2) *SmallWiki* [3] (100 classes, 1072 methods, 4347 lines of code) is a new and fully object-oriented wiki implementa- tion in Smalltalk and (3) *CodeCrawler* [4] (81 classes, 1077 methods, 4868 lines of code) is a language independent software visualization tool. These three appli- cations have different sizes and complexity. We briefly summarize some results related to coding styles. We have seen that frequency and the presence of a pat- tern in a system is besides domain specific issues of coding style. In our case studies we have seen that *CodeCrawler* has a lot of *Subclass Star* and *Facade* patterns, whereas *SmallWiki* has a lot of *Attribute Chain* and *Attribute Star* patterns. *Advance* is the only application with the *Composite* pattern.

Concrete Examples: We illustrate this case study showing two *collaboration patterns* named *Attribute Chain* and *Subclass Chain* that were found in the tool *CodeCrawler* (shown in the Figure 4).

Pattern *Subclass Chain* in order $= 3$ is described with the following properties: C_3 *is subclass of* C_2, {C_2 *is subclass of* C_1, C_1 *is abstract*}, and in order $= 4$ the property C_4 *is subclass of* C_3 is added. Pattern *Attribute Chain* in order $= 3$ is described with the following properties: {C_1 *invokes* C_2, C_2 *invokes* C_3 and C_1 *is abstract*}, and in order $= 4$ the property C_3 *invokes* C_4 is added. Pattern

---

[2] http://www.io.com/ icc/

[3] http://c2.com/cgi/wiki?SmallWiki

[4] http://www.iam.unibe.ch/ scg/Research/CodeCrawler/index.html

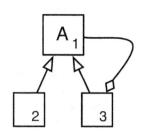

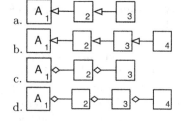

Fig. 3. Composite Pattern

Fig. 4. Subclass Chain (a)(b) and Attribute Chain (c)(d) in order = 3 and order = 4. Number in the classes indicate position in the tuple

Composite (shown in the Figure 3) is described with the following properties: $\{C_2$ is subclass of C_1, C_3 is subclass of C_1, C_3 invokes C_1, C_1 is abstract$\}$.

The first two patterns help us to analyze some coding styles used in *CodeCrawler*. We have found 25 instances of 3 classes (order = 3) of *Attribute Chain* but no instances of 4 classes (order = 4). This means that in this application the developer has used at most 3 levels of delegation between the classes. Thus, if we detect a case of delegation in a method in a class in *CodeCrawler*, we are sure that this delegation will relate 2 or 3 classes. In the case of *Subclass Chain*, we have found 11 instances of 3 classes (order = 3) but only 3 instances of 4 classes (order = 4). This means that in this application most of hierarchies have 3 levels of inheritance and only some of them have 4 levels of inheritance. This is a symptom that the application uses *flat hierarchies* in its implementation. Related to the idea of *pattern neighbourhood*, as we have said exploring the *neighbours* of a pattern is important to detect all candidates for a classical pattern. For example, we have found the *Abstract Composite* pattern of order $o = 3$ of the ADvance application. We have detected two *Abstract Composite* patterns, but in the neighborhood we find four more *Composite* patterns without an abstract composite root.

Issues: We mention some important issues about this approach.

- Choice of elements and properties. Elements are tuples of classes built from the metamodel. As our case study refines the work of Tonella and Antoniol [TA99] we use the same idea to build the elements. The choice of properties is the set of the structural relationships to characterize *Structural Design Patterns* [GHJV95]. Except the properties *is subclass of* and *is abstract* that are mapped directly from the metamodel, the rest of properties are computed from the metamodel.
- Compact representation of data. This issues is related to how the tuples of classes are generated. We avoid generating all permutations of class sequences in the tuples. For example, if the tuple (C A P) is generated, and we subsequently generate (A P C) or (C P A), we only keep one of these as being representative of all three alternatives.

- Multiple Contexts. In this case, we have used 3 lattices. Each lattice is for one order of the elements: 2, 3 and 4. All use the same set of properties.
- Performance of the algorithm. With the tuples of classes of order more than 4, we are not able to have a reasonable computation time of the algorithm. With one of the applications using tuples of classes of order 4, the computation time took around 2 days and this is not acceptable time from our viewpoint to software engineering.
- Meaningless concepts. As we said previously, each concept is a candidate for a pattern. In this case, each concept represents a graph in which the set of the elements is the set of nodes, and the set of properties define the edges that should connect the nodes, *e.g.*, if (A P C) has the properties *A is subclass of P* and *P uses C*, then we have a graph of 3 nodes with edges from A to P and from P to C. Thus we discarded all the concepts that represent graphs in which one or several nodes are not connected at all with any node in the graph.
- Mapping Partial Order of the Lattice. In this case, we map the partial order of the lattice to the definition of *neighbours* of a pattern. We can detect either missing relationships between classes needed to complete a pattern, or excess relationships between classes that extend a pattern.
- Limits of *Collaboration Patterns*. In this high level view, we consider that there are still possible new collaboration patterns to detect when applying the approach in other applications.

4   Lessons Learned

In general terms, we have seen that Formal Concept Analysis is a useful technique in reverse engineering. From our experiencies [Aré03, ABN04, ADN03] in developing this approach, several lessons learned are worthwhile mentioning.

Lack of a General Methodology. The main problem we have found in the state of the art is the lack of a general methodology for applying FCA to the analysis of software. In most publications related to software analysis, the authors only mention the FCA mapping as a trivial task, and how they interpret the concepts. With our approach, we achieved not only to identify clear steps for applying FCA to a piece of software but where we have identified different bottlenecks in using the technique.

Modelling Software Entities as FCA Components. The process of modelling software entities as FCA components is one of most difficult tasks and we consider it as one of the critical steps in the approach. Modelling is an iterative process in which we map software entities to FCA components, and we test whether they are useful enough to provide meaningful results. This task is not trivial at all, because it entails testing at least 5 small case studies (which should be representative of larger case studies). Each case study should help to refine the building of FCA components.

Performance of FCA Algorithms. Part 1 The performance of the algorithms (to build the concepts and lattice) was one of main bottlenecks. In small case studies, this factor can be ignored because the computation time is insignificant. But in large case studies this factor can cause the complete approach to fail because computing the concepts and the lattice may take several hours (eventually days).

Performance of FCA Algorithms. Part 2 The computation of the FCA algorithms is also affected by how they are implemented in a chosen language. A direct mapping of the algorithms (in pseudo-code, as they are presented in books) to a concrete programming language is not advisable. In our specific case, we took advantage of efficient data structures in *Smalltalk* language to represent the data and improve the performance.

Supporting Software Engineers. The result of our experiences must be read by software engineers. One positive issue in this point is that with the high level views the software engineer is not obliged to read the concept lattices , meaning that he need not be a FCA expert.

Interpretation of the Concepts. Although we can have an adequate choice of FCA elements and properties, the interpretation of the concepts is a difficult and time-consuming task. In most of the cases, we have tried to associate a meaning (a name) to each concept based on the conjunction of its properties. This task must be done by a software engineer applying *opportunistic code reading* to get meaningful interpretations. This process is completely subjective because it depends on the knowledge and experience of the software engineer.

Use of the Complete Lattice. Not all the concepts have a meaning in our approach, so we do not use the complete lattice in our analysis. In most of the cases, we remove meaningless concepts because from software engineering viewpoint they did not provide enough information for the analysis. We hypothesize that in certain cases it may be possible to use the complete lattice, but we did not find any.

Use of the Partial Order. Another critical factor is the interpretation of the partial order of the lattice in terms of software relationships. Only in *Collaboration Patterns* we were able to obtain a satisfactory interpretation of the partial order. So far, the interpretation is not a trivial task.

5   Related Work

Several researchers have also applied FCA to the problem of understanding object oriented software. Dekel uses CA to visualize the structure of the class in Java and to select an effective order for reading the methods and reveal the state usage [Dek03]. Godin and Mili [GMM+98] use concept analysis to maintain, understand and detect inconsistencies in the Smalltalk *Collection* hierarchy. They show how Cook's [Coo92] earlier manual attempt to build a better interface hierarchy for this class hierarchy (based on interface conformance) could be automated. In C++, Snelting and Tip [ST98] analysed a class hierarchy making the relationship between class members and variables explicit. They were able to detect design anomalies such as class members that are redundant or that can be moved into a derived class. As a result, they propose a new class hierarchy that is behaviorally equivalent to the original one. Similarly, Huchard [HDL00] applied concept analysis to improve the generalization/specialization of classes in a hierarchy. Tonella and Antoniol [TA99] use CA to detect the *structure* of Gamma-style design patterns using relationships between classes, such as *inheritance* and *composition*.

6 Conclusions and Future Work

In this paper we present a general approach for applying FCA in reverse engineering of object oriented software. We also evaluate the advantages and drawbacks of using FCA as a metatool for our reverse engineering approaches. We also identify the different bottlenecks of the approach. Thus, we are able to focus clearly on solving which and where the limitations appear (if there are some possible solutions) to draw the maximum benefit offered by FCA. From our tool builder viewpoint, we have proven that FCA is an useful technique to identify *groups* of software entities with hidden dependencies in a system. With FCA, we have built different software engineering tools that help us to generate *high level views* at different levels of abstraction of a system. We generate the *high level views* because without them, the *software engineer* should be obliged to read the lattice. This can represent a problem because in most of the cases, besides the useful information, the lattice can also have useless information that can introduce *noise* in analyzing a system.

Our future work is focused on several research directions that consist of: (1) the development of new case studies to analyze how useful the approach is or if there are still some refinements and improvements to do, (2) Testing with other concept and lattice building algorithms to see if we can improve the computation time of the ConAn engine and(3) Analysis of the partial order to get possible mappings in terms of software engineering relationships.

Acknowledgments. We gratefully acknowledge the financial support of the Swiss National Science Foundation for the projects "Tools and Techniques for Decomposing and Composing Software" (SNF Project No. 2000-067855.02), and Recast: Evolution of Object-Oriented Applications (SNF 2000-061655.00/1).

References

[ABN04]   Gabriela Arévalo, Frank Buchli, and Oscar Nierstrasz. Detecting implicit collaboration patterns. In *Proceedings of WCRE 2004*. IEEE Computer Society Press, November 2004. to appear.

[ADN03]   Gabriela Arévalo, Stéphane Ducasse, and Oscar Nierstrasz. X-Ray views: Understanding the internals of classes. In *Proceedings of ASE 2003*, pages 267–270. IEEE Computer Society, October 2003. Short paper.

[Aré03]   Gabriela Arévalo. Understanding behavioral dependencies in class hierarchies using concept analysis. In *Proceedings of LMO 2003*, pages 47–59. Hermes, Paris, January 2003.

[Aré05]   Gabriela Arévalo. *High Level Views in Object Oriented Systems using Formal Concept Analysis.* PhD thesis, University of Berne, January 2005. forthcoming.

[BDW98]   Lionel C. Briand, John W. Daly, and Jürgen Wüst. A unified framework for cohesion measurement in object-oriented systems. *Empirical Software Engineering: An International Journal,* 3(1):65–117, 1998.

[Coo92]   William R. Cook. Interfaces and specifications for the Smalltalk-80 collection classes. In *Proceedings OOPSLA '92, ACM SIGPLAN Notices,* volume 27, pages 1–15, October 1992.

[DDN02]   Serge Demeyer, Stéphane Ducasse, and Oscar Nierstrasz. *Object-Oriented Reengineering Patterns.* Morgan Kaufmann, 2002.

[Dek03]   Uri Dekel. Revealing java class structures using concept lattices. Diploma thesis, Technion-Israel Institute of Technology, February 2003.

[DLT00]   Stéphane Ducasse, Michele Lanza, and Sander Tichelaar. Moose: an extensible language-independent environment for reengineering object-oriented systems. In *Proceedings of CoSET 2000,* June 2000.

[DRW00]   Alastair Dunsmore, Marc Roper, and Murray Wood. Object-oriented inspection in the face of delocalisation. In *Proceedings of ICSE 2000,* pages 467–476. ACM Press, 2000.

[GHJV95]   Erich Gamma, Richard Helm, Ralph Johnson, and John Vlissides. *Design Patterns: Elements of Reusable Object-Oriented Software.* Addison Wesley, Reading, Mass., 1995.

[GMM+98]   Robert Godin, Hafedh Mili, Guy W. Mineau, Rokia Missaoui, Amina Arfi, and Thuy-Tien Chau. Design of class hierarchies based on concept (galois) lattices. *Theory and Application of Object Systems,* 4(2):117–134, 1998.

[GW99]   Bernhard Ganter and Rudolf Wille. *Formal Concept Analysis: Mathematical Foundations.* Springer Verlag, 1999.

[HDL00]   M. Huchard, H. Dicky, and H. Leblanc. Galois lattice as a framework to specify algorithms building class hierarchies. *Theoretical Informatics and Applications,* 34:521–548, 2000.

[LPLS96]   David Littman, Jeannine Pinto, Stan Letovsky, and Elliot Soloway. Mental models and software maintenance. In Soloway and Iyengar, editors, *Empirical Studies of Programmers, First Workshop,* pages 80–98, 1996.

[LRP95]   John Lamping, Ramana Rao, and Peter Pirolli. A focus + context technique based on hyperbolic geometry for visualising larges hierarchies. In *Proceedings of CHI '95,* 1995.

[SG95]   Raymie Stata and John V. Guttag. Modular reasoning in the presence of subclassing. In *Proceedings of OOPSLA '95,* pages 200–214. ACM Press, 1995.

[SLMD96]   Patrick Steyaert, Carine Lucas, Kim Mens, and Theo D'Hondt. Reuse contracts: Managing the evolution of reusable assets. In *Proceedings of OOPSLA '96,* pages 268–285. ACM Press, 1996.

[ST98]   Gregor Snelting and Frank Tip. Reengineering class hierarchies using concept analysis. In *ACM Trans. Programming Languages and Systems,* 1998.

[TA99]   Paolo Tonella and Giuliano Antoniol. Object oriented design pattern inference. In *Proceedings ICSM '99,* pages 230–238, October 1999.

[TDDN00]   Sander Tichelaar, Stéphane Ducasse, Serge Demeyer, and Oscar Nierstrasz. A meta-model for language-independent refactoring. In *Proceedings ISPSE 2000,* pages 157–167. IEEE, 2000.

Navigation Spaces for the Conceptual Analysis
of Software Structure

Richard Cole and Peter Becker

School of Information Technology and Electrical Engineering (ITEE),
The University of Queensland,
QLD 4072, Australia
{pbecker, rcole}@itee.uq.edu.au

Abstract. Information technology of today is often concerned with information that is not only large in quantity but also complex in structure. Understanding this structure is important in many domains – many quantitative approaches such as data mining have been proposed to address this issue. This paper presents a conceptual approach based on Formal Concept Analysis. Using software source code as an example of a complex structure we present a framework for conceptually analysing relational structures. In our framework, a browsable space of sub-contexts is automatically derived from a database of relations augmented by a rule engine and schema information. Operations are provided for the user to navigate between sub-contexts. We demonstrate how the use of these operations can lead to quick identification of an area of software source code that establishes an unecessary dependency between software parts.

1   Introduction

Many modern information systems contain not only large amounts of information, but also complex structures. From operating on large but simple tables, information technology moved on to complex models represented by object-relational structures. This can be seen in the database world in the form of *Entity-Relationship Modeling* (ER) or *Object-Role Modeling* (ORM) [Hal96], it can be found in *Object-Oriented Programming* (OOP) and it is the basis of disciplines such as *Knowledge Engineering* (KE) [Smi96].

While modeling tools have been advanced over time to accommodate the complexity found, approaches to retrieve information from such structured sources are not much more sophisticated than they were decades ago. The main methods of querying are still based on relational joins and projections, combined with statistical methods to reduce the resulting information into sizes a human can handle. But while the result of the former usually produces tabular data too large to be understood in its entirety, the latter suffers from reducing information in an opaque manner.

Formal Concept Analysis (FCA) [GW99] is a successful technique of data analysis for data that fits well into the structure of a *many-valued context*. But while the structure of a many-valued context (attribute-value data) is common, it is still quite restricting, requiring the identification of a single object set and assuming functional dependencies.

B. Ganter and R. Godin (Eds.): ICFCA 2005, LNAI 3403, pp. 113–128, 2005.

In many examples significant steps were required in order to convert relational data into the form of a many-valued context.

To address this issue, this paper proposes other approaches to use FCA in combination with relationally structured data. This does not only apply to relational databases, but also knowledge bases in the sense of Conceptual Graphs or RDF. Other data can often be mapped into a suitable structure easily; in this paper we will demonstrate the techniques with an example analysing software source code.

Part of the problem posed by complex data is that the user has to be able to select views on the whole data, which can be displayed and understood easily. Toscana systems solve this problem by predefining a number of scales to be used. Since the scales are predefined it can be the case during analysis that desired scales have not yet been constructed, or that constructed scales are never used. If the number of scales is large it can also become difficult for the user to identify which scale meeting their analysis requirements. Furthermore structural changes on the input data can invalidate scales.

To overcome these problems we propose a different approach: defining and using a *navigation space*. The navigation space consists of points and functions. A point in the navigation space corresponds to a view on the data. The user moves from point to point within the space by selecting a function to apply to their current point. This mode of goal directed interaction is akin to browsing within hypertext documents. Similar to web browsers the user should be allowed to return to points visited earlier to follow different paths of investigation.

We demonstrate that FCA can be used as a powerful visualisation technique for querying complex relational structures by applying this notion of a navigation space to a concrete example in the domain of software engineering.

Before we start the discussion of our approaches, Section 2 introduces the general notion of relational structures and our problem domain: relational structures in software. We then outline our approach in Section 3, before introducing the notion of a navigation space in Section 4. Section 5 discusses our prototypical implementation, Section 6 gives an overview of related work and Section 7 concludes the paper with an outlook on further research.

2   Relational Structures

One of the most important building blocks of modern large-scale software systems is the notion of a *relational database*. A relational database can be defined as:

> **Relational Database:** [A] collection of data items organised as a set of formally-described tables from which data can be accessed or reassembled in many different ways without having to reorganise the database tables.[1]

Quite often the notion of a relational database is connected to *relational model* as defined by E.F.Codd in [Cod70], which is historically correct, although technically modern databases do not use pure relational algebra due to the introduction of the NULL value in SQL.

---

[1] http://his.osu.edu/help/database/terms.cfm

Sometimes the notion of a relational database is extended to the notion of an *object-relational database*. In an object-relational database the relational structures are extended by introducing the notion of classes with an inheritance hierarchy: subclasses inherit the attributes of their superclasses.

In the domain of Artificial Intelligence (AI) the term *"knowledge base"* (KB) is used quite frequently. In [Smi96] a knowledge based is defined as follows:

Knowledge Base: A collection of facts and rules; the assembly of all the information and knowledge from a particular problem domain.

The underlying definition of knowledge is that "Knowledge consists of symbols, the relationships between them and rules or procedures for manipulating them." [Smi96], which is common in the area of AI.

Following these definitions the relevant distinction between a database and a knowledge base is that the former stores only facts, while the latter also stores rules. Rules can be implemented either by forward-chaining (deriving new facts from exiting facts by applying the rules) or backward-chaining (testing goals by breaking them into subgoals). For the purposes of this paper we will use the term "knowledge base" in the definition above and will not distinguish between the actual implementation of the storage system.

2.1 Relational Structures in Software

Modern software systems are a rich source of structured information. This structured information can be extracted automatically and is classified as either (i) static information, extracted from programs without running them, or (ii) dynamic information extracted from programs during their execution.

Even though modern software systems are highly structured, a significant part of the structure cannot be captured explicitly within the software code with existing tools. For example, software engineering promotes orthogonality as a desirable aspect of software structure, yet it is often hard to define the exact meaning of orthogonality in a particular software design. The process of modifying software incrementally towards a better and more orthogonal design has become increasingly interesting with larger software projects. With the notion of *refactoring* [Fow99] this process has been formalised and there is significant tool support in applying standard patterns of changes that enhance software structure without changing the functionality of the program.

Modern software tools still fail to give much guidance in where and how to refactor. Since within FCA orthogonality has a clear definition as a direct product of partial orders, we see FCA as well placed to aid humans in the task of understanding software structure in general and for the purpose of refactoring in particular.

For the purpose of this paper we consider the following artifacts of a Java program: *packages, classes, interfaces, methods,* and *fields.* Java programs follow a document model in which packages may contain packages, classes and interfaces. Classes and interfaces contain method declarations, and classes may contain method definitions and fields. Methods are identified by method signatures that define the method name and its parameters types. Classes, methods and fields additionally have modifiers such as public, private, synchronised or abstract attached to them.

The artifacts in a Java program have unique names derived from the position of that artifact within the program. Having unique names for objects in a domain greatly simplifies the task of analysis since we can use those names as referents for the artifacts and do not have to introduce artificial identifiers.

The relations between artifacts of a Java program we are interested in are: `in`, `calls`, `extends`, and `implements`. The relation `in` is derived from the document structure of the source code. The relation `calls` tracks when one method potentially makes a call to another method. If one class `extends` another class, then it inherits the fields and methods of the class it extends, similar for interfaces extending other interfaces. A class can `implement` an interface meaning that it takes the responsibility to provide the methods defined within the interface.

We will occasionally refer to some of these relations with names common for software engineering. The `calls` relation is called the *static call graph*, `in` is referred to as *containment relation* and both `extends` and `implements` are considered *inheritance* relations.

3   System Overview

Figure 1 gives an overview of our approach. A *domain analyser* extracts relational structure from the domain, in our example the domain analyser is a utility to reconstruct source code structure from Java programs.

A *rule based system* is then used to derive additional information from the data extracted by the domain analyser, thus introducing abstractions which are specific to the domain. Such additional relations can be transitive or reflexive closures or they can be defined by set functions such as unions of other relations (which can be used to generalise relations) or set-minus (which can be used to specialise relations).

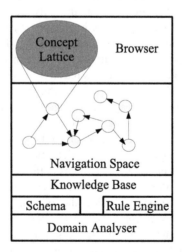

Fig. 1. System Overview: Information is extracted by a source code analyser. A knowledge base uses rules to interpret the information, provide a query interface and a definition of the schema. A navigation space of concept lattices is automatically derived and made browsable by the user

As an example, our domain analyser produces `extends` and `implements` as relations between classes and interfaces. Both model the conceptually related notion of inheritance and for our purposes it is sometimes useful to consider them as two aspects of a larger relation `is-derived-from`. This is achieved in our system by adding a rule that defines `is-derived-from` as the union of the two basic relations.

Our system applies the rules in a separate step before analysis of the data takes place, deriving additional facts using forward-chaining. We are able to do this because our rules have a finite closure that is small enough to be computed and stored on contemporary hardware. In combination with the indexing mechanisms applied, the retrieval performance of the system for derived relations allows quick navigation.

A *navigation space* is then automatically generated from the relational structure — both the relations extracted from the domain and the relations derived by the rule based engine. Each point in the navigation space corresponds to a concept lattice, and a browsing tool is used to move between points in the navigation space. Movement through the navigation space is performed by the user who moves from one point to the next via a choice of functions.

The details of the navigation space will be discussed in the next section.

4 Constructing a Navigation Space

We begin this section with a formal definition of a navigation space. We then motivate this definition within the context of our software engineering application before working through a concrete example. We then offer some discussion on how the navigation space may be expanded to employ nested line diagrams to reduce diagram complexity.

4.1 Formal Notions

The starting point for the navigation space as used in this paper is a formal context with additional orders on the objects and attributes.

Definition 1. *An **ordered context** is a triple* $((G, M, I), (G^\dagger, \leq_{G^\dagger}), (M^\dagger, \leq_{M^\dagger}))$ *where* (G, M, I) *is a formal context,* $(G^\dagger, \leq_{G^\dagger})$ *and* $(M^\dagger, \leq_{M^\dagger})$ *are partially ordered sets, with* $G \subseteq G^\dagger$ *and* $M \subseteq M^\dagger$. *These ordered sets are called the **object order** and the **attribute order** respectively.*

We then complete this context by interpreting the object and attribute orders as defining object and attribute implications, i.e. we wish that $h \leq g$ induces the object implication $h \to g$ and $n \leq m$ induces the attribute implication $n \to m$.

Definition 2. *The **completed context** of an ordered context* $((G, M, I), (G^\dagger, \leq_{G^\dagger}), (M^\dagger, \leq_{M^\dagger}))$ *is the formal context* $(G^\dagger, M^\dagger, I^\dagger)$ *where for* $g \in G^\dagger$ *and* $m \in M^\dagger$

$$(g, m) \in I^\dagger :\Leftrightarrow \exists h \in G^\dagger, n \in M^\dagger : h \leq_{G^\dagger} g, n \leq_{M^\dagger} m, (h, n) \in I$$

The user moves from point to point in the navigation space using a number of functions. In a user interface this could be achieved by allowing the user to interact with a concept, object or attribute and make a choice from a menu.

Definition 3. *A **navigation space** is a triple (P, F, X) where P is a set, and $F := \{f_j\}_{j \in J}$ is a set of functions indexed by J with $f_j : D_j \to P$ and $D_j \subseteq P \times X$ and X is a set of potential user inputs. The element of P are called **points** of the navigation space.*

Definition 4. *The **navigation space of a completed context** $C = (G^\dagger, M^\dagger, I^\dagger)$, **functions** F **and user input** X, is the navigation space $N(C, F, X) := (P, F, X)$ where $P := \mathfrak{P}(G^\dagger) \times \mathfrak{P}(M^\dagger)$.*

Each point in the navigation space is represented to the user as a concept lattice of a subcontext of the completed context. The user inputs X represent the ability of the user to select from the diagram an object, and attribute or a concept.

Definition 5. *The **subcontext of a point** (H, N) in a navigation space $N((G^\dagger, M^\dagger, I^\dagger), F, X)$ is the formal context $(H, N, I^\dagger \cap (H \times N))$.*

The choice of a set of navigation functions is a design decision. For our application we consider functions acting on the following domains:

$$D_O := \left\{ ((H, N), g) \in P \times G^\dagger \mid g \in H \right\}$$

$$D_A := \left\{ ((H, N), m) \in P \times M^\dagger \mid m \in N \right\}$$

$$D_C := \left\{ ((H, N), c) \in P \times \mathfrak{P}(G^\dagger) \times \mathfrak{P}(M^\dagger) \mid c \in \mathfrak{B}(H, N, I^\dagger \cap (H \times N)) \right\}$$

corresponding to whether the user chooses an object, an attribute or a concept.

For each of these domains a number of functions are offered. If the user has selected an object then the following functions are defined:

$$\text{move\_down}((H, N), g) := (\text{lower\_covers}(g), N)$$

$$\text{unfold\_down}((H, N), g) := (H \cup \text{lower\_covers}(g), N)$$

$$\text{fold\_down}((H, N), g) := (H \setminus \text{lower\_covers}(g), N)$$

Additionally we provide the functions zoom_up, unfold_up, and fold_up which use upper covers rather than lower covers. Six equivalent functions are available for attributes.

If the user selects a concept then the following functions are defined:

$$\text{focus\_on\_extent}((H, N), c) := (\text{extent}(c), N)$$

$$\text{focus\_on\_object\_contingent}((H, N), c) := (\text{object\_contingent}(c), N)$$

$$\text{focus\_on\_intent}((H, N), c) := (H, \text{intent}(c))$$

$$\text{focus\_on\_attribute\_contingent}((H, N), c) := (H, \text{attribute\_contingent}(c))$$

These functions are designed to strike a balance between guiding the user toward meaningful diagrams, and not overly constraining the user. By providing functions that add or remove covers the user makes strong use of the object and attribute orderings to derive meaningful subcontexts. To avoid lattices growing too large the user can focus on particular concept, but then will usually drill down by unfolding objects and attributes.

The path of the user through the navigation space is important so we additionally store this path and allow the user to move backwards and forwards along the path in a similar fashion to a web browser.

4.2 Navigation Space for Software Analysis

As is evident from our definitions, a navigation space is constructed from an ordered formal context. In our application an ordered formal context is naturally formed by the calls-t and in-rt relations. in-rt is the reflexive transitive closure of in and is used to define both the attribute and object orderings of the ordered formal context. Completing this natural ordered formal context extends the calls-t relation defined on methods to a relation defined over packages, classes and methods with the basic rule being that package *a* calls package *b* if the *a* contains a method calling a method in *b*.

Software engineers are interested in which packages contain calls to which other packages, because this introduces a dependency from one package to the other. To be used, a package requires all packages it depends on, which can affect reuse of software components. Keeping track of dependencies is also very important when software is modified because the modification of a package will potentially affect all the packages that are dependent on it.

Dependencies are also established by the is-derived-from relation and so another natural choice for an ordered formal context is a combination of the is-derived-from-t relation and the in-rt relation.

In cases where the analyst is less interested in the layered nature of the software design and instead wants to focus in the immediate dependencies between packages and classes, an ordered context formed from calls rather than calls-t is of interest.

A little less intuitive is the ordered formal context formed by the relations calls-t and the inverse relation of is-derived-from. This ordered context is related to the notions of *polymorphism* and *dynamic dispatch* in object-oriented programming and can be used to identify which method implementations are potentially executed by method calls.

By presenting the user with a navigation space, we allow browsing these complex structures via relatively simple diagrams. Each association in these diagrams can be traced to a portion of the source code. Since the names used in our application identify software artifacts, we allow navigating from the objects and attributes directly to these artifacts.

4.3 Example

To demonstrate navigation within the space of concept lattices in a realistic example we have applied our tool to the ToscanaJ source as it was on April, 1st 2002[2]. The aim of using this tool was to identify design decisions in the software and to test if these were applied consistently throughout the source code. The top level package for the program is called net.sourceforge.toscanaj. For brevity we will use tj instead, both in the diagrams and in the text.

The start point for the exploration is given on the left of Figure 2. Both the object and the attribute set are the lower covers of the top concept in the containment relation, which means the first navigation point is (lower_covers tj, lower_covers tj).

---

[2] This code can be extracted from the source code repository at http://sf.net/projects/toscanaj

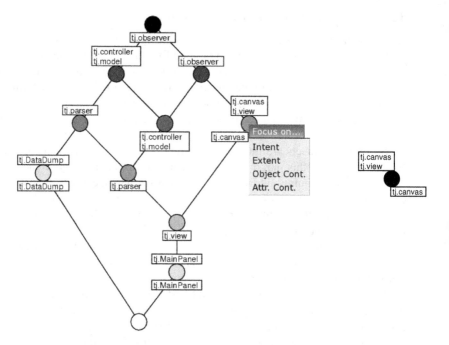

Fig. 2. First and second in a series of concept lattices generated by navigating through the space of concept lattice in which suspicious dependency of tj.canvas on tj.view is investigated

The resulting diagram shows which top level packages and classes contain method calls into which other top level packages and classes.

In this software the tj.canvas package is supposed to model a general notion of a drawing canvas, independent of the ToscanaJ software itself. In fact this package was extracted into a separate software component later in the product development. Based on this knowledge the call from tj.canvas to tj.view is considered wrong since it establishes a dependency from this supposedly general package into ToscanaJ specific code.

To examine this call in more detail we first zoom into the object concept of tj.canvas, restricting both the attribute and object set to the corresponding contingents. This produces the concept lattice on the right of Figure 2 where only one node is left. With this step we have restricted our view to the relevant aspects of the structure.

Next we unfold the lower covers of the object tj.canvas to see which parts of the canvas module contain the problematic call. Additionally we unfold the lower covers of the attribute tj.view to see which parts of the view package are called. The result is the diagram in the left of Figure 3. The central concept shows us that the dependency must be established between the two classes DrawingCanvas and ImageWriter on the canvas side and the package tj.view.diagram on the view side. Further unfolding of the attribute tj.view.diagram produces the diagram in the right of Figure 3 which identifies the dependency on the class level: the two classes mentioned above call the class DiagramSchema in tj.view.diagram.

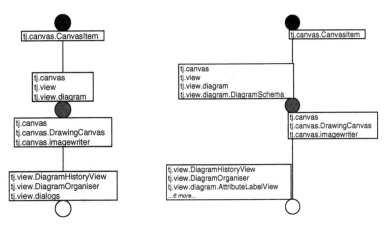

Fig. 3. Third and fourth in a series of concept lattices generated by navigating through the space of concept lattice in which suspicious dependency of tj.canvas on tj.view is investigated

In a last navigation step we identify the dependency on the method level by unfolding the object tj.canvas.DrawingCanvas to find which methods of tj.canvas.DrawingCanvas are making the relevant calls. The resulting diagram is shown in Figure 4 and shows that there are only a few dependencies between the packages. Most parts of the canvas package are now shown as independent from the view package, only two methods are marked as having dependencies. From the method names we can navigate to the source code for the methods and examine the source code. Examining the code for tj.canvas.DrawingCanvas.paintCanvas we find a call to tj.view.diagram.DiagramSchema, which in this example turns out to be not only an unwanted dependency but also a bit of code that has no functionality in the system – it was most likely left during previous changes.

To identify the dependency coming from the tj.canvas.imagewriter package we would have to unfold the package again, but in this case it is not required since this dependency turned out to be not a direct dependency, but a call to the paintCanvas method, thus fixing this method solves the problem of the invalid dependency completely. Alternatively we could have looked at the calls relation instead of its transitive closure, in which case we would not have found this dependency between the tj.canvas.imagewriter and the tj.view package.

Note that in our navigation we chose to maintain the original attributes and objects when unfolding. This was a somewhat arbitrary choice, alternatively we could have zoomed into lower covers. By unfolding, and keeping attributes from previous diagrams, we kept additional structure within the diagram that gave a hint as to the previous navigation points we visited. Keeping these attributes also allowed us to observe the object and attribute implications embedded in the lattice by the completion algorithm.

To demonstrate the generality of the approach we navigated to the concept lattice shown in Figure 5. In contrast to our earlier example this space is not based on the static

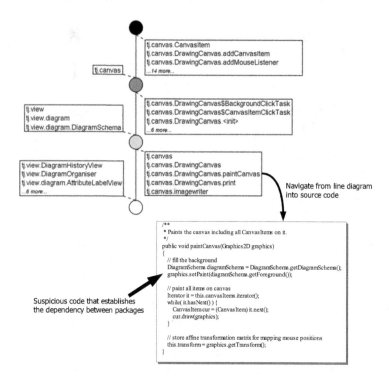

Fig. 4. Fifth in a series of concept lattices generated by navigating through the space of concept lattice in which suspicious dependency of `tj.canvas` on `tj.view` is investigated

call graph, but on the transitive and reflexive closure of the class hierarchy relation. The navigation space is again constructed from the containment relation.

The diagram shows the inheritance of interfaces and classes between the top level package and the view package on the object side and the canvas and observer package on the attribute side. The resulting lattice identifies three categories of classes: observers, observable, and observable observers. In this case the diagram reflected the expectation of the software designer. It validates a number of assumptions, which were not made explicit before: for example the class `DiagramView` is not observable because it is the outer-most view, meaning it is not contained within any other view. The classes `LineView` and `NodeView` are not observers because they don't change in response to other elements. `LabelView` objects in contrast follow `NodeView` objects around the diagram and so are observers.

It is interesting to note that even a hand rendered UML class diagram is incapable of conveying the same information — the grouping of the classes according to their inheritance — as is captured in the concept lattice. We believe that Figure 5 would be a valuable addition to a software design document accompanied by a textual description reflecting on the relationship of the diagram to the software design.

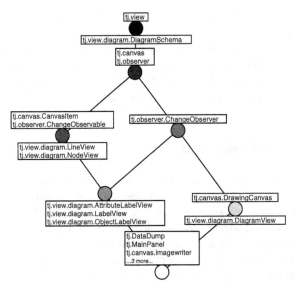

Fig. 5. Concept lattice showing inheritance of the interfaces `ChangeObserver` and `ChangeObservable` within the view package

4.4   Nested Line Diagrams

As concept lattices become large they can become difficult to layout and difficult to comprehend. A popular mechanism to reduce the complexity of concept lattice diagram is the nested line diagram. A nested line diagram essentially embeds the concept lattice of a context (G, M, I) into the direct product of the concept lattices of two sub-contexts $(G, M_1, I \cap G \times M_1)$ and $(G, M_2, I \cap G \times M_2)$ where M_1 and M_2 partition M.

In the resulting nested line diagram the grouping of attributes into the two sets M_1 and M_2 makes a very strong statement to the reader. For rhetoric reasons this grouping should be selected carefully.

In order to support nested line diagrams we extend the navigation space to include triples (H, N, m) where $m \in N$. To display a navigation point (H, N, m) we partition N to form N_{inner} and N_{outer} as:

$$N_{inner}((H, N, m)) = N \cap (\downarrow_{\leq_{M^\dagger}} m)$$
$$N_{outer}((H, N, m)) = N \setminus N_{inner}$$

The two subcontext then become $(H, N_{inner}, I^\dagger \cap H \times N_{inner})$ and $(H, N_{outer}, I^\dagger \cap H \times N_{outer})$.

This partitioning has a natural conceptual mapping in the case where the navigation relation $M^\dagger$ is some type of containment relation, since the grouping into the inner and outer diagram represents the distinction as containment. It is also very well suited for local scaling as described in [Stu96].

To include moving between nested and unnested diagrams the functions defined previously need to be modified slightly. The key property that these functions should preserve is that for any state (H, N, m) the attribute m has to be in the set N. Whenever an operation violates this condition the system has to change back into displaying an unnested diagram.

Nesting is used usually within TOSCANA systems to combine conceptual scales. The translation of this idea into the navigation space would produce a mechanism for combining two navigation points. This would be a significant extension to the approach we have outlined so far, but would most likely also add significant complexity to the user interface.

5   A Prototype Implementation

In our prototype we employ a graph based database system we implemented ourself with the aims of: (i) being comparatively fast; (ii) being comparatively succinct in expressing binary relations since our domain consists largely of these; (iii) indexing should performed automatically; (iv) simple rules should be supported to permit us to leverage background information when interpreting data.

We will briefly describe the three components of our prototype, namely: a class file analyser, a system to manage the data and the rules, and a navigation component.

5.1   Class File Analyser

Java class files contain the program in a compiled form, but still close enough to reconstruct some of the original source code structure. The main loss is the information about names of parameters and variables in the code, which is not relevant for many applications of our technique. For various technical reasons we decided to use class files instead of actual source code as the input of our prototype — later versions are planned to accept both class files and source code itself.

We used the CFPARSE library from IBM which provides functions that allow access to the contents of Java class files. The class file analyser produces a list of triples with each triple being a single relationship of the form (subject, predicate, object). An example for such a triple is (java.lang, is-a, package).

The class file analyser already extracts unique names for software artifacts which we use to determine object identity in the system.

5.2   Knowledge Base Management System

The core component for storing the knowledge base is a quadruple store which stores and indexes quadruples. Each quadruple is of the form (context, subject, predicate, object) where context is an identifier that specifies the context in which the statement represented by the triple holds. In our software analysis example the context is usually a particular version of the software analysed.

The quadruple store store provides a simple query language whereby mappings from a query graph into the quadruple store are found and returned. It can be modelled as a set of quadruples $G \subseteq C \times C \times C \times C$ where C is a set of strings[3]. A query graph, Q, is a set of quadruples $Q \subseteq W \times W \times W \times W$ where $W \subseteq V \cup C$ is a set containing strings

---

[3] a string is a sequence of characters of a given alphabet.

and variables (V is a set of variables). A mapping, $\phi : W \rightarrow C$ from the query graph Q to the data graph G is valid iff

$$\forall (x_1, x_2, x_3, x_4) \in Q : (\phi x_1, \phi x_2, \phi x_3, \phi x_4) \in G$$
$$\forall c \in C \cap W : \phi c = c$$

The rule-based system attached to this quadruple store implements simple rules composed of two graphs: a query graph as a premise and a second graph using the same variables as conclusions. The rule is applied by iterating through each valid mapping from the premise into the data graph. In each iteration the mapping is applied to the conclusion graph and the result added to the data graph.

5.3 Navigation Component

In order to make sub-contexts directly computable from the quadruple store, the current navigation point (H, N) is stored within the quadruple store. The sub-context $(H, N, I^\dagger \cap (H \times N))$ corresponding to the navigation point (H, N) is then queried directly and turned into a context representation which can be displayed by tools from the TOSCANAJ suite.

6 Related Work

There are essentially two contributions made by this paper: the idea of a navigation space as a way to structure relational data for conceptual analysis, and the application of this idea to provide a tool to aid in the analysis of software structure. We will first situate out work in relation to use of formal concept analysis as a tool in software engineering and then go on to situate our work in relation to general methods of formal concept analysis.

A significant problem facing the application of formal concept analysis to software structure is the size of the lattices constructed. Our approach of situating the user within a navigation space structured by familiar hierarchies such as the package structure hierarchy or the type system hierarchy solves this problem and is novel.

A number of authors have recommended methods using FCA as a tool to understand software structure. A significant portion of such work revolves around clustering of software artifacts with the purpose of deriving a new and improved software structure. One significant approach in this regard, is the work of Snelting et al. [ST00, Sne98] on re-engineering class hierarchies. Briefly stated, the idea is that a formal context with program artifacts (functions, common blocks, types etc.) as objects and attributes, and an incidence relation that says whether one artifact accesses another artifact, has formal concepts that can be considered candidates for program organisation. Contributions around this idea have been made by various authors [GM93, GMM+98, Ton01, KM00]. In the case considered by Snelting the context is built from variables and method calls the concepts can be considered candidate interfaces. We consider Snelting's approach to be a large lattice approach, meaning a single concept lattice from a single input context is utilised. A problem with large lattices is that they are difficult to comprehend and often contain objects and attributes that the user, for her current purpose, is not interested in. Our approach complements Snelting's work in that it allows the user to exclude some aspects of the software structure and focus on others. Being situated within a navigation space the user is able to quickly shift her focus.

Further to the work of Snelting on re-engineering class hierarchies, our work focuses on program understanding and human reflection. Our central purpose is not to re-engineer class hierarchies, but rather provide a tool for programmers to access and reflect on software structure as a way to plan his or her next move. Furthermore we believe the structure and tool we have outlined in this paper to be relevant to other domains.

The literature concerning the application of FCA to software engineering is extensive. For a more complete description of the field we refer the reader to [TCBE03] where we conducted a literature review of applications of formal concept analysis generally within the software engineering process. Our previous work [CT03] analysing software structure using a rule based system, triple store, and formal concept analysis can been seen as relevant background to the idea of a navigation space.

We now turn to the position of navigation spaces in relation to general approaches to data analysis using formal concept analysis. The notion of *conceptual scaling* [GW89] and analysis tools and techniques that have grown from it [VW95, BH04] present a way to analyse data that can be formulated as a *many valued context*. Although these tools have been very successful in a large number of domains, a difficulty that can occur when the number of *conceptual scales* becomes large. Toscana based systems require an analyst to produce conceptual scales a-priori and if there is a large number of scales this work can be extensive. Also in Toscana systems there is generally no provision for organising the conceptual scales and so as the number of scales becomes large the user tends to become lost.

The navigation space idea that we have outlined in this paper provides a way to automatically generate and structure a space of subcontexts using hierarchies. Thus it attempts to solve two difficulties associated with Toscana systems: (i) the time consuming task of generating many scales by hand, and (ii) the lack of organisation of the scales once produced. The user interacts with the system like they do with a web browser moving along links from one diagram to the next.

There have been number of attempts to move away from, or to extend, Toscana systems. Our previous work [CS00] and work by Stumme [Stu99] can be seen as a precursor to the idea of a navigation space. Stumme and Herreth [Stu00] have proposed ways to extend Toscana systems in the direction of On-Line Analytical Processing.

7   Outlook and Conclusion

When concept lattices become large they loose their ability to succinctly communicate information to the user of an analysis system as their complexity becomes overwhelming. The system presented here allows browsing a large information space and representing views on this space as concept lattices, thus mapping a larger structure into understandable parts.

In order to provide an analysis tool for complex relational data we have presented a system for automatically constructing a navigation space consisting of sub-contexts displayed as concept lattices. As opposed to Toscana systems these views can be created dynamically by the user and a navigational structure over the diagrams is provided. In order to demonstrate the utility of this navigation structure we applied it to the analysis of software structure and demonstrated its use via a simple example.

The navigation space we presented in this paper is only one of a family of browsing structures that could be constructed via a similar methodology. The use of navigation spaces is not restricted to software structure but we believe can be usefully applied to a wide range of relational structures. Ideally a software tool would be able to allow tailoring of the navigation functions to a particular application domain. It such a software tool would also abstract over different database types to make it easy to process data found in different systems.

In this paper we have not considered mechanisms to combine points from different navigation spaces. We feel that the theory of multicontexts [Wil96] will be helpful in this direction but leave it as an area of further work.

Another area of further investigation is that of temporal and modal aspects of software systems. Software programs under development undergo change – changes which are commonly stored within a version control system, which gives access to different versions of the software at different points in time (e.g. a release vs. a development version). While our prototype is specifically designed to capture information about software versions and configurations and so allow the analysis of the differences and similarities between versions or configurations, we have not yet experimented extensively with this task.

References

[BH04]   Peter Becker and Joachim Hereth Correia. The ToscanaJ suite for implementing Conceptual Information Systems. In *Formal Concept Analysis – State of the Art*. Springer, Berlin – Heidelberg – New York, 2004. To appear.

[Cod70]   E. F. Codd. A relational model of data for large shared data banks. *Communications of the ACM*, 13(6):377–387, June 1970.

[CS00]   Richard Cole and Gerd Stumme. CEM - a Conceptual Email Manager. In *Conceptual Structures; Logical, Linguistic, and Computational Issues*, number 1867 in LNAI. Springer-Verlag,, September 2000.

[CT03]   R. Cole and T. Tilley. Conceptual analysis of software structure. In *Proceedings of Fifteenth International Conference on Software Engineering and Knowledge Engineering, SEKE'03*, pages 726–733, USA, June 2003. Knowledge Systems Institute.

[Fow99]   M. Fowler. *Refactoring, Improving the Design of Existing Code*. Addison Wesley, 1999.

[GM93]   Robert Godin and Hafedh Mili. Building and maintaining analysis-level class hierarchies using galois lattices. In *Proceedings of the OOPSLA'93 Conference on Object-oriented Programming Systems, Languages and Applications*, pages 394–410, 1993.

[GMM+98]   R. Godin, H. Mili, G. W. Mineau, R. Missaoui, A. Arfi, and T.-T. Chau. Design of class hierarchies based on concept (galois) lattices. *Theory and Application of Object Systems (TAPOS)*, 4(2):117–134, 1998.

[GW89]   B. Ganter and R. Wille. Conceptual scaling. In F. Roberts, editor, *Applications of combinatorics and graph theory to the biological and social sciences*, pages 139–167. Springer-Verlag, New York, 1989.

[GW99]   Bernhard Ganter and Rudolf Wille. *Formal Concept Analysis: Mathematical Foun-dations*. Springer, Berlin – Heidelberg – New York, 1999.

[Hal96]   Terry Halpin. *Conceptual schema and relational database design (2nd ed.)*. Prentice-Hall, Inc., 1996.

[KM00]   T. Kuipers and L. Moonen. Types and concept analysis for legacy systems. Tech-nical Report SEN-R0017, Centrum voor Wiskunde en Informatica, July 2000.

[Smi96]   Peter Smith. *An Introduction to Knowledge Engineering*. International Thompson Computer Press, 1996.

[Sne98]   G. Snelting. Concept analysis - a new framework for program understanding. In *SIGPLAN/SIGSOFT Workshop on Program Analysis for Software Tools and Engi-neering (PASTE)*, pages 1–10, Montreal, Canada, June 1998.

[ST00]   G. Snelting and F. Tip. Understanding class hierarchies using concept analysis. *ACM Transactions on Programming Languages and Systems*, pages 540–582, May 2000.

[Stu96]   G. Stumme. Local scaling in conceptual data systems. In P.W. Eklund, G. El-lis, and G. Mann, editors, *Conceptual Structures: Knowledge Representation as Interlingua*, LNAI 1115, pages 308–320, Berlin, August 1996. Springer-Verlag.

[Stu99]   Stumme. Hierarchies of conceptual scales. In *Proc.Workshop on Knowledge Ac-quisition Modeling and Management*, volume 2, pages 78–95, Banff, 16'th-22'nd October 1999.

[Stu00]   G. Stumme. Conceptual On-Line Analytical Processing. In S. Ghandeharizadeh K. Tanaka and Y. Kambayashi, editors, *Information Organization and Databases*, chapter 14, pages 191–203. Kluwer, Boston–Dordrecht–London, 2000.

[TCBE03]   T. Tilley, R. Cole, P. Becker, and P. Eklund. A survey of formal concept analysis support for software engineering activities. In Gerd Stumme, editor, *Proceedings of the First International Conference on Formal Concept Analysis - ICFCA'03*. Springer-Verlag, February 2003. to appear.

[Ton01]   P. Tonella. Concept analysis for module restructuring. *IEEE Transactions on Soft-ware Engineering*, 27(4):351–363, April 2001.

[VW95]   Frank Vogt and Rudolf Wille. Toscana — a graphical tool for analyzing and ex-ploring data. In Roberto Tamassia and Ioannis G. Tollis, editors, *Graph Drawing*, pages 226–233. Springer, Berlin – Heidelberg – New York, 1995.

[Wil96]   R. Wille. Conceptual structures of multi-contexts. In P. Eklund, G. Ellis, and G. Mann, editors, *Conceptual Structures: Knowledge Representation as Interlin-gua*, number 1114 in LNAI, pages 23–39. Springer Verlag, Berlin, 1996.

Restructuring Help Systems Using Formal Concept Analysis

Peter Eklund[1] and Bastian Wormuth[2]

[1] School of Information Technology and Computer Science,
The University of Wollongong,
Northfields Avenue, Wollongong, NSW 2522, Australia
peklund@uow.edu.au
[2] Darmstadt University of Technology, Department of Mathematics,
Schloßgartenstr. 7, 64289 Darmstadt, Germany
bastian@wormuth.info

Abstract. This paper extends standard help system technology to demonstrate the suitability of Formal Concept Analysis in displaying, searching and navigating help content. The paper introduces a method for building suitable scales directly from the help system index by computing a keyword extension set. The keyword extension technique is generalisable in any document collection where a hand-crafted index of terms is available.

1 Introduction

Wille [1] writes that methods for knowledge processing presuppose an understanding of what knowledge is, we therefore begin by examining the nature of "knowledge" in the help domain.

Firstly, help systems are multipurpose. To understand their purpose, we examine help in terms of the dimensions of Conceptual Knowledge Processing (CKP) [1]. At the simplest level a help system provides instruction in the style of an on-line tutorial. In this case there is an instructional narrative where tasks are presented in a predefined order. In the same vein, help systems are also used to present more advanced "how to" information to users who have mastered the basics but are following a command sequence for the first time. This corresponds to **exploring** in CKP [1]. More advanced users have completed most command sequences in the software. For these users, the help system is used to **search** for something they can "more or less specify but not localize" [1]. The help materializes knowledge that is either beyond the present recollection or too unimportant to commit to memory.

Identifying the taxonomy of commands – or functionality – is a major problem for many software systems, particularly those based on Windows. The drill down made possible by the combination of dynamic menus, property settings, tabs and dialog boxes often leads to feature discovery through trial and error. Windows-based help systems have a particular style that supports pathfinding of this sort to reduce the "uncertainty" (as it is called in information science [2]) resulting from so many branches in a hierarchy.

B. Ganter and R. Godin (Eds.): ICFCA 2005, LNAI 3403, pp. 129–144, 2005.

Having identified the "knowledge" of help systems w.r.t. elementary methods used in CKP, the hypothesis we test is that the application of FCA to the **analysis** of help system content can be **restructured** to **improve** its presentation for the purpose of supporting inter-subjective human argumentation (**decision** support based on mixed initiative – a dominant theme in the modern practice of HCI [3]). This we believe will enhance the user's capacity to **investigate** the help content.

Two outcomes result from our study. The first is the recasting of the help system for the MAIL-SLEUTH program as a Conceptual Information System. We demonstrate how the creation of suitable scales can clarify the presentation of help content. The second is a method to extend search terms from the help system index. This method involves seeding the scale with search terms which are then expanded using the hand made index provided by the CHM file format. We describe this technique and evaluate its suitability to help systems.

2   Motivation

Fig. 1 shows the lattice that results from a context made of a set of objects as help pages (HTML documents) with attributes as URLs (`</href= />` tags to other help pages)[1]. The resulting concept lattice is rather poor. This line diagram is included because we can learn from it. First, we learn that scaling is required to isolate the help contents w.r.t. the purpose and functionality of the software being described. Second, the disarray in this concept lattice reinforces that the help system was not developed in a systematic way. The presentation of the content is *ad hoc* and is therefore likely to require restructuring.

Most help systems for Windows applications are constructed from compiled HTML files (so called CHM files) using a program called Microsoft HTML help. Compiled HTML has a number of advantages. Firstly, the compiled HTML is usually smaller than the original source. Second, compiled HTML is indexed and is therefore searchable. Third, the format for the presentation of the compiled HTML is standard in the Windows operating environment. Finally, the content of the help pages is not easily copied once compiled.

The search functionality for compiled HTML is a keyword search. Search terms can be entered into the search area, index entries relate to help pages and lack relations to other index terms; no context is given on a search term even when the term appears in multiple contexts. Although it is easy to find explanations when the correct search term has been entered, it is much harder to find the right explanation when the searcher is unsure of the appropriate search term to use. The browsing metaphor for compiled HTML is also rather limited. A tree widget is presented that requires browsing hierarchically by chapter, section and page. In short, the HTML help environment lacks the ability to semantically relate terms with pages and the search context.

---

[1] A Java program called EXLINK was written for this purpose by Shaun Domingo to automatically build the context.

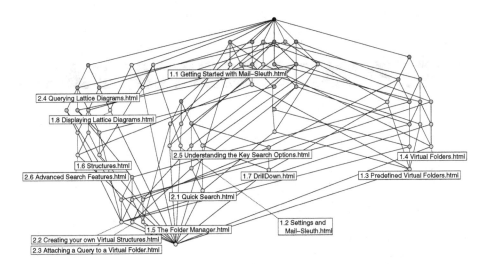

Fig. 1. Naive analysis of the help system for the MAIL-SLEUTH program using SIENA. Many of the pages towards the top of the lattice involve initial set-up, configuration or initialization tasks, done once on installation. Help pages that are insinuated across the MAIL-SLEUTH help appear lower in the lattice, the idea of "structures" (presented in the page Structure.html) is (for instance) important because it is referred to from many other pages. The poor structure and presentation are indicative of the unstructured *ad hoc* content of the help system

What attracts our attention to compiled HTML as a content platform is its potential to be improved through the application of the concept lattice for browsing and visualizing help content. In the next generation of Microsoft operating systems, the use of MAML (**M**icrosoft **A**ssistance **M**arkup **L**anguage) claims to counter this difficulty[2]. MAML will enable help to be tagged with dynamic properties that allow parts of the help system to be activated depending on the content. However, this paper addresses how FCA can value-add the current generation of help systems and compiled HTML.

Our paper follows the application thread presented by other work, specifically CEM [4] and RENTAL-FCA [5], where the use of the concept lattice has been shown to be well-suited to document browsing in the domains of email and News classifieds. More recent work by Eklund et al.[6] supports these claims through a usability trial for the MAIL-SLEUTH program. A more comprehensive usability study of MAIL-SLEUTH's Web-mail variant, MAILSTRAINER, was also recently completed [7]. In the MAILSTRAINER study, 16 users completed 7 search and document management tasks and completed a psychometric survey based

---

[2] http://www.winwriters.com/articles/longhorn

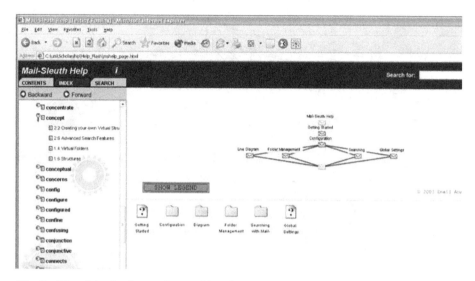

Fig. 2. The *Introduction* scale provides the starting point for the MAIL-SLEUTH help. Fig. 3 is the scale that appears when the user double clicks on the centre left vertex *Line Diagrams*. Nested-line diagrams are therefore implemented as a form of progressive drill-down rather than in the usual all-at-once nested line diagram

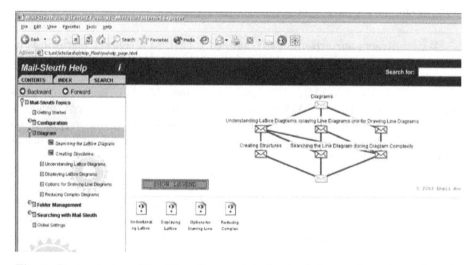

Fig. 3. Restructuring of the MAIL-SLEUTH help involved the creation of several static scales: these included a scale for searching, drawing line diagrams, Getting Started, Configuration, Folder Management and Global Settings

on 10 questions on their experience and impressions with the software. A detailed presentation of MAILSTRAINER will be presented in a subsequent paper as a synthesis of Domingo [7].

Further, Carpineto and Romano – pioneers in the application of FCA to information retrieval [8, 9] – have engineered the CREDO search engine[3]. This outcome reinforces the appeal of a lattice-based approach to search. The outcome from our paper is narrower in scope, it demonstrates the suitability of FCA to the search and navigation of help system content and characterizes the mathematical structures that support an understanding of this process.

3   Examination of Help Content Using FCA

In the following section we presume a basic knowledge of Formal Concept Analysis. For a introduction to FCA see [10], further mathematical foundations are given in [11]. Practical FCA algorithms and methods are presented in [8].

We will discuss two ways to create a Conceptual Information System (CIS) based around the functionality of help systems. The first is to take an existing help system based on the index file, the content file and the explanation files (a CHM file) and restructure it in Conceptual Knowledge Processing terms. The second is the creation of a help system from the ground up regarding principles of CKP. For both approaches, FCA is used as an analysis technique and the known advantages of line diagrams representing concept lattices are reinforced. The advantages of the concept lattice as a visual metaphor of the information help space include:

– *the line diagram simultaneously displays the results of one or more search terms;*

Search of the index in HTMLhelp CHM files is conjunctive keyword search, all search terms must be contained in the document if a match is to be found. This encourages one (or few) search terms initially with iterative search proceeding from a single keyword and then introducing additional keywords one at a time with the objective of reducing the size of the result set. Iterative search in the standard help system paradigm therefore proceeds from few-to-many keywords. In the CIS search paradigm, this is not normally the case, search encourages as many search terms as can be thought of in order to produce as rich a lattice structure as possible. This style of search isolates search terms of little value quickly, reducing the dimensionality of the search. Therefore iterative search proceeds in a many-to-few keyword direction initially. The **search** act in CIS, just described, includes **exploration**, **identification** and **investigation** of the information landscape [1].

– *the diagrams display the search results embedded in their conceptual context;*

---

[3] http://credo.fub.it

The result from search in a CIS will display the context of the pages from the search terms. In the case of DOCCO and CEM the documents retrieved are revealed along with their position in the underlying file system. This provides context for the search.

– *the diagrams at one time can offer global, general and detailed views;*

The concept lattice that emerges from search in a CIS shows either the complete boolean lattice of search terms used or alternatively the reduced line diagram showing all points on the concept lattice instantiated by the keyword used in the search. This reveals the distribution of matching documents over the entire document collection. In so doing, it provides both detailed and general views of the collection according to the dimensions of the search.

– *predefined diagrams can guide you through a domain.*

The idea of a library of predefined contextual scales is not new and has been used in the ZIT library system [12] and elsewhere [5]. The creation and re-use of conceptual scales are a fundamental basis for the TOSCANA-workflow. General CIS tools, such as CEM [4] and TOSCANAJ[4], allow the user to create their own individual search scales to provide reusable search contexts. In MAIL-SLEUTH [6], several useful predefined search scales are pre-loaded with the program and further scales may be added by the user. DOCCO[5] and CREDO[6] do not provide any scaling tools but this is consistent with the general nature of the collections they index.

3.1   Reprocessing of CHM-Based Help Systems

The CHM files can be decompiled using free decompilers like KEYTOOLS from KEYWORKS Software[7]. The result is a set of HTML files, a file containing the index and a file containing the contents of the help system. The HTML files are the explanation texts that appear in the right hand output window of a standard help system window, Fig. 4 shows such an interface. The system provides three options to find the explanation text that answers a request, the search function, the tree structured content of the help system and the index. The explanation text itself is structured in three elements: the title, the text body and a paragraph that lists related topics: "see also" or "related topics". Further, there may be hyperlinks in the body of the text that connect certain words or phrases in the explanation text to other relevant pages. All hyperlinks from the "see also" section can also address other CHM help files.

The application of Formal Concept Analysis requires the assignment of objects and attributes to the de-compiled elements. The subjects the user is looking

---

[4] http://www.toscanaj.sf.net
[5] http://www.tockit.sf.net
[6] http://credo.fub.it
[7] http://www.keyworks.net

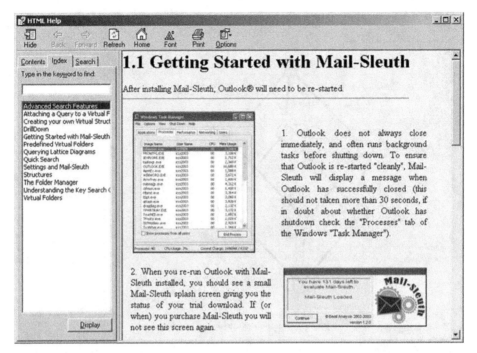

Fig. 4. The standard user interface for a help system in MICROSOFT WINDOWS systems. The user has three means to search the help content provided by tabs on the left hand of the window: the index, the contents and the search function. All these functions produce a list of potential titles, phrases of keywords to the user, selecting one item from this list displays its corresponding explanation file in the right hand frame of the window. Further navigation functionalities similar to web browsing are also provided

for are the explanation texts and the constraints for the search are the keywords the user enters. So the explanation files become the objects and the keywords entered become the attributes of the Conceptual Information System. With G as the set of explanation files, M being the set of entered keywords and the incidence relation gIm, if an explanation files contains a keyword m, we produce the Formal Context (G, M, I) that is the basis for the concept lattice displayed in the graphical search interface.

Fig. 5 shows a simple search result using the FCA based retrieval tool DOCCO[8] applied on the help content of MAIL-SLEUTH[9].

This result is produced against on the raw set of HTML files from the decompilation. DOCCO creates an index of the words contained in the explanation files and displays a diagram of a concept lattice where the search terms are the attributes and the explanation files containing any of the search terms as

---

[8] http://www.tockit.sf.net
[9] http://www.mail-sleuth.com

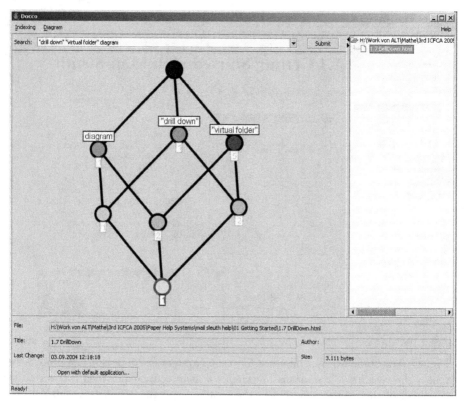

Fig. 5. Screenshot of Docco: the entered search terms are "drill down", "virtual folder" and "diagram". The relevant search result is actually the explanation file attached to the bottom node highlighted. The selected file(s) can directly be opened with the button "Open with default application..."

objects. The inherent structures of the explanation files given by the links to related topics are not considered in this approach.

The opposite approach, where the search terms are the objects and the explanation files the attributes, yields a diagram that better supports users without familiarity with FCA. This structure provides a kind of ranking with the best results, the most specific explanation files, in the top section of the diagram and more general matches below. This hierarchy is also given in the approach pursued in this paper: the best results are expected in the extent of the bottom concept of the lattice.

3.2   Search Improvements and Automatic Search Extension

A user searching a large data set for a special question would like to achieve best results with minimum effort. The process of searching has been effective and successful when the set of relevant search results resp. explanation files is small. So the objective is to minimize the number of objects in the extent of

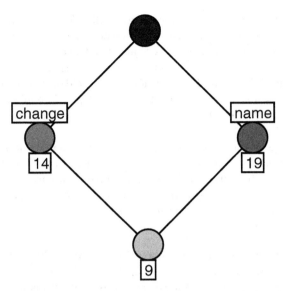

Fig. 6. Diagram representing the concept lattice of a search for the explanation for "Changing the name of a folder": the user enters "change" and "name" as search terms

the bottom concept of the line diagram. The more search terms provided, the smaller the extent and the more specific the concept. The problem here is that users do not generally enter many search terms, so some alternative support is required to improve the search functionality: through query expansion. Fig. 6 shows the result of a search for the simple question "Changing the name of a folder" in the standard help file of MICROSOFT WINDOWS. The user entered "change" and "name" as search terms, the size of extent of the bottom concept is 9, so 9 different explanation files are offered as result.

Most help systems do not support query expansion to find the relevant explanation file, so the user often has few clues on what to search for. Query expansion by the automatic extension of the search terms is one solution. The use case requires one or two keywords to act as a seed for generating extension terms. In Windows help, the search terms "change", "name" might be expanded to "change", "name" "file" and "folder" for instance.

Additional search terms can be derived from the index of the given help system. In the following we propose a simple method which leverages the human effort formalized in the hand-made index contained within the CHM file. By understanding this hand-made index as a formal context $\mathbb{K}_{CHM}$, as described in Section 3.1, we can use it to derive additional search terms using the prime operator.

Let $\mathbb{K}_{index}$ be the formal context used by DOCCO, with $G_{index} = G_{CHM}$ being the set of explanation files of the help system, M_{Index} being the automatically indexed words and gIm, if $m \in M_{index}$ appears in $g \in G_{index}$. As the index used by DOCCO is not manually created, but collects all words (except those in a stop words list) that appear in the explanation files, it is (in practice) much larger than the hand-made index. By larger we mean that $M_{CHM} \subseteq M_{index}$, the object sets are identical. It is possible that the author of the help system has added entries in the hand-made CHM index that do not occur in the explanation file, but this is a rather pathological case. Furthermore there are differences in the incidence relations I_{CHM} and I_{index}: not every incidence relation in between a term $m \in M_{index}$ and an explanation file $g \in G_{index}$ is realised in the context $\mathbb{K}_{CHM}$ and symmetrically there may be incidence relations in $\mathbb{K}_{CHM}$ that are not realised in $\mathbb{K}_{index}$.

The advantage of the hand-made index is the fact that the words or phases are assigned by a human to a certain explanation file and therefore the explanation file being "about" the phrase or keyword assigned is assured.

Let S be the set of seed search terms entered by the user, S_{ext} be the extended set of search terms and G_{CHM} the set of explanation files. S_{ext} is derived from S and G_{CHM} by the following:

1. Derive the set $S' \in G_{CHM}$ of explanation files where all instances of S occur in every instance of S'.
2. $S_{ext} = S''$, $S'' \in M_{CHM}$ is the set of words that all instances of S' have in common.

In the language of FCA this is the derivation of the attribute concept of the seed search terms in the context $\mathbb{K}_{CHM}$:

$$S_{ext} := \{S \cap M_{CHM}\}''.$$

If the CHM index on the explanation files does not remove stop words S_{ext} will contain many redundant words. This stop word filter should be applied *after* the derivation of S_{ext}, because the removal of stop words can result in substantial information loss, e.g. the words "on" and "off" in "log on" and "log off".

The resulting set S_{ext} can be offered to the user to specialize.

Applying this method on the search displayed in Fig. 6 with $S = \{change, name\}$ returns,

$$S_{ext} = \{Change, file, folder, changing, forbidden, characters, names, renaming, naming, overview\}.$$

Adding the search terms "file" and "folder" to the search yields the diagram given in Fig. 7. Due to the more specific search this line diagram has just three explanation files in the extent of the bottom concept and among those the most relevant file to the intended query, "Change the name of a file or folder" is included.

The complexity of this operation is $\mathcal{O}(|G||M|)$, $|G|$ being the number of objects and $|M|$ the number of attributes. This operation computes in an acceptable practical time frame for instantaneous keyword expansion for contexts based on indexed decompiled CHM Help files.

The method to derive an extended set of search terms relates to [13] and [14]. The following presents a short introduction to multicontexts:

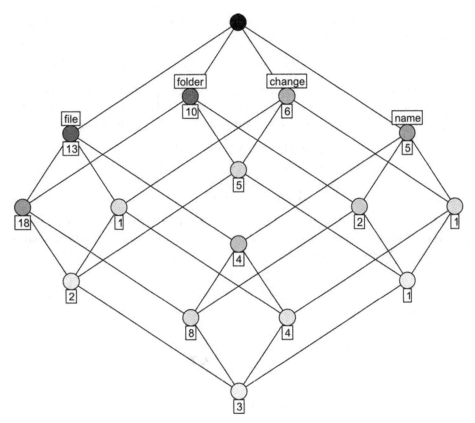

Fig. 7. A line diagram representing the concept lattice of a search for the explanation for "Changing the name of a folder" as in Fig. 6. Additional to the initially entered terms "change" and "name", the user also entered "file" and "folder": derived by automatic search term extension. The most specific results are the most relevant and represented in the three explanation files objects in the extent of the bottom concept of the line diagram

Multicontexts can be understood as a network of formal contexts describing incidence relations of a domain, each context representing a different view of the "situation". Different contexts may have different object and attribute sets, they may also share elements in common, although they do not have to. It is also possible that the object set of a context shares elements with the attribute set of another context, which causes interesting relations in between the lattices of two different contexts of a multicontext.

Following the definition of multicontexts, quoted from [13]:

Definition 1. *A multicontext of signature* $\sigma : P \rightarrow I^2$, *where* I *and* P *are nonempty sets, is defined as a pair* (S_I, R_P) *consisting of a family* $S_I := (S_i)_{i \in I}$ *of sets and a family* $R_P := (R_p)_{p \in P}$ *of binary relations with* $R_p \subseteq S_i \times S_j$ *if*

$\sigma p = (i, j)$. *A multicontext* (S_I, R_P) *of signature* $\sigma : P \to I^2$ *can be understood as a network of formal contexts* $\mathbb{K}_p := (S_i, S_j, R_p)$ *with* $\sigma p = (i, j)$.

Multicontexts provide so called coherence mappings between two contexts:

Definition 2. *Let* (S_I, R_P) *be a multicontext, let* $\mathbb{K}_p := (S_i, S_j, R_p)$ *and* $\mathbb{K}_q := (S_k, S_l, R_q)$ *be formal contexts of* (S_I, R_P), *and let* (A, B) *be a formal concept of* $\mathbb{K}_p$. *Then there are four* coherence *mappings from* $\mathfrak{B}(\mathbb{K}_p)$ *to* $\mathfrak{B}(\mathbb{K}_q)$ *defined by:*

$$\lambda_{pq}(A, B) := ((A \cap S_k)^{qq}, (A \cap S_k)^q)$$
$$\varrho_{pq}(A, B) := ((B \cap S_l)^q, (B \cap S_l)^{qq})$$
$$\varphi_{pq}(A, B) := ((B \cap S_k)^{qq}, (B \cap S_k)^q)$$
$$\psi_{pq}(A, B) := ((A \cap S_l)^q, (A \cap S_l)^{qq})$$

The concept lattices of the formal contexts of (S_I, R_P) *together with all coherence mappings form a coherence network of concept lattices which shall be denoted by* $\mathfrak{R}(S_I, R_P)$. *See also Fig. 8.*

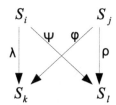

Fig. 8. Coherence mappings

Coherence mappings, shown in Fig. 8, allow the user to select a concept as a formal context of a multicontext and to map this concept to another context – simply put, the user has the option to select his objects of interest in a certain view and to compare this selection to the corresponding results in a different view.

The coherence mappings establish a means to examine the relations between two contexts – exactly the way we derived the extended set S_{ext} of search terms. In our example we took the concept of interest and mapped it to the corresponding one in the lattice of $\mathbb{K}_{CHM}$, then we transfered the intent of the mappings image back to Docco and created the sub lattice of $\mathbb{K}_{index}$ with the new set of search terms to specify the search.

Let (A, B) be the concept of $\mathbb{K}_{index}$ that seems interesting from the users point of view. Then the corresponding concept of $\mathbb{K}_{CHM}$ is derived using the mapping φ_{pq} with $p = index$ and $q = CHM$ that maps from $\mathbb{K}_{index}$ to $\mathbb{K}_{CHM}$:

$$\varphi_{pq}(A, B) = ((B \cap M_{CHM})^{qq}, (B \cap M_{CHM})^q)$$

The intent of $\varphi_{pq}(A, B)$ is the extension set S_{ext} that we used to specify the search, and because of $M_{CHM} \subseteq M_{index}$, the intent of the derived concept is still part of M_{index} and produces a promising extension of search terms.

3.3   Processing the Structure of the Help Content

Any CHM based help system is a structured document with three key elements. The first is the HTML of the actual explanation file. Sub-parts of the layout as described in Section 3.1 can easily be found in the HTML source code: the title and the body tags of the explanation file for example. The identification of the "related topic" paragraph is more difficult, because HTML provides various means to tag this paragraph. The second key element is the table of contents of the help system. Although a tree structure, the table of contents groups explanation files by their content. The third key element is given by the links from words in the body of an explanation file to other explanation files. The fact that external explanation files of other CHM files can be incorporated allows the inclusion of related files.

These elements can be used to derive sets of explanation files that are of higher importance to each explanation file: for every explanation file we define a set of related explanation files, these are those that appear in the same sub-directory of the table of contents, those listed in the "related topics" paragraph and those that are link targets from the body of the explanation file. This set, we call the *scope set* of an explanation file, because it forms a scope for every explanation file and can be used in Formal Concept Analysis to create predefined conceptual scales for the CIS. Such derived sets will not normally be disjoint, except in the case where pages indexed by the table of contents are disjoint.

The elements: title, body and "related topics" can also be used to apply constraints on the search. These provide three "search modalities": restricting search to titles, bodies, titles and bodies, an so on. Only title search can be achieved with current version of DOCCO. Fig. 9 shows the result of a title element

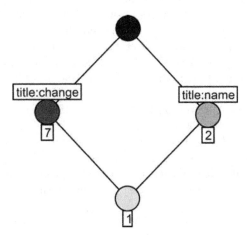

Fig. 9. A line diagram representing the concept lattice of a search for the explanation for "how to change name of a folder": the user entered "change" and "name" as search terms and the search is restricted to the title element tags of the explanation files

search using the same terms "change" and "name" as in Figure 6 and 7, but restricted to the titles. The only object in the extent of the bottom concept is the relevant explanation file.

Further possibilities to use the resolved structure of the HTML help files to improve the relevant results will be proposed in Section 4.

4   Further Research

The structure provided by the HTML files and the human effort coded in the content and index files that form part of the CHM files have not been used before to engineer a Conceptual Information System. In addition, the following paragraph proposes a means to capture the inherited structure of the explanation and the content files using multicontexts.

In addition to multicontexts, triadic contexts as introduced by Wille and Zickwolff in [15] need to be addressed also. The core difference between triadic contexts and formal contexts of a multicontext is that every incidence relation in a triadic context is valid in a certain modality. This suits our requirements, with different "search modalities" conforming to help system key elements. Triadic contexts and multicontexts are related, because every triadic context gives rise to a multicontext [13]. The problem of characterising triadic contexts to improve the exploration of help systems is that the understanding of line diagrams representing triadic contexts requires considerable experience in Formal Concept Analysis, so we focus on multicontexts.

The idea is to create a multicontext consisting of contexts based on the indexed words of the titles of the explanation texts, e.g., the bodies and the "related topics" paragraph in order to have a mathematical foundation for the search modalities as described in Section 3.3.

Additional contexts can be derived from each chapter of the table of contents. An implementation of the multicontexts would allow the user to search the data in different modalities, e.g. to restrict the search to titles and to incorporate the bodies of the explanation files after selecting a certain concept. The introduction of other content tags can enhance the search additionally by providing finer segmentation on the document. Imagine a user having problems with his network connection. A search on the standard MS Windows help for "network connection" gives about 100 hits. The Windows help has a chapter "Networking" and another "Mobile Users", a search using the "mobile" or "standard" context would focus on the related explanation files. If the user has found an explanation file that helps, but does not solve his problem, he can then compare search results to the corresponding results from another search modality.

5   Conclusion

MAIL-SLEUTH is a program that itself uses the concept lattice to dynamically organize the contents of an email collection into thematically coherent views based on the application of Formal Concept Analysis. It therefore seems fitting to experiment with the very same idea for the MAIL-SLEUTH help system. This desire for design coherence was the starting point for our study.

Our research reveals that help systems (in general) are well-suited to a content-management approach based on the concept lattice for navigation. There are several reasons for this, the first is that help systems are relatively small so computing extension keywords is realistic, the second is that help systems are a semi-structured document source with strongly inter-related content and finally help systems come equipped with a handy precomputed, hand-crafted index.

In our paper Wille's **restructuring** idea is applied by examining the content of the MAIL-SLEUTH help. The results were the creation of six predefined scales that thematically covered all functional aspects of the software. The new help contents were then re-written w.r.t. the scales with self-evident improvements to the presentation and structure of the help system. Hand crafting scales in this way is ideal but there is also scope to automatically generate scales by using seed search terms and computing the extension set. We show how this can be achieved and present examples that illustrate keyword extension.

References

1. Wille, R.: Conceptual landscapes of knowledge: A pragmatic paradigm for knowledge processing. In Gaul, W., Locarek-Junge, H., eds.: Classification in the Information Age, Springer (1999) 344–356
2. Colomb, R.: Information Spaces: the Architecture of Cyberspace. Springer (2002)
3. Horvotz, E., Kadie, C., Paek, T., Hovel, D.: Models of attention in computing and communication: From principles to applications. CACM **46** (2003) 52–59
4. Cole, R., Eklund, P., Stumme, G.: Document retrieval for email search and discovery using formal concept analysis. Journal of Experimental and Theoretical Artificial Intelligence **17** (2003) 257–280
5. Cole, R., Eklund, P.: Browsing semi-structured web texts using formal concept analysis. In: Proceedings 9th International Conference on Conceptual Structures. LNAI (2001) 319–332
6. Eklund, P., Ducrou, J., Brawn, P.: Concept lattices for information visualization: Can novices read line diagrams. In Eklund, P., ed.: Proceedings of the 2nd International Conference on Formal Concept Analysis - ICFCA'04, Springer (2004)
7. Domingo, S.: Engineering and Interactive Searching and Browsing Environment for Web-Based Email. PhD thesis, School of IT and Computer Science (2004)
8. Carpineto, C., Romano, G.: Concept Data Processing: Theory and Practice. Wiley (2004)
9. Carpineto, C., Romano, G.: Order-theoretical ranking. Journal of the American Society for Information Sciences (JASIS) **7** (2000) 587–601
10. Wille, R.: Introduction to Formal Concept Analysis. In Negrini, G., ed.: Modelli e modellazione, Models and Modelling, Roma, Consiglio Nazionale delle Ricerche, Instituto di Studi sulle Ricerca e Documentazione Scientifica (1997)
11. Ganter, B., Wille, R.: Formal Concept Analysis: Mathematical Foundations. Springer, Berlin (1999)
12. Rock, T., Wille, R.: Ein TOSCANA—Erkundungsystem zur Literatursuche. In Stumme, G., Wille, R., eds.: Begriffliche Wissensverabeitung: Merthoden und Anwendungen, Berlin-Heidelberg, Springer (2000) 239–253

13. Dörflein, S.K., Wille, R.: Coherence Networks of Concept Lattices: The Basic Theorem (2004)
14. Wille, R.: Conceptual Structures of Multicontexts. In Eklund, P.W., Ellis, G., Mann, G., eds.: Conceptual Structures: Knowledge Representation as Interlingua. Number 1115 in LNAI, Heidelberg - Berlin - New York, Springer (1996) 23–29
15. Wille, R., Zickwolff, M.: Grundlagen einer Triadischen Begriffsanalyse. In Stumme, G., Wille, R., eds.: Begriffliche Wissensverarbeitung: Methoden und Anwendungen, Heidelberg - Berlin - New York, Springer (2000) 125–150
16. Dieberger, A., Dourish, P., Höök, K., Resnick, P., Wexelblat, A.: Social navigation: Techniques for building more usable systems. ACM Transactions on Human-Computer Interaction **7** (2004) 26–58

An Application of FCA to the Analysis of Aeronautical Incidents

Nicolas Maille[1], Irving C. Statler[2], and Laurent Chaudron[3]

[1] ONERA - Centre de Salon de Provence,
Base Arienne 701 - 13661 Salon Air - France
maille@onera.fr
[2] NASA - Ames Research Center,
Code IHS, Mail Stop 262-7 - Moffett Field, CA 94035-1000
Irving.C.Statler@nasa.gov
[3] ONERA - Centre de Toulouse,
2, avenue E. Belin - 31400 Toulouse - France
chaudron@onera.fr

Abstract. This paper illustrates how a new clustering process dedicated to the analysis of anecdotal reports of aviation incidents has been designed and tested thanks to an FCA tool called *Kontex*. Special attention has been given to the necessary transcription of the data from the initial relational database to an FCA context. The graphical interface for *Kontex*, which has been specially implemented for this study, is also presented.

The study presented in this paper validates the process adopted and highlights the use of FCA to help the expert to mine the database without previous knowledge of the searched concepts. The work brought original ideas to the aviation safety community by the development of an incident model and the notion of scenario. For the FCA community, one interesting aspect of this work lies on the use of a first-order context (given by a relational database) and its translation into a classical context.

1 Introduction

1.1 Scope of This Paper

This paper describes an experiment that has been realized cooperatively by ON-ERA and NASA. Its aim was to test a new clustering methodology dedicated to the analysis of textual reports of aviation incidents. A Formal Concept Analysis (FCA) tool has been developed and used for this work.

The work brought original ideas to the aviation safety community by the development of an incident model and the notion of scenario. For the FCA community, one original aspect of this work lies on the use of a first-order context (given by a relational database) and its translation into a classical context. This point is developed in this paper in section 2.1 for the theoretical approach and in section 4.3 for its application to the aviation incident reports.

B. Ganter and R. Godin (Eds.): ICFCA 2005, LNAI 3403, pp. 145–161, 2005.

After a short overview of incident analysis, the paper describes briefly the FCA tool (*Kontex*) in section 2. Then a simplified view of the incident model and the definition of the clustering methodology are given in section 3. Finally, the experiment and a summary of the results given by the use of the FCA tool are presented in section 4.

1.2   ONERA-NASA Cooperation

Since 1995, ONERA and NASA Ames Research Center have conducted collaborative research on the analysis of aeronautical incident reports. The aim is to develop methodologies and tools that allow the experts to identify causal factors and human-factor-related issues. The ASRS[1] database [1], [2] is used as a representative data resource for this study even though the approach adopted is not designed to fit any specific incident reporting systems.

Several codification processes that have been developed and tested on ASRS reports [3] were evaluated. Then the effort have focused on the design of a new clustering process. In 2003, an experiment validating the methodology was completed [4] and is the subject of this paper. It was based on the use of the FCA tool *Kontex* (developed by ONERA) with a limited set of selected ASRS reports. Other experiments based on statistical methods are now under study.

1.3   Applicative Background

Even though air transportation is the safest mode of travel, improving the level of security is a major concern for the aeronautical community. The airlines and the authorities would like to have a proactive management of safety risk from a system-wide prospective. Such a process involves identifying hazards, evaluating causes, assessing risks and implementing appropriate solutions. It is a non-trivial task that relies on the capability of continuously monitoring the system's performance.

Some airlines have already developed quality-control strategies in which they analyze routinely their performance data. Their safety programs rely mainly on two types of data sources: (1) flight-recorded data [5], and (2) incident-reporting systems. Other techniques such as in-flight audit (LOSA) [6] are emerging but are not widely used yet[2]. While flight-data analysis provides an objective understanding of "what" happened during operations, it gives little information about the " why". This distinction between the "what" and the "why" of an incident is discussed further in the discussion of the notion of scenario in Section 3.2. An understanding of the causal factors of why an incident occurred is essential

---

[1] The ASRS (Aviation Safety Reporting System) managed by NASA and funded by the FAA Office of System Safety since 1976, is one of the world's best-known and most highly regarded repositories of safety information. The ASRS database is a collection of nearly 112,000 narratives of aviation safety incidents that have been voluntarily submitted by reporters across the aviation industry.

[2] At least when measured on the basis of the extent of the ongoing collection of data.

to formulating the appropriate intervention in proactive management of safety risk.

The search for causal factors of incidents depends mainly on the anecdotal account of the incident reporter. So, incident databases constitute the best available sources of information about why incidents happened. However, their analyses pose several challenges. First, the process is typically labor-intensive and requires high-priced domain expertise to read and interpret each report. Further, while available tools allow automated extraction from a database of reports on a specified issue, they are unable to highlight the unknown systemic issue requiring a proactive intervention. There is a need for new analytical methods and automated capabilities to help the experts mine these rich and complex textual databases for insight into the causal, contributing, and aggravating factors of an incident or event. Finally, the breakthrough in the new generation of text-analysis tools is their ability to deal with the complex semantics of these anecdotal reports. Thus, the need for formal means that could preserve and empower the symbolic content of these data is essential. A first step is to build meaningful sets of reports that identify recurrent issues and reduce the task of the experts.

2   The ONERA Kontex Tool

The ONERA team involved in the ONERA-NASA cooperation worked for several years on FCA and especially on the use of FCA with generalized contexts [7], [8]. They developed several prototypes of tools among which a Prolog III$^{©}$ program allows computation of the concept lattice when the attributes are represented by literals of a first-order language.

The investigation of real incident reports achieved during this collaborative study generated large concept lattices (over 4000 concepts) for which the exploration required a more sophisticated graphical interface. This motivated the ONERA team to develop a new tool (called *Kontex*) based on the C language and OpenGL. Unlike their previous Prolog tool, this one uses classical FCA contexts. Indeed, it is always possible to find a classical context that generates the same concept lattice as the one calculated with the first-order context (or any generalized context). The next section explains how to make such a transformation. In the specific case of a first-order context, it is better, from a computational point of view, to generate the equivalent classical context before calculating the concept lattice, as the use of first-order logic is time consuming. This explains the choices made for the new ONERA tool.

2.1   Generalized and Classical Contexts

When the knowledge about the objects to be analyzed is captured by a relational database, it is sometimes easier to characterize those objects by literals of a first-order logical language. Then the description is closer to the original information and more understandable by the domain experts. It has been shown in [7] that FCA theory can be adapted to the use of such contexts (called generalized contexts).

Definition 1.
A *generalized context* $\mathbb{K}$ is a triple $(O, (\mathfrak{L}, \leq, \sqcap, \sqcup), \zeta)$ where O is a finite set, $(\mathfrak{L}, \leq, \sqcap, \sqcup)$ is a lattice and ζ is a mapping from O to $\mathfrak{L}$.

We will not review here all the mathematical elements (i.e., generalized concepts and, generalized concept lattices) required to extend FCA to the use of such generalized contexts. As already noted, it is not really an extension of the FCA theory, but a way to use contexts represented in a more expressive way. Indeed, for each generalized context $\mathbb{K}$, there is a classical context $\mathbb{K}'$ that generates an isomorphic lattice [9].

The aim of this section is to show briefly how to build such a corresponding classical context when the generalized context $\mathbb{K}$ is based on finite lattice $\mathfrak{L}$. The demonstration relies on the use of $\vee$-irreducible elements. Let us make a short review:

Definition 2.
Let $(\mathfrak{L}, \leq, \sqcap, \sqcup)$ be a lattice. An element $x \in \mathfrak{L}$ is $\vee$-*irreducible* if:
- $x \neq 0$ (in case $\mathfrak{L}$ has a zero);
- $x = a \vee b$ implies $x = a$ or $x = b$ for all $a, b \in \mathfrak{L}$

The set of all the $\vee$-irreducible elements of $\mathfrak{L}$ is noted $J(\mathfrak{L})$.

In the case of a finite lattice $\mathfrak{L}$, $J(\mathfrak{L})$ is supremum -dense and so each element of $\mathfrak{L}$ can be represented as the supremum of elements of $J(\mathfrak{L})$ [9]. This allows us to define the decomposition of an element as follows.

Definition 3.
Let $(\mathfrak{L}, \leq, \sqcap, \sqcup)$ be a finite lattice. The decomposition of an element x of $\mathfrak{L}$ is $x^\vee = \{y \in J(\mathfrak{L}) \mid y \leq x\}$

This definition can now be used to define a classical context from a generalized one as follows.

Definition 4.
Let $\mathbb{K} = (\mathfrak{L}, \leq, \sqcap, \sqcup)$ be a finite lattice and $(O, (\mathfrak{L}, \leq, \sqcap, \sqcup), \zeta)$ be a generalized context.
$\mathbb{K}' = (G, M, I)$ is a classical context defined by $G = O$, $M = J(\mathfrak{L})$ and $(g, m) \in I$ iff $m \in g^\vee$ (for all $g, m \in G \times M$).

It is easy to check that for all $a, b \in \mathfrak{L}$ we have: (1) $a \vee b = \bigvee(a^\vee \cup b^\vee)$, (2) $a \wedge b = \bigvee(a^\vee \cap b^\vee)$, and (3) $a \leq b \Leftrightarrow a^\vee \subseteq b^\vee$

From these properties the following result can be deduced.

Proposition.
The generalized concept lattice $\mathfrak{B}(O, (\mathfrak{L}, \leq, \sqcap, \sqcup), \zeta)$ is isomorphic to the classical concept lattice $\mathfrak{B}(O, J(\mathfrak{L}), I)$.

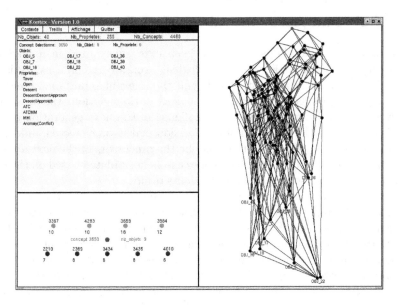

Fig. 1. The Kontex graphical interface

This property gives a theoretical way to transform any generalized context (based on a finite lattice) into a classical context, which generates an isomorphic concept lattice. Therefore, one can either choose to use the adaptation of FCA with a generalized context or choose to transform it into a classical context before calculating the concept lattice. The choice mainly depends on the complexity of the algorithm required to implement the lattice operators ($\leq, \sqcap, \sqcup$) compared to the complexity of the decomposition algorithm. When the cube lattice[3] is used, it is more efficient to transform the context as the $\leq$ operator used the subsumption algorithm. The automated transformation of a cubical context into a classical context is under development and will be integrated in *Kontex*.

2.2   The Kontex Tool

The Kontex tool has been developed by ONERA in order to explore concept lattices. The set of concepts and their relations are calculated with the algorithm of Lindig [10]. The graphical interface (shown in Figure 1) mainly enables navigation in the lattice structure and extraction of sub-lattices. It contains three windows: one (upper left) for the description of the selected concept, one (lower left) with the parents and the children of the selected concept (which is useful when the number of concepts is high), and one (right) with the current view of the lattice (it can be the full lattice or a sub-lattice). Menus allow access directly to the concept labeled by the selected attribute or by the object name.

---

[3] A cube is a conjunction of first order literals. A finite lattice, called the cube lattice, can be defined on the set of cubes and used to build a generalized context, called a cubical context [7], [8].

3   Mining an Aeronautical Incident Database

Large databases, like the ASRS, have been extensively used in a retroactive way with automated tools: when an accident occurs or when a safety issue is suspected, a dedicated request is built and the relevant reports analyzed. But, in order to enhance safety, the decision makers in the aviation system need now to develop proactive strategies[4]. The challenge of the next generation of safety tools is to bring to the attention of the decision makers unexpected, hidden, and relevant issues in the database that may be the precursors of the next accidents. Thus, the adequate intervention strategies can be formulated based on the causal factors of the precursors before the accidents occur.

Discovering relevant safety issues in an incident report database is not a trivial task. It implies the ability to extract meaningful sets of reports related to recurrent situations. Such sets of reports will then be analyzed by domain experts to assess the operational significance of the recurring event.

In order to build relevant clusters, the specificities of what is an aviation incident must be taken into account. Hence, in this section (which is completely dedicated to the application), we review briefly the content of the ASRS database and present the incident model upon which the clustering methodology is based. A detailed view of these notions can be found in [4].

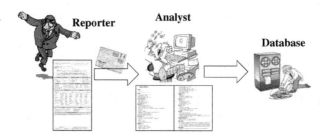

Fig. 2. From the event to the Database

3.1   An Aeronautical Incident Reporting System

Aeronautical incident reporting systems, such as the ASRS, are voluntary, non punitive and confidential. They rely on the participation of certificated personnel (pilots, air traffic controllers, dispatchers, mechanics...) who report safety concerns they observe, even if they resulted from their own errors. As we do not want to describe one particular system, we just present here some essential steps performed in order to build the database of such a system. Accordingly, an oversimplified report processing system can be described by Figure 2.

---

[4] This is sort of what the ASRS does in the triage mode when experts identify incidents worthy of alert bulletins. However this step is not at all automated and highly dependent on the analyst's memory as to whether similar incidents ever occurred before.

The report is submitted on a reporting form that contains a few fixed fields and a blank section for the narrative. When received by the experts of the reporting system, the report is analyzed and codified through a codification form that is based on a structured set of items. The resulting codification and the narrative are entered in the database.

3.2   The Incident Model

The main information of the initial report is concentrated in the narrative where reporters of aviation incidents describe problems encountered during their operations. They usually relate them as stories and focus on what happened, on the involvement and behavior of people as well as on the important features that help us to understand why these problems occurred.

Let us take an example extracted from the ASRS database. The narrative (in which the abbreviations and acronyms have been expanded) is reproduced here: *"We were on a visual approach behind a wide-body for runway 28 Right. At about 1000 feet above ground level, the tower offered us 28 Left. We changed to 28 Left and the tower cleared the wide-body to cross 28 Left ahead of us. The wide-body delayed crossing and when we were close in, the tower offered us 28 Right. We attempted to change to 28 Right but were too close in to maneuver and so we went around."*

This short story can be understood through a sequence of four states and the three transitions between them, as displayed in the Figure 3. Extracting such a decomposition from the narrative is a natural process for the expert as it allows him/her to capture quickly the essence of the evolution of the situation in a meaningful way.

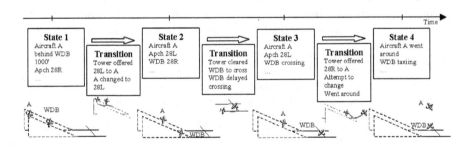

Fig. 3. The story of an incident report

In such a story, all the states and transitions do not play the same role and a story is an incident only if certain specific states are entailed. In a simplified view, we can say that a story relates an incident if it starts with a safe state, evolves to an anomalous one and returns to a safe state. Then the main features of the incident are captured by the three components of the notion of a scenario, as defined on Figure 4. We point out that the *Context*[5] (or *Initial Situation*) and the *Outcome* mainly represent the "WHAT" happened while the *Behavior* is more linked to the "WHY".

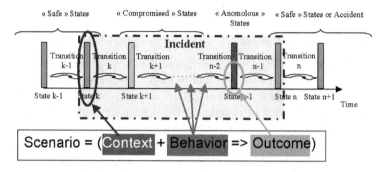

Fig. 4. Scenario and incident model

In order to generate meaningful sets of reports, the process designed will have to be based on the notions highlighted by this incident model.

3.3   The Clustering Methodology Adopted

Let us recall that we want to group reports that look similar so as to illustrate a recurrent issue. Each cluster should be meaningful for an expert who is going to decide if it is operationally significant and if there is any specific intervention to design in order to address the problem highlighted.

Our clustering process will be defined according to the scenario notion: it relies on the assumption that if two stories have similar scenarios, they fit together. As the behavior is context dependent, a three-step process has been adopted:

1. find sets of reports with similar *Outcomes*
2. for each set produced by step one search for typical *Initial Situation* and produce a subset for each identified *Initial Situation.*
3. for each subset generated by step 2, analyze the *Behavior.*

This methodology gives us the framework that should allow us to extract meaningful clusters from the incident database. However, the tools used for each step of the process are not determined yet. They will depend on the number of reports processed and on the kind of knowledge (i.e., Fixed Fields, Free text...) used to describe each of these three parts of the scenario.

---

[5] In the incident model, the *Context* refers to the state and the setting elements that are at the starting point of the compromised situation. We will, in this paper, use *"Initial Situation"* instead of *Context* in order to avoid any confusion with the meaning of the term context as it is used in the FCA. However, it should be understood that the *"Initial Situation"* is not limited to the proximate factors and may include latent factors such as organizational culture and training.

The analysis of the available codification showed that the fixed fields adequately describe the *Initial Situation* and the *Outcome* (the "WHAT"), but not the *Behavior*. Consequently, it is possible to conduct the two first steps of the process using only the information in the fixed fields. This allows us to use powerful clustering tools such as FCA for these steps. However, the third step relies mainly on free text analyses of the narrative portion of the reports. The first two steps define what happened, but the third step to identify the causal factors of the *Behavior* is critical to understanding why the incident happened and the appropriate intervention. The third step is the subject of on-going ONERA-NASA collaboration. The concept of the approach is that the automated analysis required for the third step will benefit from the focus achieved in accomplishing the first two steps.

This methodology has now to be tested on a set of ASRS reports. This first experiment is the focus of the next section.

4   Analysis of 40 Incident Reports

4.1   Aim of the Experiment

When experts analyze a set of reports, they build up, in an informal way, groups of reports that illustrate important similar issues. They use intensively the free text for this task. However, an automated tool that could achieve an initial clustering of similar reports would reduce the amount of their work. The assumption is that the designed methodology is able to build meaningful groups of reports based solely on the fixed fields recorded in the database. The underlying assumptions are that (1) it is useful to cluster reports that are similar with respect to "what" happened, (2) the factors of the *Initial Situation* and the related *Outcome* define what happened, and (3) the factors of the *Initial Situation* and the *Outcome* of an incident are adequately defined by the fixed fields of that incident report. It is important to recognize that, while these may be valid for the ASRS database, they do not necessarily maintain for all textual databases. The aim of the experiment is to test these assumptions on a small sample set of reports from the ASRS database.

Thus, we want to evaluate whether the designed process can produce meaningful clusters when used in the best possible conditions. More precisely we want to test whether we can achieve the two first steps using only the fixed fields of the codification and obtain:

– well separated categories of reports related to specific *Outcomes* and,
– meaningful *Initial Situation* associated to each category.

The groups of reports generated and the attributes that characterize them will have to be relevant and meaningful to experts. In this work, we do not intend to design the final user tool, but rather just to validate the methodology. So we decided to:

– adopt a formal representation of the data that exactly fit the expert codification entered in the database [11] (based on the cube model [8]);

- build a generalized context where each incident is described by a cube;
- produce the concept lattice and "manually" check if meaningful clusters appeared or not.

If the clusters expected by experts do not appear in the lattice structure, it will mean that the Fixed Fields entered in the database do not contain all the required knowledge: the classifications produced by the experts utilize additional information contained in the narrative.

One of the key issues here is to keep in the context all the information described in the relational database. This is why a generalized context is build even if it is not the more efficient from a computational point of view. Then, using the theoretical results of section 2.1, a classical context that gives an isomorphic concept lattice is generated and used as input to the FCA tool.

Designing an automated tool that can build similar clusters for huge sets of reports (typically 50,000 reports) will be the next step of the work and is the purpose of other experiments being conducted that are not described in this paper. The value of the study reported here is largely in establishing confidence in the approach. However, the process has more general value as a way to assess the validity (with a few representative reports) of the results obtained using fully automated tools on very large numbers of reports.

4.2   The 40 Incident Reports

As this experiment is designed to evaluate the methodology, we wanted to use a set of reports dealing with a small number of somewhat similar scenarios. The validity of the clusters generated will be determined by experts. Therefore, the global set of reports had to be small and well known. As the ASRS and NASA experts had already studied a set of reports about "In Close Approach Changes" (ICAC), we decided to randomly extract 40 of those for this study. All these reports deal with an aircraft in the approach flight phase and, thus, close to an airport. The involved *Initial Situations* of all 40 reports are not very different and the experts had identified a limited set of *Outcomes*. So we knew that this set of reports could be divided into 3 or 4 scenarios by domain experts and we wanted to test whether our two-step clustering process was able to find them.

The number of 40 reports was also taken to limit the amount of "manual" work required to transfer and format the data from the ASRS database to the ONERA Kontex tool.

4.3   From the ASRS Codification to an FCA Context

Our primary description of each incident is its ASRS codification. From that, we want to extract a context that describes both the *Initial Situation* of the incident and its *Outcome* in order to perform the two first steps of the clustering process. Figure 5 is an excerpt of the ASRS codification of the incident report number 81075.

ACN: 81075

Time

    Date : 198801

    Day : Mon

    Local Time Of Day : 0001 To 0600

Place

    Locale Reference.Airport : SFO

    State Reference : CA

    Altitude.AGL.Bound Lower : 0

    Altitude.AGL.Bound Upper : 100

Environment

    Flight Conditions : VMC

    Light : Night

    Visibility.Bound Lower : 15

    Visibility.Bound Upper : 15

Aircraft / 1

    Involvement : Unique Event

    Controlling Facilities.Tower : SFO

    Operator.Common Carrier : Air Carrier

    Make Model : Medium Large Transport, Low Wing, 2 Turbojet Eng

    Crew Size : 2

    Flight Phase.Descent : Approach

    Flight Phase.Landing : Go Around

    Airspace Occupied.Class D : SFO

Aircraft / 2

    Operator.Common Carrier : Air Carrier

    ...

Person / 1

    Involvement : Pilot Flying

    Involvement.Other : Reporter

    Affiliation.Company : Air Carrier

    Function.Oversight : PIC

    ...

Person / 4

    Involvement : Controlling

    ...

Events

    Type Of Event : Unique Event

    Anomaly.Other Anomaly.Other : Unspecified

    Independent Detector.Other.Flight CrewA : Unspecified

    Resolutory Action.Other : Flight Crew Executed Missed Approach Or Go Around

    ...

Fig. 5. Excerpt of an ASRS codification

The information encoded in such a form is complex and its meaning partly relies on the links between the fixed fields. As an example, the person 1 is the pilot of the Aircraft 1, person 4 is the controller of San Francisco tower who is in charge of Aircraft 1. A special study was conducted to highlight all these links and formally represent them in a first-order logical language [11]. Figure 6 shows an extract of the formal representation of the incident 81075.

Date(01,1988)   Aircraft(Acft_2,WDB,LW,TE,2,2)
Day(Mon)   ...
Time(1)   Person(Per_1,PilotFlying)
Place(CA(SFO(VAR1,VAR2)))   Location(Per_1,Acft_1)
Altitude(AGL(0,100))   Affiliation(Per_1,Company(AirCarrier))
Airport(CA(SFO), Open, Controlled, Parallel) Function(Per_1,Oversight(PIC))
Traffic(Samedir)   ...
Flight_Condition(VMC)   Person(Per_2,Monitoring)
Visibility(15,15)   Location(Per_2,Acft_1)
Aircraft(Acft_1,MLT)   ...
...   Person(Per_4,Controlling)
Flight_Phase(Acft_1,Descent(Approach))   Location(Per_4,Tower(CA(SFO)))
Flight_Phase(Acft_1,Landing(GoAround))   ...
Control(Acft_1,Tower(CA(SFO)))   Anomaly(Other(Other))
Airspace(Acft_1,ClassD(SFO))

Fig. 6. Formal representation of the incident 81075

The two variables, VAR1 and VAR2, indicate that the orientation and the distance between the place of the incident and the airport of San Francisco are not known. The ASRS codification has been augmented by the use of the fixed fields of two other codification forms (X-Form and Cinq-Demi form[6]). As an example the information Traffic(Samedir) (which indicates the presence of other aircraft flying in the same direction) and Parallel (parallel runways are in use) come from the fixed fields of the X-Form.

Once all the information contained in the database has been captured, the literals related to the *Initial Situation* and the *Outcome* of the incident are selected. Then, following the methodology developed in section 2.1, a classical FCA context was generated from this first-order codification. Figure 7 illustrates some of the properties used to describe our incident 81075.

Thus, the decomposition of the literal Flight_Phase(Acft_1,Descent(Approach)) in the cube lattice is a set of 6 ∨-irreducible elements:
Flight_Phase(x,y), Flight_Phase(x,Descent(y)) ,Flight_Phase(x,Descent(Approach)), Flight_Phase(Acft_1,y), Flight_Phase(Acft_1,Descent(y)) and Flight_Phase(Acft_1,Descent(Approach)).

Of these 6 elements, only two have been kept: Flight_Phase(x,Descent(y)) called Descent and Flight_Phase(x,Descent(Approach)) called DescentApproach. Indeed the other elements are not useful as all the incidents contain the ∨-irreducible element Flight_Phase(x,y) and the reference to the Aircraft Acft_1 is

---

[6] Information about these forms can be found in [4].

not meaningful by itself as it is only used to link the literals together. As the aircraft Acft_1 is of type MLT (encoded by Aircraft(Acft_1,MLT) we also generate other ∨-irreducible elements such as {Aircraft(x,MLT),Flight_Phase(x,Descent(y))} in which the two literals are linked by the variable x. This element has been called DescentMLT.

Open	DescentMLT
Parallel	DescentApproachMLT
Traffic	exp2
Samedir	Descentexp2
VMC	DescentApproachexp2
vis3	MLTexp2
...	DescentMLTexp2
Descent	...
DescentApproach	AnomalyOtherOther
MLT	AnomalyOther

Fig. 7. Some attributes of the incident 81075

The decomposition has been achieved by means of a dedicated Prolog III program. It is not a complete decomposition as some of the reducible attributes ([9] page 24) have been removed. The context describing the incidents *Initial Situation* and *Outcome* contains 210 attributes for the 40 objects and generates 2162 concepts. It has been "manually" explored and analyzed with the Kontex tool as explained in the following section.

4.4   Results of the Study

First Step: The Outcome

As explained in section 3.3 the process started with the analysis of the *Outcome*. In the description of each incident, only the attributes related to the Outcome have been kept and the lattice structure generated. The Outcomes of the forty reports were characterized by 26 attributes and 50 concepts were found (see Figure 8).

This lattice structure has been explored with a top-down approach. This top-down analysis identified discriminant concepts. A concept was considered discriminating if (1) it had none or few reports shared with other significant concepts, (2) it contained an interesting percentage of all the reports, and (3) collectively, the discriminating concepts contained nearly all of the reports.

This process highlighted two significant concepts based on the two attributes: "Track or Heading Deviation", and "Conflict". With the same top-down approach, the two sub-lattices generated respectively by these two concepts have been explored. Then, we point out that the "Conflict" concept can be decomposed into three significant sub-concepts described by the three attributes: "Airborne", "Ground", and "Near Mid-Air Collision (NMAC)". So, we are able to state two main Outcomes in the scenarios of these 40 reports; namely reports that deal

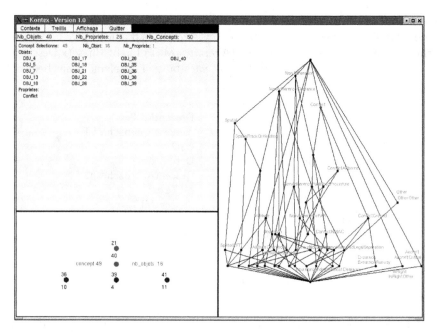

Fig. 8. The concept lattice based on the Outcome attributes

with a spatial deviation (track or heading) and reports dealing with a conflict. This first stage of analysis also showed that the attribute "Non Adherence to a Clearance" was often encountered but seemed to be a contributing factor in the four main groups of reports identified, rather than a discriminating factor.

In addition, a concept ("Other") containing all the reports that were not related to any anomaly of the ASRS taxonomy (as the incident 81075) and 2 exceptions ("InFlight": the aircraft encountered a turbulence, and "Aircraft": technical problem on the aircraft systems) were identified. The following table summarizes these results.

Table 1. Outcome classification of the reports

Report#																																								
	1	1	1	1	1	2	2	2	2	2	2	2	1	1	2	2	2	2	2	2	2	1	1	2	1	2	1	1	2	1	2		1	1	2	2				
	2	3	3	6	6	1	1	2	2	3	4	6	9	7	8	0	1	2	5	6	6	6	5	9	6	9	1	0	8	8	4	9	3	2	2	8	0	9	0	4
	1	1	6	3	4	0	3	3	9	0	0	4	9	1	3	1	5	4	4	6	9	8	3	1	3	3	2	2	1	1	0	5	6	4	6	6	4	7	4	4
	1	9	0	4	7	6	4	6	8	8	7	5	6	1	2	9	4	2	3	6	1	2	7	3	0	4	2	7	0	8	8	0	5	6	9	9	4	4	5	8
	9	1	3	6	1	9	2	7	5	9	8	9	9	7	1	7	8	9	6	2	6	9	4	7	9	1	9	2	7	7	9	8	9	5	3	0	1	5	3	5
	0	7	8	9	6	2	8	2	8	3	6	3	9	8	5	0	0	1	0	0	2	6	2	0	7	0	3	8	5	1	3	9	5	5	4	8	7	1	0	1
Track/Hd	*	*	*	*	*	*	*	*	*	*	*	*																												
Airborne													*	*	*	*	*	*	*	*	*	*																		
Ground																						*	*	*	*															
NMAC																											*	*	*											
Other																														*	*	*	*	*						
InFlight																																					*			
Aircraft																																					*			

Second Step: The Initial Situation

In this next step, we wanted to cluster on the shared *Initial Situation* within each discriminative concept. Thus for each of the discriminative concepts, an FCA context containing only their reports and the attributes related to their *Initial Situations* was built and analyzed. The main results of this second stage are summarized here.

Track or Heading Deviation: the shared *Initial Situation* of these incidents can be described by: *"An aircraft is in the approach flight phase to an open and controlled airport. The aircraft is controlled by the TRACON[7] in a Class B airspace."*

Conflict: the shared *Initial Situation* of these incidents is characterized by: *"Two aircraft are in the vicinity of an open and controlled airport. One of them is in the approach flight phase and there is some traffic."*

The *Initial Situation* of these two incidents look similar because all the incidents dealt with ICACs. Nevertheless, the differences highlighted are meaningful: Track or Heading deviations usually start far away from the airport, while conflicts often appear later, when the aircraft are closer to the airport. Conflicts require two aircraft and are associated with traffic.

When the typical *Initial Situations* of three types of Conflicts are studied, it appears that :

- for a Ground conflict, the aircraft is under the control of the Tower,
- for an Airborne conflict, the two aircraft are in the same flight phase and parallel runways are in use,
- no meaningful specific attributes for the NMAC conflict were identified.

4.5   Conclusion of the Experiment

This limited experiment shows that, within the phase of flight selected for the *Initial Situation* of these 40 reports, an initial clustering process based on the description of the *Outcome* generated well-separated groups of reports. Then, the analysis of the related *Initial Situations* was able to point out discriminating parameters. Of course, with such a small number of reports, one should be careful about the reliability and the generalization of the results. The purpose of this study was to test the value of the model and the process, and not so much to come to conclusions about the links between *Initial Situations* and *Outcomes* based on this small set of reports. It appears that an adequate description of what happened (i.e., of the *Initial Situation* and of the *Outcome* can be obtained from solely the Fixed Fields of the codification of these ASRS reports. This is important to the automated methodologies being developed that are intended to identify the causal factors of the *Behavior* entailed in a scenario of an incident.

---

[7] The TRACON (Terminal Radar Approach Control) control the aircraft when it is still far from the airport. Afterward the control is transferred to the Tower.

5   Conclusion

We presented here a new process dedicated to the analysis of reports of aviation incidents. The methodology adopted is innovative for the safety community as the design of the clustering process relied on the components of an incident model. The validity of the model has been tested on a limited set of reports with the FCA tool Kontex. The primary representation of the incident is done by conjunctions of first-order literals (the cube model) because the codification used in the relational database matches this model.

This paper demonstrates, from a theoretical point of view, how to transform the first-order context (decomposition of the elements) into a classical context and how to apply it to this particular set of reports. The decomposition function will be integrated in the Kontex tool.

Moreover this work illustrates how the FCA technique, combined with an adapted graphical interface, allows an expert to "manually" mine a textual database and extract a set of similar, relevant reports. Heuristic functionalities could be added so as to make a first selection of the discriminant concepts before validation by the expert.

The same methodology will now be tested to explore how in-flight events extracted from flight-recorded quantitative data and from subjective data of anecdotal reports could be correlated. This work has also helped us to define other experiments based on statistical methods and on natural language processing, which are currently being pursued on large numbers of reports.

Acknowledgment

We would like to thank all the ASRS and Battelle personnel who shared with us part of their experience and expertise.

This work has been partly granted by the French Ministry of Defence and conducted at the NASA Ames Research Center.

References

1. Chappell, S.: Using Voluntary Incident Reports for Human Factors Evaluations. In *Aviation Psychology in Practice*, (1997) 149–169.
2. Connell, L.: Incident Reporting: The NASA Aviation Safety Reporting System. *GSE Today*, (1999) 66–68.
3. Demonstration of a Process for Analyzing Textual Databases for Causal Factors of Human Behavior. *A Jointly Prepared Publication of NASA, ONERA, Battelle and Cinq-Demi* (2000).
4. Maille, N., Shafto, M., Statler, I.: What Happened, and Why: Towards an Understanding of Human Error Based on Automated Analyses of Incident Reports. *A Jointly Prepared Publication of NASA and ONERA* (2004).
5. Chidester, T.: Understanding Normal and Atypical Operations through Analysis of Flight Data. In *Proceedings of the 12th International Symposium on Aviation Psychology*, Ohio, (2003).

6. Helmreich, R., Klinect, J., & Wilhelm, J.: System safety and threat and error man-
    agement: The line operations safety audit (LOSA). In *Proceedings of the Eleventh
    International Symposium on Aviation Psychology*, (in press).
7. Chaudron, L., & Maille, N.: Generalized Formal Concept Analysis. In *LNAI 1867*,
    Springer, (2000) 357–370.
8. Chaudron, L., Maille, N., & Boyer, M.: The Cube Lattice Model and its Applica-
    tions. In *Applied Artificial Intelligence*, 17(3), (2003) 207–242.
9. Ganter, W., & Wille, R.: Formal Concept Analysis. Springer. (1999)
10. Lindig, C.: Fast Concept Analysis. In *Working with Conceptual Structures - Con-
    tributions to ICCS 2000*, Springer, (2000) 152–161.
11. Maille, N.: Towards a more automated analysis of data extracted from the ASRS
    database. ONERA Technical Report 3/05662 DCSD, (2001).

Characterization and Armstrong Relations for Degenerate Multivalued Dependencies Using Formal Concept Analysis*

Jaume Baixeries and José Luis Balcázar

Dept. Llenguatges i Sistemes Informàtics,
Universitat Politècnica de Catalunya,
c/ Jordi Girona, 1-3,
08034 Barcelona
{jbaixer, balqui}@lsi.upc.es

Abstract. Functional dependencies, a notion originated in Relational Database Theory, are known to admit interesting characterizations in terms of Formal Concept Analysis. In database terms, two successive, natural extensions of the notion of functional dependency are the so-called degenerate multivalued dependencies, and multivalued dependencies proper. We propose here a new Galois connection, based on any given relation, which gives rise to a formal concept lattice corresponding precisely to the degenerate multivalued dependencies that hold in the relation given. The general form of the construction departs significantly from the most usual way of handling functional dependencies. Then, we extend our approach so as to extract Armstrong relations for the degenerate multivalued dependencies from the concept lattice obtained; the proof of the correctness of this construction is nontrivial.

1   Introduction

It is well-known [19] that, from the Concept Lattice associated to a given binary relation, one can extract a number of implications that hold in the relation, for instance via the Duquenne-Guigues basis or, alternatively, by using minimal hypergraph transversals of the predecessors of each closed set ([26], [27]). Actually, the implications obtained in that way are known to characterize a Horn axiomatization of the given relation if one exists (otherwise, they provide a well-defined Horn approximation to the data; see [3], where a complete proof may be found).

Moreover, in [19] we also find an interesting application to database theory, since the syntactical similarity between implications and functional dependencies is more than a mere syntactical similarity; and there is a precise method (that we will call here "comparison-based binarization") to change a given database

---

* This work is supported in part by MCYT TIC 2002-04019-C03-01 (MOISES) and by the PASCAL-NETWORK Project.

B. Ganter and R. Godin (Eds.): ICFCA 2005, LNAI 3403, pp. 162–175, 2005.

relation r into a binary relation, or scaling, whose implications (its Horn axiomatization) provide exactly the functional dependencies that hold in r. Specifically, for each pair of tuples in r the values for each attribute are compared so as to yield a binary result, and therefore a binary relation (of quadratic size) is obtained.

There are other forms of dependencies in database theory, and we consider it interesting to find methods to obtain or characterize them on the basis of Formal Concept Analysis (FCA). Indeed, we have seen that this is possible for some of them, but the task turns out to be far from trivial. In [4] we have developed a careful semantic study of the relations or propositional theories where formulas of these sorts do hold: namely, multivalued dependencies, degenerate multivalued dependencies, and a family of propositional formulas introduced by Sagiv [28] that parallel them in the same sense as Horn clauses parallel functional dependencies. There we have identified precise meet operators, that is, various forms of combining objects to replace the standard intersections in the closure property, that characterize semantically these dependencies; but these do not readily give as yet a formal concept lattice. Here we consider instead an alternative approach based on defining Galois connections on classes, or partitions, of tuples and of attributes.

Along this line, in [1] we actually demonstrate how to define a Galois connection between attributes and partitions of tuples, inspired by the algorithms in [21] for computing functional dependencies, and thus propose a closure operator giving another precise characterization of functional dependencies in terms of FCA; and our recent, still unpublished work [2] proves that this approach, generalized to handle partitions of attributes instead of single attributes, actually does handle adequately multivalued dependencies.

Associated to each sort of dependency in databases is a notion of Armstrong relation: for a given set of rules, according to whatever syntax the sort of dependency considered allows for, an Armstrong relation is a relation that obeys that set of rules, and of course their logical consequences, but absolutely no other rule. They are useful in database design since they exemplify a set of dependencies, for instance those designed for a database schema prior to actually populating the database with tuples, and the analysis of such an Armstrong relation (seen as a potential database that the rules do allow, but where no other rules hold) is valuous for the prior understanding of the schema designed. Indeed, if the schema should have included a dependency that was forgotten, or expected to follow from the others, but actually does not, the Armstrong relation will point it out by including tuples that violate it, and this is usually a good method for the database engineer to realize the omission.

Our goal in this paper is twofold: first, we complete the existing characterizations by exhibiting an FCA-based characterization of degenerate multivalued dependencies, that did not follow from the previous study; and then, taking advantage of the crisper semantics of these dependencies in comparison with multivalued dependencies proper, we set out to explore the possibility of using FCA methods to construct appropriate Armstrong relations. For the case of functional dependencies there are methods actually very close to FCA [13]; in our case of DMVD's, we describe here a method to obtain the Armstrong relation from the associated Concept Lattice.

We choose a way of handling partitions of attributes and classes of tuples, and define a Galois connection between them; then we prove that the associated closure operator yields exactly the degenerate multivalued dependencies that hold in the original table, and move on to provide and fully verify a method to obtain an Armstrong relation from the family of the closed sets. Future goals would be to find a similarly-behaving method for multivalued dependencies proper, or for other stronger notions of dependency arising in the database theory area; and to investigate a potential, possibly strong connection between the notions of dependency that can be exemplified by Armstrong relations and those that can be characterized in terms of Formal Concept Analysis.

2   Basic Definitions

The following definitions and propositions have been taken from [14] and [31]. We keep mostly a database-theoretic notation since we are working for the most part with multivalued relations. We have a set of attributes $U := \{X_1, \ldots, X_n\}$ and, for each attribute $X_i \in U$, a domain of values $Dom(X_i)$. A tuple t is a mapping from U into the union of the domains, such that $t[X_i] \in Dom(X_i)$ for all i. Alternatively, tuples can be seen as well as elements of $Dom(X_1) \times \cdots \times Dom(X_n)$. We also use tuples over subsets X of U, naturally extending the notation to $t[X]$ for such projected sub-tuples. We will freely use the associativity (modulo a bijection) of the cartesian product to speak of tuples over a subset X of U as being obtained from cartesian products of subsets of X. A relation r over U is a set of tuples. We will use capital letters from the end of the alphabet for sets of attributes, and abuse language by not distinguishing single attributes from singleton sets of attributes; the context will allow always easy disambiguation. We denote by XY the union of the sets of attributes X and Y.

Given two tuples t_1, t_2, we say that they **agree** on a set of attributes X if their values in X are pointwise the same or, formally, $t_1[X] = t_2[X]$ seen as tuples.

2.1   Degenerate Multivalued Dependencies

This sort of database constraints, introduced in [15], is of intermediate strength between functional dependencies and multivalued dependencies. Whereas of less practical importance than either, it has been the right spot for generalizing the construction of Armstrong relations as we prove in the next section.

Definition 1. *A* **degenerate multivalued dependency** $X \to Y|Z$ *holds in a relation iff whenever two tuples t_1, t_2 agree on X, then, they agree on Y or in Z, where $X \cup Y \cup Z = U$. Formally:* $t_1[X] = t_2[X]$ *implies* $t_1[Y] = t_2[Y] \vee t_1[Z] = t_2[Z]$.

The *dependency basis* for a given set of attributes $X \subseteq U$, which is a notion deeply related to multivalued dependencies in their general form, was defined in [7]. It consists in a unique partition of the set of attributes U such that, for any (nondegenerate) multivalued dependency (MVD) that has X in the left-hand side, the right-hand side is a union of one or more of the classes of this partition. It could be defined in this way because MVD's fulfill the reflexivity property $(Y \subseteq X \models X \rightarrow Y)$ and their right-side parts are closed under union $(X \rightarrow Y, X \rightarrow Z \models X \rightarrow Y \cup Z)$.

The dependency basis of [7] for MVD's easily allows one to summarize a set of dependencies with the same left-hand side in a single expression (as it has been proposed in [12]). Since DMVD's also have the reflexivity property and are closed under union, likewise, we summarize a set of DMVD's in a single (generalized) DMVD as follows:

Definition 2. $X \rightarrow Y_1|\ldots|Y_n$ *(where $Y_1 \cup \cdots \cup Y_n$ is a partition of U) is a* **generalized DMVD** *in r if for each DMVD $X \rightarrow Z|W$ that holds in r, Z (W) can be formed by the union of some Y_i, and $\forall Y_i : X \rightarrow Y_i|U \setminus (X \cup Y_i)$ holds in r.*

We will use here vertical bars in enumerations of classes of attributes, when they make sense syntactically as a potential generalized dependency, that is, when they constitute a partition of the attributes.

3   Modeling DMVD's with FCA

We are ready to describe the first core contribution of this paper. On the basis of an input relation r, we define a pair of functions for which we prove that they constitute a Galois connection (in a slightly generalized sense); then we analyze the closure operator obtained, and show that the implications it provides correspond to the degenerate multivalued dependencies that hold in the given r. To explain our construction, we will use throughout the following running example, in which $U = \{A, B, C, D\}$ and that has four tuples: $\{1, 2, 3, 4\}$. For sake of simplicity, we will use the tuple id's to refer the tuples:

Example 1.

id	A	B	C	D
1	a	b	c	d
2	a	b	d	c
3	a	c	c	d
4	a	c	d	b

Often Galois connections are defined between two sub-semilattices of a power set lattice; however, structurally it is not necessary to restrict ourselves to power set lattices, and many other partial orderings are as appropriate [23]. We will work with the following orderings: on the one hand, partitions of the set of attributes U, ordered according to a refinement relation; on the other hand, sets of classes of tuples, ordered according to a covering relation similar to the refinement relation[1].

Thus, assume we are given a set of tuples r over the schema U. The central definition for our notion of closure is as follows:

Definition 3. *A partition P of the set of attributes U, $P := \{P_1|\ldots|P_n\}$, matches a class of tuples $\pi \subseteq r$ iff different tuples differ on a single class. We also say that P matches a set of classes of tuples Π if P matches all classes $\pi_i \in \Pi$.*

Therefore, t_1, $t_2 \in \pi$, for a matching π, requires that there is some j such that $t_1[P_j] \neq t_2[P_j]$ but $t_1[P_k] = t_2[P_k]$ for $k \neq j$. Moreover, in principle this j depends of the pair of tuples, as the definition is stated, but in fact it does not: all the different tuples that belong to the same class agree on all the values of the different classes of attributes P_i except in one and only one sole class P_j, the same for all the pairs of tuples in π as we prove in the next proposition.

Proposition 1. *Let π be a class of tuples matched by a partition of attributes P; then all different tuples differ in the same class of attributes.*

Proof. Let us suppose that given partition of attributes $P := \{P_1|\ldots|P_n\}$ there are two tuples t_1, t_2 in π such that $t_1[P_i] \neq t_2[P_i]$ and let us suppose that there is also another tuple t_3 such that $t_2[P_j] \neq t_3[P_j]$. Since each pair has to agree in the rest of the classes in which they do not differ, t_1 and t_3 will differ in P_i and in P_j, which would be a violation of the definition unless $i = j$. ∎

Intuitively, we are implicitly handling a structural binarization, or scaling, that provides us with a standard binary formal context, which in turn is able to give us the implications as we need them; the objects of this implicit scaling are the classes of tuples, of which actually only the maximal ones are relevant, whereas the implicit attributes are the partitions of the original attributes; and the binary relation is defined exactly by the *match* property. The following monotonicity property holds.

Lemma 1. *Let P, P' be two partitions of attributes, where P' is finer than (or equal to) P, that is, $\forall p' \in P' : \exists p \in P(p' \subseteq p)$. Then, each class matched by P' is also matched by P.*

Proof. Let π' be a class of tuples matched by P'. For each pair of tuples in it, there will be only one sole class of attributes from P' in which both tuples will disagree, and it will be fully contained in a class of P, which is thus the only class of P where these two tuples can disagree, which means that P will also match π'. ∎

---

[1] Our sets of attributes will be called classes since they belong to equivalence relations; to keep consistency, we call classes also the sets of tuples, and indeed their role is very similar to the equivalence classes of tuples in [1]. A set-theorist might object to our practice of calling sets (of classes) the third order objects and classes the second order objects, but note that in our absolutely finitary approach everything is (set-theoretic) sets and there are no proper (set-theoretic) classes.

We have a similar monotonicity property along the other dimension. To compare sets of classes of tuples, we use the following ordering:

Definition 4. *Let Π and Π' be sets of classes of tuples. $\Pi' \leq \Pi$ means: $\forall \pi' \in \Pi' : \exists \pi \in \Pi (\pi' \subseteq \pi)$.*

According to this ordering, the monotonicity property reads:

Lemma 2. *Let Π, Π' be sets of classes of tuples; if $\Pi' \leq \Pi$ and P matches Π, then P also matches Π'.*

Proof. Let t_1, t_2 be two tuples that are in the same class $\pi' \in \Pi'$. Since $\Pi' \leq \Pi$, then they will be in one class $\pi \in \Pi$, and they will disagree in only one class of attributes P_i. Therefore, P matches Π'. ∎

We denote by $\wp$ as the set of all possible partitions of attributes and present the definition of the two functions of the Galois connection.

Definition 5. *Operator ϕ: $\wp \to \mathcal{P}(\mathcal{P}(r))$. This operator receives a partition of the set of attributes U: $P := \{P_1 | \ldots | P_n\}$ and returns a set of classes of tuples $\Pi := \{\pi_1, \ldots, \pi_m\}$ matched by P, each of which is maximal under inclusion.*

Of course, $\phi(P)$ may not be a partition, in that a given tuple can belong to more that one class matched by P; this can be seen also in Example 1: $\phi(\{A|B|CD\}) = \{\{1, 2\}, \{3, 4\}, \{1, 3\}\}$, and $\phi(\{A|BCD\}) - \phi(\{ABCD\}) = \{\{1, 2, 3, 4\}\}$.

Definition 6. *Operator ψ: $\mathcal{P}(\mathcal{P}(r)) \to \wp$. This operator receives a set of classes of tuples $\Pi := \{\pi_1, \ldots, \pi_n\}$ and returns a partition of the set of attributes P that is the **finest** partition of attributes that matches that set of classes.*

Actually, this last definition could, in principle, be ambiguous, but fortunately there is always a unique most refined partition of attributes for a given set of classes. It is instructive to see how to prove it. We work on the basis of the following operation, which will take the role of a meet in the semilattice, and which we find most intuitive to call intersection of partitions.

Definition 7. *Let $X := \{X_1 | X_2 | \ldots | X_n\}$ and $Y := \{Y_1 | Y_2 | \ldots | Y_m\}$ be two partitions of attributes. We define the **intersection** of two partitions of attributes $X \cap Y$ as follows: $X \cap Y := \{Z_1 | Z_2 | \ldots | Z_l\}$ such that $\forall Z_i (\exists j, k : Z_i = X_j \cap Y_k)$.*

Example 2. Let $X := \{AB|CD|EFG\}$ and $Y := \{ABC|DEF|G\}$, then, $X \cap Y := \{AB|C|D|EF|G\}$.

It is easy to see that this operation is associative.

Lemma 3. *Let $X := \{X_1 | X_2 | \ldots | X_n\}$ and $Y := \{Y_1 | Y_2 | \ldots | Y_m\}$ two partitions of attributes, and let π be a class matched by both X and Y. Then, $X \cap Y$ also matches π.*

Proof. Let t_1, t_2 be two tuples in π. They disagree only in one class of attributes in X, let it be X_i, and in one sole class of attributes in Y, let it be Y_j. Then, the set of attributes in which both tuples disagree must be contained in both X_i and Y_j, and hence in $X_i \cap Y_j$. This proves that the intersection $X \cap Y$ also matches π.   ∎

Corollary 1. *(Unicity of the most refined matcher). Let π be a class of tuples. There is a unique partition of attributes that is the most refined partition of attributes that matches π.*

Proof. The trivial partition consisting of a single class of attributes matches all classes. If two or more different and incomparable partitions of attributes match the same class, according to Lemma 3 and to the associativity of the intersection, their intersection, which is more refined, also matches π.   ∎

In fact, this sort of intersection and the closure property that we have just argued correspond to the notion of intersection in the implicit binary formal context that we are using as scaling, alluded to above.

Now we can prove easily the basic property that relates both operators. To state it in the most standard terms we denote the ordering of partitions of attributes as follows:

Definition 8. *Let P, P' be partitions of a set of attributes U. We denote by $P \preceq P'$ the fact that P' is finer than P.*

Then we have:

Lemma 4. *Let P be a partition of attributes and Π' a set of classes of tuples. $P \preceq \psi(\Pi') \Leftrightarrow \Pi' \leq \phi(P)$.*

Proof. ($\Rightarrow$) Let us call $\psi(\Pi') = P'$. By Lemma 1, every class of tuples matched by P' is also matched by P. Then, every class of tuples matched by $\psi(\Pi')$ is included in some maximal class matched by P; that is: $\Pi' \leq \phi(P)$.

($\Leftarrow$) Let us call $\phi(P) = \Pi$. By Lemma 2, since $\Pi' \leq \Pi$ and and P matches Π (since $\phi(P) = \Pi$), then we have that P matches Π'. Since $\psi(\Pi')$ is the finest partition of attributes that matches Π', then it must be finer than (or equal to) P, that is, $P \preceq \psi(\Pi')$.   ∎

It is well known that the property we have just proved characterizes Galois connections; thus,

Corollary 2. *ϕ and ψ is a **Galois connection** between U and $\mathcal{P}(r)$.*

We come now to the closure operator corresponding to this Galois connection, from which we will obtain the dependencies we search for, which is $\Gamma := \psi \circ \phi$.

The informal definition is that given a partition of the set of attributes P, it returns the finest partition of attributes P' that can match the same classes as P. For instance, in Example 1 we have that $\Gamma(\{B|ACD\}) = \{A|B|CD\}$ because both match the classes $\{\{1,2\}, \{3,4\}, \{1,3\}\}$, but $\{A|B|CD\}$ is the finest; and, also, $\Gamma(\{ABCD\}) = \{A|BCD\}$. Such a composition of both directions of a Galois connection is always a closure operator [19]. Thus,

Proposition 2. Γ *is a* **closure operator**.

We are ready for the main result in this section: the closure operator so constructed corresponds, in a precise sense, to the degenerated multivalued dependencies that hold in the original relation r. Specifically, we consider pairs of partitions of attributes having the same closure; for instance, one of them could be the closure of the other. We consider pairs of partitions that differ in only one class of attributes in the following way: we assume (to simplify notation) that they have been numbered in such a way that the first k classes in both partitions coincide, and the rest of them in X are merged into a single class in X'. We prove that, in this case, the union of the common classes can be taken as left hand side of a degenerate multivalued dependency with the remaining classes of X as right hand side, and it holds in the input relation; moreover, the converse also holds, so that each true DMVD of r gives rise in the same way to two partitions X and X' having the same closure.

Theorem 1. *Let* X, X' *be partitions of attributes such that the* n *classes of* X *are* $X := \{X_1| \ldots |X_k|X_{k+1}| \ldots |X_n\}$, *whereas the* $k+1$ *classes of* X' *are* $X' := \{X_1|\ldots|X_k|X_{k+1}\ldots X_n\}$. *Then,* $\Gamma(X) = \Gamma(X') \Leftrightarrow X_1 \ldots X_k \rightarrow X_{k+1}|\ldots|X_n$.

Proof. ($\Rightarrow$) Since $\Gamma(X) = \Gamma(X')$, then for each pair of tuples t_1, t_2 we have the following cases: (a) they both belong to the same class matched by X and X'. If this is the case, then they only disagree in one class of attributes. If this class is one $X_i : i \in \{1 \ldots k\}$ then the DMVD holds because they disagree in $X_1 \ldots X_k$. If this class is one $X_i : i \in \{k+1 \ldots n\}$ then the DMVD holds as well. (b) they do not belong to the same class matched by X and X'. Then, they disagree in more than one class of attributes, let's say that at least they disagree in two classes. In X', at least, one of these classes must be in one $X_i : i \in \{1 \ldots k\}$. Then, the same must apply in X, and, since both tuples disagree in $\{X_1|\ldots|X_k\}$ the DMVD holds as well.

($\Leftarrow$) If $X_1 \ldots X_k \rightarrow X_{k+1}|\ldots|X_n$ holds, then, for each pair of tuples we can have two cases:

(a) they agree in $X_1 \ldots X_k$. In this case, in order for the DMVD to hold, they must only dissagree in one partition of attributes $X_i : i \in \{k+1 \ldots n\}$. In this case, they will also disagree only in one class of attributes in X and in X' and they will belong to the same class of tuples induced by X and X'.

(b) they disagree in $X_1 \ldots X_k$. Then, if both tuples belong to a class matched by X, it means that they disagree in only one $X_i \in \{X_1| \ldots |X_k\}$ and that, therefore, they agree in the rest of classes. Then, they will disagree in the same sole class of attributes in X' as well, and they will belong to the same class induced by X and X'. If both tuples do not belong to a class matched by X, then it means that they disagree in, at least, two classes of attributes. Since one of these classes will be one in $\{X_1| \ldots |X_k\}$, the other can also belong to the same subset of classes or to one $X_i : 1 \le i \le k$ and agree on the rest of classes. In either case, both tuples will also disagree in, at least, two classes of attributes in X' and they will not belong to the same class of tuples induced by X or X'.   ∎

4   Calculating Armstrong Relations

In this section we explain a method to construct Armstrong relations for DMVD's. As indicated above, an Armstrong relation obeys only a given set of dependencies and their logical consequences and no other dependency. In our case, there is an input relation r from which we have obtained the concept lattice via the Galois connection described in the previous section; we want to recover a relation r' only from the concept lattice, in such a way that r' satisfies exactly the same degenerate multivalued dependencies as r, which can be represented as a set of tuples or as a set of dependencies. In this section, for the sake of clarity, we will assume that r is represented as a set of tuples, but in the conclusions section we will argue that the method which is about to be presented is independent of the representation of r.

Along this section, we assume that the input relation r does not have constant attributes, that is, attributes which only take one single value along all the tuples of r. This is not a serious restriction since such attributes bear in fact no information and can be projected out and recovered later, if necessary; but, for the sake of completeness, a more detailed discussion of this case has been added to the next section of this paper.

Given a relation r, let $CL_1, CL_2, \ldots, CL_n$ be the closed partitions generated by Γ. We start with an empty relation r'. For each $CL_i : i \in \{1 \ldots n\} = \{X_1| \ldots |X_m\}$ we add to r' a new pair of tuples for each $X_j : j \in \{1 \ldots m\}$ as follows: all the values of X_j must be different in both tuples, and the values of the rest of classes must be the same. Also, each new value in this new pair of tuples must be different from the rest of existing values in r'. For each CL_i we will add $2m$ tuples, where m is the number of classes in that closed set.

Example 3. Let us suppose that, from a given relation r, the following closed sets are obtained: $\{A|BCD\}$ and $\{A|B|CD\}$. The Armstrong relation r' will be constructed as follows: for the closed set $\{A|BCD\}$ we will add two pairs of tuples: 1,2, that disagree in A and 3,4 that disagree in BCD. For the closed set $\{A|B|CD\}$ we will add three pairs: 5, 6, that disagree in A; 7, 8, that disagree in B; and 9, 10, that disagree in CD.

id	A	B	C	D
1	a_0	b_0	c_0	d_0
2	a_1	b_0	c_0	d_0
3	a_1	b_1	c_1	d_1
4	a_1	b_2	c_2	d_2
5	a_3	b_3	c_3	d_3
6	a_4	b_3	c_3	d_3
7	a_5	b_4	c_4	d_4
8	a_5	b_5	c_4	d_4
9	a_6	b_6	c_5	d_5
10	a_6	b_6	c_6	d_6

Theorem 2. *The method described calculates an Armstrong relation for a given set of DMVD's.*

Proof. According to Theorem 1, it is enough to prove that this relation produces the same set of closed partitions. Equivalently, that if a given partition of attributes is closed under r, it is also closed under r', and that if this partition is not closed under r it is not closed under r' either. We will prove it in two steps:

1. ($\Rightarrow$) If $X := \{X_1|\ldots|X_m\}$ is closed under r it is also closed under r'. Let's suppose that X is not closed in r'. Let Y be the closure of X in r' that matches the same classes of tuples as X. Since it is more refined, there will be two attributes that are in the same class in X, let it be X_i, and that are in different classes in Y, let it be Y_j and Y_k. By construction of r', there will be a pair of tuples in r' such that they will differ in all the attributes X_i, and they will be matched by X. These same pair of tuples will differ in classes Y_j and Y_k, and will not be matched by Y, but it contradicts our previous assumption.

2. ($\Leftarrow$) We assume that X is a partition closed in r' but not in r and derive a contradiction. Consider first the case where X contains a single class with all the attributes (the coarsest trivial partition). This X is always closed according to r', and would not be closed according to r only if some constant attribute exists in r; but we have explicitly assumed that this is not the case, so it is closed in both. For the rest of the argument, we consider some X that has at least two classes.

   Let Y be the closure of X according to r. Thus, Y is a partition that is strictly finer than X, since X is not closed and thus is different from its closure; and both X and Y match the same sets of tuples in r. Let X_i be a class of X that is not in Y, and let $Y_i \subset X_i$ a class of Y strictly included in X_i. Let W be the union of the rest of the classes of Y included in X_i, so that $X_i = Y_i \cup W$, and W is a union of classes of Y; also, $W \neq \emptyset$.

   We compare now Y_i to X_i according to r'. We have that X is closed in r', so that no further refinement is possible unless some class of tuples in r' is lost. We nevertheless refine X by splitting X_i into Y_i and W, and pick two tuples of r' that witness the fact that classes of tuples are thus lost. That is, let t'_1 and t'_2 be tuples in r' that do match X, but would not match it anymore if X_i is split into Y_i and W.

Given that X has at least two nonempty classes, there are attributes on which t'_1 and t'_2 have the same value. These pair of tuples, that agree in several attributes, witness in r' the existence of a closure in r, by construction of r'. Let Z be this closed partition of attributes, and let $Z_i \in Z$ be the class of attributes that differ in t'_1, t'_2. On the other hand, t'_1 and t'_2 together did match X, and differ in X_i since otherwise they would still match after refining it; thus, they cannot differ anywhere else in X. This implies that all their differing attributes, namely Z_i, belong to X_i: hence $Z_i \subseteq X_i$.

Moreover, t'_1 and t'_2 differ both in some attribute of Y_i and in some attribute of W, otherwise they would still match the split. This implies that $Z_i \cap Y_i \neq \emptyset$, and that $Z_i \cap W \neq \emptyset$ as well. Note that $Z_i \subseteq X_i = Y_i \cup W$.

We return back and reason at r. Since Z is closed according to r, if we split its class Z_i into $Z_i \cap Y_i$ and $Z_i \cap W$, then some class of tuples must be lost, and, similarly to above, there must exist tuples t_1 and t_2 in r that differ in some attribute of $Z_i \cap Y_i$ and also in some attribute of $Z_i \cap W$, and coincide everywhere outside Z_i. Since $Z_i \subseteq X_i$, they coincide everywhere outside X_i as well, and thus they match X. However, they do not match Y, because they exhibit differences both in Y_i and in some attribute of W, which belongs to some other class Y_j different from Y_i. This implies that X and Y do not match the same sets of tuples of r, and therefore Y cannot be the closure of X as assumed initially. Having reached this contradiction, we have proved that X must be closed according to r if it is closed according to r'.   ∎

5   Conclusions

In this paper we have used FCA to represent the set of DMVD's that hold in a relation. This can be useful because we can profit from the expressiveness of FCA (using the concept lattice as a graphical tool) and also because the different brands of dependencies so far examined can be expressed in function of a closure operator, thus complementing existing characterizations based on algebraic or logic-based approaches.

We also have presented a method for constructing an Armstrong relation following a similar approach to that used in [13] for functional dependencies, in which the notion of closed set (though not in an FCA context) was used: for each closed set, a witness was added to the Armstrong relation. It is worth to note that an Armstrong relation is usually constructed for an input relation, represented explicitly as a set of tuples or implicitly as a set of dependencies that the tuples coming in in the future will obey. Of course, when it is represented as a set of tuples, this same representation is an Armstrong relation for itself, but it does not prevent us from constructing a new one that can be much smaller in number of tuples, and more self-explanatory from the point of view of a database analyzer, for instance. In any case, the set of closures can be constructed from the set of DMVD's according to Theorem 1 in the same way the set of closures in [13] can be extracted from a set of FD's.

5.1   Constant Attributes

Here we discuss briefly the case where the trivial coarsest partition, consisting of a single class with all the attributes, is not closed according to the input relation r. This means that it is possible to split its only class into a finer partition, and yet the set of all the tuples in r will still match, that is, all the tuples will coincide in all classes except one, always the same by Proposition 1. Everywhere outside this class, all the tuples coincide, and therefore the attributes exhibit a constant value. The closure of the trivial coarsest partition is formed thus: each such constant attribute becomes a singleton class, and the others are kept together in a single class.

These constant attributes actually bear no information at all, and can be safely projected out of r, and added to it again later if need arises. Therefore, the construction of the Armstrong relation is to be done only in the nontrivial part, so that the argument in the previous paragraphs fully applies. Alternatively, the constant attributes of r can be maintained in r', but are to be kept constant as well, instead of following the process of creation of r' that we have explained; this process would apply only to the nonconstant attributes.

5.2   Future Extensions

The next goal is to find a method, similar in spirit, to construct Armstrong relations for multivalued dependencies proper, comparing it with the one we have described here and with somewhat similar methods existing for the construction of Armstrong relations for functional dependencies [13]. On the other hand, it is known that there are notions of dependency that do not allow for Armstrong relations, and here a higher-level question arises: what is the relationship between those sorts of dependencies that have Armstrong relations and those sorts of dependencies for which a scaling exists (based maybe on partitions or sets like our own here) that provides these dependencies as the implications corresponding to the associated concept lattice. In fact, this is the major question we intend to progress on along our current research.

References

1. Baixeries J. *A Formal Concept Analysis Framework to Model Functional Dependencies.* Mathematical Methods for Learning (2004).
2. Baixeries J., Balcázar J.L. *Using Concept Lattices to Model Multivalued Dependencies.*
3. Balcázar, J.L., Baixeries J. *Discrete Deterministic Data Mining as Knowledge Compilation.* Workshop on Discrete Mathematics and Data Mining in SIAM International Conference on Data Mining (2003).

4. Balcázar, J.L., Baixeries J. *Characterization of Multivalued Dependencies and Related Expressions* Discovery Science 2004.
5. Bastide Y., Pasquier N., Taouil R., Stumme G., Lakhal L. *Mining Minimal Non-Redundant Association Rules using Closed Itemsets.* Proc. of the 1st Int'l Conf. on Computational Logic, num. 1861, Lectures Notes in Artificial Intelligence, Springer, july 2000, pages 972-986.
6. Bastide Y., Taouil R., Pasquier N., Stumme G., Lakhal L. *Mining Frequent Patterns with Counting Inference.* SIGKDD Explorations 2(2): 66-75 (2000).
7. Beeri, C. *On the Membership Problem for Functional and Multivalued Dependencies in Relational Databases.* ACM Trans. Database Syst. 5(3): 241-259 (1980)
8. Birkhoff G. *Lattice Theory, first edition.* Amer. Math. Soc. Coll. Publ. 25, Providence, R.I. 1973.
9. Davey B.A., Priestley H.A. *Introduction to Lattices and Order.* Second edition. Cambridge University Press, 1990, 2002.
10. Elmasri R., Navathe S. B. *Fundamentals of Database Systems.* 2nd Edition. Benjamin/Cummings 1994
11. Fagin R. *Functional dependencies in a relational database and propositional logic.* IBM J. Research and Development 21, 6, Nov. 1977, pp. 534-544.
12. Fagin R. *Multivalued dependencies and a new normal form for relational databases.* ACM Trans. on Database Systems 2, 3, Sept. 1977, pp. 262-278.
13. Fagin R. *Armstrong databases.* Invited paper, Proc. 7th IBM Symposium on Mathematical Foundations of Computer Science, Kanagawa, Japan, May 1982.
14. Fagin R., Beeri C., Howard J. H. *A complete axiomatization for functional and multivalued dependencies in database relations.* Jr. Proc. 1977 ACM SIGMOD Symposium, ed. D. C. P. Smith, Toronto, pp. 47-61.
15. Fagin R., Sagiv Y., Delobel D., Stott Parker D. *An equivalence between relational database dependencies and a fragment of propositional logic.* Jr. J. ACM 28, 3, July 1981, pp. 435-453. Corrigendum: J. ACM 34, 4, Oct. 1987, pp. 1016-1018.
16. Fagin R., Vardi Y. V. *The theory of data dependencies: a survey.* Mathematics of Information Processing, Proceedings of Symposia in Applied Mathematics, AMS, 1986, vol. 34, pp. 19-72.
17. Flach P., Savnik I. *Database dependency discovery: a machine learning approach.* AI Communications,volume 12 (3): 139–160, November 1999.
18. Flach P., Savnik I. *Discovery of multivalued dependencies from relations.* Intelligent Data Analysis,volume 4 (3,4): 195–211, November 2000.
19. Ganter B., Wille R. *Formal Concept Analysis. Mathematical Foundations.* Springer, 1999.
20. Godin R., Missaoui R. *An Incremental Concept Formation Approach for Learning from Databases.* Theoretical Computer Science, Special Issue on Formal Methods in Databases and Software Engineering, 133, 387-419.
21. Huhtala Y., Karkkainen J., Porkka P., Toivonen H. *TANE: An Efficient Algorithm for Discovering Functional and Approximate Dependencies.* The Computer Journal 42(2): 100 - 111, 1999.
22. Kivinen J., Mannila H. *Approximate inference of functional dependencies from relations.* Theoretical Computer Science 149(1) (1995), 129-149.
23. Liquiere M., Sallantin J. *Structural machine learning with Galois lattice and Graphs.* International Conference in Machine Learning, ICML 1998.
24. Lopes, S., Petit J-M., Lakhal L. *Efficient Discovery of Functional Dependencies and Armstrong Relations.* Proceedings of the 7th International Conference on Extending Database Technology (EDBT 2000), Konstanz, Germany.

25. Lopes, S., Petit J-M., Lakhal L. *Functional and approximate dependency mining: database and FCA points of view.* Special issue of Journal of Experimental and Theoretical Artificial Intelligence (JETAI) on Concept Lattices for KDD, 14(2-3):93-114, Taylor and Francis, 2002.

26. Pfaltz, J.L. *Transformations of Concept Graphs: An Approach to Empirical Induction.* 2nd International Workshop on Graph Transformation and Visual Modeling Techniques. GTVM 2001, Satellite Workshop of ICALP 2001, Crete, Greece. Pages 320-326. July 2001.

27. Pfaltz, J.L., Taylor, C.M. *Scientific Discovery through Iterative Transformations of Concept Lattices.* Workshop on Discrete Mathematics and Data Mining at 2nd SIAM Conference on Data Mining, Arlington. Pages 65-74. April 2002.

28. Sagiv Y. *An algorithm for inferring multivalued dependencies with an application to propositional logic.* Journal of the ACM, 27(2):250-262, April 1980.

29. Savnik I., Flach P. *Bottom-up Induction of Functional Dependencies from Relations.* Proc. of AAAI-93 Workshop: Knowledge Discovery in Databases. 1993

30. Ullman J.D. *Principles of Database and Knowledge-Base Systems.* Computer Science Press, Inc. 1988.

31. Zaniolo C., Melkanoff M. A. *On the Design of Relational Database Schemata.* TODS 6(1): 1-47 (1981).

Formal Concept Analysis Constrained by Attribute-Dependency Formulas

Radim Bělohlávek and Vladimír Sklenář

Dept. Computer Science, Palacký University, Tomkova 40, CZ-779 00 Olomouc,
Czech Republic
{radim.belohlavek, vladimir.sklenar}@upol.cz

Abstract. An important topic in formal concept analysis is to cope
with a possibly large number of formal concepts extracted from formal
context (input data). We propose a method to reduce the number of
extracted formal concepts by means of constraints expressed by particular formulas (attribute-dependency formulas, ADF). ADF represent a
form of dependencies specified by a user expressing relative importance
of attributes. ADF are considered as additional input accompanying the
formal context $\langle X, Y, I \rangle$. The reduction consists in considering formal
concepts which are compatible with a given set of ADF and leaving out
noncompatible concepts. We present basic properties related to ADF, an
algorithm for generating the reduced set of formal concepts, and demonstrating examples.

1 Preliminaries and Problem Setting

We refer to [6] (see also [14]) for background information in formal concept
analysis (FCA). We denote a formal context by $\langle X, Y, I \rangle$, i.e. $I \subseteq X \times Y$ (object-
attribute data table, objects $x \in X$, attributes $y \in Y$); the concept deriving
operators by $^\uparrow$ and $^\downarrow$, i.e. for $A \subseteq X$, $A^\uparrow = \{y \in Y \mid$ for each $x \in A : \langle x, y \rangle \in I\}$
and dually for $^\downarrow$; a concept lattice of $\langle X, Y, I \rangle$ by $\mathcal{B}(X, Y, I)$, i.e. $\mathcal{B}(X, Y, I) = \{\langle A, B \rangle \in 2^X \times 2^Y \mid A^\uparrow = B, B^\downarrow = A\}$.

An important aspect of FCA is a possibly large number of formal concepts in
$\mathcal{B}(X, Y, I)$. Very often, the formal concepts contain those which are in a sense not
interesting for the expert. In this paper, we present a way to naturally reduce the
number of formal concepts extracted from data by taking into account information additionally supplied to the input data table (formal context). We consider
a particular form of the additional information, namely, a form of particular attribute dependencies expressed by (logical) formulas that can be supplied by an
expert/user. The primary interpretation of the dependencies is to express a kind
of relative importance of attributes. We introduce the notion of a formal concept
compatible with the attribute dependencies. The main gain of considering only
compatible formal concepts and disregarding formal concepts which are not compatible is the reduction of the number of resulting formal concepts. This leads
to a more comprehensible structure of formal concepts (clusters) extracted from

B. Ganter and R. Godin (Eds.): ICFCA 2005, LNAI 3403, pp. 176–191, 2005.

the input data. We present basic theoretical results, an algorithm for generating compatible formal concepts, and illustrate our approach by examples.

2   Constraints by Attribute Dependencies

2.1   Motivation

When people categorize objects by means of their attributes, they naturally take into account the importance of attributes. Usually, attributes which are less important are not used to form large categories (clusters, concepts). Rather, less important attributes are used to make a finer categorization within a larger category. For instance, consider a collection of certain products offered on a market, e.g. home appliances. When categorizing home appliances, one may consider several attributes like price, the purpose of the appliance, the intended placement of the appliance (kitchen appliance, bathroom appliance, office appliance, etc.), power consumption, color, etc. Intuitively, when forming appliance categories, one picks the most important attributes and forms the general categories like "kitchen appliances", "office appliances", etc. Then, one may use the less important attributes (like "price $\leq$ \$10", "price between \$15–\$40", "price $>$ \$100", etc.) and form categories like "kitchen appliance with price between \$15–\$40". Within this category, one may further form finer categories distinguished by color. This pattern of forming categories follows the rule that when an attribute y is to belong to a category, the category must contain an attribute which determines a more important characteristic of the attribute (like "kitchen appliance" determines the intended placement of the appliance). This must be true for all the characteristics that are more important than y. In this sense, the category "red appliance" is not well-formed since color is considered less important than price and the category "red appliance" does not contain any information about the price. Which attributes and characteristics are considered more important depends on the particular purpose of categorization. In the above example, it may well be the case that price be considered more important than the intended placement. Therefore, the information about the relative importance of the attributes is to be supplied by an expert (the person who determines the purpose of the categorization). Once the information has been supplied, it serves as a constraint for the formation of categories. In what follows, we propose a formal approach to the treatment of the above-described constraints to formation of categories.

2.2   Constraints by Attribute-Dependency Formulas

Consider a formal context $\langle X, Y, I \rangle$. We consider constraints expressed by formulas of the form

$$y \sqsubseteq y_1 \sqcup \cdots \sqcup y_n. \tag{1}$$

Formulas (1) will be called AD-formulas (attribute-dependency formulas). The set of all AD-formulas will be denoted by ADF. Let now $\mathcal{C} \subseteq ADF$ be a set of AD-formulas.

Definition 1. *A formal concept* $\langle A, B \rangle$ *satisfies an AD-formula (1) if we have that*

$$\text{if } y \in B \text{ then } y_1 \in B \text{ or } \cdots \text{ or } y_n \in B.$$

Remark 1. More generally, we could consider formulas $l(y) \sqsubseteq l(y_1) \sqcup \cdots \sqcup l(y_n)$ where $l(z)$ is either z or $\overline{z}$. For instance, $y \sqsubseteq \overline{y_1}$ would be satisfied by $\langle A, B \rangle$ if whenever $y \in B$ then none of $x \in A$ has y_1. For the purpose of our paper, however, we consider only (1).

The fact that $\langle A, B \rangle \in \mathcal{B}(X, Y, I)$ satisfies an AD-formula φ is denoted by $\langle A, B \rangle \models \varphi$. Therefore, $\models$ is the basic satisfaction relation (being a model) between the set $\mathcal{B}(X, Y, I)$ of all formal concepts (models, structures) and the set ADF of all AD-formulas (formulas). As usual, $\models$ induces two mappings, $\text{Mod} : 2^{ADF} \rightarrow 2^{\mathcal{B}(X,Y,I)}$ assigning a subset

$$\text{Mod}(\mathcal{C}) = \{\langle A, B \rangle \in \mathcal{B}(X, Y, I) \mid \langle A, B \rangle \models \varphi \text{ for each } \varphi \in \mathcal{C}\}$$

to a set $\mathcal{C} \subseteq ADF$ of AD-formulas, and $\text{Fml} : 2^{\mathcal{B}(X,Y,I)} \rightarrow 2^{ADF}$ assigning a subset

$$\text{Fml}(U) = \{\varphi \in ADF \mid \langle A, B \rangle \models \varphi \text{ for each } \langle A, B \rangle \in U\}$$

to a subset $U \subseteq \mathcal{B}(X, Y, I)$. The following result is immediate [12].

Theorem 1. *The mappings* Mod *and* Fml *form a Galois connection between* ADF *and* $\mathcal{B}(X, Y, I)$. *That is, we have*

$$\mathcal{C}_1 \subseteq \mathcal{C}_2 \text{ implies } \text{Mod}(\mathcal{C}_2) \subseteq \text{Mod}(\mathcal{C}_1), \tag{2}$$

$$\mathcal{C} \quad \subseteq \quad \text{Fml}(\text{Mod}(\mathcal{C})), \tag{3}$$

$$U_1 \subseteq U_2 \text{ implies } \text{Fml}(U_2) \subseteq \text{Fml}(U_1), \tag{4}$$

$$U \quad \subseteq \quad \text{Mod}(\text{Fml}(U)). \tag{5}$$

for any $\mathcal{C}, \mathcal{C}_1, \mathcal{C}_2 \subseteq ADF$, *and* $U, U_1, U_2 \subseteq \mathcal{B}(X, Y, I)$.

Definition 2. *For* $\mathcal{C} \subseteq ADF$ *we put*

$$\mathcal{B}_{\mathcal{C}}(X, Y, I) = \text{Mod}(\mathcal{C})$$

and call it the constrained (by $\mathcal{C}$) concept lattice *induced by* $\langle X, Y, I \rangle$ *and* $\mathcal{C}$.

For simplicity, we also denote $\mathcal{B}_{\mathcal{C}}(X, Y, I)$ by $\mathcal{B}_{\mathcal{C}}$. That is, $\mathcal{B}_{\mathcal{C}}(X, Y, I)$ is the collection of all formal concepts from $\mathcal{B}(X, Y, I)$ which satisfy each AD-formula from $\mathcal{C}$ (satisfy all constraints from $\mathcal{C}$).

The following is immediate.

Theorem 2. $\mathcal{B}_{\mathcal{C}}(X, Y, I)$ *is a partially ordered subset of* $\mathcal{B}(X, Y, I)$ *which is bounded from below. Moreover, if* $\mathcal{C}$ *does not contain an AD-formula (1) such that* y *is shared by all objects from* X *and none of* $y_1, \ldots, y_n$ *is shared by all objects, then* $\mathcal{B}_{\mathcal{C}}(X, Y, I)$ *is bounded from above.*

Proof. Obviously, $\langle Y^\downarrow, Y \rangle$ is the least formal concept from $\mathcal{B}(X,Y,I)$ and it is compatible with each AD-formula. Therefore, $\langle Y^\downarrow, Y \rangle$ bounds $\mathcal{B}_\mathcal{C}(X,Y,I)$ from below. Furthermore, if there is no AD-formula (1) with the above-mentioned properties then $\langle X, X^\uparrow \rangle$ is the upper bound of $\mathcal{B}_\mathcal{C}(X,Y,I)$ since in this case $\langle X, X^\uparrow \rangle$ clearly satisfies $\mathcal{C}$.

Remark 2. Note that the condition guaranteeing that $\mathcal{B}_\mathcal{C}(X,Y,I)$ is bounded from above is usually satisfied. Namely, in most cases, there is no object satisfying all attributes and so $X^\uparrow = \emptyset$ in which case the condition is fulfilled.

Let us now consider AD-formulas of the form

$$y \sqsubseteq y'. \tag{6}$$

Clearly, (6) is a particular form of (1) for $n = 1$. Constraints equivalent to (6) were considered in [1,2], see also [9] for a somewhat different perspective. In [1,2], constraints are considered in the form of a binary relation R on a set of attributes (in [1]) or objects (in [2]). On the attributes, R might be a partial order expressing importance of attributes; on the objects, R might be an equivalence relation expressing some partition of objects. Restricting ourselves to AD-formulas (6), $\mathcal{B}_\mathcal{C}(X,Y,I)$ is itself a complete lattice:

Theorem 3. *Let $\mathcal{C}$ be a set of AD-formulas of the form* (6). *Then $\mathcal{B}_\mathcal{C}(X,Y,I)$ is a complete lattice which is a $\bigvee$-sublattice of $\mathcal{B}(X,Y,I)$.*

Proof. Since $\mathcal{B}_\mathcal{C}(X,Y,I)$ is bounded from below (Theorem 2), it suffices to show that $\mathcal{B}_\mathcal{C}(X,Y,I)$ is closed under suprema in $\mathcal{B}(X,Y,I)$, i.e. that for $\langle A_j, B_j \rangle \in \mathcal{B}_\mathcal{C}(X,Y,I)$ we have $\langle (\cap_j B_j)^\downarrow, \cap_j B_j \rangle \in \mathcal{B}_\mathcal{C}(X,Y,I)$. This can be directly verified.

One can show that $\mathcal{B}_\mathcal{C}(X,Y,I)$ in Theorem 3 need not be a $\bigwedge$-sublattice of $\mathcal{B}(X,Y,I)$. Note that $\langle A, B \rangle \models (y \sqsubseteq y')$ says that B contains y' whenever it contains y. Then, $\langle A, B \rangle \models \{y \sqsubseteq y', y' \sqsubseteq y\}$ if either both y and y' are in B or none of y and y' is in B. This seems to be interesting particularly in the dual case, i.e. when considering constraints on objects, to select only formal concepts which do not separate certain groups of objects (for instance, the groups may form a partition known from outside or generated from the formal context).

In the rest of this section we briefly discuss selected topics related to constraints by AD-formulas. Due to the limited space, we omit details.

2.3   Expressive Power of AD-Formulas

An attribute may occur on left hand-side of several AD-formulas of $\mathcal{C}$. For example, we may have $y \sqsubseteq y_1 \sqcup y_2$ and $y \sqsubseteq y_3 \sqcup y_4$. Then, for a formal concept $\langle A, B \rangle$ to be compatible, it has to satisfy the following: whenever $y \in B$ then it must be the case that $y_1 \in B$ or $y_2 \in B$, and $y_3 \in B$ or $y_4 \in B$. Therefore, it is tempting to allow for expressions of the form

$$y \sqsubseteq (y_1 \sqcup y_2) \sqcap (y_3 \sqcup y_4)$$

with the intuitively clear meaning of compatibility of a formal concept and a formula of this generalized form. Note that a particular form is also e.g. $y \sqsubseteq y_2 \sqcap y_3$. One may also want to extend this form to formulas containing disjunctions of conjunctions, e.g.

$$y \sqsubseteq (y_1 \sqcap y_2) \sqcup (y_3 \sqcap y_4).$$

In general, one may consider formulas of the form

$$y \sqsubseteq t(y_1, \ldots, y_n) \tag{7}$$

where $t(y_1, \ldots, y_n)$ is a term over Y defined by: (i) each attribute $y \in Y$ is a term; (ii) if t_1 and t_2 are terms then $(t_1 \sqcup t_2)$ and $(t_1 \sqcap t_2)$ are terms. Then for a set $\mathcal{D}$ of formulas of the form (7), $\mathcal{B}_\mathcal{D}(X, Y, I)$ has the obvious meaning (formal concepts from $\mathcal{B}(X, Y, I)$ satisfying all formulas from $\mathcal{D}$). The following assertion shows that with respect to the possibility of expressing constraints, we do not gain anything new by allowing formulas (7).

Theorem 4. *For each set $\mathcal{D}$ of formulas (7) there is a set $\mathcal{C}$ of AD-formulas such that $\mathcal{B}_\mathcal{D}(X, Y, I) = \mathcal{B}_\mathcal{C}(X, Y, I)$.*

Proof. It is tedious but straightforward to show that each formula φ of the form (7) can be transformed to an equivalent formula of the form $y \sqsubseteq D_1 \sqcap \cdots \sqcap D_m$ where each D_k is of the form $y_{k,1} \sqcup \cdots \sqcup y_{k,l_k}$. Now, $y \sqsubseteq D_1 \sqcap \cdots \sqcap D_m$ is equivalent to the set $AD(\varphi) = \{y \sqsubseteq D_i \mid i = 1, \ldots, m\}$ of AD-formulas. Therefore, $\mathcal{D}$ is equivalent to $\mathcal{C} = \bigcup_{\varphi \in \mathcal{D}} AD(\varphi)$ in that $\mathcal{B}_\mathcal{D}(X, Y, I) = \mathcal{B}_\mathcal{C}(X, Y, I)$. An easy way to see this is to look at $\sqsubseteq, \sqcup, \sqcap$ as propositional connectives of implication, disjunction, and conjunction, respectively, and to observe that a formula φ of the form (7) is satisfied by a formal concept $\langle A, B \rangle$ iff the propositional formula corresponding to φ is true under a valuation $v : Y \to \{0, 1\}$ defined by $v(y) = 1$ if $y \in B$ and $v(y) = 0$ if $y \notin B$.

We will demonstrate the expressive capability of AD-formulas in Section 3.

2.4   Entailment of AD-Formulas

We now focus on the notion of entailment of AD-formulas. To this end, we extend in an obvious way the notion of satisfaction of an AD-formula. We say that a subset $B \subseteq Y$ (not necessarily being an intent of some formal concept) satisfies an AD-formula φ of the form (1) if we have that if $y \in B$ then some of $y_1, \ldots, y_n$ belongs to B as well and denote this fact by $B \models \varphi$ (we use $B \models \{\varphi_1, \ldots, \varphi_n\}$ in an obvious sense).

Definition 3. *An AD-formula φ (semantically) follows from a set $\mathcal{C}$ of AD-formulas if for each $B \subseteq Y$ we have that if $B \models \mathcal{C}$ (B satisfies each formula from $\mathcal{C}$) then $B \models \varphi$.*

We are going to demonstrate an interesting relationship between AD-formulas and so-called attribute implications which are being used in formal concept analysis and have a strong connection to functional dependencies in databases, see [6, 10]. Recall that an attribute implication is an expression of the form $A \Rightarrow B$ where $A, B \subseteq Y$. We say that $A \Rightarrow B$ is satisfied by $C \subseteq Y$ (denoted by $C \models A \Rightarrow B$) if $B \subseteq C$ whenever $A \subseteq C$ (this obviously extends to $\mathcal{M} \models T$ for a set $\mathcal{M}$ of subsets of Y and a set T of attribute implications). The connection between AD-formulas and attribute implications is the following.

Lemma 1. *For a set $B \subseteq Y$ we have*

$$B \models y \sqsubseteq y_1 \sqcup \cdots \sqcup y_n \qquad \text{iff} \qquad \overline{B} \models \{y_1, \ldots, y_n\} \Rightarrow y$$

where $\overline{B} = Y - B$. Furthermore,

$$B \models \{y_1, \ldots, y_n\} \Rightarrow \{z_1, \ldots, z_m\} \qquad \text{iff}$$
$$\overline{B} \models z_i \sqsubseteq y_1 \sqcup \cdots \sqcup y_n \quad \text{for each } i = 1, \ldots, m.$$

Proof. The assertion follows from definition by a moment's reflection.

Now, this connection can be used to reducing the notion of semantical entailment of AD-formulas to that of entailment of attribute implications which is well studied and well known [6, 10]. Recall that an attribute implication φ (semantically) follows from a set $\mathcal{C}$ of attribute implications ($\mathcal{C} \models \varphi$) if φ is true in each $B \subseteq Y$ which satisfies each attribute implication from $\mathcal{C}$. For an AD-formula $\varphi = y \sqsubseteq y_1 \sqcup \cdots \sqcup y_n$, denote by $I(\varphi)$ the attribute implication $\{y_1, \ldots, y_n\} \Rightarrow y$. Conversely, for an attribute implication $\varphi = \{y_1, \ldots, y_n\} \Rightarrow \{z_1, \ldots, z_m\}$, denote by $A(\varphi)$ the set $\{z_i \sqsubseteq y_1 \sqcup \cdots \sqcup y_n \mid i = 1, \ldots, m\}$ of AD-formulas. Then it is immediate to see that we have the following "translation rules":

Lemma 2. *(1) Let φ, φ_j, $(j \in J)$ be AD-formulas. Then $\{\varphi_j \mid j \in J\} \models \varphi$ (entailment of AD-formulas) iff $\{I(\varphi_j) \mid j \in J\} \models I(\varphi)$ (entailment of attribute implications). (2) Let φ, φ_j, $(j \in J)$ be attribute implications. Then $\{\varphi_j \mid j \in J\} \models \varphi$ (entailment of attribute implications) iff $\cup_{j \in J} A(\varphi_j) \models A(\varphi)$ (entailment of AD-formulas).*

Lemma 2 gives a possibility to answer important problems related to entailment like closedness, completeness, (non)redundancy, being a base, etc. Due to the limited space, we leave a detailed discussion to our forthcoming paper.

2.5   Trees from Concept Lattices

A particularly interesting structure of clusters in data is that of a tree, see [3, 5]. In a natural way, trees can be extracted from concept lattices by AD-formulas. Here, we present a criterion under which $\mathcal{B}_\mathcal{C}(X, Y, I)$ is a tree. Since $\mathcal{B}_\mathcal{C}(X, Y, I)$ has always the least element (see Theorem 2), we will call $\mathcal{B}_\mathcal{C}(X, Y, I)$ a tree if $\mathcal{B}_\mathcal{C}(X, Y, I) - \{\langle Y^\downarrow, Y \rangle\}$ (deleting the least element) is a tree.

Theorem 5. *If for each $y_1, y_2 \in Y$ which are not disjoint (i.e. $\{y_1\}^\downarrow \cap \{y_2\}^\downarrow \neq \emptyset$) $\mathcal{C}$ contains an AD-formula $y_1 \sqsubseteq \cdots \sqcup y_2 \sqcup \cdots$ or $y_2 \sqsubseteq \cdots \sqcup y_1 \sqcup \cdots$ such that the attributes on the right hand-side of each the formulas are pairwise disjoint then $\mathcal{B}_\mathcal{C}(X, Y, I)$ is a tree.*

Proof. If $\mathcal{B}_\mathcal{C}(X, Y, I)$ is not a tree, there are distinct and noncomparable $\langle A_1, B_1 \rangle$, $\langle A_2, B_2 \rangle \in \mathcal{B}_\mathcal{C}(X, Y, I)$ such that $A_1 \cap A_2 \neq Y^\downarrow$ (their meet in $\mathcal{B}_\mathcal{C}(X, Y, I)$) is greater than the least formal concept $\langle Y^\downarrow, Y \rangle$. Then there are $y_1 \in B_1 - B_2$ and $y_2 \in B_2 - B_1$. But y_1 and y_2 cannot be disjoint (otherwise $A_1 \cap A_2 = \emptyset$ which is not the case). By assumption (and without the loss of generality), $\mathcal{C}$ contains an AD-formula $y_1 \sqsubseteq \cdots \sqcup y_2 \sqcup \cdots$. Now, since there is some $x \in A_1 \cap A_2$, we have $\langle x, y_1 \rangle \in I$ and $\langle x, y_2 \rangle \in I$. Since $\langle A_1, B_1 \rangle$ satisfies $y_1 \sqsubseteq \cdots \sqcup y_2 \sqcup \cdots$, B_1 must contain some y' which appears on the right hand-side of the AD-formula, from which we get $\langle x, y' \rangle$. But $\langle x, y_2 \rangle \in I$ and $\langle x, y' \rangle \in I$ contradict the disjointness of y_2 and y'.

Remark 3. Note that the assumption of Theorem 5 is satisfied in the following situation: Attributes from Y are partitioned into subsets $Y_1, \ldots, Y_n$ of Y such that each $Y_i = \{y_{i,1}, \ldots, y_{i,n_i}\}$ corresponds to some higher-level attribute. E.g., Y_i may correspond to color and may contain attributes red, gree, blue, ... Then, we linearly order Y_i's, e.g. by $Y_1 < \ldots Y_n$ and for each $i < j$ add a set $Y_i \sqsubseteq Y_j$ of AD-formulas of the form $y_{i,j} \sqsubseteq y_{j,1} \sqcup \cdots \sqcup y_{j,n_j}$ for each $y_{i,j} \in Y_i$. In fact, we may add only $Y_i \sqsubseteq Y_{i+1}$ and omit the rest with the same restriction effect. Then the assumptions of Theorem 5 are met. Such a situation occurs when one linearly orders higher-level attributes like color < price < manufacturer and want to see the formal concepts respecting this order. The just described ordering of attributes is typical of so-called decision trees [13].

Remark 4. Note that formulas $y_1 \sqsubseteq \cdots \sqcup y_2 \sqcup \cdots$ and $y_2 \sqsubseteq \cdots \sqcup y_1 \sqcup \cdots$ in Theorem 5 need not belong to $\mathcal{C}$. It is sufficient if they are entailed by $\mathcal{C}$, i.e. if $\mathcal{C} \models y_1 \sqsubseteq \cdots \sqcup y_2 \sqcup \cdots$ or $\mathcal{C} \models y_2 \sqsubseteq \cdots \sqcup y_1 \sqcup \cdots$.

2.6   Algorithm for Computing $\mathcal{B}_\mathcal{C}(X, Y, I)$

In this section we present an algorithm for computing $\mathcal{B}_\mathcal{C}(X, Y, I)$. The algorithm computes both the formal concepts and their ordering, and is a modification of the incremental algorithms (see [8]), particularly of [11]. The purpose of our algorithm is to compute $\mathcal{B}_\mathcal{C}(X, Y, I)$ without the need to compute the whole $\mathcal{B}(X, Y, I)$ and then test which formal concepts $\langle A, B \rangle \in \mathcal{B}(X, Y, I)$ satisfy $\mathcal{C}$. For the reader's convenience, we list the whole pseudocode. For explaining of the idea of the background algorithm we refer to [11]. In what follows, we comment only on the modification related to taking into account the constraining AD-formulas. In addition to `CreateLatticeIncrementally` (computes the constrained lattice) and `GetMaximalConcept` (auxiliary function) which are the same as in [11], our modification consists in introducing a new function `GetValidSubIntent` and a new function `ADFAddIntent` which is a modified version of `AddIntent` from [11].

The algorithm is a kind of so-called incremental algorithms. The main procedure, called `CreateLatticeIncrementally`, calls for each object $x \in X$ recursive function `ADFAddIntent`. `ADFAddIntent` checks whether the intent of the input object x satisfies AD-formulas from $\mathcal{C}$. If yes, it passes the intent. If not, it passes the largest subset of the intent which satisfies $\mathcal{C}$. The largest subset B is obtained by `GetValidSubIntent` and need not be an intent. Nevertheless,

ADFAddIntent, in the further processing, can create new intents from B. This way, "false" concepts may be created. Therefore, after processing of all the objects, we need to check and remove the "false" concepts. The algorithm plus its illustration follow.

```
01:Procedure CreateLatticeIncrementally (X,Y,I)
02: BottomConcept := (Y',Y)
03: L := {BottomConcept}
04: For each x in X
05:  ObjectConcept := AddIntent(x', BottomConcept, L)
06:  Add x to the extent of ObjectConcept and all concepts above
07: End For
```

```
01:Function ADFAddIntent(inputIntent, generator, L)
02: if IsValidIntent(inputIntent)
03:      intent := inputIntent
04: else
05:      intent := GetValidSubIntent(inputIntent)
06: End If
07:      GeneratorConcept = GetMaximalConcept(intent, generator,L)
08:      If GeneratorConcept.intent = intent
09:       Return GeneratorConcept
10:      End If
11:      newParents := Empty Set
12:      GeneratorParents = GetParents(GeneratorConcept, L)
13:      For each Candidate in GeneratorParents
14:         if Candidate.Intent Is Not SubSet intent
15:          Candidate:=
16:          ADFAddIntent(Intersection(Candidate.Intent,intent),
17:                            Candidate, L)
18:         End if
19:          addParent = true
20:          for each Parent in NewParents
21:             if Candidate.Intent Is Subset Parent.Intent
22:                addParent := false
23:                Exit For
24:             Else If Parent.Intent Is Subset Candidate.Intent
25:                Remove Parent from newParents
26:             End If
27:          End For
28:        if addParent
29:          Add candidate to newParents
30:        End If
31:         End For
32:      newConcept := (GeneratorConcept.Extent,intent)
33:      Add NewConcept to L
```

```
34:     For each Parent in newParents
35:         RemoveLink(Parent, GeneratorConcept, L)
36:         SetLink(Parent, NewConcept, L)
37:     End For
38:     SetLink(NewConcept, GeneratorConcept, L)
39: Return NewConcept
```

```
01:Function GetValidSubIntent(intent, L)
02: notValid := true
03: While notValid and intent <> EmptySet
04:     conflictADF := GetConflictADF (intent, L)
05:     For each adf in conflictADF
06:       Remove adf.LeftSide from intent
07:     End For
08:     notValid := IsValidIntent(intent)
09:   End If
10: End While
11:return intent
```

```
01:Function GetMaximalConcept(intent, GeneratorConcept, L)
02: parentIsMaximal := true
03: While parentIsMaximal
04:   parentIsMaximal := false
05:   Parents = GetParents(GeneratorConcept, L)
06:   For each Parent in Parents
07:     If intent is subset Parent.Intent
08:       GeneratorConcept := Parent
09:       parentIsMaximal := true
10:       Exit For
11:     End If
12:   End For
13: End While
14: return GeneratorConcept
```

We now illustrate the algorithm on an example which is a modification of an example from [8].

First, we add object 6 with attributes b, c, f, i to (already computed) concept lattice depicted in Fig. 1. First, if we do not apply any constraints, we obtain 4 new concepts and get the concept lattice in Fig. 2.

Second, we add 6 under an AD-formula $f \sqsubseteq h \sqcup i$ (note that all concepts from the concept lattice in Fig. 1 satisfy this AD-formula). In this case, the intent of 6 satisfies $f \sqsubseteq h \sqcup i$. As a result, we get the constrained concept lattice in Fig. 3.

Third, we add object 6 under an AD-formula $i \sqsubseteq a$ (note again that all concepts from the concept lattice in Fig. 1 satisfy this AD-formula). In this case, the intent of 6 does not satisfy $i \sqsubseteq a$. As a result, we get the constrained concept lattice in Fig. 4.

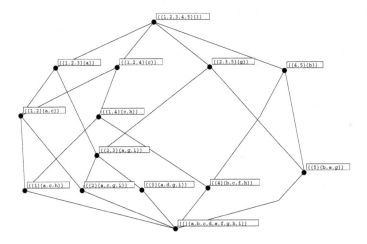

Fig. 1. A concept lattice to which we wish to add object 6 having attributes b, c, f, i, under different constraints

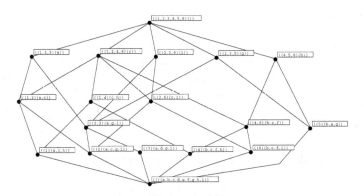

Fig. 2. Adding object 6 to concept lattice from Fig. 1 with no constraints

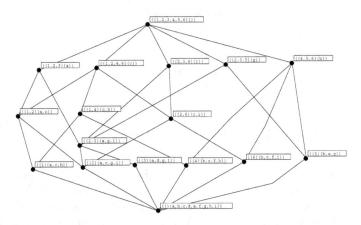

Fig. 3. Adding object 6 to concept lattice from Fig. 1 under constraint f ⊑ h ⊔ i

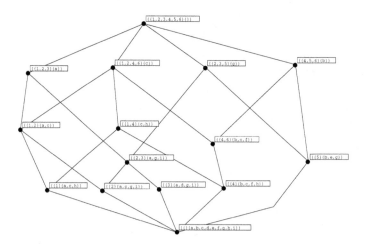

Fig. 4. Adding object 6 to concept lattice from Fig. 1 under constraint i ⊑ a

3   Examples

We now present illustrative examples. We use Hasse diagrams and label the nodes corresponding to formal concepts by boxes containing concept descriptions. For example, $(\{1, 3, 7\}, \{3, 4\})$ is a concept with extent $\{1, 3, 7\}$ and intent $\{3, 4\}$. Consider a formal context described in Tab. 1. The context represents data about eight car models (1–8) and their selected attributes (1: diesel engine,..., 8: ABS).

Table 1. Formal context given by cars and their properties

	1	2	3	4	5	6	7	8
car 1	1	0	1	0	0	1	0	1
car 2	1	0	1	0	1	1	0	1
car 3	0	1	1	0	0	0	0	1
car 4	0	1	0	1	1	0	0	0
car 5	0	1	1	0	1	1	0	0
car 6	0	1	0	1	0	1	1	0
car 7	0	1	0	1	1	1	1	1
car 8	0	1	0	1	0	0	0	1

attributes: 1 - diesel engine, 2 - gasoline engine, 3 - sedan, 4 - hatchback, 5 - air-conditioning, 6 - airbag, 7 - power stearing, 8 - ABS

The attributes can be partitioned into three groups: $\{1, 2\}$ (engine type), $\{3, 4\}$ (car type), $\{5, 6, 7, 8\}$ (equipment). The concept lattice $\mathcal{B}(X, Y, I)$ corresponding to formal concept $\langle X, Y, I \rangle$ contains 27 formal concepts and is depicted in Fig. 5. The formal concepts of $\mathcal{B}(X, Y, I)$ represent, in the sense of FCA, all meaninfull concepts-clusters present in the data. Suppose we want to use a (part

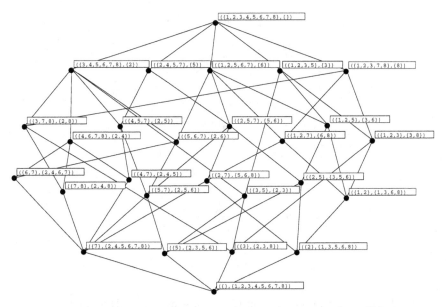

Fig. 5. Concept lattice corresponding to the context from Tab. 1

of a) concept lattice to provide a conceptual view of the data table and suppose we know that the customers find most important the type of engine (diesel, gasoline) and then the type of the bodywork. Such a situation is described by a set $\mathcal{C}$ of AD-formulas from (8).

$$
\begin{aligned}
\text{air-conditioning} &\sqsubseteq \text{hatchback} \sqcup \text{sedan} \\
\text{power stearing} &\sqsubseteq \text{hatchback} \sqcup \text{sedan} \\
\text{airbag} &\sqsubseteq \text{hatchback} \sqcup \text{sedan} \\
\text{ABS} &\sqsubseteq \text{hatchback} \sqcup \text{sedan} \\
\text{hatchback} &\sqsubseteq \text{gasoline engine} \sqcup \text{diesel engine} \\
\text{sedan} &\sqsubseteq \text{gasoline engine} \sqcup \text{diesel engine}
\end{aligned}
\tag{8}
$$

The corresponding constrained $\mathcal{B}_{\mathcal{C}}(X, Y, I)$ contains 13 formal concepts and is depicted in Fig. 6. Second, consider a set $\mathcal{C}$ of AD-formulas (9). Contrary to the previous example, the importance of the type of a bodywork and the type of the engine are reversed.

$$
\begin{aligned}
\text{air-conditioning} &\sqsubseteq \text{diesel engine} \sqcup \text{gasoline engine} \\
\text{power stearing} &\sqsubseteq \text{diesel engine} \sqcup \text{gasoline engine} \\
\text{airbag} &\sqsubseteq \text{diesel engine} \sqcup \text{gasoline engine} \\
\text{ABS} &\sqsubseteq \text{diesel engine} \sqcup \text{gasoline engine} \\
\text{gasoline engine} &\sqsubseteq \text{hatchback} \sqcup \text{sedan} \\
\text{diesel engine} &\sqsubseteq \text{hatchback} \sqcup \text{sedan}
\end{aligned}
\tag{9}
$$

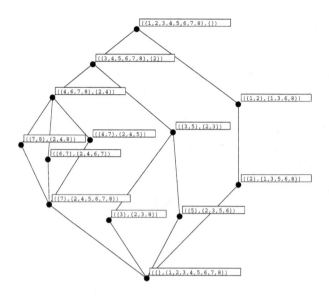

Fig. 6. Concept lattice from Fig. 5 constrained by AD-formulas 8

The corresponding constrained $\mathcal{B}_{\mathcal{C}}(X, Y, I)$ contains 13 formal concepts and is depicted in Fig. 7. We can see that the bottom parts in Fig. 6 and Fig. 7 are the same. The lattices differ according to the selection of importance of the attributes.

Third, suppose the customer changes the preferences and finds the most important car property to be safety and requires ABS and airbag. The situation is described by a set $\mathcal{C}$ of AD-formulas from 10.

$$\text{air-conditioning} \sqsubseteq \text{diesel engine} \sqcup \text{gasoline engine} \qquad (10)$$
$$\text{power stearing} \sqsubseteq \text{diesel engine} \sqcup \text{gasoline engine}$$
$$\text{gasoline engine} \sqsubseteq \text{hatchback} \sqcup \text{sedan}$$
$$\text{diesel engine} \sqsubseteq \text{hatchback} \sqcup \text{sedan}$$
$$\text{sedan} \sqsubseteq \text{ABS}$$
$$\text{hatchback} \sqsubseteq \text{ABS}$$
$$\text{ABS} \sqsubseteq \text{airbag}$$
$$\text{airbag} \sqsubseteq \text{ABS}$$

The resulting $\mathcal{B}_{\mathcal{C}}(X, Y, I)$ corresponding to (10) contains 6 formal concepts and is depicted in Fig. 8. In general, if an attribute $y' \in Y$ is required, it is sufficient to have $\mathcal{C}$ such that $\mathcal{C} \models \{y \sqsubseteq y'\}$ for each $y \in Y$. In particular, for $\mathcal{C} = \{y \sqsubseteq y' \mid y \in Y\}$ we have $\mathcal{B}_{\mathcal{C}}(X, Y, I) = \{\langle A, B \rangle \in \mathcal{B}(X, Y, I) \mid \langle \{y'\}^{\downarrow}, \{y'\}^{\downarrow\uparrow} \rangle \leq \langle A, B \rangle\}$, i.e. $\mathcal{B}_{\mathcal{C}}(X, Y, I)$ is a main filter in $\mathcal{B}(X, Y, I)$ corresponding to attribute concept of y'.

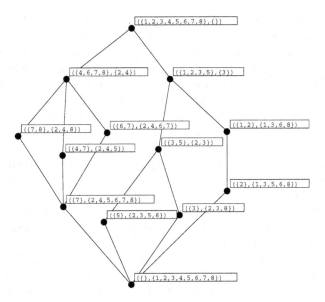

Fig. 7. Concept lattice from Fig. 5 constrained by AD-formulas 9

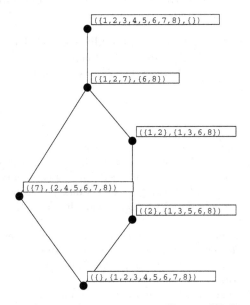

Fig. 8. Concept lattice corresponding to AD-formulas 10

4   Future Research

In the next future, we focus on the following:

– As one of the referees pointed out, AD-formulas are what is called surmise relationships in [4]. In [7], even more general formulas (clauses) are studied. A future research will be focused on constraints by more general formulas, e.g. to allow negations (see Section 2.2), and on studying relationships to earlier work on formulas over attributes, see [7] and the papers cited therein.

- Entailment. In Section 2.4, we showed basic observations which make it possible to reduce the problems related to entailment of AD-formulas to the corresponding problems of entailment of attribute implications which are well described. In our forthcoming paper, we elaborate more on this.
- Algorithms. We will study further possibilities to generate concepts satisfying the constraints directly, see Section 2.6.
- Experiments with large datasets. In a preliminary study we used a database of mushrooms avaliable at UCI KDD Archive (http://kdd.ics.uci.edu/). We transformed the database to a formal context with 8124 objects and 119 attributes. The corresponding concept lattice contains 238710 formal concepts and took about 10 minutes to compute it. We run experiments with several sets of AD-formulas which seemed to express natural constraints. The resulting constrained concept lattices were considerably smaller (from tens to thousands of formal concepts, depending on the sets of AD-formulas). The computation of the constrained concept lattices took a correspondingly smaller amount of time (from 2 seconds up). We will run further experiments on this database, possibly with an expert in the field and compare our results with other clustering methods.

Acknowledgment. Radim Bělohlávek gratefully acknowledges support by grant No. 201/02/P076 of the Grant Agency of the Czech Republic. Also, we acknowledge support by grant No. 1ET101370417 of the GA AV CR.

References

1. Bělohlávek R., Sklenář V., Zacpal J.: Formal concept analysis with hierarchically ordered attributes. *Int. J. General Systems* **33**(4)(2004), 283–294.
2. Bělohlávek R., Sklenář V., Zacpal J.: Concept lattices constrained by equivalence relations. *Proc. CLA 2004*, Ostrava, Czech Republic, pp. 58–66.
3. Bock H. H.: *Automatische Klassifikation.* Vandenhoeck & Ruprecht, Göttingen, 1974.
4. Diognon J.-P., Falmagne J.-C.: *Knowledge Spaces.* Springer, Berlin, 1999.
5. Everitt, Brian S.: *Cluster Analysis, 4th ed.* Edward Arnold, 2001.
6. Ganter B., Wille R.:   *Formal Concept Analysis. Mathematical Foundations.* Springer-Verlag, Berlin, 1999.
7. Ganter, Wille: Contextual attribute logic. In: Tepfenhart W., Cyre W. (Eds.): *Proceedings of ICCS 2001*, Springer, 2001.
8. Godin R., Missaoui R., Alaoui H.: Incremental concept formation algorithms based on Galois (concept) lattices. *Computational Intelligence* **11**(2) (1995), 246–267.
9. Kent R. E.: Rough concept analysis. In: Ziarko W. P. (Ed.): *Rough Sets, Fuzzy Sets, and Knowledge Discovery.* Proc. of the Intern. Workshop RSKD'93, Springer-Verlag, London, 1994.
10. Maier D.: *The Theory of Relational Databases.* Computer Science Press, Rockville, 1983.

11. Van der Merwe, F.J., Obiedkov, S., and Kourie, D.G.: AddIntent: A new incremental lattice construction algorithm. In: Concept Lattices. Proc. of the 2nd Int. Conf. on Formal Concept Analysis, Sydney, Australia, Lecture Notes in Artificial Intelligence, vol. 2961, pp. 372–385.
12. Ore O.: Galois connections. *Trans. Amer. Math. Soc.* 55:493–513, 1944.
13. Quinlan J. R.: *C4.5: Programs for Machine Learning.* Morgan-Kaufmann, San Francisco, CA, 1993.
14. Wille R.: Restructuring lattice theory: an approach based on hierarchies of concepts. In: Rival I.: *Ordered Sets.* Reidel, Dordrecht, Boston, 1982, 445–470.

On Computing the Minimal Generator Family for Concept Lattices and Icebergs

Kamal Nehmé[1], Petko Valtchev[1], Mohamed H. Rouane[1], and Robert Godin[2]

[1] DIRO, Université de Montréal, Montréal (Qc), Canada
[2] Département d'informatique, UQAM, Montréal (Qc), Canada

Abstract. Minimal generators (or *mingen*) constitute a remarkable part of the closure space landscape since they are the antipodes of the closures, i.e., minimal sets in the underlying equivalence relation over the powerset of the ground set. As such, they appear in both theoretical and practical problem settings related to closures that stem from fields as diverging as graph theory, database design and data mining. In FCA, though, they have been almost ignored, a fact that has motivated our long-term study of the underlying structures under different perspectives. This paper is a two-fold contribution to the study of mingen families associated to a context or, equivalently, a closure space. On the one hand, it sheds light on the evolution of the family upon increases in the context attribute set (e.g., for purposes of interactive data exploration). On the other hand, it proposes a novel method for computing the mingen family that, although based on incremental lattice construction, is intended to be run in a batch mode. Theoretical and empirical evidence witnessing the potential of our approach is provided.

1 Introduction

Within the closure operators/systems framework, *minimal generators*, or, as we shall call them for short, *mingen*, are, beside closed and pseudo-closed elements, key elements of the landscape. In some sense they are the antipodes of the closed elements: a mingen lays at the bottom of its class in the closure-induced equivalence relation over the ground set, whereas the respective closure is the unique top of the class. This is the reason for mingen to appear in almost every context where closures are used, e.g., in fields as diverging as the database design (as *key* sets [7]), graph theory (as *minimal transversals* [2]), data analysis (as *lacunes irréductibles*[1], the name given to them in French in [6]) and data mining (as minimal premises of association rules [8]). In FCA, mingen have been used for computational reasons, e.g., in TITANIC [11], where they appear explicitly, as opposed to their implicit use in NextClosure [3] as canonical representations (prefixes) of concept intents.

Despite the important role played by mingen, they have been paid little attention so far in the FCA literature. In particular, many computational problems

---

[1] Irreducible gaps, translation is ours.

B. Ganter and R. Godin (Eds.): ICFCA 2005, LNAI 3403, pp. 192–207, 2005.
© Springer-Verlag Berlin Heidelberg 2005

related to the mingen family are not well understood, let alone efficiently solved. This observation has motivated an ongoing study focusing on the mingen sets in a formal context that considers them from different standpoints including batch and incremental computation, links to other remarkable members of the closure framework such as pseudo-closed, etc. Recently, we proposed an efficient method for maintaining the mingen family of a context upon increases in the context object set [16]. The extension of the method to lattice merge has been briefly sketched as well. Moreover, the mingen-related part of the lattice maintenance method from [16] was proved to easily fit the iceberg lattice maintenance task as in [10].

In this paper, we study the mingen maintenance problem in dual settings, i.e., upon increases in the attribute set of the context. The study has a two-fold motivation and hence contributes in two different ways to the FCA field. Thus, on the one hand, the evolution of the mingen is given a characterization, in particular, with respect to the sets of stable/vanishing/newly forming mingen. To assess the impact of the provided results, it is noteworthy that although in lattice maintenance the attribute/object cases admit dual resolution, this does not hold for mingen maintenance, hence the necessity to study the attribute case separately. On the other hand, the resulting structure characterizations are embedded into an efficient maintenance method that can, as all other incremental algorithms, be run in a batch mode. The practical performances of the new method as batch iceberg-plus-mingen constructor have been compared to the performances of TITANIC, the algorithm which is reportedly the most efficient one producing the mingen family and the frequent part of the closure family[2]. The results of the comparison proved very encouraging: although our algorithm produces the lattice precedence relation beside concepts and mingen, it outperformed TITANIC when run on a sparse data set. We tend to see this as a clear indication of the potential the incremental paradigm has for mingen computation.

The paper starts with a recall of basic results about lattices, mingen, and incremental lattice update (Section 2). The results of the investigation on the evolution of the mingen family are presented in Section 3 while the proposed maintenance algorithm, INCA-GEN, is described in Section 4. In Section 5, we design a straightforward adaptation of INCA-GEN to iceberg concept lattice maintenance. Section 6 discusses preliminary results of the practical performance study that compared the algorithm to TITANIC.

2   Background on Concept Lattices

In the following, we recall basic results from FCA [18] that will be used in later paragraphs.

---

[2] Other algorithms include CLOSE and A-CLOSE [9].

2.1   FCA Basics

Throughout the paper, we use standard FCA notations (see [4]) except for the elements of a formal context for which English-based abbreviations are preferred to German-based ones. Thus, a formal context is a triple $\mathcal{K} = (O, A, I)$ where O and A are sets of objects and attributes, respectively, and I is the binary incidence relation.

We recall that two derivation operators, both denoted by $'$ are defined: for $X \subseteq O$, $X' = \{a \in A | \forall o \in X, oIa\}$ and for $Y \subseteq A$, $Y' = \{o \in O | \forall a \in Y, oIa\}$. The compound operators $''$ are closure operators over 2^O and 2^A, respectively. Hence each of them induces a family of *closed* subsets, $\mathcal{C}_\mathcal{K}^o$ and $\mathcal{C}_\mathcal{K}^a$, respectively. A pair (X, Y) of sets, where $X \subseteq O$, $Y \subseteq A$, $X = Y'$ and $Y = X'$, is called a *(formal) concept* [18].

Furthermore, the set $\mathcal{C}_\mathcal{K}$ of all concepts of the context $\mathcal{K}$ is partially ordered by extent/intent inclusion and the structure $\mathcal{L} = \langle \mathcal{C}_\mathcal{K}, \leq_\mathcal{K} \rangle$ is a complete lattice. In the remainder, the subscript $_\mathcal{K}$ will be avoided whenever confusion is impossible. Fig. 1 shows a sample context where objects correspond to lines and attributes to columns. Its concept lattice is shown next.

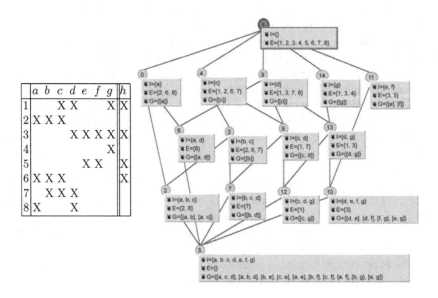

Fig. 1. Left: Binary table $\mathcal{K}_1 = (O = \{1, 2, ..., 8\}, A_1 = \{a, b, ..., g\}, I_1)$ and the attribute h. **Right:** The Hasse diagram of the lattice $\mathcal{L}_1$ of $\mathcal{K}_1$. Concepts are provided with their respective intent (I), extent (E) and mingen set (G)

Within a context $\mathcal{K}$, a set $G \subseteq A$ is a *minimal generator* (mingen) of a closed set $Y \subseteq A$ (hence of the concept (Y', Y)) iff G is a minimal subset of Y such that $G'' = Y$. As there may be more than one mingen for a given intent Y, we define the set-valued function *gen*. Formally,

Definition 1. *The function associating to concepts their mingen sets, gen(c)* :
$C \to 2^{2^A}$, *is defined as follows:*

$$gen(Y', Y) = \{G \subseteq Y \mid G'' = Y \text{ and } \forall F \subset G, \ F'' \subset Y\}.$$

In Fig. 1, the concept $c_{\#2} = (26, abc)$ has two mingen: $gen(c_{\#2}) = \{ab, ac\}$. In the remainder, *gen* will be used both on individual concepts and on concept sets with a straightforward interpretation.

2.2   Incremental Lattice Update, a Recall

Assume that $\mathcal{K}_1$ and $\mathcal{K}_2$ are two contexts diverging by only one attribute, i.e., $\mathcal{K}_1 = (O, A_1, I_1)$ and $\mathcal{K}_2 = (O, A_2, I_2)$ with $A_2 = A_1 \cup \{a\}$ and $I_2 = I_1 \cup \{a\} \times a'$. In the following, to avoid confusion, we shall denote the derivation operators in $\mathcal{K}_i$ ($i = 1, 2$), by ^{i}. Similarly, mingen functions will be subscripted. Let now $\mathcal{L}_1$ and $\mathcal{L}_2$ be the two concept lattices of $\mathcal{K}_1$ and $\mathcal{K}_2$, respectively. If $\mathcal{L}_1$ is already available, say, as a data structure in the main memory of a computer, then, according to the incremental lattice construction paradigm [5], it can be transformed at a relatively low cost into a structure representing $\mathcal{L}_2$. Hence there is no need to construct $\mathcal{L}_2$ from scratch, i.e., by looking on $\mathcal{K}_2$.

In doing the minimal reconstruction that yields $\mathcal{L}_2$ from $\mathcal{L}_1$ and (a, a^2), all the incremental methods rely on basic property of closure systems: C^o is closed under set intersection [1]. In other words, if $c = (X, Y)$ is a concept of $\mathcal{L}_1$ then $X \cap a^2$ is closed object set and corresponds to an extent in $\mathcal{L}_2$. Hence, the transformation of $\mathcal{L}_1$ into $\mathcal{L}_2$ via the attribute a is mainly aimed at computing all the concepts from $\mathcal{L}_2$ whose extent is not an extent in $\mathcal{L}_1$. Those concepts are called the *new* concepts in [5] (here denoted $\mathbf{N}(a)$). As to $\mathcal{L}_1$, its concepts are partitioned into three categories. The first one is made of *modified* concepts (labeled $\mathbf{M}(a)$): their extent is included in a^2, the extent of a, hence they evolve from $\mathcal{L}_1$ into $\mathcal{L}_2$ by integrating a into their intents while extents remain stable. The second category is made of *genitor* concepts (denoted $\mathbf{G}(a)$) which help create new concepts but remain themselves stable (changes appear in the sets of neighbor concepts). Finally, *old* concepts (denoted $\mathbf{U}(a)$) remain completely unchanged. As concepts from $\mathcal{L}_1$ have their counterparts in $\mathcal{L}_2$, we shall use subscripts to distinguish both copies of a set. Thus, $\mathbf{G}_1$, $\mathbf{U}_1$ and $\mathbf{M}_1$ will refer to $\mathcal{L}_1$, while $\mathbf{G}_2$, $\mathbf{U}_2$, $\mathbf{M}_2$ and $\mathbf{N}_2$ will refer to $\mathcal{L}_2$.

A characterization of each of the above seven concept categories is provided in [17]. It relies on two functions which map concepts to the intersection of their extents with a^2: $\mathcal{R}_i : \mathcal{C}_i \to 2^O$ with $\mathcal{R}_i(c) = extent(c) \cap a^2$. Each $\mathcal{R}_i$ induces an equivalence relation on $\mathcal{C}_i$ where $[c]_{\mathcal{R}_i} = \{\bar{c} \in \mathcal{C}_i \mid \mathcal{R}_i(c) = \mathcal{R}_i(\bar{c})\}$. Moreover, following [15], $\mathbf{G}_1(a)$ and $\mathbf{M}_1(a)$ are the minimal concepts in their respective equivalence classes in $\mathcal{L}_1$.

Example 1. Assume $\mathcal{L}_1$ is the lattice induced by the attribute set $abcdefg$ (see Fig. 1 on the right) and consider h the new attribute to add to $\mathcal{K}_1$. Fig. 2 shows the resulting lattice $\mathcal{L}_2$. The sets of concepts are as follows: $\mathbf{U}_2(h) =$

$\{c_{\#0}, c_{\#3}, c_{\#6}, c_{\#7}, c_{\#8}, c_{\#9}, c_{\#14}\}; \mathbf{M}_2(h) = \{c_{\#5}, c_{\#10}, c_{\#11}, c_{\#12}, c_{\#13}\}; \mathbf{G}_2(h)$
$= \{c_{\#1}, c_{\#2}, c_{\#4}\}$ and $\mathbf{N}_2(h) = \{c_{\#15}, c_{\#16}, c_{\#17}\}$.

Following [17], three mappings will be used to connect $\mathcal{L}_1$ into $\mathcal{L}_2$. First, σ maps a concept in $\mathcal{L}_1$ to the concept with the same extent in $\mathcal{L}_2$. Second, γ projects a concept from $\mathcal{L}_2$ on A_1, i.e., returns the concept having the same attributes but a. Finally, χ^+ returns for a concept c in $\mathcal{L}_1$ the minimal element of the equivalence class $[]_{\mathcal{R}_2}$ for its counterpart $\sigma(c)$ in $\mathcal{L}_2$.

Definition 2. *We define the following mappings between $\mathcal{L}_1$ and $\mathcal{L}_2$:*

- $\sigma: \mathcal{C}_1 \to \mathcal{C}_2$, $\sigma(X, Y) = (X, X^2)$,
- $\gamma: \mathcal{C}_2 \to \mathcal{C}_1$, $\gamma(X, Y) = (\bar{Y}^1, \bar{Y})$ *where* $\bar{Y} = Y - \{a\}$,
- $\chi^+: \mathcal{C}_1 \to \mathcal{C}_2$, $\chi^+(X, Y) = (\bar{X}, \bar{X}^2)$ *where* $\bar{X} = X \cap a^2$.

3   Structure Characterization

We clarify here the evolution of the mingen family between $\mathcal{L}_1$ and $\mathcal{L}_2$. First, we prove two properties stating, respectively, that no generator vanishes from $\mathcal{L}_1$ to $\mathcal{L}_2$ and that only modified and new concepts in $\mathcal{L}_2$ contribute to the difference $gen_2(\mathcal{C}_2) - gen_1(\mathcal{C}_1)$. Then, we focus on the set of new mingen that forms in each of the two cases and prove that a new mingen is made of an old one augmented by the attribute a.

3.1   Global Properties

Let us first find the relation between the mingen of a concept c in $\mathcal{L}_1$ and those of its counterpart $\sigma(c)$ in $\mathcal{L}_2$. The following property shows that the mingen of c are also mingen for $\sigma(c)$.

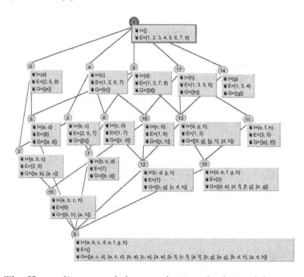

Fig. 2. The Hasse diagram of the new lattice $\mathcal{L}_2$ derived from context $\mathcal{K}_2$

Property 1. $\forall \, c \in \mathcal{L}_1 \, gen_1(c) \subseteq gen_2(\sigma(c))$.

Proof. If $G \in gen_1(c)$ then $G^1 = G^2$, hence, $G^{22} = \bar{Y}$. Moreover G is minimal for $\bar{Y}$, otherwise it could not be a mingen of c.

Consequently, if G is a mingen in $\mathcal{L}_1$ it can only be a mingen in $\mathcal{L}_2$.

Corrolary 1. $gen(\mathcal{C}_1) \subseteq gen(\mathcal{C}_2)$

Proof. Given a concept $c = (X, Y)$ *in* $\mathcal{L}_1$, whatever is the category of $\sigma(c)$ in $\mathcal{L}_2$(*old, genitor* or *modified*), we always have $gen_1(c) \subseteq gen_2(\sigma(c))$.

For example, consider the concept $c_{\#12} = (1, cdg)$ in $\mathcal{L}_1$ whose mingen cg is also a mingen of $\sigma(c)_{\#12} = (1, cdgh)$ in $\mathcal{L}_2$. However, the concept $c_{\#12}$ in $\mathcal{L}_2$ has another mingen, cdh, which it does not share with its image in $\mathcal{L}_1$. This case can only happen with modified concepts from $\mathcal{L}_1$ because, as the following property states it, for old and genitor concepts in $\mathcal{L}_1$, the mingen of their $\sigma(c)$-counterpart in $\mathcal{L}_2$ are exactly the same as their own mingen.

Property 2. $\forall \, c = (X, Y) \in \mathcal{C}_1$, *if* $\sigma(c) = c$ *then* $gen_1(c) = gen_2(\sigma(c))$

Proof. Following Property 1, $gen_1(c) \subseteq gen_2(\sigma(c))$. Then, $gen_2(\sigma(c) \subseteq gen_1(c))$ comes from the fact that if $G \in gen_2(\sigma(c)$ and $c = \sigma(c)$ then $G^1 = G^2$ and $G^{11} = G^{22} = Y$. Moreover, G is minimal for c otherwise it would not be a mingen of $\sigma(c)$.

For example, consider the concept $c_{\#2} = (26, abc)$ in $\mathcal{L}_1$. It is easily seen that the set of its mingen, $\{ab, ac\}$, is the same as the mingen set of $\sigma(c_{\#2})$ in $\mathcal{L}_2$.

3.2 Characterizing the New Mingen

Now that we know that all mingen in $\mathcal{L}_1$ stay mingen in $\mathcal{L}_2$, the next step consists in finding the new mingen in $\mathcal{L}_2$. As the modified concepts and the genitor ones are the minimal elements of their equivalence classes in $\mathcal{L}_1$, we express the evolution of these classes from $\mathcal{L}_1$ to $\mathcal{L}_2$. In fact, the class of a new concept c in $\mathcal{L}_2$ is exactly the image of the class of its genitor in $\mathcal{L}_1$ to which we add c. The class of a modified concept is identical in both $\mathcal{L}_1$ and $\mathcal{L}_2$.

Property 3. *The equivalence classes of new and modified concepts in* $\mathcal{L}_2$ *are composed as follows:*

- $\forall c \in \mathbf{N}_2(a)$, $[c]_{\mathcal{R}_2} = [\gamma(c)]_{\mathcal{R}_1} \cup c$
- $\forall c \in \mathbf{M}_2(a)$, $[c]_{\mathcal{R}_2} = [\gamma(c)]_{\mathcal{R}_1}$

In summary, it was established that the equivalence classes $[]_{\mathcal{R}_2}$ differ by at most one element from their counterparts $[]_{\mathcal{R}_1}$ and that new mingen may only appear at the minimal element of each class. The next question to ask is how mingen of concepts in $[\gamma(c)]_{\mathcal{R}_1}$ are related to those of the minimal element of $[c]_{\mathcal{R}_2}$. The first step is to notice that whenever a mingen of a concept c from $\mathcal{L}_1$

is augmented with the new attribute a, the closure of the resulting set in $\mathcal{K}_2$ is the intent of the minimal element in the respective class $[\sigma(c)]_{\mathcal{R}_2}$.

Assume $c_{\min} = (X, Y) \in \mathcal{C}_2$ is the minimal element of its class in $\mathcal{L}_2$ and let $\bar{c} = (\bar{X}, \bar{Y})$ a concept of that class while $\bar{G}$ is a mingen of $\bar{c}$. According to the definition of $[c]_{\mathcal{R}_i}$, we have: $\bar{X} \cap a^2 = X$ and hence $\bar{G}^2 \cap a^2 = Y^2$. Moreover, it is known that for $A, B \in 2^O$, $(A \cup B)^2 = A^2 \cap B^2$. Thus, $\bar{G}^2 \cap a^2 = Y^2$ can be written as $(\bar{G} \cup a)^2 = Y^2$ and consequently $(\bar{G} \cup a)^{22} = Y^{22}$.

In summary, for every mingen G of a concept in a class $[c]_{\mathcal{R}_1}$, its superset obtained by adding the new attribute a, $\bar{G} \cup a$, has as closure the intent of the minimal element of the corresponding class $[\sigma(c)]_{\mathcal{R}_2}$. The following property states that $(\bar{G} \cup a)$ is a mingen of c iff $\bar{G}$ is minimal among the mingen of the entire equivalence class $[c]_{\mathcal{R}_1}$.

Property 4. $\forall c \in \mathbf{M}_2(a)$ then :

$$gen_2(c) = gen_1(\gamma(c)) \cup min\left(\bigcup_{\hat{c} \in [c]_{\mathcal{R}_i} \ and \ \hat{c} \neq c} gen_1(\hat{c}) \right) \times \{a\}$$

For example, consider the concept $c_{\#13} = (13, dgh)$ in $\mathcal{L}_2$. Here $\mathcal{R}_2(c_{\#13}) = 13$ and $[c_{\#13}]_{\mathcal{L}_2} = \{c_{\#9}, c_{\#14}, c_{\#13}\}$. Clearly, $c_{\#13}$ is minimal in its class, and more precisely, it is a modified concept. Since $gen_1(c_{\#9}) = \{d\}$ and $gen_1(c_{\#9}) = \{g\}$ whereas both mingen d and g are incomparable, the newly formed mingen of $c_{\#13}$ in $\mathcal{L}_2$ are $\{dh, gh\}$. Consequently, the entire set of mingen for $c_{\#13}$ in $\mathcal{L}_2$ is $gen_2(c_{\#13}) = \{dg, dh, gh\}$. Indeed, the correctness of that result can be checked upon Fig. 2.

A similar result can be proved for the complementary case for the minima in a class $[]_{\mathcal{R}_2}$, i.e., for new concepts. The following property states that the mingen of a new concept $c = (X, Y) \in \mathbf{N}_2(a)$ are exactly the sets produced by adding a to each of the mingen that are minimal in the entire class $[\gamma(c)]_{\mathcal{R}_1}$.

Property 5.

$$\forall c \in \mathbf{N}_2(a), gen_2(c) = min\left(\bigcup_{\hat{c} \in [\gamma(c)]_{\mathcal{R}_1}} gen_1(\hat{c}) \right) \times \{a\}$$

For example, consider the concept $c_{\#15} = (6, abch)$ in $\mathcal{L}_2$. $\mathcal{R}_2(c_{\#15}) = 6$ and $[c_{\#15}]_{\mathcal{R}_2} = \{c_{\#0}, c_{\#3}, c_{\#2}, c_{\#15}\}$. Clearly, $[\gamma(c_{\#15})]_{\mathcal{R}_1} = \{c_{\#0}, c_{\#3}, c_{\#2}\}$. Moreover, $gen_1(c_{\#0}) = \{a\}$, $gen_1(c_{\#3}) = \{b\}$, $gen_1(c_{\#2}) = \{ab, ac\}$. Thus, $min(\bigcup_{\hat{c} \in [c_{\#15}]_{\mathcal{R}_1}} gen_1(\hat{c})) = min\{a, b, ab, ac\} = \{a, b\}$ and hence $gen(c)_{\#15} = \{ah, bh\}$.

Following Properties 4 and 5, we state that every new mingen in $\mathcal{L}_2$ is obtained by adding a to a mingen from $\mathcal{L}_1$.

Corrolary 2. $gen(\mathcal{C}_2) - gen(\mathcal{C}_1) \subseteq gen(\mathcal{C}_1) \times \{a\}$.

In summary, to compute the mingen in $\mathcal{L}_2$, one only needs to focus on new and modified elements. In both cases, the essential part of the calculation is the detection of all the mingen G of concepts from the underlying equivalence classes which are themselves minimal, i.e., there exists no other mingen in the class that is strictly included in G. Obviously, this requires the equivalence classes to be explicitly constructed during the maintenance step.

4   Mingen Maintenance Method

The results presented in the previous section are transformed into an algorithmic procedure, provided in Algorithm 1, that updates both the lattice and the mingen sets of the lattice concepts upon the insertion of a new attribute into the context. The key task of the method hence consists in computing/updating the mingen of the minimal concept in each class $[]_{\mathcal{R}_2}$ in $\mathcal{L}_2$. Such a class will be explicitly represented by a variable, θ, which is a structure with two fields: the minimal concept, *min-concept*, and minimal mingen, *min-gen*. Moreover, the variable for all classes are gathered in a index, called *Classes*, where each variable is indexed by the $\mathcal{R}_1$ value for the respective class.

```
 1: procedure INCA-GEN(In/Out: L = ⟨C, ≤⟩ a Lattice, In: a an attribute)
 2: Local : Classes : an indexed structure of classes
 3:
 4:   COMPUTE-CLASSES(C,a)
 5:   for all θ in Classes do
 6:     c ← θ.min-concept
 7:     if |R(c)| = |extent(c)| then
 8:       intent(c) ← intent(c) ∪ {a}      {c is modified}
 9:     else
10:       ĉ ← newConcept(R(c),Intent(c) ∪ {a})      {c is genitor}
11:       L ← L ∪ {ĉ}
12:       updateOrder(c,ĉ)
13:       gen(ĉ) ← ∅
14:       θ.min-concept ← ĉ
15:       c ← θ.min-concept
16:       gen(c) ← gen(c) ∪ θ.min-gen × {a}
```

Algorithm 1: Insertion of a new attribute in the context

It is noteworthy that the lattice update part of the work is done in a way which is dual to the object-wise incremental update. Thus, the Algorithm 1 follows the equivalent of the basic steps for object incrementing described in [14]. The work starts with a pre-processing step that extracts the class information from the lattice and stores it in the *Classes* structure (primitive COMPUTE-CLASSES, see Algorithm 2). At a second step, the variables θ corresponding to each class are explored to restructure the lattice and to compute the mingen (lines 6 to 16). First, the kind of restructuring, i.e., modification of an intent versus creation of a new concept, is determined (line 7). Then, the standard update procedures are carried out for modified (line 8) and genitor concepts (lines 10 to 12). In the second case, the computation specific to the mingen family update is limited to lines 13 and 14. Finally, the mingen set of a minimal concept is updated in a uniform manner that strictly follows the Properties 4 and 5.

The preprocessing step as described in Algorithm 2 basically represents a traversal of the lattice during which the content of the *Classes* structure is gradually collected.

At each concept, the intersection of the extent and the object set of the attribute a is computed (line 5). This provides an entry point for the concept to its equivalence class which is tentatively retrieved from the *Classes* structure using the intersection as a key (line 6). If the class is not yet present in the structure (line 7), which means that the current concept is its first encountered member, the corresponding variable θ is created (line 8), initialized with the information found in c (line 9), and then inserted in *Classes* (line 10). The current concept is also compared to the current minimum of the class (lines 11 and 12) and the current minima of the total mingen set of the class are updated (line 13). At the end, the structure *Classes* comprises the variables of all the equivalence classes in $\mathcal{R}_1$ with the accurate information about its minimal representative and about the minima of the global mingen set.

```
 1: procedure COMPUTE-CLASSES(In/Out: C concept set, In: a an attribute)
 2:
 3: for all c in C do
 4:     E ← extent(c) ∩ a²
 5:     θ ← lookup(Classes, E)
 6:     if (θ = NULL) then
 7:         θ ← newClass()
 8:         θ.min-concept ← c
 9:         put(Classes, θ, E)
10:     if (θ.min-concept < c) then
11:         θ.min-concept ← c
12:     θ.min-gen ← Min(θ.min-gen ∪ gen(c))
```

Algorithm 2: Computation of the equivalence classes $[]_{\mathcal{R}}$ in the initial lattice

5   Iceberg Lattice Variant

Let $c = (X, Y)$ a concept in $\mathcal{L}_1$. The frequency of c, denoted *freq(c)*, is defined as the ratio of its extent and the size of the object set: $freq(c) = \|X\|/\|O\|$). Given α a minimal threshold of support defined by the user, the concept c is frequent if $freq(c) \geq \alpha$. The *iceberg concept lattice* generated by α, $\mathcal{L}_1^{\alpha}$, is made of all frequent concepts. An iceberg is thus a join-semi-lattice, a sub-semi-lattice of the complete concept lattice [11].

For example, the *iceberg* $\mathcal{L}_1^{0.20}$ obtained from the complete lattice of Fig. 1 with $\alpha \geq 0.20$ is shown in Fig. 3.

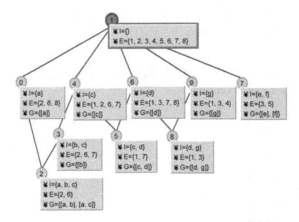

Fig. 3. The iceberg lattice $\mathcal{L}_1^{0.20}$ of the context $\mathcal{K}_1$ with $A_1 = \{a, .., g\}$

Unlike the object-wise incremental iceberg update [10], when a new attribute is added, the maintenance of the iceberg follows strictly that of the complete lattice. This is due to the invariance of concept frequency: Once a concept is generated, its frequent status remains unchanged. Assume c is a concept in $\mathcal{L}_1^\alpha$. The $freq(\sigma(c))$ in $\mathcal{L}_2^\alpha$ will be $\frac{|extent(\sigma(c))|}{|O|}$. As $extent(\sigma(c)) = extent(c)$ and the number of objects does not vary along the transition from $\mathcal{L}_1^\alpha$ to $\mathcal{L}_2^\alpha$ it follows that $freq(c) = freq(\sigma(c))$.

Moreover, the frequency of the intersection $\mathcal{R}_1(c)$ is monotonously non-decreasing with respect to lattice order since it is the composition of two monotonous non-decreasing functions.

Property 6. $\forall c \in \mathcal{L}_1^\alpha$, if $|\mathcal{R}_1(c)| \leq \alpha|O|$ then $\forall \underline{c}$ s.t $c \prec \underline{c}$, $|\mathcal{R}_1(\underline{c})| \leq \alpha|O|$.

Exploring the monotonicity of the frequency function and restricting the $\mathcal{R}_i$ functions $(i = 1, 2)$ from section 2.2 to icebergs, we obtain the fact that a class may be only partially included in the iceberg. Furthermore, as for any concept $c = (X, Y)$ in $\mathcal{L}_1^\alpha$, $extent(\chi^+(c))$ is the extent of a new or a modified concept in $\mathcal{L}_2^\alpha$, only the frequent intersections could be considered since the concepts corresponding to infrequent ones do not belong to the iceberg.

The above observation could potentially invalidate our mingen calculation mechanism since it relies on the presence of the entire mingen set for a class in $\mathcal{R}_1$. However, observe that only the minima of the equivalence classes require some calculation. Moreover, because of the monotonicity, the minima are also the less frequent concepts of a class and thus cannot be present in the iceberg $\mathcal{L}_2^\alpha$ if the class is only partially covered by $\mathcal{L}_2^\alpha$. Consequently, infrequent intersections could simply be ignored.

In summary, the icebergs could be dealt with in a way similar to that for complete lattices. The only thing to be added is a filter for infrequent intersections. Thus, a concept producing such an intersection should be discarded from the preprocessing step. As a result, only the classes corresponding to frequent intersections, or, equivalently, having frequent minima, will be sent to the main algorithm for further processing.

```
1: procedure COMPUTE-F-CLASSES(In/Out: L_α an iceberg lattice, In: a an attribute)
2: Local : cQ : a queue of concepts
3:
4: in(cQ, top(L_α))
5: while nonempty(cQ) do
6:    c ← out(cQ)
7:    E ← extent(c) ∩ a²
8:    if (|E| ≥ α|O|) then
9:       θ ← lookup(Classes, E)
10:      if (θ = NULL) then
11:         θ ← newClass()
12:         θ.min-concept ← c
13:         put(Classes, θ, E)
14:      if (θ.min-concept > c) then
15:         θ.min-concept ← c
16:      θ.min-gen ← Min(θ.min-gen ∪ gen(c))
17:      for all ĉ ∈ Cov^u(c) do
18:         in(cQ, ĉ)
```

Algorithm 3: Computation of the frequent equivalence classes $[]_R$ in the initial iceberg lattice

Luckily enough, the difference between the processing of icebergs and that of complete lattices can be confined in the class construction step. Thus, the frequency-aware traversal of the iceberg relies on the monotonicity of the intersection function to prune unnecessary lattice paths. More specifically, it explores the concept set following the lattice order and in a top-down manner whereby the

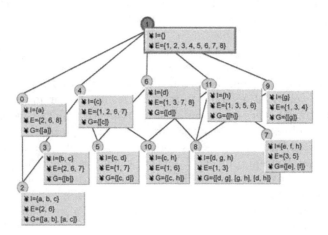

Fig. 4. The iceberg lattice $\mathcal{L}_2^{0.20}$ of the context $\mathcal{K}_2$ with $A_2 = A_1 \cup \{h\}$

exploration of a new path stops with the first concept producing an infrequent intersection. These differences are reflected in the code of Algorithm 3. The algorithm uses a queue structure to guide the top-down, breadth-first traversal of the iceberg (lines 4 to 6 and 17 to 18). The remaining noteworthy difference with Algorithm 2 is that infrequent concepts are ignored whenever pulled out of the queue (line 8).

Finally, to obtain an iceberg-based version of the lattice algorithm INCA-GEN, it would suffice to replace in Algorithm 1 the call of COMPUTE-CLASSES by COMPUTE-F-CLASSES on line 4. For example, the iceberg lattice $\mathcal{L}_2^{0.20}$ resulting from the addition of the attribute h to the iceberg lattice $\mathcal{L}_1^{0.20}$ in Fig. 3 is depicted in Fig. 4.

6   Experiments and Performance Evaluation

The algorithm INCA-GEN has been implemented in Java and a version thereof is available within the Galicia[3] platform [13]. The platform version is designed for portability and genericity and therefore is not optimized for performances.

A stand-alone version of the algorithm, called MAGALICE-A, was devised for use in experimental studies. Its performances have been examined on a comparative basis. In a preliminary series of tests, MAGALICE-A was confronted to the TITANIC algorithm [11]. TITANIC is a batch method which solves a similar problem, known as *frequent closed itemset mining*, and produces comparable results, i.e., the set of frequent concept intents and the corresponding mingen sets. The choice of TITANIC was further motivated by the reported efficiency of the method and its status of reference algorithm in the FCA community. We used our own Java implementation of TITANIC for the experiments since at the time of the study, no code was publicly available[4].

The experiments were carried out on a Pentium IV 2 GHz workstation running Windows XP, with 1 GB of RAM. The comparisons were performed on two types of datasets:

- subsets of MUSHROOM, a real-world dataset which is also a dense one ($8,124$ objects, 119 attributes, average of 23 attributes per object), and
- subsets of T25.I10.D10K, a sparse synthetic dataset popular with the data mining community ($10,000$ objects, 1000 attributes, average of 25 attributes per object).

The choice of both datasets was motivated by the following observation. It is now widely admitted that incremental lattice algorithms perform well on sparse datasets but lag behind batch methods when applied to dense ones. Our goal was to test whether this trend persists when the mingen families are fed into the computation process. Indeed, the experimental results seem to confirm the

---

[3] http://galicia.sourceforge.net
[4] The authors would like to thank G. Stumme for the valuable information he provided about TITANIC.

hypothesis that incremental methods may be used as efficient batch procedure whenever dealing with low-density data tables.

More concretely, two types of statistics have been gathered. On the one hand, the efficiency of both algorithms has been directly related to the CPU time that was required to solve the task. On the other hand, we have recorded the memory consumption as an important secondary indicator of how suitable the algorithm is for large dataset analysis.

Fig. 5 depicts the CPU time of the analysis of both datasets: the dense one, on the left, and the sparse one, on the right. The first diagram indicates that TITANIC outperforms MAGALICE-A by far: it runs 2 to 7 times faster. However, the reader should bear in mind the fact that beside the concept set and the mingen, MAGALICE-A also maintains the order in the iceberg lattice while TITANIC does not.

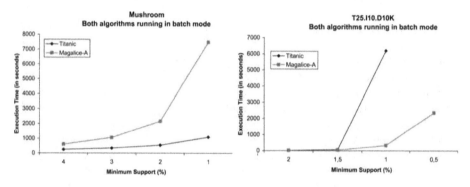

Fig. 5. CPU-time for MAGALICE-A and TITANIC. The tests involved the entire dataset from which only the frequent concepts had to be computed using a range of support threshold values

The second diagram reverses the situation: MAGALICE-A beats TITANIC by a factor going up to 18. A more careful analysis would be necessary to explain such a dramatic shift in performances. However, the main reason seems to lay in the fact that the performance of the incremental algorithm is strongly impacted by the actual number of concepts in the iceberg. Thus, with a higher number of concepts as with the MUSHROOM dataset, the algorithm suffers a significant slow-down whereas with T25.I10.D10K where the number of concepts is low, it performs well. The closure computation used by TITANIC to find concept intents depends to a much lesser degree on the number of concepts in the iceberg and therefore the performances of the algorithm vary in a narrow interval.

Fig. 6 visualizes the results about the memory usage for both algorithms. The same trends as with CPU-time seem to appear here. Thus, while dense datasets increase substantially the memory consumption of our algorithm, TITANIC remains at a reasonable usage rate. With sparse datasets however, the figures are mirrored: MAGALICE-A leads by far with a much smaller memory demand. Once again, the number of frequent concepts might be the key to the interpretation

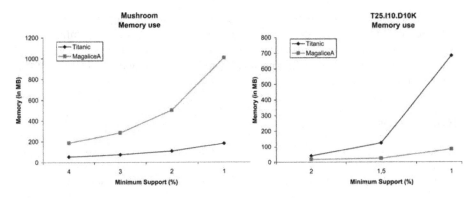

Fig. 6. Memory consumption for MAGALICE-A and TITANIC. The tests involved the entire dataset from which only the frequent concepts had to be computed using a range of support threshold values

of the results: with larger number of concepts, the overhead due to additional computations in MAGALICE-A, i.e., equivalence class constitution, extent and order computation/storage, etc., takes over the core tasks of computing intents and mingen. Conversely, with a small number of frequent intents, the number of mingen is (proportionally) larger. MAGALICE-A does well since only small amount of computing is performed on a mingen, whereas TITANIC wastes time in computing a large number of closures.

It is noteworthy that for support values of 1% and less, both algorithms exhaust the main memory capacity and relied on swapping to continue their work. For TITANIC, this happens when working on the sparse dataset while for MAGALICE-A it is the case with the dense one.

7   Conclusion

Minimal generators of concept intents are intriguing members of the FCA landscape with strong links to practical and theoretical problems from neighbor areas. Because of their important role, it is worth studying their behavior under different circumstances, in particular their evolution upon small changes in the input context. In this paper we studied the evolution of the mingen family of a context upon increases of the attribute set. The operation has a certain practical value since in many FCA tools, dynamic changes in the set of "visible" attributes are admitted. However, in this study we looked at the incrementing of the attribute set as a pure computational technique and examined its relative merits compared to those of an existing batch method.

The results up to date suggest that the incremental paradigm has its place in this particular branch of FCA algorithmic practices. They also motivate the research on a second generation algorithms that would improve on the design of the initial, rather straightforward procedures, INCA-GEN and MAGALICE-A.

The presented research is a first stage in a broader study on the dynamic behavior of FCA-related subset families: closures, pseudo-closed, mingen, etc. An even more intriguing subject is the cross-fertilization between methods for computing separate families in the way it is done in TITANIC or in the MERGE algorithm described in [12].

References

1. M. Barbut and B. Monjardet. *Ordre et Classification: Algèbre et combinatoire.* Hachette, 1970.
2. C. Berge. *Hypergraphs: Combinatorics of Finite Sets.* North Holland, Amsterdam, 1989.
3. B. Ganter. Two basic algorithms in concept analysis. preprint 831, Technische Hochschule, Darmstadt, 1984.
4. B. Ganter and R. Wille. *Formal Concept Analysis, Mathematical Foundations.* Springer-Verlag, 1999.
5. R. Godin, R. Missaoui, and H. Alaoui. Incremental Concept Formation Algorithms Based on Galois (Concept) Lattices. *Computational Intelligence*, 11(2):246–267, 1995.
6. J.L. Guigues and V. Duquenne. Familles minimales d'implications informatives résultant d'un tableau de données binaires. *Mathématiques et Sciences Humaines*, 95:5–18, 1986.
7. D. Maier. *The theory of Relational Databases.* Computer Science Press, 1983.
8. N. Pasquier. Extraction de bases pour les règles d'association à partir des itemsets fermés fréquents. In *Proceedings of the 18th INFORSID'2000*, pages 56–77, Lyon, France, 2000.
9. N. Pasquier, Y. Bastide, R. Taouil, and L. Lakhal. Discovering frequent closed itemsets for association rules. In *Proceedings, ICDT-99*, pages 398–416, Jerusalem, Israel, 1999.
10. M. H. Rouane, K. Nehmé, P. Valtchev, and R. Godin. On-line maintenance of iceberg concept lattices. In *Contributions to the 12th ICCS*, page 14 p., Huntsville (AL), 2004. Shaker Verlag.
11. G. Stumme, R. Taouil, Y. Bastide, N. Pasquier, and L. Lakhal. Computing Iceberg Concept Lattices with Titanic. *Data and Knowledge Engineering*, 42(2):189–222, 2002.
12. P. Valtchev and V. Duquenne. Implication-based methods for the merge of factor concept lattices (32 p.). *submitted to Discrete Applied Mathematics.*
13. P. Valtchev, D. Grosser, C. Roume, and M. Rouane Hacene. GALICIA: an open platform for lattices. In B. Ganter and A. de Moor, editors, *Using Conceptual Structures: Contributions to 11th Intl. Conference on Conceptual Structures (ICCS'03)*, pages 241–254, Aachen (DE), 2003. Shaker Verlag.
14. P. Valtchev, M. Rouane Hacene, and R. Missaoui. A generic scheme for the design of efficient on-line algorithms for lattices. In A. de Moor, W. Lex, and B. Ganter, editors, *Proceedings of the 11th Intl. Conference on Conceptual Structures (ICCS'03)*, volume 2746 of *Lecture Notes in Computer Science*, pages 282–295, Berlin (DE), 2003. Springer-Verlag.
15. P. Valtchev and R. Missaoui. Building concept (Galois) lattices from parts: generalizing the incremental methods. In H. Delugach and G. Stumme, editors, *Proceedings of the ICCS'01*, volume 2120 of *Lecture Notes in Computer Science*, pages 290–303, 2001.

16. P. Valtchev, R. Missaoui, and R. Godin. Formal Concept Analysis for Knowledge Discovery and Data Mining: The New Challenges. In P. Eklund, editor, *Concept Lattices: Proceedings of the 2nd Int. Conf. on Formal Concept Analysis (FCA'04)*, volume 2961 of *Lecture Notes in Computer Science*, pages 352–371. Springer-Verlag, 2004.

17. P. Valtchev, R. Missaoui, R. Godin, and M. Meridji. Generating Frequent Itemsets Incrementally: Two Novel Approaches Based On Galois Lattice Theory. *Journal of Experimental & Theoretical Artificial Intelligence*, 14(2-3):115–142, 2002.

18. R. Wille. Restructuring lattice theory: An approach based on hierarchies of concepts. In I. Rival, editor, *Ordered sets*, pages 445–470, Dordrecht-Boston, 1982. Reidel.

Efficiently Computing a Linear Extension of the Sub-hierarchy of a Concept Lattice

Anne Berry[1], Marianne Huchard[2], Ross M. McConnell[3],
Alain Sigayret[1], and Jeremy P. Spinrad[4]

[1] LIMOS (Laboratoire d'Informatique de Modélisation et d'Optimisation des Systèmes), CNRS UMR 6158, Université Clermont-Ferrand II, Ensemble scientifique des Cézeaux, 63177 Aubière Cedex, France
{berry, sigayret}@isima.fr

[2] LIRMM (Laboratoire d'Informatique de Robotique et de Micro-électronnique de Montpellier), CNRS UMR 5506, Université Montpellier II, 161 rue Ada, 34392 Montpellier cedex 5, France
huchard@lirmm.fr

[3] EECS Department, Vanderbilt University, Nashville, TN 37235 USA
spin@vuse.vanderbilt.edu

[4] Computer Science Department, Colorado State University, Fort Collins, CO 80523-1873 USA
rmm@cs.colostate.edu

Abstract. Galois sub-hierarchies have been introduced as an interesting polynomial-size sub-order of a concept lattice, with useful applications. We present an algorithm which, given a context, efficiently computes an ordered partition which corresponds to a linear extension of this sub-hierarchy.

1   Introduction

Formal Concept Analysis (FCA) aims at mining concepts in a set of entities described by properties, with many applications in a broad spectrum of research fields including knowledge representation, data mining, machine learning, software engineering or databases. Concepts are organized in concept (Galois) lattices where the partial order emphasizes the degree of generalization of concepts and helps to visually apprehend sets of shared properties as well as groups of objects which have similarities.

The main drawback of concept lattices is that the number of concepts may be very large, or even exponential in the size of the relation.

One of the options for dealing with this problem is to use a polynomial-size representation of the lattice which preserves the most pertinent information. A way of doing this is to restrict the lattice to the concepts which introduce a new object or property, leading to two similar structures called the 'Galois sub-hierarchy' (GSH) and the 'Attribute Object Concept poset' (AOC-poset). GSH has been introduced in the software engineering field by Godin and Mili in 1993

B. Ganter and R. Godin (Eds.): ICFCA 2005, LNAI 3403, pp. 208–222, 2005.

([11]) for class hierarchy reconstruction and successfully applied in later research works ([12, 20, 14, 6]).

Recent work has shown interest of GSH in an extension of FCA to Relational Concept Analysis (RCA); RCA has been tested to identify abstractions in UML (Unified Modeling Language, see [19]) diagrams allowing to improve such diagrams in a way that had not been explored before ([7]). AOC-poset has been used in applications of FCA to non-monotonic reasoning and domain theory ([13]) and to produce classifications from linguistic data ([17, 16]). Considering AOC-poset or GSH is interesting from two points of view, namely the algorithmic and the conceptual (human perception), because the structure which is used has only a restricted number of elements.

Several algorithms have been proposed to construct the Galois sub-hierarchy, either incrementally or globally. Incremental algorithm ARES [8] and ISGOOD [10] add a new object given with its property set in an already constructed GSH. The best worst-case complexity is in $O(k^3 n^2)$ for ARES (in $O(k^4 n^2)$ for ISGOOD) where k is the maximal size of a property set and n is the number of elements in the initial GSH. Note that best pratical results are nevertheless obtained by ISGOOD. The global algorithm CERES ([15]) computes the elements of the GSH as well as the order and has worst case complexity in $O(|O| (|O| + |P|)^2)$.

In this paper, we present an algorithm which outputs the elements of the GSH in a special order, compatible with a linear extension of the GSH: roughly speaking, we decompose each element of the GSH into an extent (which is the set of objects of this element) and an intent (which is the set of properties of this element), and we output a list of subsets of objects and properties, such that if E_1 is a predecessor of E_2 in the GSH, then both the intent and the extent of E_1 are listed before the intent and extent of E_2 in our output ordering.

To do this, we use a partition refinement technique, inspired by work done in Graph Theory, which can easily be implemented to run in linear time. We used this in previous works to improve Bordat's concept generation algorithm (see [4]), in order to rapidly group together the objects (or, dually, the properties) which are similar. Partition refinement has also been used to reorder a matrix ([18]).

One of the interesting new points of the algorithm presented here is that it uses the objects and properties at the same time, instead of just the objects or just the properties.

The other interesting development is that the orderings on the properties and objects created by our algorithm define a new representation of the input relation, essentially by re-ordering its rows and columns, creating a zone of zeroes in its lower right-hand corner.

The paper is organized as follows: in Section 2, we give a few necessary notations, and present a running example which we will use throughout the paper to illustrate our work. Section 3 presents some results from previous papers, and explains the general algorithmic process which is used. Section 4 gives the algorithm, as well as some interesting properties of the output. The algorithm is proved in Section 5.

2   Notations and Example

It is assumed that the reader is familiar with classical notions of partial orderings and lattices, and is referred to [5], [1] and [9]. We will need a few preliminary notations and definitions.

Given a context $(\mathcal{O}, \mathcal{P}, R)$, $\mathcal{O}$ is a set of objects and $\mathcal{P}$ a set of properties, for $X \subseteq \mathcal{O}, Y \subseteq \mathcal{P}$, we will denote by $R(X, Y)$ the subrelation $R \cap (X \times Y)$. In the algorithms we present in this paper, objects and properties can be interchanged, so we will need to unify notations x' and x'' from [9] into $R[x]$ as follows: we will denote by $R[x]$ the set $\{y \in \mathcal{P}, (x, y) \in R\}$ if $x \in \mathcal{O}$ and $\{y \in \mathcal{O}, (y, x) \in R\}$ if $x \in \mathcal{P}$. $\bar{R}$ will denote the complement of relation R: $(x, y) \in \bar{R}$ iff $(x, y) \notin R$.

We will say that a concept $A' \times B'$ is a **successor** of concept $A \times B$ if $A \subset A'$ and there is no intermediate concept $A'' \times B''$ such that $A \subset A'' \subset A'$. A concept $A' \times B'$ is a **descendant** of concept $A \times B$ if $A \subset A'$. The notions of **predecessor** and **ancestor** are defined dually.

In the rest of this paper, we will use the same running example to illustrate our definitions and algorithms.

Example 1. Binary relation R:

Set of objects:
$\mathcal{O} = \{1, 2, 3, 4, 5, 6\}$,

Set of properties:
$\mathcal{P} = \{a, b, c, d, e, f, g, h\}$.

	a	b	c	d	e	f	g	h
1		×	×	×	×			
2	×	×	×				×	×
3	×	×				×	×	×
4				×	×			
5			×	×				
6	×							×

The associated concept lattice $\mathcal{L}(R)$ is shown in Figure 1.

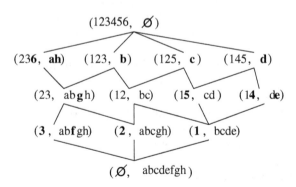

Fig. 1. Concept lattice $\mathcal{L}(R)$ of Example 1. For example, $(12, bc)$ denotes concept $\{1, 2\} \times \{b, c\}$

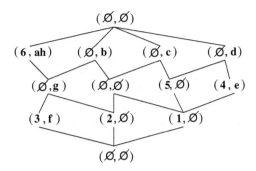

Fig. 2. Simplified labelling of Lattice of Figure 1, first step to generate Galois sub-hierarchy of Figure 3

Definition 1. *An object (resp. a property) x is said to be* introduced *by a concept if x is in the extent (resp. intent) of this concept and no ancestor (resp. descendant) of this concept contains x in its extent (resp. intent). An element of the lattice is said to be an* **introducer** *if it introduces a property or an object.*

In [9], an introducer of an object is called an 'Object Concept' and an introducer of a property is called a 'Property Concept'.

Definition 2. *([11]) The* **Galois sub-hierarchy** *(GSH) of a concept lattice is the partially ordered set of elements $X \times Y$, $X \cup Y \neq \emptyset$, such that there exists a concept where X is the set of objects introduced by this concept and Y is the set of properties introduced by this concept. The ordering of the elements in the GSH is the same as in the lattice.*

Let E_1 and E_2 be two elements of the GSH. We will denote $E_1 < E_2$ if E_1 is an ancestor (i.e. represented below in the GSH) of E_2 and $E_1 \leq E_2$ if $E_1 = E_2$ or $E_1 < E_2$.

The **Attribute Object Concept Poset** (AOC-Poset) is a GSH with addition of the top and bottom elements of the lattice, if they are not present in the GSH ([16]).

Example 2.
For example, $23 \times abgh$ is the introducer of g, as the descendants of this concept: $236 \times ah$, $123 \times b$ and $123456 \times \emptyset$ do not have g in their intent.

Simplifying the labeling of Figure 1 to the introduced objects and properties leads to the lattice represented in Figure 2. Then, removing trivial nodes and nodes labeled with empty sets leads to the Galois sub-hierarchy represented in Figure 3.

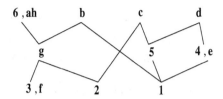

Fig. 3. Galois sub-hierarchy of Example 1, with the simplified labeling: b, for example, corresponds to $(\emptyset, \{b\})$. This hierarchy is obtained by removing trivial nodes from simplified lattice of Figure 2

3   Previous Results and Algorithmic Process

In order to explain our algorithm, we will need some extra notions which we introduced in previous papers to improve concept-related algorithms.

First, we will need the notion of *maxmod*, which is a set of objects or properties which share the same row or column. These are useful, because when generating concepts using Bordat's algorithm for example, the elements of a maxmod of the relation or subrelation which is used are always grouped together in the concepts. Note that if the relation is reduced ([1]), all maxmods are of cardinality one; however, reducing a relations costs $O(|\mathcal{O} + \mathcal{P}|^3)$ time, whereas our algorithm is in $O(|\mathcal{O} + \mathcal{P}|^2)$, so we cannot assume that the relation is reduced.

Definition 3. *([2]) Let $(\mathcal{O}, \mathcal{P}, R)$ be a context. For $x, y \in \mathcal{O}$ or $x, y \in \mathcal{P}$, x and y belong to the same* **maxmod** *if $R[x] = R[y]$. We refer to maxmods as* object maxmods *or* property maxmods, *according to the nature of their elements.*

The maxmods define a partition of $\mathcal{O} + \mathcal{P}$. The GSH, with its simplified labeling, contains exactly the maxmods of $\mathcal{O} + \mathcal{P}$. For any maxmod M, we will denote $E(M)$ the element of the GSH which contains M in its label.

Example 3. For example, $\{a, h\}$ defines a maxmod, as a and h have identical columns in the matrix of R. a and h always appear together, both in the lattice and in the GSH. The set of maxmods is:
$\{\{a, h\}, \{b\}, \{c\}, \{d\}, \{e\}, \{f\}, \{1\}, \{2\}, \{3\}, \{4\}, \{5\}, \{6\}\}$.
These are exactly the labels of the GSH of Figure 1.

Another important notion, which stems from Graph Theory, is that of *domination*:

Definition 4. *([2]) Let $(\mathcal{O}, \mathcal{P}, R)$ be a context. For $x, y \in \mathcal{O}$ or $x, y \in \mathcal{P}$, x is said to* **dominate** *y if $R[x] \subseteq R[y]$. A maxmod X is said to dominate another maxmod Y if for $x \in X$, for $y \in Y$, x dominates y.*

In the rest of this paper, we will often use maxmods, and denote them informally without columns, for example we will use ah instead of $\{a, h\}$; matrices

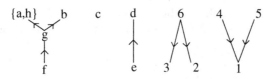

Fig. 4. Domination order of Example 1. For example, we see that e dominates d and f dominates b, whereas 6 dominates 2

will be represented with maxmods instead of lone objects or properties, which is equivalent to keeping in the matrix only one representative for every set of lines which are identical.

The notion of domination leads us to a decomposition of the GSH into an object-GSH and a property-GSH, as introduced in [3]. We will use the following simple terms in referring to this:

Definition 5. *([3]) Let $(\mathcal{O}, \mathcal{P}, R)$ be a context. The partial order on the property maxmods is called the* property domination order, *and likewise the partial order on the object maxmods is called the* object domination order.

Note that both partial orders on objects and properties are compatible with the GSH:

Theorem 1. *([3]) Let P_1 and P_2 be two property maxmods ; then P_1 dominates P_2 iff $E(P_1) < E(P_2)$. Let O_1 and O_2 be two object maxmods, then O_1 dominates O_2 iff $E(O_2) < E(O_1)$.*

Example 4. Figure 4 shows the domination orders on our example.

We showed in [4] that a linear extension of the object or property ordering can be very efficiently computed in linear time, using another Graph Theory tool: partition refinement. The partition algorithm is given in Section 4.

Our aim here is to compute a linear extension of the GSH. Since the object and property orderings are preserved in the GSH, both the corresponding linear extensions are compatible with the GSH. Our idea here is to use the object linear extension and the property linear extension and merge them into a linear extension of the GSH.

4   The Algorithm

We present an algorithm which takes a context as input, and outputs a linear extension of the GSH.

Our algorithm uses as a primitive a partition refinement technique (based on a process presented in [4]), called Algorithm MAXMOD-PARTITION, which computes a linear extension of the domination order. We use this primitive twice in our main algorithm (Algorithm Tom Thumb): the first time, we use Algorithm MAXMOD-PARTITION with an arbitrary ordering L on $\mathcal{P}$; the second time, we use in the input the ordering on $\mathcal{O}$ output by the first pass of Algorithm MAXMOD-PARTITION.

Algorithm MAXMOD-PARTITION
Input: A context $(\mathcal{O}, \mathcal{P}, R)$, a set S which is either $\mathcal{O}$ or $\mathcal{P}$, and an ordered partition L of $(\mathcal{O} + \mathcal{P}) - S$.
Output: An ordered partition of the maxmods of S.
PART is a queue, initialized with S;
for each class of partition L taken in the input ordering **do**
    choose a representative x of the class;
    for each class K of PART such that $|K| > 1$ **do**
        $K' \leftarrow K \cap R[x]$;
        $K'' \leftarrow K - R[x]$;
        if $K' \neq \emptyset$ and $K'' \neq \emptyset$ **then**
            In PART, replace K by K' followed by K'';
return PART.

Note that the execution can be stopped if all classes trivially contain a single element.

This process has the remarkable property that a given maxmod taken in the output partition can dominate only maxmods which lie to its left:

Theorem 2. *([4]) Algorithm* MAXMOD-PARTITION *outputs an ordered list of maxmods such that if maxmod A dominates maxmod B, then B is before A in this list.*

Example 5. Let us apply Algorithm MAXMOD-PARTITION to our running example, using $S = \mathcal{O}$, L is the total ordering (a, b, c, d, e, f, g, h) of $\mathcal{P}$ (each partition is trivially formed of a single element).

$(\{1,2,3,4,5,6\})$
    $\downarrow$ $R[a] = \{2,3,6\}$
$(\{2,3,6\}; \{1,4,5\})$
    $\downarrow$ $R[b] = \{1,2,3\}$
$(\{2,3\}; \{6\}; \{1\}; \{4,5\})$
    $\downarrow$ $R[c] = \{1,2,5\}$
$(\{2\}; \{3\}; \{6\}; \{1\}; \{5\}; \{4\})$.

Our algorithm for computing a linear extension of the GSH first uses Algorithm MAXMOD-PARTITION to compute a linear extension of the object maxmods, using any ordering of the property set, then uses the output ordering on object maxmods to find a special linear extension of the property maxmods. In a third step, the algorithm will merge the two extensions. The result is a list LINEXT of maxmods, which represent a linear extension of the GSH, in which for any element of the GSH formed by both an object maxmod and a property maxmod, these appear consecutively.

Algorithm TOM THUMB
Input: A context $(\mathcal{O}, \mathcal{P}, R)$.
Output: A linear extension of the GSH, where object maxmods and property maxmods are separated (if an element of the GSH contains an object maxmod O and a property maxmod P then P and O appear consecutively in the linear extension).

1. Apply MAXMOD-PARTITION to $(\mathcal{O}, \mathcal{P}, R)$, with $S = \mathcal{O}$ and L an arbitrary ordering of $\mathcal{P}$, resulting in an ordered partition $(Y_1, ..., Y_q)$ of **object** maxmods;
2. Apply MAXMOD-PARTITION, with $S = \mathcal{P}$, and using for L partition $(Y_q, ..., Y_1)$, i.e. the partition obtained at step one in reverse order, resulting in an ordered partition $(X_1, ..., X_r)$ of **property** maxmods;
3. LIST is an empty queue.
   $j \leftarrow q$; $i \leftarrow 1$;
   while j>0 and i≤r **do**
        choose representatives: $y \in Y_j$ and $x \in X_i$;
        if $(y, x) \in R$ **then** add X_i to queue LIST; $i \leftarrow i+1$;
        else add Y_j to queue LIST; $j \leftarrow j-1$;
   // *At this point, there may remain either objects or properties, which are added to* LIST
   while j>0 **do** add Y_j to queue LIST; $j \leftarrow j-1$;
   while i≤r **do** add X_i to queue LIST; $i \leftarrow i+1$;
   LINEXT $\leftarrow$ LIST in reverse; **Return** LINEXT.

Note that LIST computes a linear extension which is in reverse order with respect to the orientation of the GSH we have chosen for this paper, as illustrated in our running example. LIST thus has to be output in reverse as LINEXT. In the final LINEXT output, the list of properties output by MAXMOD-PARTITION is reversed, while the list of objects is preserved.

Example 6. An execution of Algorithm TOM THUMB on Relation R from Example 1.

Step 1: Partition the object set, using any ordering on the property set. With ordering (a, b, c, d, e, f, g, h) used in Example 5, (2,3,6,1,5,4) is obtained.

Step 2: Partition property set using objects ordering (4,5,1,6,3,2).
({a,b,c,d,e,f,g,h})
      ↓   R[4]={d,e}
({d,e}; {a,b,c,f,g,h})
      ↓   R[5]={c,d}
({d}; {e}; {c,}; {a,b,f,g,h})
      ↓   R[1]={b,c,d,e}
({d}; {e}; {c,}; {b}; {a,f,g,h})
      ↓   R[6]={a,h}

$(\{d\}; \{e\}; \{c,\}; \{b\}; \{a,h\}; \{f,g\})$
 $\downarrow$ $R[3]=\{a,b,f,g,h\}$ let the partition unchanged
 $\downarrow$ $R[2]=\{a,b,c,g,h\}$
$(\{d\}; \{e\}; \{c,\}; \{b\}; \{a,h\}; \{g\}; \{f\})$

Step 3: Merge ordered partitions (4,5,1,6,3,2) and (d,e,c,b,ah,g,f).
LIST=();
 current object: 4, current property: d, $(4, d) \in R$, d is chosen next,
LIST=(d);
 current object: 4, current property: e, $(4, e) \in R$, e is chosen next,
LIST=(d; e);
 current object: 4, current property: c, $(4, c) \notin R$, 4 is chosen next,
LIST=(d; e; 4);
 current object: 5, current property: c, $(5, c) \in R$, c is chosen next,
LIST=(d; e; 4; c);
 current object: 5, current property: b, $(5, b) \notin R$, 5 is chosen next,
LIST=(d; e; 4; c; 5);
 current object: 1, current property: b, $(1, b) \in R$, b is chosen next,
LIST=(d; e; 4; c; 5; b);
 current object: 1, current properties: ah, $(1, a) \notin R$, 1 is chosen next,
LIST=(d; e; 4; c; 5; b; 1);
 current object: 6, current properties: ah, $(6, a) \in R$, ah is chosen next,
LIST=(d; e; 4; c; 5; b; 1; ah);
 current object: 6, current property: g, $(6, g) \notin R$, 6 is chosen next,
LIST=(d; e; 4; c; 5; b; 1; ah; 6);
 current object: 3 current property: g, $(3, g) \in R$, g is chosen next,
LIST=(d; e; 4; c; 5; b; 1; ah ;6 ;g);
 current object: 3 current property: f, $(3, f) \in R$, f is chosen next,
LIST=(d; e; 4; c; 5; b; 1; ah; 6; g; f);
 no property left, add objects 3 then 2,
LIST=(d; e; 4; c; 5; b; 1; ah; 6; g; f; 3; 2).
The resulting list of maxmods is $(d; e; 4; c; 5; b; 1; ah; 6; g; f; 3; 2)$.
In the GSH of Example 2, ah and 6 are in the same element, and so are e and
4, as well as f and 3.
The output LINEXT represents linear extension
$(\{2\}, \{3, f\}, \{g\}, \{6, ah\}, \{1\}, \{b\}, \{5\}, \{c\}, \{4, e\}, \{d\})$
of the GSH.

Note that, as usual, $\mathcal{O}$ and $\mathcal{P}$ could be interchanged in the algorithms above.

Time Complexity:
Algorithm MAXMOD-PARTITION can be implemented to run in $O(|R|)$ time,
thus Step 1 and Step 2 cost $O(|R|)$. In Step 3, each time we enter the while-loop
$i + (q - j)$ is incremented; as $i + (q - j) \leq r + q \leq |\mathcal{P}| + |\mathcal{O}|$, Step 3 costs
$O(|\mathcal{P}| + |\mathcal{O}|)$. Algorithm TOM THUMB has then a complexity in $O(|R|)$.

Algorithm Tom Thumb runs in linear time, but does not explicitly compute the elements of the GSH. Computing these elements, using a brute-force approach, costs $O((|\mathcal{P}| + |\mathcal{O}|)^3)$ time, and no better process is known for this. However, in order to compute these elements, we could use the output of Algorithm Tom Thumb in the following fashion: if an element of the GSH has both a non-empty extent and a non-empty intent, call them O and P, then in list LINEXT output by the algorithm, P comes first and O is just after P; thus we need to test all pairs of the list where an object maxmod immediately follows a property maxmod, to find out whether they together form an element of the GSH. This test can be performed by taking an element o_1 in O and an element p_1 in P; (O, P) is an element of the GSH iff $R[o_1] \times R[p_1]$ is a rectangle, i.e. the corresponding cartesian product is a subset of R. The test costs $O((|\mathcal{P}| + |\mathcal{O}|)^2)$ time for each pair which is tested. The overall cost of computing all the elements of the GSH may become lower than $O((|\mathcal{P}| + |\mathcal{O}|)^3)$ in some cases.

Algorithm Tom Thumb turns out to have a variety of interesting properties, due to the fact that it computes a very special linear extension of the GSH, as Theorem 3 for which will will define the notion of 'staircase':

Definition 6. *Let $(\mathcal{O}, \mathcal{P}, R)$ be a context. Let $\alpha = (o_1, ..., o_q)$ be a total ordering of $\mathcal{O}$ and $\beta = (p_1, ..., p_r)$ be a total ordering of $\mathcal{P}$. Let M be the matrix of R, with the rows ordered by α and the columns ordered by β.*

We will say that M has a lower-right-hand staircase of zeroes *if there exist an total function φ from an interval $[o_h, o_q]$ of α to $\mathcal{P}$ such that:*

- *for o_i and o_j objects of $[o_h, o_q]$, if $o - i$ before o_j in α then $\varphi(o_i)$ is after $\varphi(o_j)$ in β, and*
- *for each i in $[h, q]$, the rectangle $S_i = \{o_i\} \times [\varphi(o_i), p_r]$ is a rectangle of zeroes (i.e. $\forall y \in [\varphi(o_i), p_r], M[o_i, y] = 0$).*

We will say that the union Z_2 of all these rectangles S_i of zeroes is a lower-right-hand staircase of zeroes *of M. We will denote by Z_1 the other part $M - Z_2$ of M.*

Theorem 3. *Using an arbitrary ordering α of $\mathcal{O}$ to compute with Algorithm* MAXMOD-PARTITION *an ordered partition α of the maxmods of $\mathcal{P}$ (as in Step 2 of the Tom Thumb Algorithm), and reordering the rows and columns of the matrix with α in reverse and β results in a matrix which has a lower-right-hand staircase of zeroes.*

The proof follows the definition of the partition process: at each step, 'ones' are put to the left, and 'zeroes' are left at the right; this is repeated one step higher, to partition the zone above the previous zeroes zone; this results in a staircase of zeroes, in any matrix defined in this fashion.

Proof. Let $(\mathcal{O}, \mathcal{P}, R)$ be a context. Let $\alpha = (o_1, ..., o_q)$ be a total ordering of $\mathcal{O}$ and $\beta = (p_1, ..., p_r)$ be a total ordering of $\mathcal{P}$. Let M be the matrix of R with the rows ordered by α in reverse: $(o_q, ..., o_1)$ and the columns ordered by $\alpha = (p_1, ..., p_r)$.

At the first step of the algorithm, if $\overline{R}[o_1] \neq \emptyset$, set $\mathcal{P}$ will be split into $K' = R[o_1]$ and $K'' = \overline{R}[o_1]$; thus $\forall y \in K''$, $M[o_1, y] = 0$, and K'' is the righmost class of PART at the end of step 1. If $\overline{R}[o_1] = \emptyset$, let $k = 0$; note that this only occurs if $R[o_1] = \mathcal{P}$. Suppose that at the end of step i of the algorihm, the rightmost class of PART is $S_i = \bigcap_{j=1}^{i} \overline{R}[o_i]$. If, at step $i+1$, $\overline{R}[o_{i+1}] \neq \emptyset$, $S_{i+1} = \bigcap_{j=1}^{i+1} \overline{R}[o_k]$ will be the new rightmost class in PART. If $\overline{R}[o_{i+1}] = \emptyset$, let $k = i$.

Thus we recursively construct a list $\{o_1\} \times S_1, ..., \{o_k\} \times S_k$ of rectangles of zeroes, with inclusions $S_1 \supseteq ... \supseteq S_k$. This will define the successive elements of the ordered partition on $\mathcal{P}$, which w.l.o.g. are: $R[o_1], R[o_2] \cap S_1, ... , R[o_k] \cap S_{k-1}, S_k$.

We can thus define function φ from $[o_1, o_k]$ to $\mathcal{P}$ by $\varphi(o_i) = S_i$ with the different rectangles of zeroes S_i of Definition 6. Thus M has a lower-right-hand staircase of zeroes $Z_2 = \cup_{i=1}^{k}(\{o_i\} \times S_i)$. Note that the use of a reverse ordering on $\mathcal{O}$ makes the object indices different from those of Definition 6.

Example 7. Let us use (2,3,6,1,5,4) and (d,e,c,b,ah,g,f) as output by Steps 1 and 2 of the execution of Algorithm Tom Thumb of Example 6. The resulting matrix is:

R	d	e	c	b	ah	g	f
2			×	×	×	×	
3				×	×	×	×
6					×		
1	×	×	×	×			
5	×		×				
4	×	×					

The lower-right-hand part of the matrix contains only zeroes. The limit of this zone is defined by the succession of queries on R given in Step 3 of the Tom Thumb Algorithm: $(x, y) \notin R$, $(4, c) \notin R$, $(5, b) \notin R$, $(1, a) \notin R$, $(6, g) \notin R$. Thus $Z_2 = \{4\} \times \{c, b, ah, g, f\} \cup \{5\} \times \{b, ah, g, f\} \cup \{1\} \times \{ah, g, f\} \cup \{6\} \times \{g, f\}$.

5   Proof of the Algorithm

The TOM THUMB Algorithm does a traversal of the GSH in a such fashion that an element of the GSH is reached – and then put in the list – only after all its descendants in the GSH are reached:

Theorem 4. *Algorithm* TOM THUMB *gives a linear extension of the GSH, where object maxmods and property maxmods are separated: if a property maxmod P and an object maxmod O are in the same element of the GSH then O appears just before P in the linear extension.*

In order to prove this, we will use Theorem 3, as well as the following lemmas.

Lemma 1. *In the course of the execution of Algorithm Tom Thumb,*

1. *when a* property *maxmod P is added to* List, *then for any* object *maxmod O added after P, (O, P) is in Z_1.*
2. *when an* object *maxmod O is added to* List, *then for any* property *P added after O, (O, P) is in Z_2.*

Proof. Let $(\mathcal{O}, \mathcal{P}, R)$ be a context. Let $\alpha = (O_1, ..., O_q)$ be the partition in object maxmods obtained at the first step of Algorithm Tom Tumb, let $\beta = (P_1, ..., P_r)$ be the partition in property maxmods obtained at the second step of Algorithm Tom Tumb, using α in reverse: $(O_q, ..., O_1)$, as ordered partition L in the input. Let M be the matrix of R with rows in order α and columns in order β. By Theorem 3, M has a lower-right-hand staircase of zeroes.

1. Let O be an object maxmod and let P be a property maxmod such that O is after P in List. If P and O have been compared in the third step of Algorithm Tom Tumb, then $(O, P) \in R$ and (O, P) may not be in Z_2. On the other hand, if P has been inserted in List without being compared to O, this means there exists another object maxmod O' which is after O in α, which has been compared to P, and which has been put after P as $(O', P) \in R$. Thus $(O', P) \in Z_1$ and, by Definition 6 of M, (O, P) is in Z_1.
2. We will prove by induction that for each object maxmod O_i of α, all the properties P_j put after O_i in List verify $(O_i, P_j) \in Z_2$.
   The first step of Algorithm Tom Thumb begins by comparing O_q and P_1. All the property maxmods of $R[O_q]$ are put before O_q in List, as for $P_j \in R[O_q]$, $(O_q, P_j) \in R$. The property maxmods of $\overline{R}[O_q]$ will be put after O_q in List and will constitute the first 'step' S_1 of the lower-right-hand staircase in matrix M. Thus for all $P_j \in \overline{R}[O_q]$, $(O_1, P_j) \in Z_2$.
   Suppose that for object maxmod O_i, all the property maxmod P_j that are after O_i in List verify $(O_i, P_j) \in Z_2$.
   When O_{i-1} is processed, the set $\mathcal{B}$ of property maxmods which have yet to be processed can be split into $\mathcal{B} \cap R[O_{i-1}]$ and $\mathcal{B} \cap \overline{R}[O_{i-1}]$. Then, the property maxmods of $\mathcal{B} \cap R[O_{i-1}]$ will be put before O_{i-1} in List and these of $\mathcal{B} \cap \overline{R}[O_{i-1}]$ will be put after O_{i-1}. By Definition 6, $O_{i-1} \times (\mathcal{B} \cap \overline{R}[O_{i-1}])$ is an element of the staircase of zeroes of matrix M. Thus, all P_j in $\mathcal{B} \cap \overline{R}[O_{i-1}]$ put after O_{i-1} in List will verify $(O_{i-1}, P_j) \in Z_2$.

Lemma 2. *Let O be an object maxmod associated with element $E(O)$ of the GSH, let P be a property maxmod associated with $E(P)$; then $(O, P) \in R$ iff $E(O) \leq E(P)$.*

Proof.

$\Rightarrow$ This follows directly from the definitions of concept lattice and GSH: the introducer of P is concept $R[P] \times R[R[P]]$. If $(O, P) \in R$, O is in this concept and in all its predecessors, one of which is the introducer of O. Then $E(O) \leq E(P)$ if $(O, P) \in R$.

$\Leftarrow$ If $E(O) \leq E(P)$, the introducer of P is a descendant of the introducer of O and thus will have O in its extent. Then $(O, P) \in R$.

Theorem 5. *For any pair (O, P) of maxmods such that $(O, P) \notin R$ and $(O, P) \in Z_1$, O and P belong to non-comparable elements $E(O)$ and $E(P)$ of the GSH.*

Proof. Let (O_1, P_1) be a zero in Z_1. Suppose by contradiction that P_1 and O_1 belong to comparable elements of the GSH.

Since $(O_1, P_1) \notin R$, by Lemma 2, we must have $E(P_1) < E(O_1)$. Let us consider the moment when P_1 is added to LIST: by Lemma 1, since $(O_1, P_1) \in Z_1$, O_1 has not yet been added. Let O_2 be the object which is queried by Step 3 of Algorithm Tom Thumb, and which results in adding P_1 to LIST, let $E(O_2)$ be the element of the GSH which contains O_2: we have $(O_2, P_1) \in R$. Clearly, the algorithm will insert O_2 after P_1 in LIST; by Lemma 2, $E(O_2) \leq E(P_1)$.

Combining the above remarks together, we obtain $E(O_2) \leq E(P_1) < E(O_1)$, so by Theorem 1, O_1 dominates O_2. But this is impossible, by Theorem 2, as the ordering used ensures that O_1 can dominate only objects which are output before it by Algorithm MAXMOD-PARTITION; since this ordering is used in reverse by Step 3 of Algorithm Tom Thumb, O_2, which is used first, cannot be dominated by O_1 - a contradiction.

We are now ready to prove Algorithm TOM THUMB:

Proof. (of Theorem 4) Let A and B be two different maxmods, belonging to elements $E(A)$ and $E(B)$ of the GSH respectively. There will be two cases:

- Suppose that $E(B) < E(A)$. We will show that B is placed before A in LINEXT, which is equivalent to saying that A is placed before B in LIST.
  1. If A and B are both property maxmods: by Theorem 1, B dominates A. Clearly, the ordering output by Step 2 of Algorithm Tom Thumb is preserved in LIST. By Theorem 2, A is before B in LIST.
  2. If A and B are both object maxmods: by Theorem 1, A dominates B. The ordering output by Step 1 of Algorithm Tom Thumb is reversed in LIST. By Theorem 2, A is before B in LIST.
  3. If A is a property maxmod and B is an object maxmod: as $E(B) < E(A)$, by Lemma 2, $(B, A) \in R$, which implies $(B, A) \notin Z_2$. By contraposition of the second item of Lemma 1, A must be before B in LIST.
  4. If A is an object maxmod and B is a property maxmod: as $E(B) < E(A)$, by contraposition of Lemma 2, $(A, B) \notin R$. By Theorem 5, $(A, B) \notin Z_1$, as $E(A)$ and $E(B)$ are comparable. Finally, by contraposition of the first item of Lemma 1, object maxmod A is before property maxmod B in LIST.
- Suppose that $E(B) = E(A)$. This corresponds to the case where object maxmod A and property maxmod B appear in the same element of the GSH. We will show that B is placed after A in LINEXT, which is equivalent to saying that A is placed after B in LIST.

If $E(A) = E(B)$ then, by Lemma 2, $(A, B) \in R$ and thus $(A, B) \notin Z_2$. By contraposition of the second item of Lemma 1, object maxmod A is after property maxmod B in LIST. Finally, as a direct consequence of the precedent case, no other maxmod will appear between A and B in LIST.

6   Conclusion

In this paper, we present a new algorithm to efficiently compute an ordering on the object and property maxmods which is compatible with a linear extension of the Galois sub-hierarchy.

It turns out that the linear extension we compute has very special properties, which will require further investigation, both as useful for dealing with Galois sub-hierarchies, and as interesting in the more general context of handling a binary relation. For example, the way the algorithm traverses the sub-hierarchy is interesting, as well as the definition of some zones of the matrix with non-comparable elements.

Moreover, the family of reorderings computed by our Tom Thumb algorithm turns out to often lead to cases where concept generation can be accomplished faster than in the general case. Experimentation on this is being pursued.

References

1. Barbut M., Monjardet B.: *Ordre et classification*. Classiques Hachette, (1970).
2. Berry A., Sigayret A.: Representing a concept lattice by a graph. *Proc. Discrete Maths and Data Mining Workshop, 2nd SIAM Conference on Data Mining (SDM'02), Arlington (VA, USA)*, (April 2002). *Discrete Applied Mathematics, special issue on Discrete Maths and Data Mining*, **144(1-2)** (2004) 27–42.
3. Berry A., Sigayret A.: Maintaining class membership information. *Advances in Object-Oriented Information Systems - OOIS 2002 Workshops*, LNCS, **2426** (Sept. 2002) 13–23.
4. Berry A., Bordat J-P., Sigayret A.: Concepts can't afford to stammer. *INRIA Proc. International Conference "Journées de l'Informatique Messine" (JIM'03), Metz (France)*, (Sept. 2003). Submitted as 'A local approach to concept generation.'
5. Birkhoff G.: *Lattice Theory*. American Mathematical Society, 3rd Edition, (1967).
6. Dao M., Huchard M., Leblanc H., Libourel T., Roume C.: A New Approach of Factorization : Introducing Metrics. *Proc. 8th IEEE international Software Metrics Symposium (METRICS'02), Ottawa (Canada)*, (June 2002) 227–236.
7. Dao M., Huchard M. Rouane Hacene M., Roume C., Valtchev P.: Improving Generalization Level in UML. *LNCS Proc. 12th International Conference on Conceptual Structures (ICCS'04)*, (2004) 346–360.
8. Dicky H., Dony C., Huchard M., Libourel T.: ARES, Adding a class and RESstructuring Inheritance Hierarchies. Proc. 11me *Journées Bases de Données Avancées*, Nancy (France), (1995) 25–42.
9. Ganter B., Wille R.: *Formal Concept Analysis*. Springer, (1999).
10. Godin R., Chau T. T.: Comparaison d'algorithmes de construction de hiérarchies de classes. *L'Objet*, **5(3)** (2000) 321–338.

11. Godin R., Mili H.: Building and Maintaining Analysis-Level Class Hierarchies Using Galois Lattices. *Proc. OOPSLA'93, Washington (DC, USA)*, Special issue of Sigplan Notice, **28(10)** (1993) 394–410.
12. Godin R., Mili H, Mineau G., Missaoui R., Arfi A., Chau T. T.: Design of Class Hierarchies Based on Concept (Galois) Lattices. *Theory and Practice of Object Systems.* **2(4)** (1998) 117–134.
13. Hitzler P.: Default Reasoning over Domains and Concept Hierarchies. *LNAI Proc. 27th German Conference on Artificial Intelligence, (KI'04), Ulm (Germany).* (Sept. 2004).
14. Huchard M., Dicky H., Leblanc H.. Galois lattice as a framework to specify building class hierarchies algorithms. *Theoretical Informatics and Applications*, EDP Science ed., **34** (Jan. 2000) 521–548.
15. Huchard M., Leblanc H.: From Java classes to Java interfaces through Galois lattices. *Proc. 3rd International Conference on Orders, Algorithms and Applications (Ordal'99)*, Montpellier (France), (1999) 211–216.
16. Osswald R., Pedersen W.: Induction of Classifications from Linguistic Data. *Proc. ECAI-Workshop on Advances in Formal Concept Analysis for Knowledge Discovery in Databases (FCAKDD'02), Lyon (France)*, (July 2002).
17. Pedersen W.: A Set-Theoretical Approach for the Induction of Inheritance Hierarchies. *Electronic Notes in Theoretical Computer Science*, **51** (July 2001).
18. Spinrad J. P.: Doubly Lexical Orderings of Dense 0-1 Matrices. *Information Processing Letters*, **45** (1993) 229-235.
19. UML 2.0 superstructure Specification. *OMG Final Adopted Specification ptc/03-08-02*. URL http://www.omg.org/
20. Yahia A., Lakhal L., Cicchetti R., Bordat J-P.: iO2, An Algorithmic Method for Building Inheritance Graphs in Object Database Design *Proc. 15th Conf. on Conceptual Modeling (ER'96), Cottbus (Germany)*, LNCS **1157** (1996) 422–437.

A Generic Algorithm for Generating Closed Sets of a Binary Relation

Alain Gély

LIMOS, Université Blaise Pascal,
Campus des Ceźeaux, 63170 Aubière Cedex,
France
gely@isima.fr

Abstract. In this paper we propose a "divide and conquer" based generating algorithm for closed sets of a binary relation. We show that some existing algorithms are particular instances of our algorithm. This allows us to compare those algorithms and exhibit that the practical efficiency relies on the number of invalid closed sets generated. This number strongly depends on a choice function and the structure of the lattice. We exhibit a class of lattices for which no invalid closed sets are generated and thus reduce time complexity for such lattices. We made several tests which illustrate the impact of the choice function in practical efficiency.

Keywords: Generation algorithm, Closure operator, Galois or concept lattice.

1 Introduction

Closure systems or lattices are mathematical structures which are used for many applications in computer science and in particular, for discovering information in databases [1, 2].

In this paper, we give a generic divide and conquer algorithm which generates closed sets of a given binary relation. The main idea of this approach is to generate closed sets which contain an element a, then closed sets which do not contain a. This approach is then recursively applied. Moreover, we show that existing algorithms, in the spirit of generating algorithms, are particular instances of this algorithm, by only modifying the choice function. Time complexity of our algorithm is the same as that of Ganter's algorithm [5]. However, our algorithm has some extra features which can be of interest: It is simple, general, well suited for particular classes of lattices and well suited for distributed computation. It can be used as a basis for comparing different algorithms or to conceive new ones, using polynomial space.

The aim of this paper is to generate all the closed sets associated to a binary relation. This is a classical problem which could be equivalently formulated as "Generate all the maximal bicliques of a bipartite graph" or "Generate all the maximal 1-rectangles of a Boolean matrix". Number of closed sets of a binary relation may be exponential in the size of the relation.

B. Ganter and R. Godin (Eds.): ICFCA 2005, LNAI 3403, pp. 223–234, 2005.

Complexity of a generating algorithm is based on the size of the input as well as the size of the output (i.e. the number of closed sets). In this paper, when time complexity is given without mention of the output size, we talk about the time complexity *per closed set*.

We only want to generate all the closed sets, and not build the lattice structure. As we do not need to store all the closed sets, we do not allow an exponential use of memory. Thus, the algorithm presented here uses, at worst, a polynomial space in the size of the binary relation.

To describe our algorithm, we first recall some notations we will use in this paper. For definitions and proofs not given here, we refer to [7].

In this paper, we consider a binary relation $R = (J, M, I)$ where elements of J are usually called the objects, those of M the attributes of R. For $j \in J$, $m \in M$ one has jIm iff there exists an edge between j and m, i.e. if the object j possesses the attribute m. We denote usually the attributes by numbers, and the objects by small letters.

Recall Galois connection is given by the following derivation operators:

- For a set $A \subseteq J$, $A' = \{m \in M \mid jIm \text{ for all } j \in A\}$.
- For a set $B \subseteq M$, $B' = \{j \in J \mid jIm \text{ for all } m \in B\}$.

We consider the classical closure operator $'' : 2^J \to 2^J$, defined by applying twice the derivation operators. Let X be a set, then X'' is the closure of X.

2   A Generic Algorithm

The main idea of our algorithm is to generate closed sets which contain element a and closed sets which do not contain a. And then apply this principle recursively. However, input data of our algorithm is a binary relation R. Main difficulty is thus to determine two sub-relations R_1 and R_2 of R such that closed sets containing a can be generated from R_1, while closed sets which do not contain a can be generated from R_2.

Let $\mathcal{F}(R)$ be the set of all closed subsets of J with respect to the binary relation $R = (J, M, <)$. Given $a \in J$, the closed sets of R may be decomposed according to whether they contain a or not.

- $\mathcal{F}_a(R) = \{F \in \mathcal{F}(R) \mid a \in F\}$, i.e. closed sets which contain a.
- $\mathcal{F}_{\bar{a}}(R) = \{F \in \mathcal{F}(R) \mid a \notin F\}$, i.e. closed sets which do not contain a.

Clearly $\mathcal{F}_a(R)$ is a closure system on J, but the set $\mathcal{F}_{\bar{a}}(R)$ may not be a closure system since it may not contain a top element (see Figure 1).

Since $\mathcal{F}_a(R)$ is a closure system on J, we can find a sub-relation $R_1 \subseteq R$ such that $\mathcal{F}(R_1) = \mathcal{F}_a(R)$. More formally $R_1 = (J, a', <)$ since a closed set X of R that contains a must satisfy $X' \subseteq a'$. Note that, in a sub-relation, $<$ denotes the restriction of $<$ in R for present elements in the sub-relation.

On the other hand, since $\mathcal{F}_{\bar{a}}(R)$ is not always a closure system, we need to find $R_2 \subseteq R$ such that $\mathcal{F}_{\bar{a}}(R) \subseteq \mathcal{F}(R_2)$. We define R_2 by $R_2 = (J \backslash \uparrow a, M, <)$

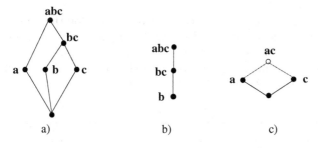

Fig. 1. A decomposition of the closure system in Figure 2 when b is chosen. Notice that the set $\{ac\}$ is not a closed set of R

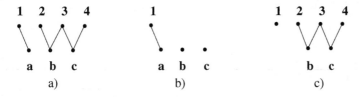

Fig. 2. a) A binary relation $R = (J, M, <)$, b) $R_1 = (J, a', <)$, c),$R_2 = (J \backslash \uparrow a, M, <)$

since a closed set of R which does not contain a must not contain any element of $\uparrow a$, where $\uparrow a = \{j \in J \mid j' \subseteq a'\}$.

Figure 2 illustrates this decomposition of a binary relation to sub-relations.

Definition 1. *Invalid closed set.*
Let R the initial context, and R_i a context derived from R by a sequence of transformations of R in R_1 or R_2. We say X is an invalid closed set if $X \in \mathcal{F}(R_i)$ but $X \notin \mathcal{F}(R)$.

We say that a set X is valid if $X \in \mathcal{F}(R)$. $\mathcal{F}(R_2)$ may contain more closed sets than $\mathcal{F}_{\overline{a}}(R)$. We denote by $NV(R) = \mathcal{F}(R_2) \backslash \mathcal{F}_{\overline{a}}(R)$ the set of closed sets of R_2 which are invalid closed sets in R. Thus the whole difficulty with our approach is first to determine from R the two sub-relations which contain the valid closed sets, and second to avoid generating too many invalid closed sets.

The following Lemma shows that if a set is invalid then all its upper sets are invalid.

Lemma 1. *Let $R(J, M, <)$ be a binary relation and $X \in NV(R)$. Then for all $Y \in \mathcal{F}(R_2)$ such that $X \subset Y$, one has $Y \in NV(R)$.*

Proof. Let $X \in NV(R)$ and X'' its closed set in R. Then $a \in X''$. Moreover, for all $Y \in \mathcal{F}(R_2)$: $X \subset Y$, and we have $X'' \subseteq Y''$. And thus, $a \in Y''$.
We conclude $Y \in NV(R)$. $\qquad\qquad\qquad\qquad\qquad\qquad\qquad\qquad\qquad\qquad\square$

Thus using Lemma 1, we can avoid generating too many invalid closed sets since we know that upper sets of an invalid closed sets are also invalid. We can thus deduce the following divide and conquer algorithm for the generation of closed sets. In the algorithms below, R is the binary relation considered, P is the set of elements of J which can be chosen to separate the closure system. So, when the algorithm runs from R to R_1, we delete a'' from P, and when the algorithm runs from R to R_2, we delete $\uparrow a$ from P.

Algorithm 1: Generating closed sets

> **Data**   : A binary relation $R = (J, M, <)$.
> **Result** : $\mathcal{F}(R)$
> **begin**
> > Output $\emptyset''$;
> > Closure(R, $J \setminus \emptyset''$);
>
> **end**

Algorithm 2: Closure(R,P)

> **begin**
> > **if** $P \neq \emptyset$ **then**
> > > 1   Choose an element $a \in P$;
> > > Compute a'' in the current relation R;
> > > **if** a'' *is valid* **then**
> > > > 2   Output a'';
> > > > $R_1 \leftarrow (J, a', <)$;
> > > > 3   Closure(R_1, $P \setminus a''$);
> > >
> > > $R_2 \leftarrow (J \setminus \uparrow a, M, <)$;
> > > 4   Closure(R_2, $P \setminus \uparrow a$);
>
> **end**

Theorem 1. *Algorithm 2 generates the closed sets of a binary relation $R = (J, M, <)$ using a polynomial space and $O(|J|^2 * |M|)$ time per closed set.*

Proof. We first explain how to construct Closure(R_i, P_i). If $P_i = \emptyset$ then our list of choices is empty and the procedure stops.

Suppose $P_i \neq \emptyset$ and let $a \in P_i$ be chosen. The procedure computes the closed set a'' in the binary relation R_i and tests if it is valid or not. It suffices to test if a'' is a closed set on the sub-relation $(J, M_i, <)$, where M_i is the set of attributes of the context R_i considered at the present time. It costs at most $O(|J|.|M|)$.

Indeed, the closure operator is applied only once, on the relation $R_i' = (J, M_i, <)$. Then, a closed set $X \subset J$ is valid if $\forall x \in J \setminus J_i$, we have $x \notin X$.

If a'' is valid, then the procedure computes R_1 and R_2 from R and calls Closure($R_1, P_i \setminus a''$) and Closure($R_2, P_i \setminus \uparrow a''$). It costs at most $O(|J|.|M|)$.

But if a'' is not valid then only R_2 is computed and only one call is realized with the same time complexity.

It remains to show that the number of invalid closed sets is bounded by $O(|J|.|\mathcal{F}(R)|)$.

Consider the computation tree corresponding to the procedure Closure(R, P). Notice that the computation tree is a binary tree since the procedure contains two recursive calls. To each invalid closed set there corresponds a node of degree 1 in the computation tree and to each valid closed set there corresponds a node of degree 2. It is known that for any binary tree the number of leaves is at most twice the number of nodes of degree 2. And, since the height of the search tree is at most $|J|$, then any path (containing only nodes of degree 1) from a leaf to a node of degree 2 contains at most $|J|$ invalid closed sets. We charge complexity of computing these invalid closed sets to the corresponding node of degree 2. Moreover, to any node of degree 2 we associate at most two leaves. We conclude that the number of invalid closed sets is at most $2.|J|.|\mathcal{F}(R)|$.

Actually, it is easy to see that the complexity can be expressed as a function of the number of valid and invalid closed sets: $O(|J||M|\mathcal{F}(R))+O(|J||M||NV(R)|)$.

As we have seen above $|NV(R)| \leq 2.|J|.\mathcal{F}(R)$, thus we obtain the classical complexity of $O(|J|^2.|M|)$ per closed set.

$\square$

The proof of Theorem 1 shows that time complexity of the generic algorithm deeply relies on the generation of invalid closed sets: In $O(|J|.|M|.\mathcal{F}(R))+O(|J|.|M|.|NV(R)|)$, the second term represents the impact of invalid closed sets.

Clearly, one can ask if the number of invalid closed sets is bounded by $K.|\mathcal{F}(R)|$, where K is a constant. We made several tests on different binary relations and all results seem to confirm this. However, experimental results cannot validate theoretical bounds and thus this question remains open. Answering this question may improve time complexity of our algorithm.

We now discuss two particular instances of this algorithm, in goal to deal with two approaches: The approach of Ganter's [5, 6], which uses a property on the label elements, and an approach like the one of Bordat's [4] (Or Lindig [10]) algorithms, which use lattice property (computation of the immediate successors of a closed set).

Other existing algorithms might be seen as particular instances of our generic algorithm. Indeed, depending on how element a is to be chosen in line 1 and the order of recursive calls, we can simulate the behaviour of well-known algorithms.

2.1   Example of Particular Instance: Ganter's Algorithm

In Ganter's algorithm [5], closed sets are generated in a lectic order as defined below:

Definition 2. *Let A, B be two subsets of a set of ordered elements. We say the set A is lectically smaller than the set B if the smallest element distinct in A and B belongs to B.*

More formally, $\exists i \in B \backslash A$ such that $A \cap \{1, 2, \ldots, i-1\} = B \cap \{1, 2, \ldots, i-1\}$

To generate closed sets in a lectic order using our algorithm, we first suppose that J is totally ordered. Choosing element a is then reduced to picking up the smallest element of J (line 1). Moreover, the two recursive calls are swapped in order to respect the lectic order definition (line 2 and 5). Ganter's algorithm thus looks like as follows:

Algorithm 3: Closure(R,P) (Ganter's instance)

---
    begin
        if $P \neq \emptyset$ **then**
1   Choose the smallest element $a \in P$;
            $R_2 \leftarrow (J\backslash \uparrow a,\ M,\ <)$;
2   Closure($R_2,\ P\backslash \uparrow a$);
            Compute a'';
            if a'' *is valid* **then**
3   Output a'';
4                $R_1 \leftarrow (J,\ a',\ <)$;
5   Closure($R_1,\ P\backslash a''$);
    end

One can note this version is not totally equivalent to the Next Closure algorithm: This version computes less invalid closed sets than the original one.

One of the main advantages of using Next Closure algorithm is that the only data needed are J and a closure operator. For Ganter's algorithm, the closure operator may be used as a black box. As a consequence, Next Closure does not use informations whose can be deduced from a particular closure operator. Here, we use such informations since from the binary relation, we deduce the partial order of J.

As the closed sets are generated in the lectic order, our version can be considered as a particular implementation of Next Closure.

It is possible to obtain a behaviour similar to the original one by replacing $R_2 = (J\backslash \uparrow a, M, <)$ by $R_2 = (J\backslash a, M, <)$. In this case, we do not use the additional knowledge about the partial order on J given to us by the binary relation. Number of closure computed is then equal to the number of closure computed by Next Closure.

Time complexity remains unchanged. However, even this implementation using extra-informations on the poset of J elements generates more invalid closed sets than an implementation based on Bordat's method (See tests in Section 4). A possible explanation will be given in the section dealing with choice functions.

2.2 Another Particular Instance: Bordat's Algorithm

Bordat's algorithm [4] realizes a depth first search of the closed sets lattice by choosing at each step element a such that a'' is minimal. In a similar way, Lindig [10] computes from a concept all the covers of this concept.

These algorithms have been designed to build the lattice. Here, we only want to generate the set of closed sets, without structural informations.

So, of course, implementation below is not strictly equivalent to the one of Bordat (or Lindig). The principle, nevertheless, is the same: go from a closed set to another closed set which covers it. Here, we use a bottom-up approach for generation problem where Bordat and Lindig used such an approach for the diagram-building problem.

Thus, to use a bottom-up approach, it is sufficient to modify the choice function in our algorithm to obtain the same output order as that of Bordat.

Algorithm 4: Closure(R,P) (Bordat's instance)

---

  begin
     if $P \neq \emptyset$ **then**
1   Choose $a \in P$ such that a'' is minimal;
         Compute a'';
         if a'' *is valid* **then**
2   Output a'';
3           $R_1 \leftarrow (J, a', <)$;
4   Closure(R_1, $P \backslash a''$);
         $R_2 \leftarrow (J \backslash \uparrow a, M, <)$;
5   Closure(R_2, $P \backslash \uparrow a$);
  end

Again, time complexity remains unchanged. However, this instance generates less invalid closed sets than Ganter's instance as shown by our tests (see Section 4).

It is easy to see it is also possible to generate closed sets in the lexicographic order, or, more generally, using arbitrary property on the labels of elements. We can also generate the closed sets using some structural properties on the poset J. In all cases, testing if a closed set is invalid may be viewed as a test of canonicity [9], since it is equivalent to say that a closed set is invalid if it contains some forbidden elements.

However, for all instances, same remark applies: practical efficiency of the algorithms depends on the number of invalid closed sets generated. Thus, a question arises: does there exist a class of lattices for which no invalid closed set is generated?

The next Section answers this question.

3   Decomposition of Closure Systems

We have previously seen that invalid closed sets might be generated since $\mathcal{F}_{\overline{a}}(R)$ is not a closure system. However, we show that whenever a is a $\vee$-prime element then $\mathcal{F}_{\overline{a}}(R)$ is a closure system. We first recall the definition of a $\vee$-prime element:

Definition 3. $a \in J$ *is* $\vee$-*prime if for all* $A, B \in \mathcal{F}(R)$, $a \notin A \cup B$ *implies* $a \notin (A \cup B)''$.

∨-prime elements were first introduced by G. Markowsky in [11] to characterize certain types of lattices. Here, we use property of decomposition of ∨-prime
elements to deduce a generation algorithm.

This leads to the following property:

Property 1. Let $a \in J$ be an element of $(J, \leq)$. Then

1. $\mathcal{F}_a(R)$ is a closure system.
2. $\mathcal{F}_{\bar{a}}(R)$ is a closure system iff a is ∨-prime.

Figure 3 illustrates this property.

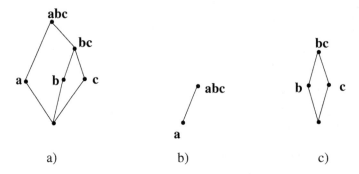

Fig. 3. A closure system and its decomposition based on the ∨-prime element a.
a)$\mathcal{F}(R)$, b)$\mathcal{F}_a(R)$, c) $\mathcal{F}_{\bar{a}}(R)$

As a consequence, using the same sub-relations R_1 and R_2 of R as defined in
Section 2, we obtain the following corollary:

Corollary 1. *Let $\mathcal{F}(R)$ be a closure system and a be a ∨-prime element. Then*

1. $\mathcal{F}_a(R) = \mathcal{F}(R_1)$
2. $\mathcal{F}_{\bar{a}}(R) = \mathcal{F}(R_2)$

Thus, whenever there exists a ∨-prime element, $\mathcal{F}(R)$ can be decomposed
in two closure systems $\mathcal{F}_a(R)$ and $\mathcal{F}_{\bar{a}}(R)$. If this property is recursively verified
on both closure systems, then $\mathcal{F}(R)$ is called ∨-separable. For instance, ∧-semi-
distributive lattices are ∨-separable [12].

Definition 4. *[12] A closure system $\mathcal{F}$ is called ∨-separable if one of the following condition is satisfied:*

1. $|\mathcal{F}| \leq 1$.
2. $\mathcal{F}$ contains a ∨-prime element a and both $\mathcal{F}_a(R)$ and $\mathcal{F}_{\bar{a}}(R)$ are ∨-separable.

As a consequence, for such class of lattices, a simplicial elimination scheme
can be used at each step by choosing a ∨-prime element and decomposing the
relation in two closure systems. Moreover, according to Corollary 1, no invalid

closed set will be generated. Thus to respect this elimination scheme it is sufficient to modify the choice function in our algorithm in order to choose at each step a $\vee$-prime element (line 1). Moreover, we can also drop the validity test.

Algorithm 5: Closure(R, P)

---

   begin
      if $P \neq \emptyset$ **then**

1   Choose a $\vee$-prime element $a \in P$;
2   Compute a'';
3   Output a'';
4          $R_1 \leftarrow (J,\ a',\ <)$;
5   Closure$(R_1,\ P \backslash a'')$;
6          $R_2 \leftarrow (J \backslash \uparrow a,\ M,\ <)$;
7   Closure$(R_2,\ P \backslash \uparrow a)$;

   end

The next proposition shows how to determine if an element a is $\vee$-prime in a relation R.

Proposition 1. *Let $R = (J, M, <)$ be a binary relation and $a \in J$. Then a is $\vee$-prime iff there exists $m \in M$ such that $m' = J \backslash \uparrow a$.*

Proof. Suppose that for all $m \in M$, $m' \neq J \backslash \uparrow a$. Then, there exists $m_1, m_2 \in M$ such that $a \in (m_1' \cup m_2')''$, with $a \notin m_1' \cup m_2'$ and $m_1', m_2' \in \mathcal{F}(R)$

Conversely, let $m \in M$ with $m' = J \backslash \uparrow a$. We show that a is $\vee$-prime. Let $A, B \in \mathcal{F}(R)$ such that $a \notin A \cup B$. Then $A \cup B \subseteq J \backslash \uparrow a = m'$, since A and B are ideals of $(J, \leq)$. Since the operator $''$ is extensive, we have $(A \cup B)'' \subseteq (J \backslash \uparrow a)'' = (m')'' = m'$. This implies that $a \notin (A \cup B)''$, since $a \notin m'$.

□

Detection of $\vee$-prime element can be done in $O(|J| \times |M|)$ time. Thus this choice function does not interfere with the overall complexity of our algorithm.

To be convinced it is possible to detect a $\vee$-prime element, note that we can use counters over elements of J and M. For each element $j \in J$ (resp. $m \in M$), a counter is associated to $|j'|$ (resp. $|m'|$).

When computing the closure of an element a, it is possible, at the same time, to compute $\uparrow a$. So, it is possible to compute $|J| - |\uparrow a|$. Then a is $\vee$-prime if exists $m \in M$ such that $|m'| = |J| - |\uparrow a|$ and $j \not{I} m$. This test may be done in $O(M)$.

To keep the $O(|J| \times |M|)$ complexity, the idea is to pre-compute and test, at each step, $|J|$ closed sets. Space remains polynomial since the call depth is no more than $|J|$. At most, we store $|J|^2$ closed sets at the same time.

However, since now no invalid closed sets are generated, the overall complexity of the algorithm is lowered as shown by the next theorem.

Theorem 2. *Algorithm 5 generates the closed sets of $\vee$-separable closure system in $O(|J| \times |M|)$ time per concept using polynomial space.*

The proof of Theorem 2 is contained in the proof of Theorem 1, considering $|NV(R)| = 0$.

4   On the Impact of the Choice Function

All along this paper we have shown that the choice function is essential for practical efficiency, since it will determine the number of invalid closed sets generated. This has been possible since our algorithm is generic and since existing algorithms are just particular instances. The only difference which modifies the practical efficiency is the way element a is chosen.

We made different tests by implementing several choice functions in order to have a hint on what could be a "good" choice function. The choice functions we decided to implement are the following:

- "Random": Choose an arbitrary element of $a \in J$. It may serve as a reference for other choices.
- "Maximize $|a''|$": Choose $a \in J$ such that $|a''|$ is maximum.
- "Minimize $|a''|$": Choose $a \in J$ such that $|a''|$ is minimum.
- "Maximize $|a'|$": Choose $a \in J$ such that $|a'|$ is maximum.
- "Minimize $|a'|$": Choose $a \in J$ such that $|a'|$ is minimum.

We have tested our implementations on binary relations generated randomly with different densities. All binary relations are tables of 50×50 size and densities $50\%, 60\%$ and 70%. For each density, we have made several tests on several binary relations, and the average numbers of closed sets of these binary relations are 77515 for density 50, 512779 for density 60 and 4715307 for density 70. Figure 4 shows the results (number of invalid closed sets) for these different choices.

The random choice may be viewed as an execution of the Next Closure instance. In effect, taking element randomly is equivalent to taking a particular total order on the label of elements.

So, tests let think that Next Closure implementation is less efficient than an implementation in bottom-up style (which minimizes $|a''|$). A partial answer is the following: Tests were done on randomly generated contexts. These contexts are not reduced, and so, some labelled elements may not be irreducible elements. Furthermore, even if the initial context was reduced, there are no warranty that sub-context R_i are reduced. Closed sets which are not $\vee$-irreducible are bad choices to separate the closure system. In effect, in this case, it is sure that the family of closed sets which do not contain this element is not a closure system. On the contrary, using a bottom-up approach warrants us to choose, at each step, a $\vee$-irreducible of the closure system.

Our conclusion is that the better implementation is to choose an element with minimum closure. Moreover, we confirm the observation in [9] that Bordat

	#Invalid	#not valid/#valid
Density 50		
Random	94629,1	1,22
Min $\|a''\|$	52446,6	**0,67**
Max $\|a''\|$	160432	2,06
Min $\|a'\|$	70816,3	0,91
Max $\|a'\|$	142385	1,83
Density 60		
Random	509295	0,99
Min $\|a''\|$	252623	**0,49**
Max $\|a''\|$	952471,9	1,85
Min $\|a'\|$	361749,9	0,70
Max $\|a'\|$	858963	1,65
Density 70		
Random	3389429	0,71
Min $\|a''\|$	1544736	**0,32**
Max $\|a''\|$	6902954	1,46
Min $\|a'\|$	2264626	0,48
Max $\|a'\|$	6366286	1,35

Fig. 4. Binary Relations 50x50, for density 50%, 60% and 70%. The average numbers of closed sets of these binary relations are 77515 for density 50, 512779 for density 60 and 4715307 for density 70

algorithm is better for dense binary relations, as shown in Figure 4 for invalid closed sets.

In this paper we have considered that the choice function is defined "a priori". However one could imagine having several choice functions and use the best one at each step. For instance, if at a given step there exists a $\vee$-prime element then this element should be chosen. In the same way, if at some steps the lattice is of a particular class for which an efficient algorithm exists, then the divide and conquer approach should be dropped and the efficient algorithm preferred. For instance, in the case of distributive lattice an efficient algorithm exists [8] and testing distributivity can be done in $O(|J| \times |M|)$ time.

5   Conclusion

In this paper we presented a generic "divide and conquer" algorithm which generates closed sets induced by a binary relation. We have shown that some existing algorithms can be viewed as particular instances of our algorithm. This allowed us to exhibit the key role played by the "choice function" for practical efficiency. This "choice function" can be viewed as a heuristic which tries to minimize the number of invalid closed sets generated. For instance, we showed that when there exists a $\vee$-prime element, no invalid closed set is generated. It is important to notice that the time complexity of the "choice function" has to be lower or equal to $|J| \times |M|$ in order to keep the overall complexity unchanged.

Indeed, this factor corresponds to the cost of computing the closure of a set which has to be done anyway. One could imagine that proposing new closed sets generation algorithms is reduced to proposing a new choice function for which the number of invalid closed set can be bounded. The overall complexity could then be somewhere between $|J| \times |M|$ and $|J|^2 \times |M|$ per closed set. The $|J| \times |M|$ corresponding to the choice of a $\vee$-prime element, while $|J|^2 \times |M|$ being the choice of a random element. Based on our tests, we strongly believe that the number of invalid closed sets generated is bounded by a constant factor. However, this has still to be proved formally and remains our next challenge.

Acknowledgment

I wish to thank the referees for their remarks and suggestions. Particularly thanks to turn my attention to an imprecision in the Ganter's instance. I also thank Raoul Medina for his assistance during the writing of this paper.

References

1. R. Agrawal, T. Imielinski, and A. Swami. Mining association rules between sets of items in large database. In *ACM SIGMOD conf. Management of data*, pages 265–290, 1993.
2. K. Bertet, R. Medina, L. Nourine, and O. Raynaud. Algorithmique combinatoire dans les bases de données massives. In *Actes du workshop 'Usage des treillis de Galois pour l'intelligence artificielle' AFIA03*, Laval, 2003.
3. G. Birkhoff. *Lattice Theory, third edition*, volume XXV. American Mathematical Colloquium Publications, Providence, 1967.
4. J.P. Bordat. Calcul pratique du treillis de galois d'une correspondance. *Journal of Math. Sci. Hum.*, 96:31–47, 1986.
5. B. Ganter. Two basic algorithms in concept analysis. Technical report, Technische Hoschule Darmstadt, 1984.
6. B. Ganter and K. Reuter. Finding all closed sets: a general approach. *Order*, 8, 1991.
7. B. Ganter and R. Wille. *Formal Concept Analysis, Mathematical Foundations*. Springer-Verlag Berlin, 1996.
8. M. Habib, R. Medina, L. Nourine, and G. Steiner. Efficient algorithms on distributive lattices. *Discrete Applied Mathematics*, 110:169–187, 2001.
9. S. Kuznetsov and S. Obiedkov. Comparing performance of algorithms for generating concept lattices. *Journal of Experimental and Theoretical Artificial Intelligence(JETAI)*, 2/3(14):189–216, 2002.
10. Christian Lindig. Fast concept analysis. In Gerd Stumme, editor, *Working with Conceptual Structures, Contributions to ICCS 2000*, 2000.
11. G. Markowsky. Primes, irreducibles and extremal lattices. *Order*, 9:265–290, 1992.
12. L. Nourine. *Une structuration algorithmique de la théorie des treillis*. Habilitation à diriger des recherches, Université Montpellier II, 2000.

Uncovering and Reducing Hidden Combinatorics in Guigues-Duquenne Bases

A. Gély, R. Medina, L. Nourine, and Y. Renaud

LIMOS, Université Blaise Pascal,
Campus des Cézeaux, 63173 Aubière,
France
{gely, medina, nourine, renaud}@isima.fr

Abstract. Mannila and Räihä [5] have shown that minimum implicational bases can have an exponential number of implications. Aim of our paper is to understand how and why this combinatorial explosion arises and to propose mechanisms which reduce it.

Keywords: Guigues-Duquenne base, closure systems, clone attributes.

1   Introduction

One of the most important open problems in formal concept analysis is the generation of a minimum implicational base from a context. This problem has two major practical difficulties.

First, there is no known polynomial time algorithm which computes such a base. This is still an open problem and many ongoing researches try to classify this problem for particular cases (for instance, finding the keys of a multi-valued context [2]). The other critical problem is the size of the result. A minimum implicational base might have an exponential size (see Mannila and Räihä [5]).

In this paper, we explain why such combinatorial explosion arises and then try to reduce it. For this study, we consider only Guigues-Duquenne bases since they are well defined.

In the example given by Mannila and Räihä, one can notice that some attributes play similar roles in the pseudo-closed sets, i.e. some pseudo-closed sets can be obtained from others by simply exchanging one attribute by another one. We say that a and b are *P-clone* attributes if all pseudo-closed sets containing attribute b can be obtained by exchanging a for attribute b in all pseudo-closed sets containing attribute a, and reciprocally.

We believe that the combinatorial explosion of the Guigues-Duquenne base is due to the presence of P-clone attributes. Aim of our ongoing work is either to prove or invalidate this belief. In this paper we present some results which could help in achieving this aim.

Medina and Nourine [6] introduced the notion of clone attributes, which is a relaxed definition of P-clone attributes. Indeed, clone attributes are attributes

B. Ganter and R. Godin (Eds.): ICFCA 2005, LNAI 3403, pp. 235–248, 2005.

having a similar role on closed sets[1] rather than on pseudo-closed sets. With this notion, it has been shown that the combinatorial explosion of Mannila and Räihä example is due to clone attributes. Moreover, a clone reduction operator which drastically reduces the size of the Guigues-Duquenne base has been proposed. The Mannila and Räihä example reduces to only one implication on its clone-free reduced context.

However, in spite of this reduction, one can easily find a new example of a base with exponential size. In this example, combinatorial explosion is due to the presence of implications having a single attribute as premise. We thus propose a new context transformation operator, called *atomization*, which computes a new context that preserves pseudo-closed sets having more than one attribute in their premise. We thus obtain a new relaxed definition of P-clone attributes: the *A-clone* attributes which are the clone attributes present in an *atomized* context. We do not know if there exists an example of *A-clone free context* with an exponential minimum base. This is still an open problem.

2   Notations, Definitions and Main Problem Statement

2.1   Definitions

In this paper, minimum implicational bases are noted by Σ and are supposed to be in the Guigues-Duquenne form. We abusively use the notation $\mathcal{F}$ to denote a closure system and its corresponding lattice. The classical closure operator (Galois operator) over a context $R = (J, M, I)$ is noted by $^-$. Thus, the closure of x is noted $\overline{x}$.

We suppose that small letters are used to represent attributes of the context and two attributes are not identical in the context; i.e. for any pair of attributes (a, b), $\overline{a} \neq \overline{b}$. We denote by J the set of attributes and M the objects of the context $R = (J, M, I)$. We consider that objects present in the context are the meet-irreducible closed sets of the closure system associated to R. A closed set is said to be meet-irreducible in a closure system if it has exactly one cover.

We briefly recall here the main property of the Guigues-Duquenne base.

Definition 1. *Quasi-closed and Pseudo-closed set*
Let $\mathcal{F}$ be a closure system and $^-$ the closure operator associated to $\mathcal{F}$. Let $P \in 2^J$ and $P \notin \mathcal{F}$.

 - *P is a Quasi-closed set iff for all $Q \subset P$, $\overline{Q} \subset P$ or $\overline{Q} = \overline{P}$.*
 - *A quasi-closed set P is a pseudo-closed set if there is no quasi-closed set $Q \subset P$ with $\overline{Q} = \overline{P}$.*

Theorem 1. *Duquenne-Guigues base [1].*
Let $\mathcal{F}$ be a closure system. The set $\Sigma_{\mathcal{F}} = \{P \to \overline{P} \backslash P \mid P$ is pseudo closed$\}$ is an implicational base of $\mathcal{F}$ and has a minimum number of implications.

---

[1] This idea was used in Ganter [3] to generate closed sets under symmetry, i.e. generate only one closed set in each class of closed sets.

2.2   P-Clone Attributes

As said in the introduction, there are two ways to cope with the efficiency of the Guigues-Duquenne base generation. The first way is to produce effective algorithms to compute the base. Many researches are done in this direction but it is still an open problem even for particular cases [2].

The second way is to deal with the size of the Guigues-Duquenne base. But since there exists an example of context having an exponential base [5], this approach looks like a dead-end. Let us try to understand on an example why this combinatorial explosion might arise.

Consider the following Guigues-Duquenne base:

$$\Sigma \begin{cases} ac \rightarrow bd, ae \rightarrow bf, ce \rightarrow df \\ ad \rightarrow bc, af \rightarrow be, cf \rightarrow de \\ bc \rightarrow ad, be \rightarrow af, de \rightarrow cf \\ bd \rightarrow ac, bf \rightarrow ae, df \rightarrow ce \end{cases}$$

One can remark that for each implication where attribute a appears as premise, there is the same implication obtained by swapping attributes a and b. We say that attributes a and b are P-clone. The same remark applies for attributes c and d as well as for attributes e and f. We thus obtain a family of P-clone attributes: $\mathcal{C} = \{\{a, b\}, \{c, d\}, \{e, f\}\}$. When keeping only one attribute per class of P-clone attributes, we obtain a new base $\Sigma' = \{ac \rightarrow bd, ae \rightarrow bf, ce \rightarrow df\}$. Computing the original base Σ from Σ' and $\mathcal{C}$ is straightforward. Thus P-clone attributes seem to play a key role in the possibles combinatorial explosion of minimum bases.

One can notice that implications which premises are a single attribute do not introduce any combinatorial explosion. Indeed, their number is polynomial and they can trivially be computed from the context. Thus, we will not consider pseudo-closed sets of size one in the P-clone definition.

Let us now define formally the notion of P-clone.

Consider $\varphi_{a,b}$ be the mapping $\varphi_{a,b} : 2^J \rightarrow 2^J$ which associates to each subset F of $\mathcal{F}$ its image by swapping a and b.

$$\varphi_{a,b}(F) = \begin{cases} (F \backslash \{a\}) \cup \{b\}, & \text{if } b \notin F \text{ and } a \in F; \\ (F \backslash \{b\}) \cup \{a\}, & \text{if } a \notin F \text{ and } b \in F; \\ F, \text{otherwise.} \end{cases}$$

Definition 2 (P-Clone Attributes). *Let $R = (J, M, I)$ be a context and $a, b \in J$. We say that a and b are P-clone if for any pseudo-closed set P of R (| P |> 1), $\varphi_{a,b}(P)$ is a pseudo-closed set of R.*

This leads to the following decision problem.

Problem 1 (P-Clone Problem).
Data : A context $R = (J, M, I)$ and $a, b \in J$.
Question : Are a and b P-clone in R?

This problem remains open in this paper.

3   Clone Attributes

3.1   Motivations

Clone attributes were initially proposed in [6]. Definition of clone attributes is a relaxation of the P-clone definition: we now consider similarities on closed sets rather than on pseudo-closed sets.

Let us illustrate the notion of clone attributes through an example (see figure 1).

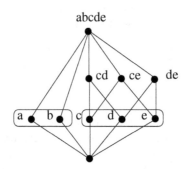

Fig. 1. Elements with a similar behaviour

When observing attributes a and b in the closure system in figure 1, one can notice that these attributes can be swapped (a becoming b and b becoming a) without changing the closure system. A same phenomenon arises with attributes c, d and e: any of those attributes can be swapped with any of the two others, and the closure system will remain unchanged. On the other hand, swapping attributes a and c changes the closure system; indeed, we loose the closed sets cd and ce and we obtain extra closed sets ac and ae. We will say that a and b are clone, as well as attributes c, d and e. Formally, two attributes can be swapped if the closure system remains unchanged.

Once this information discovered, it is possible to reduce the size of the closure system by keeping one attribute for each clone class. A possible application of clone attributes is thus the reduction of a closure system.

Moreover, one can also notice that a and b, as well as c, d and e, play symmetrical roles in the implicational base of the context (see figure 2). For any implication where a is in the premise, there exists the same implication where a and b have been swapped. Similarly, for any implication containing c in the premise, there is another one where c is swapped with d and another one obtained by swapping c and e. If several attributes belonging to a clone class are present in the premise of an implication, their swapping resumes to the identity function.

We can now define more formally the notion of clone attributes.

Definition 3 (Clone Attributes). *Let $R = (J, M, I)$ be a context and $a, b \in J$. We say that a and b are clone if for any closed set F of R, $\varphi_{a,b}(P)$ is a closed set of R.*

	a	b	cd	ce	de
a	X				
b		X			
c			X	X	
d			X		X
e				X	X

$$\begin{cases} ab \rightarrow cde, \\ ac \rightarrow bde, ad \rightarrow bce, ae \rightarrow bcd, \\ bc \rightarrow ade, bd \rightarrow ace, be \rightarrow acd, \\ cde \rightarrow ab \end{cases}$$

Fig. 2. Context (lines correspond to join-irreducible elements, columns to meet-irreducible elements) and its implicational base corresponding to the closure system in figure 1

First, let us show through the example in figure 3 that P-clone attributes generalize clone attributes.

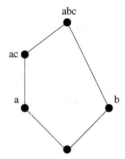

Fig. 3. $\Sigma = \{c \rightarrow a, \; ab \rightarrow c\}$, $\mathcal{F} = \{\emptyset, a, b, ac, abc\}$

Let examine the following implicational base: $\{c \rightarrow a, \; ab \rightarrow c\}$ (cf. figure 3). Here, a and b are P-clone since they appear as premise in the same implication: $(ab \rightarrow c)$. The closure system associated to this minimum base is $\mathcal{F} = \{\emptyset, a, b, ac, abc\}$. If we consider this closure system, permutation of a and b cannot be done without changing the closure system. Indeed, the set bc obtained from ac by exchanging a by b does not belong to the closure system $\mathcal{F}$. Thus, a and b are not clone. The notion of P-clone is thus more general than the notion of clone.

What is the impact of clone attributes on pseudo-closed sets computation? A first answer is given by the following lemma.

Lemma 1. *[6] Let $a, b \in J$ be two clone attributes. Then $P = \bar{a} \cup \bar{b}$ is either a closed-set or a pseudo-closed set.*

Thus, some pseudo-closed sets can be found by simply checking for all pairs (a, b) of clone attributes if $\bar{a} \cup \bar{b}$ is not a closed set.

Next lemma shows that the search space can be reduced when clone attributes are present in the context.

Lemma 2. *[6] Let $\mathcal{F}$ be a closure system and P a pseudo-closed set of $\mathcal{F}$. If $a, b \in J$ are clone attributes then $\varphi_{a,b}(P)$ is a pseudo-closed set of $\mathcal{F}$.*

In other words, if a and b are clone, for any pseudo-closed set P containing a, its φ mapping is also a pseudo-closed set (and reciprocally). From the above, we conclude that *clone* attributes are *P-clone* attributes.

3.2   Clone Detection

In this section we deal with the following decision problem (which is a particular case of Problem 1).

Problem 2 (Clone Problem).
Data : A context $R = (J, M, I)$ and $a, b \in J$.
Question : Are a and b clone in R?

In the following, we show that this problem is polynomial.

Proposition 1. *[6] Let $\mathcal{F}$ be a closure system and $a, b \in J$. Then a and b are clone attributes iff for any meet-irreducible M of $\mathcal{F}$, we have $\varphi_{a,b}(M)$ a meet-irreducible of $\mathcal{F}$.*

We note by $\mathcal{F}_{(a)}$ the set of closed sets containing a. Previous proposition indicates that in order to detect clone attributes, it is sufficient to consider only meet-irreducible elements. Indeed, a and b are clone if there exists an isomorphism between the closure systems $\mathcal{F}_{(a)}$ and $\mathcal{F}_{(b)}$. Thus, if such isomorphism exists between those closure systems, then it also exists for their meet-irreducible elements, since any closed set is the intersection of meet-irreducible elements. Thus, only meet-irreducible elements need to be considered to detect clone. And meet-irreducible elements are necessarily in the context. An example of such behaviour of the meet-irreducible elements can be seen for clone attributes c and d in figure 1.

Thus, testing if two attributes a and b are clone can be done in polynomial time by simply testing for each meet-irreducible M that $\varphi_{a,b}(M)$ is also a meet-irreducible element.

3.3   Clone Reduction

Once clone attributes have been detected, we can do a clone reduction of the input context. Indeed, according to Lemma 2, if two attributes a and b are clone then pseudo-closed sets containing b can be deduced from pseudo-closed sets containing a. Thus, closed sets containing attribute b without attribute a are no longer necessary. When removing those closed sets, we obtain a new closure system $\mathcal{F}_{a \backslash b}$. Which are the meet-irreducible elements belonging to the new closure system $\mathcal{F}_{a \backslash b}$? Since only closed sets containing b without a are removed from the system, it is straightforward to see that:

$$\{M \in \mathcal{M}_{\mathcal{F}} \mid b \notin M \text{ or } a \in M\} \subseteq \mathcal{M}_{\mathcal{F}_{a \backslash b}}$$

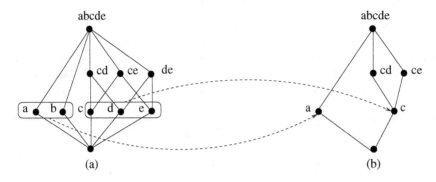

Fig. 4. Closure system after clone-reduction of $\{a, b\}$ and $\{c, d, e\}$

New meet-irreducible elements might appear during the reduction phase since some closed sets are removed from the closure system. A closed set becomes a meet-irreducible, if all closed sets that cover it are removed but one. Lemma 3 characterizes closed sets which are new meet-irreducible elements in the new closure system $\mathcal{F}_{a \backslash b}$.

Lemma 3. *[6] Let $M \subset J$ be a new meet-irreducible in $\mathcal{F}_{a \backslash b}$. Then $M \cup \{b\} \in \mathcal{M}_{\mathcal{F}}$*

From previous results we deduce that:

$$\mathcal{M}_{\mathcal{F}_{a \backslash b}} \subseteq \{M \in \mathcal{M}_{\mathcal{F}} \mid b \notin M \text{ or } a \in M\} \cup \{M \backslash \{b\} \mid M \in \mathcal{M}_{\mathcal{F}_{(b)}}\}$$

All those results can be applied on the context (via the meet-irreducible elements). All the operations, such as the removal or the insertion of meet-irreducible elements are done in polynomial time. Clone reduction can thus be done in polynomial time. According to the previous lemma, the size of the new context is lesser or equal to the size of the original context.

Finally, note that during the clone reduction phase, new clone attributes might appear. For instance, in figure 4(b), attributes d and e are clone, and thus the whole process can be applied again.

	a	cd	ce
a	X		
c		X	X
d		X	
e			X
b			

$$\left\{ \begin{array}{l} ac \rightarrow bde, \\ d \rightarrow c, e \rightarrow c, b \rightarrow acde \\ cde \rightarrow ab \end{array} \right\}$$

Fig. 5. Context and Guigues-Duquenne base after clone-reduction

3.4   Restoring Guigues-Duquenne Base

After clone reduction of the context, the size of the Guigues-Duquenne base of the new context is smaller. Thus, after several clone reductions, we obtain a clone-free context. Now let us show how to retrieve the Guigues-Duquenne base of the original context from that of the clone-free context and classes of clone attributes. The reconstruction process relies on the following theorem:

Theorem 2. *Let $\mathcal{F}$ be a closure system and a and b two clone attributes. If $\Sigma_{\mathcal{F}_{a\setminus b}}$ is a Guigues-Duquenne base of $\mathcal{F}_{a\setminus b}$ then*

$$\Sigma_{\mathcal{F}} = \begin{cases} \Sigma' = \Sigma_{\mathcal{F}_{a\setminus b}} \setminus \{b \to \overline{b}\} \\ \bigcup \{\overline{a} \cup \overline{b} \to \overline{\overline{a} \cup \overline{b}}, \ if \ \overline{a} \cup \overline{b} \ is \ not \ closed\} \\ \bigcup \{\varphi_{a,b}(P) \to \varphi_{a,b}(\overline{P}) \mid P \to \overline{P} \in \Sigma_{\mathcal{F}_{a\setminus b}} \setminus \{b \to \overline{b}\}\} \end{cases}$$

is a Guigues-Duquenne base of $\mathcal{F}$.

Here, we explain Theorem 2 by a commented example. We note Σ the Guigues-Duquenne base of the original closure system, and Σ_r the Guigues-Duquenne base of the clone-reduced context.

Let consider the minimum base of figure 5 and see how to come back to the minimum base of figure 2. The input data of the process is the base Σ_r of figure 5 as well as the clone classes: $\{a, b\}, \{c, d, e\}$.

There are three major steps to obtain Σ from Σ_r. Those steps are detailed for both clone classes detected (first $\{a, b\}$, and then $\{c, d, e\}$).

Restoration Process for Clone Class $\{a, b\}$

1. *Extra implication removal*
   New implications appear during the clone reduction operation. Let C be a clone class and let a be its representative attribute. Then, for any $x \in C$ and $x \neq a$, the implication $x \to \overline{a} \cup \overline{x}$ has been artificially added by the clone reduction, since closed sets containing x are kept only if they also contain attribute a. Thus $\overline{a} \subset \overline{x}$, which is expressed by the new implication.
   Thus, the first step is to remove such implications from Σ_r. Here, implication $\{b \to acde\}$ is removed.

$$\Sigma_r = \begin{cases} ac \to bde, \\ d \to c, e \to c, b \to acde \\ cde \to ab \end{cases} \Rightarrow \begin{cases} ac \to bde, \\ d \to c, e \to c, \\ cde \to ab \end{cases} = \Sigma_1$$

2. *Restoring pseudo-closed sets*
   In a second step, it is mandatory to check if some pseudo-closed sets have been removed from the closure system during the clone reduction. Let C be a clone class and a be its representative attribute. For any $x \in C$, $x \neq a$, we have to check if $\overline{a} \cup \overline{x}$ is a closed set of the original closure system. If it is not

a closed set, then it is a pseudo-closed set (see lemma 1) and thus we need to restore the implication $\overline{a} \cup \overline{x} \rightarrow \overline{\overline{a} \cup \overline{x}}$.

In our example, when processing the clone class $\{a, b\}$, we obtain the new implication $ab \rightarrow cde$ since ab is not a closed set in the initial closure system. The new minimum base is now:

$$\left.\begin{cases} ac \rightarrow bde, \\ ab \rightarrow cde, \\ d \rightarrow c, e \rightarrow c, \\ cde \rightarrow ab \end{cases}\right\} = \Sigma_2$$

3. *Applying the φ mapping*

   The last step is to apply the φ mapping, in order to recover the initial implications (direct application of Lemma 2):

$$\begin{array}{lcl} ac \rightarrow bde & \xrightarrow{\varphi_{a,b}} & bc \rightarrow ade \\ d \rightarrow c & \xrightarrow{\varphi_{a,b}} & d \rightarrow c \ (identity) \\ e \rightarrow c & \xrightarrow{\varphi_{a,b}} & e \rightarrow c \ (identity) \\ cde \rightarrow ab & \xrightarrow{\varphi_{a,b}} & cde \rightarrow ab \ (identity) \end{array}$$

The new minimum base is then:

$$\left.\begin{cases} ac \rightarrow bde \\ bc \rightarrow ade \\ ab \rightarrow cde \\ d \rightarrow c \\ e \rightarrow c \\ cde \rightarrow ab \end{cases}\right\} = \Sigma_3$$

Once the three steps have been done for the clone class $\{a, b\}$, we restart the whole process for the clone class $\{c, d, e\}$. Beware that the input data is now the minimum base obtained after restoring the clone class $\{a, b\}$. Thus, potentially, the size of the minimum base might double at each global step, which explains the combinatorial explosion that might appear in some minimum bases.

Restoration Process for Clone Class $\{c, d, e\}$

1. *Extra implications removal*

   During this step, two implications are removed: $\{d \rightarrow c\}$ and $\{e \rightarrow c\}$. The new minimum base is then:

$$\Sigma_3 = \begin{cases} ac \rightarrow bde \\ bc \rightarrow ade \\ ab \rightarrow cde \\ d \rightarrow c \\ e \rightarrow c \\ cde \rightarrow ab \end{cases} \Rightarrow \left.\begin{cases} ac \rightarrow bde \\ bc \rightarrow ade \\ ab \rightarrow cde \\ cde \rightarrow ab \end{cases}\right\} = \Sigma_4$$

2. *Restoring pseudo-closed sets*

   The sets cd, de, ce are closed sets of the initial closure system. Thus, no pseudo-closed set was removed by the clone reduction. Thus, no implication is restored.

3. *Applying the φ mapping*

   Here are the different mappings we apply:

$$
\begin{array}{lll}
ac \to bde & \xrightarrow{\varphi_{c,d}} & ad \to bce \\
 & \xrightarrow{\varphi_{c,e}} & ae \to bcd \\
 & \xrightarrow{\varphi_{d,e}} & (identity) \\
bc \to ade & \xrightarrow{\varphi_{c,d}} & bd \to ace \\
 & \xrightarrow{\varphi_{c,e}} & be \to acd \\
 & \xrightarrow{\varphi_{d,e}} & (identity) \\
ab \to cde & \xrightarrow{\varphi_{c,d}} & (identity) \\
 & \xrightarrow{\varphi_{c,e}} & (identity) \\
 & \xrightarrow{\varphi_{d,e}} & (identity) \\
cde \to ab & \xrightarrow{\varphi_{c,d}} & (identity) \\
 & \xrightarrow{\varphi_{c,e}} & (identity) \\
 & \xrightarrow{\varphi_{d,e}} & (identity)
\end{array}
$$

Finally, we thus obtain the minimum base of the initial closure system:

$$
\left\{
\begin{array}{l}
ab \to cde, \\
ac \to bde, ad \to bce, ae \to bcd, \\
bc \to ade, bd \to ace, be \to acd, \\
cde \to ab
\end{array}
\right\} = \Sigma
$$

4   A-Clone Attributes

4.1   Motivation

Consider again the example in figure 3. Clearly, the attributes a and b are not clone, but they are P-clone. This is due to attribute c, since ac is a closed set but not bc. Thus, if we add bc and c we obtain a new closure system which preserves pseudo-closed sets with premise size greater than one, and, where a and b are clone.

In a more general way, consider two clone attributes a and b and insert a new implication $x \to a$. Then a and b are no longer clone attributes, but they remain P-clone attributes. The existence of an order between attributes might break the possibility of permuting two closed sets at the closed set level... but not at minimum base level. Thus, one can deduce from Mannila and Räihä example a new example of clone-free context which has an exponential base.

Knowing this, the straightforward idea is to get rid of any order between attributes. Implications having a single element as premise define exactly the

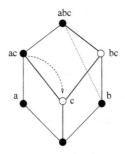

Fig. 6. Lattice $\mathcal{F}$, associated with $\Sigma' = \Sigma \backslash \{c \to a\}$

order between attributes. Those implications can easily be computed in polynomial time from the context. The idea is then to remove those implications from the Guigues-Duquenne base, and thus obtain a new closure system, where attributes are pairwise incomparable.

Consider the set of implications in Σ which premises are a single attribute, denoted by $\Sigma_J = \{a_1 \to \overline{a_1}, a_2 \to \overline{a_2}, ..., a_n \to \overline{a_n}\}$. The atomization process will remove $a_1 \to \overline{a_1}, a_2 \to \overline{a_2}$, and so on... We denote by $\mathcal{F}_i$ the closure system corresponding to $\Sigma_i = \Sigma \setminus \{a_j \to \overline{a_j}, j \leq i\}$ and $R_i = (J, M_i, I_i)$ its context. This process is called atomization and the resulting closure system $\mathcal{F}_n$ is said to be atomistic.

Figure 6 shows how the closure system in figure 3 evolves after atomization, i.e. when removing the implication $\{c \to a\}$.

One can check that, now, the attribute c appears in $\mathcal{F}_{(a)}$ with the set ac and in $\mathcal{F}_{(b)}$ with the set bc which was not present in the previous closure system. After this operation, there is now an isomorphism between the two closure systems $\mathcal{F}_{(a)}$ and $\mathcal{F}_{(b)}$. Thus, attributes a and b are clone in the new closure system.

Definition 4 (A-Clone Attributes). *Let $R = (J, M, I)$ be a context and $R_n = (J, M_n, I_n)$ be the corresponding* **atomized** *context. We say that a and b are A-clone in R if they are clone in R_n. In other words, A-clone attributes are the clone attributes that can be found after the atomization process.*

This leads to the new decision problem.

Problem 3 (A-Clone Problem).
Data : A context $R = (J, M, I)$ and $a, b \in J$.
Question : Are a and b A-clone in R?

In the remaining of this section, we study the impact of the atomization process on the closure system $\mathcal{F}_n$, the context R_n and the closure operator of $\mathcal{F}_n$.

4.2   The Impact of Atomization on Closure Systems

Here, we present some results around the atomization process when handling the closure system rather than the context.

Problem 4 (Atomization Problem 1).
Input : A closure system $\mathcal{F}$ on J and Σ_J.
Output : The closure system $\mathcal{F}_n$.

Let Σ be the Guigues-Duquenne base of the closure system $\mathcal{F}$. We examine how the closure system $\mathcal{F}$ evolves when the first implication $a_1 \rightarrow \overline{a_1} \in \Sigma$ is removed; i.e. the closure system $\mathcal{F}_1$ corresponding to $\Sigma_1 = \Sigma \setminus \{a_1 \rightarrow \overline{a_1}\}$. Intuitively, there are new sets which are not closed sets in $\mathcal{F}$ and become closed sets in $\mathcal{F}_1$.

Recall, that a set F satisfies an implication $P \rightarrow \overline{P}$, if $P \subseteq F$ implies $\overline{P} \subseteq F$. Thus a closure system associated to a base Σ is the set of all subsets that satisfy any implication in Σ.

First, we have $\mathcal{F} \subset \mathcal{F}_1$ since any closed set of $\mathcal{F}$ that satisfies Σ also satisfies Σ_1. Intuitively, the closed sets in $\mathcal{F}_1 \setminus \mathcal{F}$ are all sets that satisfy Σ_1 but $a_1 \rightarrow \overline{a_1}$. The following lemma characterizes the new closed sets.

Lemma 4. *A set $F \subseteq J$ belongs to $\mathcal{F}_1 \setminus \mathcal{F}$ if and only if $a_1 \in F$, $\overline{a_1} \not\subseteq F$ and $F \setminus \{a_1\} \in \mathcal{F}$.*

According to lemma 4, removing an implication $a_1 \rightarrow \overline{a_1}$ will consist in adding new closed sets obtained by copying any closed set $F \in \mathcal{F}$ such that $\overline{a_1} \setminus a_1 \not\subset F$. Thus, we deduce that $\mathcal{F}_1 = \mathcal{F} \cup \{F \cup \{a_1\} \mid F \in \mathcal{F} \text{ and } \overline{a_1} \setminus a_1 \not\subset F\}$.

Inductively, we can construct the closure system $\mathcal{F}_i$ from the closure system $\mathcal{F}_{i-1}$. As a consequence, there is a polynomial time algorithm to compute the closure system $\mathcal{F}_n$.

Moreover, the atomization process preserves pseudo-closed sets which premise size is greater than one, since when deleting any implication in a Guigues-Duquenne base we obtain a Guigues-Duquenne base. Thus, if two attributes are A-clone then they are P-clone. We can also show that atomization process preserves clone attributes.

4.3   The Impact of Atomization on Contexts

In this section, we study how the context R evolves during the atomization process, and more precisely how the meet-irreducible closed sets of $\mathcal{F}$ evolve.

This leads to the second "atomization problem".

Problem 5 (Atomization Problem 2).
Input : A context $R = (J, M, I)$ and Σ_J.
Output : $R_n = (J, M_n, I_n)$ the context corresponding to $\mathcal{F}_n$.

To solve this problem, we need to know which meet-irreducible closed sets will disappear and which meet-irreducible closed sets will appear.

Lemma 5. *Let $F \subseteq J$ be a meet-irreducible closed set in $\mathcal{F}_1$. Then either F is a meet-irreducible closed set of $\mathcal{F}$ or $F \setminus \{a_1\}$ has at most one cover not containing $\overline{a_1} \setminus \{a_1\}$.*

From previous lemma, we thus know that some of the meet-irreducible closed sets of $\mathcal{F}$ will remain meet-irreducible closed sets in $\mathcal{F}_1$.

Now, since a closed set $F \setminus \{a_1\}$ does not contain $\overline{a_1} \setminus \{a_1\}$, it cannot have two covers containing $\overline{a_1} \setminus \{a_1\}$. Thus $F \setminus \{a_1\}$ has at most one cover containing $\overline{a_1} \setminus \{a_1\}$ and at most one cover not containing $\overline{a_1} \setminus \{a_1\}$. Thus $F \setminus \{a_1\}$ can be obtained as the intersection of two meet-irreducible closed sets (possibly the same) in $\mathcal{F}$. As a consequence, we can state that:

$$\mathcal{M}(\mathcal{F}_1) \subseteq \mathcal{M}(\mathcal{F}) \cup \{M_1 \cap M_2 \cup \{a_1\} \mid M_1, M_2 \in \mathcal{M}(\mathcal{F}) \text{ and } \overline{a_1} \setminus \{a_1\} \not\subseteq M_1 \cap M_2\}$$

Thus, we can polynomially compute R_1. Inductively, we can compute in polynomial time R_n.

However, the following question remains open.

Question 1. Is R_n polynomial in the size of R?

A positive answer to this question would imply that problem 3 is polynomial.

4.4   The Impact of Atomization on Closure Operators

In this section, we exhibit a closure operator of the atomized closure system $\mathcal{F}_n$. This closure operator is derived from the initial context R and the set of implications Σ_J.

Proposition 2. *The mapping* $C_{\mathcal{F}_n} : 2^J \rightarrow 2^J$, *with* $C_{\mathcal{F}_n}(X') = \overline{X} \cup Y$, *where* $X = X' \setminus Y$ *and* $Y = \{a \mid a \rightarrow \overline{a} \in \Sigma_J, a \in X' \text{ and } \overline{a} \not\subseteq X'\}$ *is a closure operator for* $\mathcal{F}_n$.

According to proposition 2, there exists a polynomial time algorithm (in the size of R) that computes the closure $C_{\mathcal{F}_n}(X'), X' \subset J$, since computing $\overline{X}$ is polynomial.

5   Conclusion

We showed that P-clone attributes might induce a combinatorial explosion of minimum bases. But we do not know how to detect P-clone attributes from the context.

We thus proposed a relaxed definition of P-clone attributes: the clone attributes, which exhibit similarities on closed sets rather than on pseudo-closed sets. Their detection can be done in polynomial time. Moreover, we proposed a clone reduction operator which can be interesting as a pre-processing for practical improvement of various other algorithms. Unfortunately, even after clone reduction, some combinatorial explosion could remain.

We then introduced a new context transformation operator: atomization. More clone attributes can be found on an atomized context: the A-clone attributes. We could not find an example of A-clone free context such that its minimum base is exponential. This leads to the following question:

Question 2. What is the maximal number of implications of a minimum base of an A-clone free context?

We now have two context transformation operators: atomization and clone reduction. Those operations can be applied several times on a context before obtaining an A-clone free context. This leads to this other question:

Question 3. Is the operators order significant when repeatedly applying them? In other words, does there exist a fixed point?

Even, if many questions remain open, we hope this study gives a better understanding of combinatorics explosion in a Guigues-Duquenne base.

Acknowledgments. The authors are very grateful to the referees for their helpful comments and suggestions.

References

1. J.-L. Guigues and V. Duquenne : Famille minimale d'implications informatives résultant d'un tableau de donneés binaires, Mathématiques et sciences humaines,24, 1986.
2. T. Eiter and G. Gottlob: Hypergraph transversal computation and related problems in logic and AI. In European Conference on Logics in Artificial Intelligence (JELIA'02), pages 549–564, 2002.
3. B. Ganter: Finding closed sets under symmetry, FB4-Preprint 1307, TH Darmstadt, 1990.
4. B. Ganter and R. Wille: Formal Concept Analysis, Mathematical Foundations. Springler, 1999.
5. H. Mannila and K.J. Räihä: On the complexity of inferring functional dependencies. Discrete Applied Mathematics, 40(2):237-243, 1992.
6. R. Medina and L. Nourine: Clone items: a pre-processing information for knowledge discovery, Submitted.

A Parallel Algorithm for Lattice Construction

Jean François Djoufak Kengue[1], Petko Valtchev[2], and
Clémentin Tayou Djamegni[3]

[1] Département d'informatique, Faculté de Sciences, Université de Yaoundé 1, Cameroun
[2] DIRO, Université de Montréal, Canada
[3] Laboratoire d'informatique, Faculté de Sciences, Université de Dschang, Cameroun

Abstract. The construction of the concept lattice of a context is a time consuming process. However, in many practical cases where FCA has proven to provide theoretical strength, e.g., in data mining, the volume of data to analyze is huge. This fact emphasizes the need for efficient lattice manipulations. The processing of large datasets has often been approached with parallel algorithms and some preliminary studies on parallel lattice construction exist in the literature. We propose here a novel divide-and-conquer (D&C) approach that operates by data slicing. In this paper, we present a new parallel algorithm, called DAC-PARALAX, which borrows its main operating primitives from an existing sequential procedure and integrates them into a multi-process architecture. The algorithm has been implemented using a parallel dialect of the $C++$ language and its practical performances have been compared to those of a homologue sequential algorithm.

1 Introduction

FCA [19] has already found a wide range of applications in various domains, in particular in data mining and information retrieval where the volume of data to analyze is usually huge. However, the construction of the concept lattice or even the extraction of the concept set can be a time consuming task because of the potentially exponential growth of the lattice size in the number of data items. Therefore, there is a room for the design of efficient manipulation methods for concept lattices and derived structures such as iceberg lattices and implication bases.

Utilization of parallel processing is a typical approach for dealing with large datasets [2]. It allows the work load to be divided among a set of computing units which communicate in the process of constructing the solution of the initial sequentially defined problem. To that end, these units may establish various modes of collaboration such as data sharing, remote procedure calls, message sending, etc.

Unlike previous studies of parallel lattice construction, we follow a data-centered approach. The approach amounts to slicing the input context into disjoint fragments and assigning each fragment to a different processing unit. Once the processing of a particular unit is finished, an assembly of the results for two neighbor fragments, i.e., the concept lattices of the respective subcontexts, takes place. The assembly task is repeated until a single global lattice is obtained. The approach represents a parallel homologue of an existing sequential algorithm for lattice construction of divide-and-conquer (D&C) type [17].

B. Ganter and R. Godin (Eds.): ICFCA 2005, LNAI 3403, pp. 249–264, 2005.
© Springer-Verlag Berlin Heidelberg 2005

In this paper, we present a concrete parallel algorithm, called DAC-PARALAX, which implements the D&C approach. DAC-PARALAX relies on a multi-task architecture made of three different sorts of processes: concept servers, shared data servers, and concept assemblers. Each sort plays specific role in the global collaboration: while servers provide access to data and partial results, assemblers use those information chunks to create new concepts and link them in the factor lattice under construction. The entire set of processes is divided into blocks: Each block is assigned a specific fragment of the initial table whereby the aim is to construct the lattice of the fragment. Moreover, at the end of a parallel assembly round, blocks assigned to neighbor fragments are merged.

The algorithm has been implemented in a parallel dialect of the C++ language using the *STL* and the *MPI* standard libraries. Experiments has been carried out on a cluster of 16 CPUs running Linux and related by a Myrinet-type network. A comparison of the sequential and the parallel versions of the algorithm along the performance axis is provided together with a discussion of the observed strengths and weaknesses of our D&C approach.

The paper is organized as follows. In section 2 the basic principles of the sequential lattice assembly are recalled. Section 3 describes the transition from the sequential to a parallel design of the lattice assembly approach. The current realization of that design, the DAC-PARALAX algorithm, is presented in section 4. Finally, section 5 summarizes the results of an experimental study on the performances of the algorithm.

2   The Sequential D&C Algorithm

In its sequential version, the D&C lattice construction [17] is composed of a series of assembly tasks performed along a recursive binary split of the initial context.

2.1   Global Construction Process

The global construction process has three steps:

1. The initial context $C = (O, A, I)$ is recursively split into two parts until contexts of singleton attribute sets are obtained. The result is a (strictly) binary tree of contexts, further termed D&C-tree, where the leafs correspond to single-column tables while the context at each inner node is the *apposition*[1] of the contexts at the children nodes.
2. The lattices for each leaf context are constructed in a direct manner.
3. The lattices of all the inner contexts are constructed by assembling the lattices corresponding to children contexts. The process is itself a multi-step one: At each step, the nodes of a particular depth in the D&C-tree are processed. The lattices for inner nodes of depth $i - 1$ are obtained from the lattices for nodes at depth i. The final result is provided by the lattice of the root node.

The lattice constructing tasks involved in the above process, i.e., in steps two and three, are described with further details in the following paragraphs.

---

[1] Apposition is the horizontal concatenation of contexts sharing the same set of objects [5].

2.2   Lattice Construction for Single-Attribute Contexts

With a single-attribute context $C = (O, A = \{a\}, I)$, the corresponding lattice may have at most two concepts. Actually, the following cases may occur:

1. All the objects have the attribute a, i.e., $a' = O$. In this case, the lattice is reduced to a single node since its top and bottom concept coincide $((O, O') = (A', A))$.
2. Some of the objects are not incident to a, i.e., $a' \subset O$. There are two distinct concepts in the lattice: the top $(O, \emptyset)$ and the bottom (a', a).

The recognition of the specific case and the construction of the corresponding lattice are straightforward tasks (see [17]).

2.3   Assembly of Two Lattices Corresponding to the Fragments of a Global Context

The task is to construct the lattice L of the context $C = (O, A, I)$ from the lattices L_1 and L_2 corresponding to two complementary fragments of C, $C_1 = (O, A_1, I_1)$ and $C_2 = (O, A_2, I_2)$, respectively. Here $A = A_1 \cup A_2$ and $I_k = I \cap O \times A_k$ ($k = 1, 2$). L_1 and L_2 are called factor lattices [18].

The direct product of the factor lattices, $L_{1,2} = L_1 \times L_2$, provides the search space for the target lattice L. Indeed, L is a join sub-semi-lattice of $L_{1,2}$ while there is an order morphism φ from $L_{1,2}$ to L. The morphism φ is realized through extent intersection: The extent of the image concept in L is the intersection of the extents of the component concepts in every member of the equivalence class. The φ morphism induces an equivalence relation on $L_{1,2}$ in which two nodes belong to the same class whenever they are mapped to the same concept of L. Each class has a unique canonical representative which is the minimal node of the class. The canonical representatives are called *genitors* in the remainder as they provide all the information required for the creation of the corresponding concept in L. Actually, beside the extent of the image concept which is defined by an arbitrary member of the equivalence class, the intent is the union of both genitor intents.

The precedence order in L is also computed with respect to genitors. Indeed, the sub-order induced by the genitor set in $L_{1,2}$ is isomorphic to L. Therefore, the computation of precedence links for an image concept only requires the information about the respective genitor node in $L_{1,2}$. In particular, the lower covers of the genitor are used to compute a set of candidate lower covers of the image concept. The actual lower covers in L are the maximal concepts in that set.

The algorithm ASSEMBLY described in [17] implements a straightforward approach for lattice merge. Given two factor lattices L_1 and L_2, it performs a bottom-up traversal of $L_{1,2}$ with a canonicity test at each node. The information available at genitor nodes are then used in the creation of the respective global concepts. Lower cover computation for a new concept relies on a materialization of the φ morphism, called *Embed*. *Embed* is a two-way indexed structure which maps the pairs of factor concepts behind the nodes of $L_{1,2}$ to the corresponding image concepts from L. Algorithmically speaking, the structure is constructed "on the fly', i.e., simultaneously with the construction of L. Thus, at each moment, it comprises correct information for all the nodes of $L_{1,2}$ which have already undergone the canonicity test.

3   Design of DAC-PARALAX Algorithm

Here we design a parallel version, called DAC-PARALAX, of the sequential algorithm presented in section 2. This is done following the $PCAM^2$ methodology [2]. This methodology organizes the design of a parallel algorithm from a sequential algorithm into four phases. The starting phase dealts with partitioning of the total work into tasks and the second one with communications among tasks. The third and fourth phase study the possible agglomerations of tasks and the mapping of the resulting processes to CPU, also called allocation. The partitioning is concerned with a fine-grain decomposition [14, 16] of the sequential algorithm into tasks, called fine-grain tasks, to be executed. The study of communications involves the identification of data to be transferred between tasks as well as the definition of data structures and of reliable communication protocols (if possible optimal) for data exchanges between tasks. The study of agglomerations leads to a coarse-grain decomposition of the sequential algorithm stemming from the fine-grain decomposition of the partitioning phase. In this phase, fine-grain tasks are gathered to obtain coarse-grain tasks so as to reduce the number of data transfers between tasks as much as possible. The mapping [14] consists in assigning coarse-grain tasks to processors so as to minimize communication costs and the sum of idle times of all the processors used in the parallel execution of the algorithm.

At first glance, the sequential algorithm presented in section 2 may be decomposed in terms of factor lattice assembly. In this approach, a task is defined as the assembly of factor lattices. Therefore, in each level of the D&C-tree associated with the sequential algorithm, one may perform all the assemblies of factor lattices simultaneously. However, this approach will not minimize the idle time of available processors as the number of available processors is greater than the number of tasks in many levels of the D&C-tree. In this paper, the processors utilization problem is tackled by parallelizing the assembly of factor lattices. Following this strategy, the parallel assembly of two factor lattices is performed by all the processors that have constructed them.

3.1   Parallel Assembly of Two Factor Lattices

Here we design an algorithm devoted to the parallel assembly of two factor lattices following the PCAM methodology. Assume that the factor lattices L_1 and L_2 are to be assembled. Let L denotes the resulting lattice. Let also c_1, c_2 and c denote the concepts of lattices L_1, L_2 and L, respectively. Concept c_1 will be referred to as *1-genitor concept*, concept c_2 as *2-genitor concept* and concept c as *new concept*. Let (c_1, c_2) be a couple concept of the direct product of lattices L_1 and L_2. Given a 1-genitor, all the couples (c_1, c_2) where c_2 is an arbitrary 2-genitor concept will be referred to as *1-genitor line*. In other words, a 1-genitor line is a direct product of the form $\{c_1\} \times L_2$.

The Partitioning. The tasks obtained from this phase are provided by basic treatments related to the assembly of two factor lattices. For instance, the question of knowing if an arbitrary couple of concepts (c_1, c_2) is the minimum of its equivalence class is considered as a basic treatment. If this is the case, the corresponding new concept is

---

[2] Partitioning, Communication, Agglomeration and Mapping.

generated, then the *Embed* structure is used to compute its lower covers and the *Embed* entry corresponding to (c_1, c_2) is updated.

The different types of tasks emerge from the two following remarks:

1. All new concepts generated from a 1-genitor line need the same line of *Embed* for the computation of their respective lower covers.
2. A 1-genitor line is used to construct the corresponding line of *Embed* (indexed by the 1-genitor concept rank).

The first type of tasks that comes from these remarks consist in performing the treatments related to a 1-genitor line. Such a task will be referred to as *assembler*. Each assembler computes new concepts from a 1-genitor line, participates in the computation of a line of table *Embed* and constructs the lower cover of each new concept. To do this, each assembler must have a sorted copy of lattice L_2 [17], suitable lines of *Embed* and a 1-genitor concept. We will explain how each assembler gets a copy of lattice L_2 in section 3.2. There are as many assemblers as there are concepts in the lattice L_1. The execution graph of assemblers corresponds to the Hasse diagram of lattice L_1.

Another task, the SDS[3], is given the responsibility to supply each assembler with suitable *Embed* lines for the computation of the lower covers of new concepts. The 1-genitor concepts are provided to each assembler by a task called CS[4] The CS should ensure that the concurrent execution of assemblers is correct: No two assemblers related by a precedence link in the execution graph should be executed simultaneously. This can be done by supplying assemblers with 1-genitor concepts in a suitable order. In particular, all the assemblers of a same level in the assemblers execution graph can be executed simultaneously. More generally, one can consider simultaneous executions of assemblers that are not related with a precedence link. This is exploited here by making the CS compute and sort the concepts of L_1 according to the order proposed in [17]. The CS also collects lists of concepts computed by assemblers and returns the result of the parallel assembly. Moreover, CS informs the SDS about the number of active assemblers, i.e., assemblers that are being executed.

The SDS supplies each of the active assemblers with a suitable *Embed* lines and receives the lines of *Embed* that each assembler has computed. Based on the received lines, the SDS updates *Embed* and invites the CS to supply other assemblers with 1-genitor concept This is the beginning of a new assembly pass.

The Study of Communications. A pass corresponds to a concurrent execution of one CS, one SDS and a number of assemblers. A new pass begins when the CS chooses a number of assemblers and supplies each of them with a 1-genitor concept. These assemblers will be referred to as *elected* assemblers or *active* assemblers. A pass ends when all the elected assemblers accomplish their tasks. Each pass is identified by an integer number. A parallel assembly consists of a successive execution of a number of passes, the tasks of each pass being concurrently executed. Each parallel assembly starts with an initialization phase. During it the different sorts of tasks obtain the number of the

---

[3] for Shared Data Server.

[4] for Concept Server.

first pass and the sizes of lattices L_1 and L_2. The former two parameters are computed by the CS whereas the size of L_2 is calculated by each assembler. During each pass, three types of data are exchanged by the tasks: concepts, lists of concepts and 1-genitor concept ranks. A concept is represented by a data record storing its intent, its extent, the list of lower covers and its rank.

At the beginning of each pass, the CS sends the number of elected assemblers to the SDS. After the reception of the corresponding message, the SDS are ready to supply *Embed* lines to assemblers. When an assembler receives a 1-genitor concept from the CS it asks the SDS for all the lines from *Embed* that are required for the computation of the related new concepts. The query is made up of the ranks of all the lower covers of the 1-genitor concept. The SDS sends back a list of concepts (*Embed* lines). Furthermore, each *Embed* line starts with the concept it corresponds to.

Each assembler sends to the CS the list of the new concepts resulting from the computation of its 1-genitor line. It also sends to the SDS its contribution to the update of the *Embed* structure. i.e., a list of concepts in which the first one indicates the line number. Once the SDS received the contributions of all the elected assemblers, it sends to the CS a pass termination message. This is an invitation for the CS to launch the next pass. Once all the passes of a parallel assembly have been successfully executed, the assembly itself terminates. To that end, the CS sends a concept of rank -1 to assemblers and the integer number -1 to the SDS. Fig.1. summarizes the communications between a CS, a SDS and assemblers during a parallel assembly.

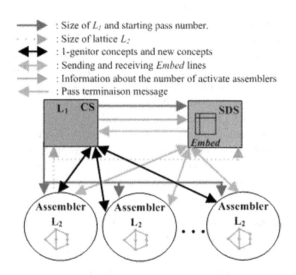

Fig. 1. Communications in a parallel assembly block

The Study of Agglomerations. Here, the granularity of assemblers is increased by assigning to each of them a set of 1-genitor lines. This leads to an increase in the granularity of a pass, i.e. the number of the processed 1-genitor lines. Hence, that number is no longer limited to the number of elected assemblers. Now, a pass involves all the lines of 1-genitors of the same level[5] in L_1. Each pass is numbered by the level of its 1-genitors.

Due to constraints related to the handling of $Embed$ (see [17]) a pass numbered p should be executed after the passes of numbers greater or equal to $p+1$. As a consequence, the passes of a parallel assembly following a decreasing order of their pass numbers. The maximal number of elected assemblers is fixed at the beginning of a parallel assembly. It depends on the number of available processors.

The increase in the granularity of assemblers also leads to an increase in the granularity of messages. At the beginning of a pass each assembler receives from the CS a block of 1-genitor concepts. The load balancing between assemblers is obtained by supplying them with blocks of substantially the same size. Once an assembler completes its execution, the new concepts that have been computed are sent to the CS and its contribution to the update of $Embed$ to the SDS.

The Mapping. The target architecture is a star network that consists of a cluster of computers connected by a Myrinet switch. Each task is assigned to a distinct processor. Let P denote the number of available processors. This means that the maximum number of elected assemblers in a pass is $P - 2$.

Optimization. The 1-genitor lines assigned to a given assembler are processed in a row. The treatment of a 1-genitor concept involves a set of lines from $Embed$. Because of the potentially large size of lines from $Embed$, their transfers from the SDS to assemblers may generate heavy communications traffic. Therefore, a well known strategy called communication-computation overlap was applied to $Embed$ line transfer. More precisely, an anticipation mechanism similar to that used in [15] was implemented. It allows assemblers to anticipate the requests of $Embed$ lines. Here, the request of $Embed$ lines related to the next 1-genitor is initiated once the $Embed$ lines related to the current 1-genitor are received. The number of such requests can be reduced by allowing each assembler to keep a copy of each $Embed$ line received and a copy of its contribution to the update of $Embed$.

Another optimization is the minimization of the idle time of assemblers due to the tranfers of concepts from the CS. This problem is tackled here by introducing another anticipation mechanism [15] that enables assemblers to launch the processing of the 1-genitor lines of the next pass once the current pass is completed. To that end, during the execution of a pass all the buckets of concepts related to the next pass are sent to the assemblers by the CS.

3.2   Parallel Construction of Concept Lattice with Several Assembly Blocks

We call *Assembly block* any set of tasks that consists of a CS, a SDS and at least two assemblers. The tasks of an assembly block are assigned to distinct processes that concurrently run on distinct processors.

Segmentation of the Initial Context. Assume that we have $B = 2^n$ assembly blocks that can be executed simultaneously. Here, these assembly blocks are numbered from 0

---

[5] The level of a 1-genitor is given by the cardinal of its intent.

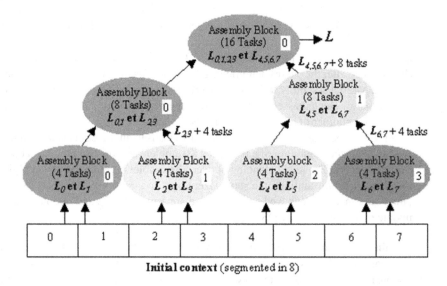

Fig. 2. Parallel Construction of Concept Lattice with four assemblies bloc

to $B - 1$. To compute the concept lattice associated with context $C = (O, A, I)$, we first perform a segmention of context C into $2B$ partial contexts that have approximatively the same size. The first partial context is made up of the $\lceil \frac{\|A\|}{2B} \rceil$ first attributes of context $C = (O, A, I)$, the next one is made up of the $\lceil \frac{\|A\|}{2B} \rceil$ next attributes. More generaly, let C_i, $i = 0, 1, ...2B - 1$ denotes the ith partial context resulting from that segmentation. C_i, $0 \leq i < 2B - 1$ is made up of the $(i + 1)$th block of $\lceil \frac{\|A\|}{2B} \rceil$ attributes while C_{2B-1} is made up of the $\|A\| - (2B - 1) \lceil \frac{\|A\|}{2B} \rceil$ last attributes. The first and the second partial contexts are assigned to the first assembly block, the third and the fourth ones to the second assembly block and so on. More generaly, the partial contexts C_{2i} and C_{2i+1} are assigned to the assembly block numbered i. In what follows L_i denotes the ordered concept lattice [17] corresponding to the partial context C_i. The ordered concept lattice resulting from the parallel assembly of lattices L_{2i} and L_{2i+1} is denoted by $L_{2i,2i+1}$.

Interactions Between Assembly Blocks. In each assembly block numbered i, the CS constructs the concept lattice L_{2i} and meanwhile each assembler constructs the concept lattice L_{2i+1}. After that, the parallel assembly of the two concept lattices L_{2i} and L_{2i+1} is performed. This leads to the larger concept lattice $L_{2i,2i+1}$. Then, the assembly blocks numbered $2i$ and $2i + 1$ merge to obtain a new and larger assembly block. The new block performs the parallel assembly of the concepts lattices $L_{2i,2i+1}$ and $L_{2i+2,2i+3}$ built by the assembly blocks which merged. To that end, all the processes of the former block $2i + 1$ play the assembler role in the new block which is numbered i. Its CS and SDS processes are respectively the CS and the SDS processes of the former block $2i$.

The merge of assembly blocks continues until a single block remains, i.e., the one corresponding to the root context (referenced by 0). The CS of this assembly block returns the concept lattice related to context $C = (O, A, I)$. Fig.2. illustrates the evolution of the tasks in a DAC-PARALAX execution which started with four assembly blocks each made up of four tasks. The assembly block merge involves important data transfers: after the merging of blocks $2i$ and $2i + 1$, all the assemblers of the new assembly block numbered i need a copy of the concept lattice constructed by the block $2i + 1$. This may lead to heavy communication costs and important idle times, especially if a whole concept lattice is transferred at once. Once more, a communication-computation overlap strategy is used to tackle the problem. To that end, we introduced the notion of diffusion groups.

The Diffusion Groups. Here we tackled the data transfer problem that arises when two assembly blocks, say $2i$ and $2i + 1$, are merged. Our solution is based, once again. on an anticipation mechanism following the communication-computation overlap strategy. To that end, we consider the group of tasks concerned by diffusion of the final result of a pass. The group comprises the assemblers in both assembly blocks and the SDS of the block $2i + 1$. At the end of a pass, every assembler of block $2i + 1$ sends a copy of the new concepts it has computed to the other tasks of its respective diffusion group. Thus, it is ensured that, when the execution of the assembly is completed, all the tasks in the diffusion group have a copy of the concept lattice computed by the block $2i + 1$. Note that, as stated in section 3.1 the CS of the block $2i + 1$ also has a copy of the lattice at the same moment. Consequently, at the end of the current assembly, every assembler of the new block i has the necessary information for the next step.

4   DAC-ParaLaX Algorithm

DAC-ParaLaX is a SPMD[6] algorithm, i.e. each process performs the same program on distinct data. Here data are either partial contexts assigned to each assembly block or the concept lattice produced by a block. Each process executes a task of an assembly block , i.e. a CS, a SDS or an assembler. The processes are identified by positive integer numbers. Thus, given an execution of DAC-ParaLaX with B blocks of α tasks each, one has $B\alpha$ processes addressed by integer numbers ranging from 0 to $B\alpha - 1$. Let $n \in \mathbb{N}$ be the number of an assembly block. The block includes the processes whose addresses range from $n\alpha$ to $n\alpha - 1$ whereby $n\alpha$ and $n\alpha + 1$ correspond to the CS and the SDS of the block, respectively. The remaining adresses indentify assemblers.

The behavior of DAC-PARALAX algorithm can be summarized as follows. It gathers processes in blocks (initAssemblyGroups), creates diffusion groups (createDiffGroups), determines the data to be used (partial context or concept lattice) by each process, defines the task executed by each process (executeTask) and handles the merging of assembly blocks (mergeOfAssemblyBlocks). This corresponds to the following algorithm.

```
program DAC-ParaLaX (L: Lattice)
 Input
    c: Context;
```

---

[6] Single Program and Multiple Data.

```
    sizeOfGroups: Integer
 var
   vc: Vector of Context;
   diff: Diffusion Group
   adr, whichTask: Integer
 begin
  vc := splitContext(c);
  adr=getAdress();
  nbOfGroups := initAssemblyGroups(adr, sizeOfGroups);
  while(nbOfGroups >0)
    diff := createDiffGroups(adr, nbOfGroups);
    whichTask := getExecutionCode(adr, nbOfGroups);
    L := executeTask(adr, whichTask, vc, L, diff);
    nbOfGroups := mergeOfAssemblyBlocks(adr);
 end.
```

Depending on the process address, `executeTask` will launch the execution of a CS, a SDS or an assembler. The algorithms for each type of task are given in the following paragraphs.

4.1   CS Execution Algorithm

During each pass, the CS supplies assemblers (`serviceOfConcepts`) with concepts, receives the computed new concepts from the assemblers (`receiveNewConcepts`) and a pass termination message from the SDS (`receiveTermMessage`). At the begining of a parallel assembly the CS starts by configuring itself (`initCS`). When all the passes have been accomplished, the CS informs the other tasks by an assembly-end message (`endOfWork`). This behavior is summarized by the algorithm below.

```
program CS (L: Lattice)
 Input
   L1: Lattice or Cj: Context
 var
   nbAss, level, i: Integer;
 begin
  level := initCS(L1 or Cj);
  for i := level downto 0 do
    nbAss := serviceOfConcepts(i, i-1);
    L := L + receiveNewConcepts(nbAss, i);
    receiveTermMessage();
  endOfWork();
 end.
```

4.2   Assembler Execution Algorithm

Each assembler configures itself at the beginning of each parallel assembly (*initAss*). During the assembly process, the assembler receives a list of 1-genitor concepts from the

CS (recieveConceptList), computes both the new concepts generated by those 1-genitors and the respective lines in the *Embed* structure of the assembly (treatmentOfRecievedList), sends its contribution to the update of *Embed* lines to the SDS (sendEmbedLines), sends the new concepts to the CS and takes part in the diffusion process within its group (bcastResults). As a mandatory part of the diffusion, an assembler receives new concepts from other members of its diffusion group. Moreover, if it belongs to an assembly block of an odd number it sends the computed new concepts to the rest of the group. In the later case, its execution ends when it receives an assembly-end message from the CS (messType = -1). The assembler then sends a diffusion-end message within the diffusion group (endOfDiffusionMessage). An assembler from an even-number block, say $2i$, ends its execution only after the reception of all the concepts computed by the block $2i + 1$ and an assembly-end message from its own CS.

```
program Assembler (L: Lattice)
 Input
   L2: Lattice or Ck: Context
 var
   messType, level: Integer;
   cwork: Boolean;
   lineToSend: list of Integer;
   mess, lcToSend: List of Concepts;
 begin
  level := initAss(L2 or Ck);
  cwork := true;
  while(cwork = true)
    mess := recieveConceptList(level);
    messType := getMessageType(mess);
    if (messType <> -1)
     treatmentOfRecievedList(mess, lcToSend, lineToSend);
     L := L + bcastResults(lcToSend, level, messtype);
     sendEmbedLines(lineToSend);
    else
      endOfDiffusionMessage();
      cwork := false;
    level := level-1;
 end.
```

4.3   SDS Execution Algorithm

The SDS starts a parallel assembly by configuring itself (initSDS). At the beginning of each pass the SDS receives the required parameters, i.e., the number of elected assemblers and the number of concepts in the pass, from its CS (recievePassParam). During its execution, the SDS serves *Embed* lines to the assemblers (*serviceOfLines*), takes part in diffusions of concepts whenever its assembly block has an odd number (bcastResults), receives the lines computed by the assemblers and updates *Embed* (receptionOfLines). When the SDS receives an assembly-end message from the CS (messType = -1), its execution ends if it belongs to an assembly block of even number. Otherwise, it needs also a diffusion-end message from each assembler in its diffusion group (endOfDiffusionMessage). The algorithm below provides a summary of the SDS behavior.

```
program SDS (L: Lattice)
   var
      Embed: 2D Array of Concepts
      mess: Message;
   begin
      level := initSDS();
      cwork := true;
      while(cwork = true)
        mess := recievePassParam(level);
        messType := getMessageType(mess);
        if (messType <> -1)
          serviceOfLines(mess, level, Embed);
          L := L + bcastResults(level);
          receptionOfLines(mess, level, Embed);
        else
          endOfDiffusionMessage();
          cwork := false ;
   end.
```

5   Implementation and Experiments

We have implemented DAC-ParaLaX in C++ for Linux. For this purpose, we used the STL[7] [13] and the MPI[8] [6] standard libraries.

STL was chosen since it implements operations on sets and lists with satisfactory complexity. Its choice allowed the extent, intent and the lower covers of a concept to be represented and handled as integer sets. MPI is a message-passing library specification for parallel programming. Since its version 2, MPI provides an object-oriented specification compatible with the C++ language. The choice of MPI was influenced by our target architecture, a cluster of computers equipped with LAM/MPI [8] and a Myrinet version of MPICH [9] which are two implementations of the MPI specification. For the current study we used MPICH.

In MPI, a message is limited to using the basic data types in C/C++[9]. For composite data types of fixed size such as C/C++ data structures that do not contain pointers, MPI offers possibilities for representing them on top of admitted basic data types [6]. These could not be used in our context as concept intents, extents, and lists of lower covers are

---

[7] Standard Template Library.
[8] Message Passing Interface.
[9] byte, integer, Boolean, double etc.

modeled as STL sets, i.e., using pointers. Therefore, we had to solve the data conversion problem which arises when concepts are transferred between tasks. This gap between STL and MPI was tackled by converting each concept or list of concepts into a vector of integers before sending them through MPI. The reverse transformation is done upon the reception of an encoded structure.

The obtained program was used to reach two experimentations goals. The first goal was to evaluate the data conversion time whereas the second goal was to evaluate the time saving when DAC-PARALAX is executed at the place of the sequential D&C algorithm in [17].

To reach the first goal, we consider contexts with a large number of concepts. DAC-PARALAX was then executed with two assembly blocks of four tasks each (DAC-ParaLaX 2x4) on five contexts with 117 attributes, extracted from the *Mushroom* context. The number of concepts in these contexts ranges from 3460 to 17782. The CPU times related to data conversion and to DAC-PARALAX execution on these contexts are consigned in Fig.3. We tend to see the results of the first experiment as an indication that for contexts

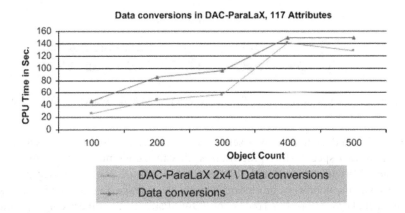

Fig. 3. The time spent by DAC-PARALAX on concept transformation vs the rest of its total time

involving a large number of concepts, the data conversion operations may take a large part of the total CPU time and hence cause a substantial slow-down of the algorithm. In fact, the CPU time spent on that task amounts to more than 50% of the total time of DAC-PARALAX.

To reach the second goal, it was therefore necessary to consider contexts for which the data conversion time could be neglected. Thus, for this initial stage of our study we examined contexts with 34 attributes, extracted from the *Mushroom* context, the number of objects ranging from 500 to 4000. DAC-ParaLaX was executed on these contexts in four different configurations:

- one assembly block of four tasks each (DAC-ParaLaX 1x4),
- one assembly block of eight tasks each (DAC-ParaLaX 1x8),
- two assemblies blocks of four tasks each (DAC-ParaLaX 2x4)
- two assemblies blocks of eight task each (DAC-ParaLaX 2x8).

For each context we computed the speedup of DAC-PARALAX with respect to our STL-based implementation of the sequential D&C algorithm in [17]. Recall that a speedup is obtained by dividing the execution time of the sequential algorithm by the execution time of DAC-PARALAX. Also recall that a speedup $S \succ 1$ means that DAC-PARALAX goes S time more quickly than the sequential algorithm. An acceleration is obtained by deviding a speedup with the number of processors used. The various speedups are depicted in Fig.4. Obviously, the speedup of DAC-PARALAX depends on both the number

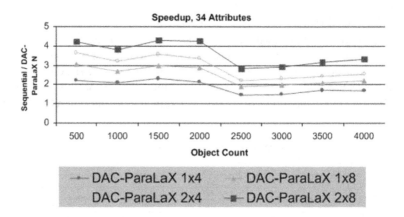

Fig. 4. Speedup obtained by DAC-PARALAX with respect to the sequential D&C algorithm

of processors and the number of assembly blocks. Indeed, with the contexts used for the tests, the average speedup for 8 processors is 2.5 with one assembly block (DAC-ParaLaX 1x8), and 2.9 with two assembly blocks (DAC-ParaLaX 2x4). The average speedup growths with the number of processors whereas the average acceleration decreases with the number of processors. For 4 processors (DAC-ParaLaX 1x4), the average speedup and acceleration are respectively 1.9 and 0.475. This case provides the worst average speedup and the best average acceleration. For 16 processors (DAC-ParaLaX 2x8), the average speedup is 3.60 whereas the average acceleration is 0.225. This case provides the best average speedup and the worst average acceleration. The average acceleration decreases with the number of processors due to the fact that the number of communications related to the handling of the table $Embed$ seriously growths with the number of processors.

The two experiments confirm our hypothesis that DAC-PARALAX realizes significant computation time gains with respect to the sequential version. However, the current implementation of DAC-PARALAX is not yet ready for the computation of the concept lattices from real-world contexts because of the heavy communication costs related to the handling of the table $Embed$. Indeed, the concept redundancy unnecessarily increases the size of the messages related to the transfers of $Embed$ lines. To improve the handling of the table $Embed$, data conversions should first be avoided. To this end, the handling of concepts lists should be implemented using C/C++ vectors of basic data types, that are recognized by MPI, rather than STL set data structure. Secondly, a study of communications will be necessary to determine the size of messages, made of $Embed$

lines, that will maximize communication-computation overlap. In particular, the size of a message should be determined in such a way that its transfer time do not exceed the time spent by the receiver of the corresponding message to perform computations without needing that message.

6   Related Work

In [10], the standard methodology for the parallelization of nested loops is used to design a parallel version of Bordat algorithm [1]. In that parallel version the parallelization is confined to the handling of redundancies that appear in the generation of new concepts. Because of this, that parallelization is incompleted as stated in [11] where a better fine-grain parallelization is proposed. In [3] a parallel version of Ganter algorithm [4] for large context is proposed. This parallelization is based on a partitioning of the search space. The main drawback of that approach stems from the fact that a large search space may provide a few number of concepts. Moreover, it is important to point out that for each of these previous works no effective implementation of the parallel algorithm designed is provided.

7   Conclusion

In this paper, we have proposed a parallel version of the sequential D&C algorithm proposed in [17] for the computation of concept lattices. The obtained parallel algorithm, DAC-PARALAX, is based on coarse-grain tasks. It was designed following the PCAM methodology to which we have added an optimization phase. Although the algorithm is at an initial stage of parallel optimization, its practical performances show that the approach has a potential for further development. In fact, the current bottlenecks are mainly due to technological constraints rather than to design faults.

A subject of ongoing research is the minimization of communications costs in DAC-PARALAX for an arbitrary number of processors. This study should lead to a better implementation of DAC-PARALAX that scales over real-world contexts. Another research track to follow will be to integrate in DAC-PARALAX the improvements made to the initial sequential D&C algorithm in [18]. This should lead to a parallel algorithm for the computation of both the lattice and its Guigues-Duquenne implication base [7].

Acknowledgments

This work was made possible by a grant from the *AUF* (Agence Universitaire de la Francophonie) that financed the short-term visit by the first author at DIRO, University of Montréal. The first author will also like to thank the members of the *GéLo Team* from DIRO for their hospitality during the design and implementation phases of this work, the technical support team of DIRO for the help provided during the experimental stage and the anonymous reviewers for useful comments which help in improving the quality of this paper.

References

1. Bordat, J.P.: Calcul pratique du treillis de Galois d'une correspondance. Mathématiques et Sciences Humaines, Vol. 96 (1986) 31-47
2. Foster, I.: Designing and Building Parallel Program. Addison Wesley (1995)
3. Fu, H., Mephu Nguifo, E.: A Parallel Algorithm to Generate Formal Concepts for Large Data. In Proc. of the Second International Conference on Formal Concept Analysis (ICFCA04), Springer LNCS, Sydney Australia (2004)
4. Ganter, B.: Two basic algorithms in concept analysis. Preprint 831, Technische Hochschule, Darmstadt (1984)
5. Ganter, B, Wille, R.: Formal Concept Analysis, Mathematical Foundations. Springer-Verlag (1999)
6. Gropp, W., Lusk, E., Skjellum, A.: Using MPI: Portable parallel programming with the Message Passing Interface. Cambridge, Mass.: MIT Press (1994)
7. Guigues, J.L., Duquenne, V.: Familles minimales d'implications informatives résultant d'un tableau de données binaires. Mathématiques et Sciences Sociales, Vol. 95 (1986) 5-18
8. LAM/MPI Team: LAM/MPI User Guide, version 7.0.4. Pervasive Technology labs, Indiana University (2004)
9. MPICH home page: http://www.mpich.org. (2004)
10. Ndoundam, R., Njiwoua, P., Mephu Nguifo, E.: Une étude comparative de la parallélisation d'algorithmes de construction de treillis de Galois. Atelier francophone de la plate forme de l'AFIA: Usage des treillis de Galois pour l'intelligence artificielle, ESIEA Recherche (2003)
11. Njiwoua, P., Mephu Nguifo, E.: A Parallel Algorithm to build Concept Lattice. In proc. of the 4th Groningen Int. Information Tech. Conf. for students (1997)
12. Nourine, L., Raynaud, O.: A Fast Algorithm for Building Lattices. Information Processing Letters, Vol. 71 (1999) 199-204
13. Robson, R.: Using the STL: The C++ Standard Template Library, 2nd Edition. New York Springer (2000)
14. Tayou Djamegni, C.: Mapping Rectangular Mesh Algorithm on to Asymptotically Space-Optimal Arrays. Journal of Parallel and Distributed Computing, Vol. 64 **3** (2004) 345-359
15. Tayou Djamegni, C., Tchuenté, M.: A cost-optimal pipeline algorithm for permutation generation in lexicographic order. Journal of Parallel and Distributed Computing, Vol. 44 (1997) 153-159.
16. Tchuenté, M.: Parallel computation on rectangular arrays. Manchester University Press, Manchester and Willey, NY, USA (1992)
17. Valtchev, P., Missaoui, R., Lebrun, P.: A partition-based approach towards building Galois (concept) lattices. Discrete Mathematics, Vol. 256. **3** (2002) 801-829
18. Valtchev, P., Duquenne, V.: Towards scalable divide-and-conquer methods for computing concepts and implications. Preprint submitted to Discrete Applied Mathematics (2004)
19. Wille, R.: Restructuring lattice theory: An approach based on hierarchies of concepts. In I. Rival, editor, *Ordered sets*, Dordrecht-Boston (1982) 445-470

Using Intermediate Representation Systems to Interact with Concept Lattices

Peter Becker[1]

School of Information Technology and Electrical Engineering (ITEE)
The University of Queensland
QLD 4072, Australia
pbecker@itee.uq.edu.au

Abstract. Automated layout of line diagrams for concept lattices is a hard problem as it requires not only asthetical but also semantic considerations. While many layout approaches have been proposed to produce line diagrams that are perceived as good for many applications, a general approach that suits all applications has not yet been found. Instead of proposing another specific layout approach we propose a framework that allows modelling layout constraints that are not only applied for automated layout, but also during manipulation of the diagram layout.

1   Introduction

Laying out concept lattices is a complex problem since there not only asthetical considerations, but the layout of the diagram can also be considered the rhetorical structure of the lattice presentation, an aspect which can be very important for understanding of the underlying information.

Additionally the need for means to interact with the diagrams arises in computer-based tools. For example, the tool ANACONDA offers the user to change diagram layout with the mouse by moving either single nodes, or the nodes representing the corresponding concept's downset or upset.

The program CERNATO introduced the notion of using an intermediate representation to offer manipulation methods that ensure certain layout properties are retained [Bec01]. Experience has shown that the restrictions induced by these constraints were rarely perceived as negative by the intended target audience (business analysts), while people experienced in FCA perceived some of the constraints as too limiting.

This paper gives an overview of the model behind the TOSCANAJ suite [BH04], which is based on the notion of representation systems. It allows modelling different constraints for the manipulation of the diagrams. Basic knowledge of Formal Concept Analyis is assumed throughout the paper.

2   Issues in Lattice Layout

While one can consider a number of asthetical goals for lattice layout, this paper is focused on semantic considerations. The layout of a concept lattice can be

B. Ganter and R. Godin (Eds.): ICFCA 2005, LNAI 3403, pp. 265–268, 2005.

seen as a rhetorical structure, emphasizing certain aspects and de-emphasizing others. In the worst case a wrong layout can be strongly misleading.

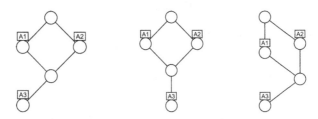

Fig. 1. Three different diagram layouts for the same lattice

Consider Figure 1, where three diagrams represent the same lattice. While they are technically equivalent and all attribute-additive, the left diagram emphasizes the $A3 \rightarrow A2$ implication, the center one the $A3 \rightarrow A1 \wedge A2$ implication and the rightmost diagram does not emphasize any particular implication.

We believe that in many cases finding a suitable rhetorical structure is beyond the capabilities of an automated system. But an automated system can provide some guidance in the process.

3   Representation Systems

Ganter and Wille present a framework to model additive line diagrams using set representations and grid projections [GW99]. To create an attribute-additive line diagram of a concept lattive, the set representation used is a mapping from the concept onto the power set of the irreducible attributes, i.e. the attributes of the purified context. These attributes then get projected using a grid projection.

The general approach to manipulate such a diagram without breaking attribute additivity is to map every movement of a node into a change of the grid projection such that the new position of the node matches the target position. A simple approach to achieve this in the attribute-additive case is to distribute the movement evenly through the projection vectors of all irreducible attributes matching the corresponding upset.

Sometimes additional constraints on the layout should be enforced, especially when additional information is available, such as an order on the attributes. Consider an interordinal scale as an example, where the two ordinal scales should be represented by only two directions in the diagram, not more.

The core idea of our approach is to project the set representation into the target space $\mathbb{R}^2$ by going through two independent steps: first projecting into $\mathbb{R}^n_+$, then applying a parallel projection onto the plane. While this is equivalent to a direct projection in the static case, it allows us to distinguish modifiable and unmodifiable aspects in the interactive case. We split the projection into one part we consider to be fixed and another which we allow the user to change. The intermediate representation in $\mathbb{R}^n_+$ allows us to model a notion of *dimensionality* explicitly which is intuitively found in many line diagrams, for example those representing interordinal or boolean scales.

4   Layout Manipulations

We want the user to be able to move any node excepts for the top node in a diagram by dragging it with the mouse. To fit into the model proposed, the movement of a node has to map into a change of the parallel projection, while at the same time the user will expect the node to move to the location underneath the mouse pointer.

To enforce nothing but attribute-additivity we can assign a unit vector of each dimension to the irreducible attributes. The result for the lattice in Figure 1 is shown in Figure 2, with the vectors in the resulting three-dimensional space attached as labels. Movement along the arrows in the left diagram results in either of the other two as labelled.

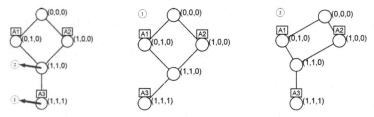

Fig. 2. Diagram manipulation when constrained to attribute additive layouts

If we assume that $A3$ is related to $A1$ in a way that we want to force their vectors to be aligned (for example if we know that generally A3 implies A1 in the system), then we can assign both the same vector as shown in Figure 3.

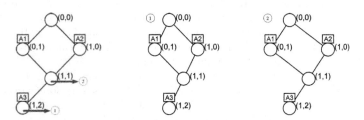

Fig. 3. Diagram manipulation with a chain constraint

Exactly the same layout can be achieved with the first representation system – Figure 2 shows this in the middle. By adding the constraint we enforce that this alignment can not be broken through the manipulations. For example trying to move the bottom node to achieve the layout shown in the left of Figure 2 results in the diagram shown in Figure 3 instead. But moving the center node behaves exactly as with the first layout – the projection of all nodes but the bottom stayed the same, no constraints are enforced on this part of the diagram.

Another constraint possible for diagrams of this lattice is enforcing that the vector assigned to $A3$ is always extending the vector straight through the diamond shape above. In that case the second representation and layout system has to be changed to assign a vector such as $(\sqrt{2}, \sqrt{2})$ instead of $(0, 1)$ to $A3$.

In some situations the user laying out a diagram might want to ignore the constraints given by the framework proposed. In the case of highly non-distributive lattices such as M_n with large n, breaking a constraint such as attribute-additivity is highly desirable. To achieve this, another set of vectors can be added to the concepts, which is applied as an additional offset when determining the position of the concept.

5   Conclusion

We have presented a short overview of a framework that enables software users to interact with concept lattices while maintaining certain layout constraints. The framework allows modelling a range of different constraints, such as attribute-additivity and chain layouts. These constraints can be application-specific, they can be derived automatically from additional information or they can be explicitly given by a user.

Further work will have to investigate which constraint systems are useful for different applications and different types of users. The connection to related work in general lattice layout (e.g. [Sko92]) and to force-based lattice layout (e.g. [Col00]) should be elaborated.

References

[Bec01]   Peter Becker. Multi-dimensional representations of conceptual hierarchies. In G. Stumme and G. Mineau, editors, *Proceedings of the 9th International Conference on Conceptual Structures*, Supplementary Proceedings ICCS, pages 33–46. Department of Computer Science, University Laval, 2001.

[BH04]   Peter Becker and Joachim Hereth Correia. The ToscanaJ suite for implementing Conceptual Information Systems. In *Formal Concept Analysis – State of the Art*. Springer, Berlin – Heidelberg – New York, 2004. To appear.

[Col00]   Richard Cole. Automatic layout of concept lattices using force directed placement and genetic algorithms. In Jenny Edwards, editor, *23th Australiasian Computer Science Conference*, volume 22 of *Australian Computer Science Communications*, pages 47–53. IEEE Computer Society, 2000.

[GW99]   B. Ganter and R. Wille. *Formal Concept Analysis: Mathematical Foundations*. Springer-Verlag, Berlin, 1999.

[Sko92]   Martin Skorsky. *Endliche Verbände — Diagramme und Eigenschaften*. PhD thesis, TH Darmstadt, 1992. Verlag Shaker, Aachen, 1992.

Crisply Generated Fuzzy Concepts

Radim Bělohlávek, Vladimír Sklenář, and Jiří Zacpal

Dept. Computer Science, Palacký University,
Tomkova 40, CZ-779 00 Olomouc, Czech Republic
{radim.belohlavek, vladimir.sklenar, jiri.zacpal}@upol.cz

Abstract. In formal concept analysis of data with fuzzy attributes, both the extent and the intent of a formal (fuzzy) concept may be fuzzy sets. In this paper we focus on so-called crisply generated formal concepts. A concept $\langle A, B \rangle \in \mathcal{B}(X, Y, I)$ is crisply generated if $A = D^{\downarrow}$ (and so $B = D^{\downarrow\uparrow}$) for some crisp (i.e., ordinary) set $D \subseteq Y$ of attributes (generator). Considering only crisply generated concepts has two practical consequences. First, the number of crisply generated formal concepts is considerably less than the number of all formal fuzzy concepts. Second, since crisply generated concepts may be identified with a (ordinary, not fuzzy) set of attributes (the largest generator), they might be considered "the important ones" among all formal fuzzy concepts. We present basic properties of the set of all crisply generated concepts, an algorithm for listing all crisply generated concepts, a version of the main theorem of concept lattices for crisply generated concepts, and show that crisply generated concepts are just the fixed points of pairs of mappings resembling Galois connections. Furthermore, we show connections to other papers on formal concept analysis of data with fuzzy attributes. Also, we present examples demonstrating the reduction of the number of formal concepts and the speed-up of our algorithm (compared to listing of all formal concepts and testing whether a concept is crisply generated).

1 Problem Setting and Preliminaries

1.1 Problem Setting

Formal concept analysis (FCA) [12] deals with object-attribute data tables (objects and attributes corresponding to table rows and columns, respectively). In the basic setting, attributes are assumed to be binary, i.e. table entries are 1 or 0 according to whether an attribute applies to an object or not. If the attributes under consideration are fuzzy (like "cheap", "expensive"), each table entry contains a truth degree to which an attribute applies to an object. The degrees can be taken from some appropriate scale containing 0 (does not apply at all) and 1 (fully applies) as bounds. The most popular choice is some subinterval of $[0, 1]$, but in general, degrees need not be numbers. A data table with truth degrees can be considered a many-valued context and can be transformed to a binary data table via so-called conceptual scaling [12]. Alternatively, the table with truth degrees can be approached using the apparatus of FCA generalized

B. Ganter and R. Godin (Eds.): ICFCA 2005, LNAI 3403, pp. 269–284, 2005.

to fuzzy setting (generalization of FCA from the point of view of fuzzy logic). A general discussion about the relationship between conceptual scaling in the sense of FCA and membership functions in the sense of fuzzy set can be found in [21].

In the present paper, we are interested in FCA of data with fuzzy attributes (FCAf) in the framework of fuzzy logic and fuzzy set theory. Probably the first paper on this was [11]. Later on, FCAf was developed by Pollandt [18] and, independently, by the first author of this paper, e.g. [1, 2, 3, 7]. An important aspect of FCA in general is the possibly large number of formal concepts extracted from data. In this paper, we propose and study what we call crisply generated formal fuzzy concepts. These are particular formal fuzzy concepts which can be considered "more important" than the others (non-crisply generated). Considering only crisply generated concepts, the main practical effect is the reduction of the number of formal concepts extracted from data. In the rest of this section, we present preliminaries on fuzzy logic and FCAf. In Section 2 we present our approach and theoretical results. Section 3 contains examples and experiments studying mainly the reduction of the number of extracted concepts.

1.2   Preliminaries

Fuzzy Sets and Fuzzy Logic. We assume basic familiarity with fuzzy logic and fuzzy sets [16, 13, 6]. An element may belong to a fuzzy set in an intermediate degree not necessarily being 0 or 1. Formally, a fuzzy set A in a universe X is a mapping assigning to each $x \in X$ a truth degree $A(x) \in L$ where L is some partially ordered set of truth degrees containing at least 0 (full falsity) and 1 (full truth). L needs to be equipped with logical connectives, e.g. $\otimes$ (fuzzy conjunction), $\rightarrow$ (fuzzy implication), etc. L together with logical connectives forms a structure $\mathbf{L}$ of truth degrees. We assume that $\mathbf{L}$ forms a so-called complete residuated lattice. Recall that a complete residuated lattice [6, 13, 14] is a structure $\mathbf{L} = \langle L, \wedge, \vee, \otimes, \rightarrow, 0, 1 \rangle$ such that (1) $\langle L, \wedge, \vee, 0, 1 \rangle$ is a complete lattice (with the least element 0, greatest element 1), i.e. a partially ordered set in which arbitrary infima ($\bigwedge$) and suprema ($\bigvee$) exist; (2) $\langle L, \otimes, 1 \rangle$ is a commutative monoid, i.e. $\otimes$ is a binary operation satisfying $x \otimes (y \otimes z) = (x \otimes y) \otimes z$, $x \otimes y = y \otimes x$, and $x \otimes 1 = x$; (3) $\otimes, \rightarrow$ satisfy $x \otimes y \leq z$ iff $x \leq y \rightarrow z$. In what follows, $\mathbf{L}$ always denotes a fixed complete residuated lattice.

The most applied set of truth degrees is the real interval $[0, 1]$; with $a \wedge b = \min(a, b)$, $a \vee b = \max(a, b)$, and with three important pairs of fuzzy conjunction and fuzzy implication: Łukasiewicz ($a \otimes b = \max(a + b - 1, 0)$, $a \rightarrow b = \min(1 - a + b, 1)$), minimum ($a \otimes b = \min(a, b)$, $a \rightarrow b = 1$ if $a \leq b$ and $= b$ else), and product ($a \otimes b = a \cdot b$, $a \rightarrow b = 1$ if $a \leq b$ and $= b/a$ else). Another possibility is to take a finite chain $\{a_0 = 0, a_1, \ldots, a_n = 1\}$ ($a_0 < \cdots < a_n$) equipped with Łukasiewicz structure ($a_k \otimes a_l = a_{\max(k+l-n, 0)}$, $a_k \rightarrow a_l = a_{\min(n-k+l, n)}$) or minimum ($a_k \otimes a_l = a_{\min(k, l)}$, $a_k \rightarrow a_l = a_n$ for $a_k \leq a_l$ and $a_k \rightarrow a_l = a_l$ otherwise).

The set of all fuzzy sets (or $\mathbf{L}$-sets) in X is denoted L^X. For a fuzzy set $A \in L^X$, the 1-cut $^1 A$ of A is an ordinary set $^1 A = \{x \in X \mid A(x) = 1\}$. A is called crisp if $A(x) \in \{0,1\}$. By $\{a/x\}$ we denote a fuzzy set A for which $A(x) = a$ and $A(y) = 0$ for $y \neq x$. For fuzzy sets A, B in X we put $A \subseteq B$ (A is a subset of B) if for each $x \in X$ we have $A(x) \leq B(x)$. More generally, the degree $S(A, B)$ to which A is a subset of B is defined by $S(A, B) = \bigwedge_{x \in X}(A(x) \rightarrow B(x))$. Then, $A \subseteq B$ means $S(A, B) = 1$.

Formal Concept Analysis of Data with Fuzzy Attributes. Let X and Y be sets of objects and attributes, respectively, I be a fuzzy relation between X and Y. That is, $I : X \times Y \rightarrow L$ assigns to each $x \in X$ and each $y \in Y$ a truth degree $I(x, y) \in L$ to which object x has attribute y (L is a support set of some complete residuated lattice **L**). The triplet $\langle X, Y, I \rangle$ is called a formal fuzzy context (corresponds to a data table with fuzzy attributes). For fuzzy sets $A \in L^X$ and $B \in L^Y$, consider fuzzy sets $A^\uparrow \in L^Y$ and $B^\downarrow \in L^X$ (denoted also $A^{\uparrow I}$ and $B^{\downarrow I}$) defined by

$$A^\uparrow(y) = \bigwedge_{x \in X} (A(x) \rightarrow I(x, y)), \tag{1}$$

$$B^\downarrow(x) = \bigwedge_{y \in Y} (B(y) \rightarrow I(x, y)) \tag{2}$$

for $y \in Y$ and $x \in X$. Using basic rules of predicate fuzzy logic, one can see that $A^\uparrow(y)$ is the truth degree of the proposition "y is shared by all objects from A" and $B^\downarrow(x)$ is the truth degree of "x has all attributes from B". Putting

$$\mathcal{B}(X, Y, I) = \{\langle A, B \rangle \mid A^\uparrow = B, \ B^\downarrow = A\},$$

$\mathcal{B}(X, Y, I)$ is the set of all pairs $\langle A, B \rangle$ such that (a) A is the collection of all objects that have all the attributes of (the intent) B and (b) B is the collection of all attributes that are shared by all the objects of (the extent) A. Elements of $\mathcal{B}(X, Y, I)$ are called formal concepts of $\langle X, Y, I \rangle$ (formal fuzzy concepts, formal **L**-concepts); $\mathcal{B}(X, Y, I)$ is called the concept lattice given by $\langle X, Y, I \rangle$ (fuzzy concept lattice, **L**-concept lattice). Both the extent A and the intent B of a formal concept $\langle A, B \rangle$ are in general fuzzy sets. This corresponds to the fact that in general, concepts apply to objects and attributes to various intermediate degrees, not only 0 and 1.

Putting

$$\langle A_1, B_1 \rangle \leq \langle A_1, B_1 \rangle \text{ iff } A_1 \subseteq A_2 (\text{iff } B_1 \supseteq B_2) \tag{3}$$

for $\langle A_1, B_1 \rangle, \langle A_2, B_2 \rangle \in \mathcal{B}(X, Y, I)$, $\leq$ models the subconcept-superconcept hierarchy in $\mathcal{B}(X, Y, I)$.

The following is a version of the main theorem for fuzzy concept lattices (see [7, 18]).

Theorem 1. *The set $\mathcal{B}(X, Y, I)$ is under $\leq$ a complete lattice where infima and suprema are given by*

$$\bigwedge_{j \in J} \langle A_j, B_j \rangle = \langle \bigcap_{j \in J} A_j, (\bigcup_{j \in J} B_j)^{\downarrow \uparrow} \rangle, \tag{4}$$

$$\bigvee_{j \in J} \langle A_j, B_j \rangle = \langle (\bigcup_{j \in J} A_j)^{\uparrow \downarrow}, \bigcap_{j \in J} B_j \rangle. \tag{5}$$

Moreover, an arbitrary complete lattice $\mathbf{V} = \langle V, \wedge, \vee \rangle$ *is isomorphic to some* $\mathcal{B}(X, Y, I)$ *iff there are mappings* $\gamma : X \times L \to V$, $\mu : Y \times L \to V$ *such that* $\gamma(X, L)$ *is* $\bigvee$-*dense in* V, $\mu(Y, L)$ *is* $\bigwedge$-*dense in* V; $a \otimes b \leq I(x, y)$ *iff* $\gamma(x, a) \leq \mu(y, b)$.

Note that Theorem 1 can be proved by reduction (see [18, 4]) to the main theorem of ordinary concept lattices [20] or directly in the framework of fuzzy logic [7]. Note also that Theorem 1 is concerned with bivalent order, there is still a more general version [7] dealing with many-valued (fuzzy) order.

Taking $L = \{0, 1\}$ (two truth degrees; bivalent case), the notions of formal fuzzy context, formal fuzzy concept, and fuzzy concept lattice coincide with the ordinary notions [12]. In the following we denote $\text{Ext}(I) = \{A \mid \langle A, B \rangle \in \mathcal{B}(X, Y, I)$ for some $B\}$ (extents of concepts) and $\text{Int}(I) = \{B \mid \langle A, B \rangle \in \mathcal{B}(X, Y, I)$ for some $A\}$ (intents of concepts). Recall [3] that $^\uparrow$ and $^\downarrow$ satisfy $S(A_1, A_2) \leq S(A_2^\uparrow, A_1^\uparrow)$; $S(B_1, B_2) \leq S(B_2^\downarrow, B_1^\downarrow)$; $A \subseteq A^{\uparrow\downarrow}$; $B \subseteq B^{\downarrow\uparrow}$. As a consequence, $A^\uparrow = A^{\uparrow\downarrow\uparrow}$ and $B^\downarrow = B^{\downarrow\uparrow\downarrow}$.

2   Crisply Generated Formal Concepts

2.1   Motivation and Definition

A formal concept $\langle A, B \rangle$ consists of a fuzzy set A and a fuzzy set B such that $A^\uparrow = B$ and $B^\downarrow = A$. Due to (1) and (2), and the basic rules of predicate fuzzy logic, this directly captures the verbal definition of a formal concept inspired by Port-Royal logic. Nevertheless, this definition actually allows for formal fuzzy concepts $\langle A, B \rangle$ such that, for example, for any $x \in X$ and $y \in Y$ we have $A(x) = 1/2$ and $B(y) = 1/2$. A verbal description of such a concept is "a concept to which each attribute belongs to degree 1/2". Such a concept, although satisfying the verbally described condition $A^\uparrow = B$, $B^\downarrow = A$, will probably be considered "not the important one". This is because people expect concepts to be determined by "some attributes", i.e. by an ordinary set of attributes. This leads to the following definition.

Definition 1. *A formal fuzzy concept* $\langle A, B \rangle \in \mathcal{B}(X, Y, I)$ *is called* crisply generated *if there is a crisp set* $B_c \subseteq Y$ *such that* $A = B_c^\downarrow$ *(and thus* $B = B_c^{\downarrow\uparrow}$*).*

We say that B_c crisply generates $\langle A, B \rangle$. Let $\mathcal{B}_c(X, Y, I)$ denote the collection of all crisply generated formal concepts in $\langle X, Y, I \rangle$, i.e.

$$\mathcal{B}_c(X, Y, I) = \{\langle A, B \rangle \in \mathcal{B}(X, Y, I) \mid \text{ there is } B_c \subseteq Y : A = B_c^\downarrow\}.$$

If $\langle A, B \rangle$ is a crisply generated concept with $A = B_c^\downarrow$, it might be actually more informative to write $\langle A, B_c \rangle$ instead of $\langle A, B \rangle$. Doing so, no information is lost since the corresponding fuzzy concept $\langle A, B \rangle$ can be obtained from $\langle A, B_c \rangle$ by taking $B = B_c^{\downarrow\uparrow}$. In general, there may be several crisp B_c's with $A = B_c^\downarrow$. To remove this ambiguity, we can always take the greatest B_c:

Lemma 1. *For a crisply generated formal concept* $\langle A, B \rangle$, ^1B *is the largest crisp set* B_c *for which* $A = B_c^\downarrow$.

Proof. Let $\langle A, B \rangle$ be crisply generated by some B_c, i.e. $A = B_c^{\downarrow}$. Since $B_c \subseteq B_c^{\downarrow\uparrow} = B$, we have $B_c = {}^1B_c \subseteq {}^1B$. That is, ${}^1B$ contains any crisp B_c which generates $\langle A, B \rangle$. Moreover, ${}^1B$ itself is a crisp set which generates $\langle A, B \rangle$. Indeed, take some crisp B_c which generates $\langle A, B \rangle$. We know that $B_c \subseteq {}^1B$, from which we get $B = B_c^{\downarrow\uparrow} \subseteq ({}^1B)^{\downarrow\uparrow}$. On the other hand, ${}^1B \subseteq B$ gives $({}^1B)^{\downarrow\uparrow} \subseteq B^{\downarrow\uparrow} = B$ which shows $({}^1B)^{\downarrow\uparrow} = B$.

Crisply generated formal concepts can be alternatively defined as maximal rectangles $\langle A, B \rangle$ contained in I for which A is the projection of ${}^1B$. Call a fuzzy relation $I' \in L^{X \times Y}$ a *rectangular relation* if there are $A \in L^X$, $B \in L^Y$ such that $I'(x, y) = A(x) \otimes B(y)$, written $I' = A \otimes B$ (call then $\langle A, B \rangle$ a *rectangle*). $\langle A, B \rangle$ is said to be *contained* in $I'' \in L^{X \times Y}$ if $A \otimes B \subseteq I''$. We put $\langle A_1, B_1 \rangle \trianglelefteq \langle A_2, B_2 \rangle$ if for each $x \in X$, $y \in Y$ we have $A_1(x) \leq A_2(x)$ and $B_1(y) \leq B_2(y)$. By an I-projection of a subset $C \subseteq Y$ on X we mean a fuzzy set A in X defined by $A(x) = \bigwedge_{y \in C} I(x, y)$.

Lemma 2. *$\langle A, B \rangle$ is a crisply generated concept iff $\langle A, B \rangle$ is a maximal (w.r.t. $\trianglelefteq$) rectangle contained in I such that A is the projection of ${}^1B$.*

Proof. The assertion follows from [6, Theorem 5.7], the fact that $\langle A, B \rangle$ is crisply generated iff $A = ({}^1B)^{\downarrow}$ (see Lemma 1), and from $({}^1B)^{\downarrow}(x) = \bigwedge_{y \in {}^1B} I(x, y)$.

2.2   Independence of the Choice of Fuzzy Logical Connectives

The next step is to observe that restricting ourselves to crisply generated concepts, one is no more dependent (almost) on the logical connectives defined on the scale L of truth degrees. To formulate this precisely, let us denote the concept lattice over the structure $\mathbf{L}$ of truth degrees by $\mathcal{B}^{\mathbf{L}}(X, Y, I)$ and denote $\mathcal{B}_c^{\mathbf{L}}(X, Y, I)$ the set of all crisply generated concepts of $\mathcal{B}^{\mathbf{L}}(X, Y, I)$. Suppose we have two structures $\mathbf{L}_1$ and $\mathbf{L}_2$ with a common set L of truth degrees, i.e. $\mathbf{L}_1 = \langle L, \otimes_1, \rightarrow_1, \ldots \rangle$ and $\mathbf{L}_2 = \langle L, \otimes_2, \rightarrow_2, \ldots \rangle$, and a data table (formal fuzzy context) $\langle X, Y, I \rangle$ which is filled with truth degrees from L.

Lemma 3. *Let $\mathbf{L}_1$ and $\mathbf{L}_2$ have a common set L of truth degrees, let $\langle X, Y, I \rangle$ be a formal fuzzy context with truth degrees from L. Then there is an isomorphism between $\mathcal{B}_c^{\mathbf{L}_1}(X, Y, I)$ and $\mathcal{B}_c^{\mathbf{L}_2}(X, Y, I)$ such that for the corresponding formal concepts $\langle A_1, B_1 \rangle \in \mathcal{B}_c^{\mathbf{L}_1}(X, Y, I)$ and $\langle A_2, B_2 \rangle \in \mathcal{B}_c^{\mathbf{L}_2}(X, Y, I)$ we have $A_1 = A_2$ and ${}^1B_1 = {}^1B_2$.*

Proof. Denote by $\downarrow_i$ and $\uparrow_i$ the operators generated by $\rightarrow_i$ ($i = 1, 2$). Recall that for each residuated implication connective $\rightarrow$ we have $1 \rightarrow a = a$. Therefore, for each crisp $B \subseteq Y$ we have $B^{\downarrow_i}(x) = \bigwedge_{y \in Y}(B(y) \rightarrow_i I(x, y)) = \bigwedge_{y \in B}(1 \rightarrow_i I(x, y)) = \bigwedge_{y \in B} I(x, y)$. That is, $B^{\downarrow_i}$ does not depend on $\rightarrow_i$. Therefore, if $\langle A, B \rangle \in \mathcal{B}_c^{\mathbf{L}_1}(X, Y, I)$ is crisply generated then from Lemma 1 we have $({}^1B)^{\downarrow_1} = A$ and since ${}^1B$ is crisp, also $({}^1B)^{\downarrow_2} = A$. This shows that $\langle A, D \rangle$, for $D = A^{\uparrow_2}$, is a crisply generated formal concept from $\mathcal{B}_c^{\mathbf{L}_2}(X, Y, I)$. Clearly, ${}^1B \subseteq {}^1D$. If ${}^1B \subset {}^1D$, i.e. D is larger than B, then Lemma 1 gives $({}^1D)^{\downarrow_2} = A$ and so $({}^1D)^{\downarrow_1} = A$

which is impossible since by Lemma 1, ^1B is the largest one with $(^1B)^{\downarrow_1} = A$. In a similar way one shows that if $\langle A, B \rangle \in \mathcal{B}_c^{\mathbf{L}_2}(X, Y, I)$ is crisply generated then $\langle A, A^{\uparrow_1} \rangle \in \mathcal{B}^{\mathbf{L}_1}(X, Y, I)$ is crisply generated as well and $^1B = \, ^1A^{\uparrow_1}$. The assertion then immediately follows.

Note that in general, $\mathcal{B}^{\mathbf{L}_1}(X, Y, I)$ and $\mathcal{B}^{\mathbf{L}_2}(X, Y, I)$ may have different number of formal concepts, i.e. the choice of fuzzy logical connectives matters. Lemma 3 shows that their crisply generated parts $\mathcal{B}_c^{\mathbf{L}_1}(X, Y, I)$ and $\mathcal{B}_c^{\mathbf{L}_2}(X, Y, I)$ are isomorphic. That is, if we consider only crisply generated concepts, the choice of fuzzy logical connectives, in a sense, does not matter.

2.3   Computing All Crisply Generated Formal Concepts

We now present an algorithm for generating $\mathcal{B}_c(X, Y, I)$. Going directly by definition, i.e. creating $\langle B^{\downarrow}, B^{\downarrow\uparrow} \rangle$ for each crisp $B \in 2^X$, has exponential time complexity and thus, cannot be used. Our algorithm is inspired by Ganter's Next Closure algorithm [12, p. 67] for generating an ordinary concept lattice, i.e. generating all formal concepts in lexicographic order. This idea can be adopted to fuzzy setting to generate all crisply generated formal fuzzy concepts.

The idea of our algorithm is to introduce a linear ordering $<$ on $\mathcal{B}_c(X, Y, I)$ such that for a given $\langle A, B \rangle \in \mathcal{B}_c(X, Y, I)$, we can compute its immediate successor w.r.t. to $<$. Since a formal concept $\langle A, B \rangle$ is uniquely given by its intent B, it is sufficient to generate all intents B. By $\mathrm{Int}_c(I)$ we denote all intents of crisply generated fuzzy concepts, i.e. $\mathrm{Int}_c(I) = \{B \mid \langle A, B \rangle \in \mathcal{B}_c(X, Y, I) \text{ for some } A \in L^X\}$. We suppose that $Y = \{1, \ldots, n\}$; $L = \{0 = a_1, a_2, \ldots, a_k = 1\}$ such that if $a_i \leq a_j$ in $\mathbf{L}$ then $i \leq j$ (that is, the ordering of elements of L by indices extends their ordering in $\leq$ in $\mathbf{L}$; such an indexing is always possible and is automatically satisfied if $\mathbf{L}$ is linearly ordered and we index the elements in L using this order from the least to the greatest element, i.e. $a_1 \leq a_2 \leq \cdots \leq a_k$). For $i = 1, \ldots, n$, introduce a relation $<_i$ on L^Y by

$$B_1 <_i B_2 \text{ iff } (^1B_1)(i) = 0, (^1B_2)(i) = 1, \text{ and } (^1B_1)(j) = (^1B_2)(j) \text{ for } j < i.$$

Furthermore, we put

$$B_1 < B_2 \text{ iff } B_1 <_i B_2 \text{ for some } i.$$

That is, $B_1 < B_2$ iff the first element of Y on which $^1B_1$ and $^1B_2$ differ, belongs to B_2; i.e. $B_1 < B_2$ means that $^1B_1$ is lexicographically smaller than $^1B_2$.

Lemma 4. $<$ is a strict total order on $\mathrm{Int}_c(I)$ which extends $\subset$.

Proof. Easy to see since every $B_1, B_2 \in \mathrm{Int}_c(I)$ with common 1-cuts (i.e. with $^1B_1 = \, ^1B_2$) are equal. Indeed, By Lemma 1, if $^1B_1 = \, ^1B_2$ then $B_1 = (^1B_1)^{\downarrow\uparrow} = (^1B_2)^{\downarrow\uparrow} = B_2$.

Furthermore, for $B \in L^Y$ and $i \in \{1, \ldots, n\}$, we put

$$B \oplus_c i := ((^1B \cap \{1, \ldots, i-1\}) \cup \{i\})^{\downarrow\uparrow}.$$

That is, we obtain $B \oplus_c i$ by taking the 1-cut of B, cutting off the elements $i, \ldots, n$, joining with element i and applying the closure $^{\downarrow\uparrow}$.

Lemma 5. *The following assertions are true.*

(1) If $B <_i D_1$, $B <_j D_2$, and $i < j$ then $D_2 <_i D_1$;
(2) if $B <_i D$ and $D = D^{\downarrow\uparrow}$ then $(^1B) \oplus_c i \subseteq D$;
(3) if $B <_i D$ and $D = D^{\downarrow\uparrow}$ then $B <_i (^1B) \oplus_c i$.

Proof. (1) follows directly from definition. (2) From $B <_i D$ we have $D(i) = 1$ and $^1B \cap \{1, \ldots, i-1\} \subseteq D$. Putting $C_1 = {}^1B \cap \{1, \ldots, i-1\}$, $C_2 = \{\,^1/i\,\}$, we thus have $C_1 \cup C_2 \subseteq D$, whence $(^1B) \oplus_c i = (C_1 \cup C_2)^{\downarrow\uparrow} \subseteq D^{\downarrow\uparrow} = D$. (3) From $B <_i D$ we have $^1B \cap \{1, \ldots, i-1\} = {}^1D \cap \{1, \ldots, i-1\}$. Using (2) we get $(^1B) \oplus_c i \subseteq D$, and so $^1(^1B \oplus_c i) \cap \{1, \ldots, i-1\} \subseteq {}^1D \cap \{1, \ldots, i-1\} = {}^1B \cap \{1, \ldots, i-1\}$. On the other hand, $^1(^1B \oplus_c i) \cap \{1, \ldots, i-1\} \supseteq {}^1(^1B \cap \{1, \ldots, i-1\})^{\downarrow\uparrow} \cap \{1, \ldots, i-1\} \supseteq {}^1(^1B \cap \{1, \ldots, i-1\}) \cap \{1, \ldots, i-1\} \supseteq {}^1B \cap \{1, \ldots, i-1\}$. Therefore, $^1B \cap \{1, \ldots, i-1\} = {}^1(^1B \oplus_c i) \cap \{1, \ldots, i-1\}$. Finally, by (2), $1 = (^1B \oplus_c i)(i) \leq D(i)$, i.e. $D(i) = 1 = a_k$ proving $B <_i (^1B) \oplus_c i$.

Theorem 2. *For $B \in L^Y$, the least crisply generated intent $B^{+c} \in \mathrm{Int}_c(I)$ which is greater than B is given by*

$$B^{+c} = B \oplus_c i$$

where i is the greatest element with $B <_i B \oplus_c i$.

Proof. Let B^{+c} be the required successor of B w.r.t. $<$. We have $B < B^{+c}$, i.e $B <_i B^{+c}$ for some i. By Lemma 5 (3), $B <_i {}^1B \oplus_c i$. By Lemma 5 (2), $^1B \oplus_c i \subseteq B^{+c}$ and thus $^1B \oplus_c i \leq B^{+c}$ (i.e. $^1B \oplus_c i < B^{+c}$ or $^1B \oplus_c i = B^{+c}$), and so $B <_i {}^1B \oplus_c i \leq B^{+c}$. Since B^{+c} is the successor of B, we have $B^{+c} = {}^1B \oplus_c i$. It remains to show that i is the greatest element with $B <_i {}^1B \oplus_c i$, i.e. $^1B <_i {}^1(^1B \oplus_c i)$. If $^1B <_j {}^1(^1B \oplus_c j)$ for $i < j$ then Lemma 5 (1) yields $^1(^1B \oplus_c j) <_i {}^1(^1B \oplus_c i)$, i.e. $^1B \oplus_c j < {}^1B \oplus_c i$ which is a contradiction to $^1B \oplus_c i = B^{+c} < {}^1B \oplus_c j$ (since B^{+c} is the immediate of B).

Theorem 2 leads to the following algorithm.

INPUT: $\langle X, Y, I \rangle$, OUTPUT: $\mathrm{Int}_c(I)$

```
store(B)
while B ≠ Y do
   B := B⁺ᶜ
   store(B)
```

The time complexity of computing from B the next crisply generated intent B^{+c} is $O(|X| \cdot |Y|^2)$. Therefore, our algorithm has polynomial time delay complexity [15] (generating crisply generated intents, one generates the successor B^{+c} of B in polynomial time $O(|X| \cdot |Y|^2)$). The time complexity of the algorithm is thus $O(|\mathrm{Int}_c(I)| \cdot |X| \cdot |Y|^2)$.

Remark 1. Note that in [8] we presented an algorithm for generating all formal fuzzy concepts of $\mathcal{B}(X, Y, I)$. This algorithm is inspired by Ganter's Next Closure which is its particular case. Using this algorithm, we can generate $\mathcal{B}_c(X, Y, I)$ in the following way: Generate all $\langle A, B \rangle \in \mathcal{B}(X, Y, I)$ and for each such $\langle A, B \rangle$, test by Lemma 1 whether $\langle A, B \rangle$ is crisply generated. Comprated to this, the algorithm presented here generates $\mathcal{B}_c(X, Y, I)$ directly, going from one crisply generated concept to the next one. We demonstrate the speed-up in Section 3.

2.4 Crisply Generated Fuzzy Concepts as Fixed Points of Fuzzy Galois-Like Mappings

It is well-known that ordinary formal concepts $\langle X, Y, I \rangle$ are exactly the fixed points of a Galois connection formed by (the concept derivation operators) $^{\uparrow} : 2^X \to 2^Y$ and $^{\downarrow} : 2^Y \to 2^X$ induced by I [19, 12]. Moreover, each Galois connection between X and Y is induced by some relation $I \in 2^{X \times Y}$. In [2], this fact was generalized to the setting of fuzzy logic: Call a fuzzy Galois connection between X and Y any pair $\langle ^{\uparrow}, ^{\downarrow} \rangle$ of mappings $^{\uparrow} : L^X \to L^Y$ and $^{\uparrow} : L^Y \to L^X$ satisfying

$$S(A_1, A_2) \le S(A_2^{\uparrow}, A_1^{\uparrow}) \tag{6}$$

$$S(B_1, B_2) \le S(B_2^{\downarrow}, B_1^{\downarrow}) \tag{7}$$

$$A \subseteq A^{\uparrow\downarrow} \tag{8}$$

$$B \subseteq B^{\downarrow\uparrow}, \tag{9}$$

for each $A, A_1, A_2 \in L^X$ and $B, B_1, B_2 \in L^Y$. It was proved in [2] that given $\langle X, Y, I \rangle$, the pair $\langle ^{\uparrow}, ^{\downarrow} \rangle$ defined by (1) and (2) is a fuzzy Galois connection and, conversely, each fuzzy Galois connection is induced by some $\langle X, Y, I \rangle$ by (1) and (2). The relationship between fuzzy Galois connections and fuzzy relations between X and Y is one-to-one.

A natural question arises as to whether crisply generated fuzzy concepts can be thought of as fixed points of suitable mappings, possibly axiomatically definable. In the following, we present a positive answer. In fact, what we are going to present is a special case of a more general case of so-called (fuzzy) Galois connections with hedges [10]. However, to keep our discussion in the framework of crisply generated concepts, we do not go to the more general notions of [10] and present the results with proofs for our special case.

Consider mappings $^{\triangle} : L^X \to L^Y$ and $^{\triangledown} : L^Y \to L^X$ resulting from $\langle X, Y, I \rangle$ by

$$A^{\triangle}(y) = \bigwedge_{x \in X} (A(x) \to I(x, y)) \tag{10}$$

and

$$B^{\triangledown}(x) = \bigwedge_{y \in Y} (^1 B(y) \to I(x, y)). \tag{11}$$

Note that we have $A^{\triangle} = A^{\uparrow}$ and $B^{\triangledown} = (^1 B)^{\downarrow}$ where $^{\uparrow}$ and $^{\downarrow}$ are defined by (1) and (2). Now, denote by $\mathcal{B}(X, {}^1Y, I)$ the set of all fixed points of $\langle ^{\triangle}, ^{\triangledown} \rangle$, i.e.

$$\mathcal{B}(X, {}^1Y, I) = \{\langle A, B \rangle \in L^X \times L^Y \mid A^{\triangle} = B, B^{\triangledown} = A\}.$$

Theorem 3. $\mathcal{B}\left(X, {}^1Y, I\right) = \mathcal{B}_c\left(X, Y, I\right)$, *i.e. crisply generated fuzzy concepts are exactly the fixed points of $^\triangle$ and $^\triangledown$.*

Proof. "$\subseteq$": If $\langle A, B \rangle \in \mathcal{B}\left(X, {}^1Y, I\right)$ then $A^\triangle = B$ and $B^\triangledown = A$, i.e. $A^\uparrow = B$ and $({}^1B)^\downarrow = A$. Therefore, $\langle A, B \rangle \in \mathcal{B}_c\left(X, Y, I\right)$, by definition.

"$\supseteq$": Let $\langle A, B \rangle \in \mathcal{B}_c\left(X, Y, I\right)$, i.e. $A^\uparrow = B$, $B^\downarrow = A$, and $A = D^\downarrow$ for some crisp $D \subseteq Y$. We need to verify $A^\triangle = B$ and $B^\triangledown = A$. By Lemma 1, it clearly suffices to check $({}^1B)^\downarrow = A$, i.e. $B^\triangledown = A$. As $A = D^\downarrow$ and $B = B^{\downarrow\uparrow}$, we need to verify $D^\downarrow = ({}^1D^{\downarrow\uparrow})^\downarrow$. But $D^\downarrow = ({}^1D)^\downarrow = ({}^{11}D^{\downarrow\uparrow})^\downarrow$. Indeed, the first equality follows from the fact that D is crisp and thus ${}^1D = D$. For the second equality, $({}^1D)^\downarrow \subseteq ({}^1({}^1D)^{\downarrow\uparrow})^\downarrow$ follows from $F \subseteq ({}^1F^\uparrow)^\downarrow$ for $F = ({}^1D)^\downarrow$ (easy), and $({}^1D)^\downarrow \supseteq ({}^1({}^1D)^{\downarrow\uparrow})^\downarrow$ follows from $D = {}^1D$, from ${}^1D \subseteq ({}^1D)^{\downarrow\uparrow}$, and from the fact that if $E \subseteq F$ then ${}^1E^\downarrow \supseteq {}^1F^\downarrow$ (just put $E = D$ and $F = ({}^1D)^{\downarrow\uparrow}$). Hence, $\langle A, B \rangle \in \mathcal{B}\left(X, {}^1Y, I\right)$.

Now, we turn to the investigation of the properties of $^\triangle$ and $^\triangledown$ and the problem of axiomatization of these properties.

Lemma 6. $^\triangle$ *and* $^\triangledown$ *defined by (10) and (11) satisfy*

$$S(A, B^\triangledown) = S({}^1B, A^\triangle) \tag{12}$$

$$\left(\bigcup_{j \in J} A_j\right)^\triangle = \bigcap_{j \in J} A_j^\triangle \tag{13}$$

for every $A, A_j \in L^X$ and $B \in L^Y$.

Proof. We have

$$S(A, B^\triangledown) = \bigwedge_{x \in X} \left(A(x) \to \left(\bigwedge_{y \in Y} {}^1B(y) \to I(x, y)\right)\right) =$$

$$= \bigwedge_{y \in Y} \bigwedge_{x \in X} \left({}^1B(y) \to (A(x) \to I(x, y))\right) =$$

$$= \bigwedge_{y \in Y} \left({}^1B(y) \to \left(\bigwedge_{x \in X} A(x) \to I(x, y)\right)\right) = S({}^1B, A^\triangle),$$

proving (12). (13) is a consequence of properties of fuzzy Galois connections [2].

Definition 2. *A pair $\langle ^\triangle, ^\triangledown \rangle$ of mappings satisfying (12) and (13) is called a c-Galois connection between X and Y.*

The following are some consequences of (12).

Lemma 7. *If $^\triangle : L^X \to L^Y$ and $^\triangledown : L^Y \to L^X$ satisfy (12) then*

$$\left(\bigcup_{j \in J} {}^1B_j\right)^\triangledown = \bigcap_{j \in J} B_j^\triangledown \tag{14}$$

$$B^\triangledown = ({}^1B)^\triangledown \tag{15}$$

$${}^a\!/x\}^\triangle(y) = a \to \{{}^1\!/x\}^\triangle(y) \tag{16}$$

$${}^a\!/y\}^\triangledown(x) = a \to \{{}^1\!/y\}^\triangledown(x) \tag{17}$$

for any $B, B_j \in L^Y$, $x \in X$, $y \in Y$, $a \in L$.

Proof. (14): We show that $S(A, (\bigcup_i {}^1B_i)^\nabla) = 1$ iff $S(A, \bigcap_i({}^1B)_i^\nabla) = 1$ for each $A \in L^X$. Using (12) we have $S(A, (\bigcup_i {}^1B_i)^\nabla) = S({}^1(\bigcup_i {}^1B_i), A^\triangle) = S((\bigcup_i {}^1B_i), A^\triangle)$. As a result, we have $S(A, (\bigcup_i {}^1B_i)^\nabla) = 1$ iff $S((\bigcup_i {}^1B_i), A^\triangle) = 1$ iff for each i we have ${}^1B_i \subseteq A^\triangle$ iff for each i we have $S({}^1B_i, A^\triangle)$ iff for each i we have $S(A, B_i^\nabla)$ iff $S(A, \bigcap_i B_i^\nabla)$.

(15) is just (14) for $|J| = 1$.

(16) and (17) follow from ${}^1b \to \{ {}^a/x \}^\triangle(y) = a \to \{ {}^b/y \}^\nabla(x)$ and $\{ {}^1/x \}^\triangle(y) = \{ {}^1/y \}^\nabla(x)$ which we now verify. First, ${}^1b \to \{ {}^a/x \}^\triangle(y) = S(\{ {}^1b/y \}, \{ {}^a/x \}^\triangle) = S(\{ {}^1 {}^b/y \}, \{ {}^a/x \}^\triangle) = S(\{ {}^a/x \}, \{ {}^b/y \}^\nabla) = a \to \{ {}^b/y \}^\nabla(x)$ (here ${}^1b = 1$ for $b = 1$ and ${}^1b = 0$ otherwise). For $\{ {}^1/x \}^\triangle(y) = \{ {}^1/y \}^\nabla(x)$ just put $a = b = 1$ in the foregoing equality.

Lemma 8. *Let $\langle \triangle, \nabla \rangle$ be a c-Galois connection. Then there is a fuzzy relation $I \in L^{X \times Y}$ such that $\langle \triangle, \nabla \rangle = \langle \triangle_I, \nabla_I \rangle$ where $\triangle_I$ and ∇_I are induced by I by (10) and (11).*

Proof. Let I be defined by $I(x, y) = \{ {}^1/x \}^\triangle(y) = \{ {}^1/y \}^\nabla(x)$. Then using (16), it is straightforward to show $A^\triangle = A^{\triangle_I}$. Furthermore, using (14) and (15), and (17) we get

$$B^\nabla(x) = {}^1B^\nabla(x) = (\bigcup_{y \in Y} \{ {}^1B(y)/y \})^\nabla(x) =$$

$$= (\bigcup_{y \in Y} {}^1\{ B(y)/y \})^\nabla(x) = (\bigcap_{y \in Y} \{ B(y)/y \}^\nabla)(x) = \bigwedge_{y \in Y} \{ B(y)/y \}^\nabla(x) =$$

$$= \bigwedge_{y \in Y} {}^1B(y) \to \{ {}^1/y \}^\nabla(x) = \bigwedge_{y \in Y} {}^1B(y) \to I(x, y) = B^{\nabla_I}(x).$$

Next, we have the desired one-to-one correspondence between fuzzy relations and c-Galois connections.

Theorem 4. *Let $I \in L^{X \times Y}$ be a fuzzy relation, let $\triangle_I$ and ∇_I be defined by (10) and (11). Let $\langle \triangle, \nabla \rangle$ be a c-Galois connection. Then*

(1) $\langle \triangle_I, \nabla_I \rangle$ satisfy (12) and (13).
(2) $I_{\langle \triangle, \nabla \rangle}$ defined as in the proof of Lemma 8 is a fuzzy relation and we have
(3) $\langle \triangle, \nabla \rangle = \langle \triangle_{I_{\langle \triangle, \nabla \rangle}}, \nabla_{I_{\langle \triangle, \nabla \rangle}} \rangle$ and $I = I_{\langle \triangle_I, \nabla_I \rangle}$.

Proof. Due to the previous results, it remains to check $I = I_{\langle \triangle_I, \nabla_I \rangle}$. We have $I_{\langle \triangle_I, \nabla_I \rangle}(x, y) = \{ {}^1/x \}^{\triangle_I}(y) = \bigwedge_{z \in X} \{ {}^1/x \}(z) \to I(z, y) = I(x, y)$, completing the proof.

Coming back to conditions (6)–(9), one can easily see that they are in general not satisfied by a c-Galois connection $\triangle, \nabla$. The next lemma shows properties of c-Galois connections which are analogous to (6)–(9).

Lemma 9. *For a c-Galois connection* $\langle^{\triangle}, ^{\triangledown}\rangle$, *we have*

$$S(A_1, A_2) \leq S(A_2^{\triangle}, A_1^{\triangle}) \tag{18}$$

$$S(^1B_1, ^1B_2) \leq S(B_2^{\triangledown}, B_1^{\triangledown}) \tag{19}$$

$$A \subseteq A^{\triangle\triangledown} \tag{20}$$

$$^1B \subseteq B^{\triangledown\triangle} \tag{21}$$

Proof. By direct verification.

2.5   Main Theorem for $\mathcal{B}_c(X, Y, I)$

Now we present a version of main theorem of concept lattices for $\mathcal{B}_c(X, Y, I)$. Due to the limited scope of the paper, we present only sketch of proof.

Note (see [6]) that for a fuzzy set $E \in L^U$, $\lfloor E \rfloor$ is a subset of $U \times L$ defined by $\lfloor E \rfloor = \{\langle u, a \rangle \in U \times L \mid a \leq E(u)\}$. Conversely, for $F \subseteq U \times L$, $\lceil F \rceil$ is a fuzzy set in U defined by $\lceil F \rceil(u) = \bigvee\{a \mid \langle u, a \rangle \in F\}$. Now, for a fuzzy relation $I \in L^{X \times Y}$, define an ordinary relation I^+ between $X \times L$ and Y by $\langle\langle x, a \rangle, y\rangle \in I^+$ iff $a \leq I(x, y)$.

Theorem 5. *The set* $\mathcal{B}_c(X, Y, I)$ *equipped with* $\leq$ *is a complete lattice where infima and suprema are given by* (4) *and*

$$\bigvee_{j \in J} \langle A_j, B_j \rangle = \langle \lceil (\bigcup_{j \in J} \lfloor A_j \rfloor)^{\uparrow\downarrow} \rceil, \lceil \bigcap_{j \in J} \lfloor ^1B_j \rfloor \rceil \rangle . \tag{22}$$

Moreover, an arbitrary complete lattice $\mathbf{V} = \langle V, \wedge, \vee \rangle$ *is isomorphic to some* $\mathcal{B}_c(X, Y, I)$ *iff there are mappings* $\gamma : X \times L \to V$, $\mu : Y \to V$ *such that* $\gamma(X, L)$ *is* $\bigvee$-*dense in* V, $\mu(Y)$ *is* $\bigwedge$-*dense in* V, *and* $a \leq I(x, y)$ *iff* $\gamma(x, a) \leq \mu(y)$.

Proof. Sketch: Analogously as in [4], we can find a bijection between c-Galois connections between X and Y, and ordinary Galois connections between $X \times L$ and Y. Under this bijection, I (fuzzy relation corresponding an c-Galois connection) corresponds to I^+ (ordinary relation corresponding to a Galois connection) and the corresponding c-Galois connection and Galois connection have isomorphic lattices of fixed points. One of them is our $\mathcal{B}_c(X, Y, I)$, the other one is $\mathcal{B}(X \times L, Y, I^+)$. Now, $\mathcal{B}(X \times L, Y, I^+)$ is an ordinary concept lattice, and thus obeys Wille's Main Theorem [20]. Translating the Main Theorem to $\mathcal{B}_c(X, Y, I)$ then gives our theorem.

Corollary 1. $\mathcal{B}_c(X, Y, I)$ *is a* $\bigwedge$-*subsemilattice of* $\mathcal{B}(X, Y, I)$.

Remark 2. $\mathcal{B}_c(X, Y, I)$ need not be a $\bigvee$-subsemilattice of $\mathcal{B}(X, Y, I)$. One can verify by taking $X = \{x_1, x_2\}$, $Y = \{y_1, y_2, y_3\}$, and $I(x_1, y_1) = 0.3$, $I(x_1, y_2) = 0.5$, $I(x_1, y_3) = 0.4$, $I(x_2, y_1) = 0.2$, $I(x_2, y_2) = 0.6$, $I(x_2, y_3) = 0.1$.

2.6 Crisply Generated Concepts and One-Sided Fuzzy Concepts

In [22], the authors deal with the following. Let $\langle X, Y, I \rangle$ be a fuzzy context (with $L = [0, 1]$). Define mappings $f : 2^X \to L^X$ (assigning a *fuzzy set* $f(A) \in L^Y$ of attributes to a *set* $A \subseteq X$ of objects) and $h : L^X \to 2^X$ (assigning a *set* $h(B) \subseteq$ of objects to a *fuzzy set* $B \in L^Y$ of attributes) by

$$f(A)(y) = \bigwedge_{x \in A} I(x, y) \tag{23}$$

and

$$h(B) = \{x \in X \mid \text{for each } y \in Y : B(y) \leq I(x, y)\}. \tag{24}$$

The same definition was later "rediscovered" by Krajči [17]. Pairs $\langle A, B \rangle \in 2^X \times L^Y$ are called one-sided fuzzy concepts (A is a set, B is a fuzzy set) in [17]. By direct computation one can verify that $\langle A, B \rangle$ is a one sided fuzzy concept iff it is of the form $\langle A, B \rangle = \langle {}^1 A', B \rangle$ for some fuzzy concept $\langle A', B \rangle \in \mathcal{B}(X, Y, I)$ which is "crisply generated by extents", i.e. such that for some $\mathbf{L}$ we have $B = C^\uparrow$ (and $A' = C^{\uparrow\downarrow}$) for some set $C \subseteq X$. Therefore, up to exchanging roles of extents and intents, [22, 17] in fact deal with particular formal fuzzy concepts (crisply generated by extents) from $\mathcal{B}(X, Y, I)$, only that instead of $\langle A, B \rangle$ they consider $\langle {}^1 A, B \rangle$. As a consequence, our result presented in this paper apply in an appropriate modification to one-sided fuzzy concepts of [22, 17].

3 Examples and Experiments

Tab. 1 describes economic indexes of selected countries, transformed to $[0, 1]$ to get a formal fuzzy context. Using minimum-based fuzzy logical operations, the corresponding concept lattice $\mathcal{B}(X, Y, I)$ contains 304 formal concepts and is depicted in Fig. 1. The corresponding set $\mathcal{B}_c(X, Y, I)$ of all crisply generated fuzzy concepts contains 27 formal concepts and is depicted in Fig. 2. As we are interested only in the reduction of the size of the concept lattice, we omit the descriptions of formal concepts.

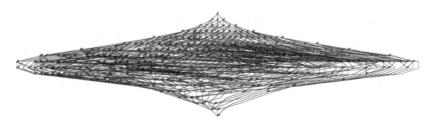

Fig. 1. Concept lattice corresponding to data from Tab. 1

Next we show results of experiments demonstrating the factor of reduction. That is, we are interested in the ratio $r = |\mathcal{B}_c(X, Y, I)|/|\mathcal{B}(X, Y, I)|$ (the smaller, the larger the reduction). Tab. 2 shows the values of r for 10 experiments

Table 1. Economic indexes: data table with fuzzy attributes

	1	2	3	4	5	6	7
1 Czech	0.4	0.4	0.6	0.2	0.2	0.4	0.2
2 Hungary	0.4	1.0	0.4	0.0	0.0	0.4	0.2
3 Poland	0.2	1.0	1.0	0.0	0.0	0.0	0.0
4 Slovakia	0.2	0.6	1.0	0.0	0.2	0.2	0.2
5 Austria	1.0	0.0	0.2	0.2	0.2	1.0	1.0
6 France	1.0	0.0	0.6	0.4	0.4	0.6	0.6
7 Italy	1.0	0.2	0.6	0.0	0.2	0.6	0.4
8 Germany	1.0	0.0	0.6	0.2	0.2	1.0	0.6
9 UK	1.0	0.2	0.4	0.0	0.2	0.6	0.6
10 Japan	1.0	0.0	0.4	0.2	0.2	0.4	0.2
11 Canada	1.0	0.2	0.4	1.0	1.0	1.0	1.0
12 USA	1.0	0.2	0.4	1.0	1.0	0.2	0.4

attributes: 1 - high gross domestic product per capita (USD), 2 - high consumer price index (1995=100) , 3 - high unemployment rate (percent - ILO), 4 - high electricity production per capita (kWh), 5 - high energy consumption per capita (GJ), 6 - high export per capita (USD), 7 - high import per capita (USD)

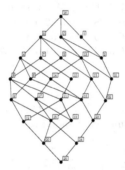

Fig. 2. Crisply generated formal concepts corresponding to data from Tab. 1

(columns) run over randomly generated formal contexts (rows) with the number of objects equal to the number of attributes (from 5 to 25 objects/attributes) and with $|L| = 11$ (11 truth degrees). Moreover, we show average and dispersion of r. We can see that the dispersion is low and that r decreases with growing size of data. Further experiments need to be run to show in more detail the behavior of r. In the second experiment, we randomly generated tables with 20 objects and 20 attributes, $|L| = 11$ with minimum-based fuzzy conjunction, each object with 10 attributes with a degree > 0 (and 10 attributes with a degree $=0$); of the ten attribute with nonzero degrees, we varied the number of attributes, from 1 to 10 (rows), with degree $= 1$; columns represent experiments; we consider average and dispersion of r, see Tab. 3. In the third experiment, we randomly generated tables with 20 objects and 20 attributes, $|L| = 11$ with minimum-based fuzzy conjunction, each object with varying number of attributes with a

Table 2. Behavior of r (average Av, dispersion Var) in dependence on the size of input data table (rows); columns correspond to experiments

	1	2	3	4	5	6	7	8	9	10	Av	Var
5	0.58	0.4	0.38	0.53	0.48	0.38	0.43	0.41	0.48	0.33	0.441	0.0733
6	0.31	0.31	0.38	0.43	0.38	0.32	0.43	0.42	0.36	0.38	0.372	0.0443
7	0.46	0.37	0.31	0.48	0.45	0.27	0.41	0.43	0.4	0.37	0.395	0.0635
...	...	...	...	...	...	...	...	...	...	...	...	...
23	0.09	0.11	0.1	0.1	0.1	0.1	0.1	0.09	0.1	0.11	0.099	0.0066
24	0.1	0.09	0.08	0.1	0.09	0.1	0.09	0.09	0.09	0.08	0.090	0.0079
25	0.08	0.07	0.09	0.08	0.09	0.09	0.08	0.07	0.09	0.08	0.081	0.0074

Table 3. Dependence of r (average Av, dispersion Var) on the number of 1's in object attributes (rows); columns correspond to experiments

	1	2	3	4	5	6	7	8	9	10	Av	Var
1	0.1	0.08	0.07	0.08	0.08	0.09	0.1	0.08	0.08	0.08	0.084	0.0100
2	0.09	0.1	0.11	0.11	0.08	0.09	0.1	0.08	0.09	0.09	0.094	0.0086
3	0.12	0.11	0.11	0.1	0.11	0.11	0.12	0.1	0.1	0.11	0.107	0.0084
4	0.12	0.13	0.11	0.14	0.14	0.13	0.13	0.12	0.14	0.14	0.130	0.0093
5	0.18	0.15	0.15	0.16	0.16	0.16	0.16	0.17	0.16	0.17	0.162	0.0095
6	0.18	0.21	0.17	0.17	0.2	0.2	0.2	0.18	0.2	0.22	0.193	0.0151
7	0.24	0.24	0.26	0.26	0.28	0.26	0.22	0.27	0.25	0.28	0.256	0.0193
8	0.36	0.33	0.34	0.36	0.35	0.35	0.33	0.34	0.37	0.35	0.347	0.0121
9	0.54	0.59	0.55	0.56	0.53	0.48	0.51	0.56	0.52	0.55	0.539	0.0298
10	1	1	1	1	1	1	1	1	1	1	1,000	0.0000

Table 4. Dependence of r (average Av, dispersion Var) on the number of nonzero values in attributes (rows); columns correspond to experiments

	1	2	3	4	5	6	7	8	9	10	Av	Var
1	0.59	0.59	0.64	0.56	0.62	0.56	0.58	0.63	0.61	0.63	0.600	0.0273
2	0.55	0.55	0.47	0.49	0.59	0.43	0.51	0.47	0.55	0.44	0.504	0.0503
3	0.62	0.43	0.57	0.5	0.48	0.54	0.47	0.48	0.61	0.45	0.514	0.0642
...	...	...	...	...	...	...	...	...	...	...	...	...
13	0.06	0.07	0.08	0.07	0.08	0.07	0.07	0.08	0.06	0.05	0.067	0.0089
14	0.03	0.05	0.04	0.06	0.04	0.07	0.04	0.04	0.04	0.03	0.044	0.0113
15	0.04	0.05	0.06	0.07	0.05	0.04	0.06	0.08	0.05	0.07	0.058	0.0117

degree > 0 (the number varies from 1 to 15, rows); columns represent experiments; we consider average and dispersion of r, see Tab. 4. Next, we randomly generated input data tables with 20 objects and 20 attributes with varying $|L|$ for $|L| = 3, 6, 11, 16, 21, 31$ (rows), see Tab. 5; columns represent experiments; we consider average and dispersion of r.

In the last experiment, we observed the speed-up of the algorithm described in Section 2.3 compared to just using [8] and testing which concepts are crisply generated, see Remark 1. We randomly generated several data tables with di-

Table 5. Dependence of r (average Av, dispersion Var) on the number $|L|$ of truth degrees (rows); columns correspond to experiments

	1	2	3	4	5	6	7	8	9	10	Av	Var
3	0.36	0.34	0.34	0.34	0.33	0.3	0.34	0.42	0.33	0.36	0.346	0.0286
6	0.18	0.21	0.19	0.26	0.23	0.21	0.24	0.17	0.18	0.18	0.205	0.0312
11	0.17	0.23	0.16	0.21	0.17	0.18	0.2	0.17	0.19	0.22	0.187	0.0244
16	0.13	0.15	0.22	0.14	0.18	0.17	0.17	0.15	0.16	0.16	0.163	0.0220
21	0.14	0.18	0.15	0.15	0.15	0.15	0.16	0.13	0.13	0.16	0.150	0.0151
26	0.14	0.2	0.14	0.14	0.13	0.16	0.16	0.11	0.11	0.18	0.147	0.0280
31	0.17	0.14	0.13	0.1	0.15	0.14	0.13	0.15	0.16	0.21	0.147	0.0276

mensions 50 objects × 50 attributes to 70 objects × 70 attributes with $|L| = 6$ under a constraint that each object has 10 attributes with a nonzero degree and 4 of these equal 1. The graph in Fig. 3 demonstrates the speed-up in dependence on the size of input data (50 to 70), i.e. the ratio T/T_c where T is the time needed for computing the whole $\mathcal{B}(X, Y, I)$ and using Lemma 1 to test whether each concept is crisply generated, and T_c is the time needed by the algorithm from Section 2.3.

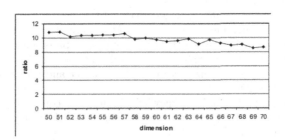

Fig. 3. Speed-up of algorithm from Section 2.3, see Remark 1

Acknowledgment. The authors acknowledge support by grant No. 1ET101370417 of the GA AV CR. Radim Bělohlávek gratefully acknowledges support by grant No. 201/02/P076 of GA CR.

References

1. Bělohlávek R.: Fuzzy concepts and conceptual structures: induced similarities. In *Proc. Joint Conf. Inf. Sci. '98*, Vol. I, pages 179–182, Durham, NC, 1998.
2. Bělohlávek R.: Fuzzy Galois connections. *Math. Log. Quart.* **45**(4)(1999), 497–504.
3. Bělohlávek R.: Similarity relations in concept lattices. *J. Logic Comput.* 10(6):823–845, 2000.
4. Bělohlávek R.: Reduction and a simple proof of characterization of fuzzy concept lattices. *Fundamenta Informaticae* **46**(4)(2001), 277–285.
5. Bělohlávek R.: Fuzzy closure operators. *J. Math. Anal. Appl.* **262**(2001), 473–489.

6. Bělohlávek R.: *Fuzzy Relational Systems: Foundations and Principles.* Kluwer, Academic/Plenum Publishers, New York, 2002.
7. Bělohlávek R.: Concept lattices and order in fuzzy logic. *Ann. Pure Appl. Logic* **128**(2004), 277–298.
8. Bělohlávek R.: *Proc. Fourth Int. Conf. on Recent Advances in Soft Computing.* Nottingham, United Kingdom, 12–13 December, 2002, pp. 200–205.
9. Bělohlávek R.: What is a fuzzy concept lattice (in preparation).
10. Bělohlávek R., Funioková T., Vychodil V.: Galois connections with hedges (submitted). Preliminary version to appear in Proc. 8-th Fuzzy Days, Dortmund, September 2004.
11. Burusco A., Fuentes-Gonzáles R.: The study of the L-fuzzy concept lattice. *Mathware & Soft Computing*, 3:209–218, 1994.
12. Ganter B., Wille R.: *Formal Concept Analysis. Mathematical Foundations.* Springer, Berlin, 1999.
13. Hájek P.: *Metamathematics of Fuzzy Logic.* Kluwer, Dordrecht, 1998.
14. Höhle U.: On the fundamentals of fuzzy set theory. *J. Math. Anal. Appl.* **201**(1996), 786–826.
15. Johnson D. S., Yannakakis M., Papadimitrou C. H.: On generating all maximal independent sets. *Inf. Processing Letters* **15**(1988), 129–133.
16. Klir G. J., Yuan B.: *Fuzzy Sets and Fuzzy Logic. Theory and Applications.* Prentice Hall, Upper Saddle River, NJ, 1995.
17. Krajči S.: Cluster based efficient generation of fuzzy concepts. *Neural Network World* **5**(2003), 521–530.
18. Pollandt S.: *Fuzzy Begriffe.* Springer, Berlin, 1997.
19. Ore O.: Galois connections. *Trans. Amer. Math. Soc.* **55** (1944), 493–513.
20. Wille R.: Restructuring lattice theory: an approach based on hierarchies of concepts. In: Rival I.: *Ordered Sets.* Reidel, Dordrecht, Boston, 1982, 445–470.
21. Wolff K. E.: Concepts in fuzzy scaling theory: order and granularity. *Fuzzy Sets and Systems* **132**(2002), 63–75.
22. Yahia S., Jaoua A.: Discovering knowledge from fuzzy concept lattice. In: Kandel A., Last M., Bunke H.: *Data Mining and Computational Intelligence*, pp. 167–190, Physica-Verlag, 2001.

Triadic Concept Graphs and Their Conceptual Contents

Lars Schoolmann

Technische Universität Darmstadt, Fachbereich Mathematik AG 1,
Schloßgartenstr. 7, D–64289 Darmstadt
schoolmann@mathematik.tu-darmstadt.de

Abstract. *Concept graphs* are mathematizations of asserting propositions consisting of (dyadic) concepts and objects. In order to take conditions or modalities in consideration *triadic concept graphs* are introduced as a straightforward generalization based on the basic notions of *Triadic Concept Analysis*. Then concept implications are discussed and *conceptual contents* of triadic concept graphs are introduced. It turns out that this approach can be reduced to a dyadic view; and the Basic Theorem on Conceptual Contents is obtained as a consequence of that. Finally, triadic concept graphs are generalized by introducing a subdivision, i.e. concept graphs with a more complex "rhetoric structure" are considered.

Contents

1 Triadic Concept Graphs

In this section we will recall some of the basic notions of *Contextual Concept and Judgment Logic* ([Wi04]). For an overview about *Contextual Logic*, its philosophical roots and its aims see [Wi00]. Here, judgments are understood as assertional combinations of concepts. Since the semantics should be based on *Formal Concept Analysis*, formal contexts are extended to *power context families* in order to express k-ary relation concepts (cf. [Wi02]):

Definition 1. *A sequence* $\vec{\mathbb{K}} := (\mathbb{K}_0, \mathbb{K}_1, \mathbb{K}_2, \ldots)$ *of contexts* $\mathbb{K}_k := (G_k, M_k, I_k)$ *with* $G_k \subseteq (G_0)^k$ *is called a* power context family. *The concepts of the contexts* $\mathbb{K}_k$ *($k \geq 1$) are called* relation concepts *since their extents are relations on the set* G_0. □

The elementary judgments which we consider have the form: An object (or a sequence of objects) is in the extent of a concept (or relation concept). Concept

B. Ganter and R. Godin (Eds.): ICFCA 2005, LNAI 3403, pp. 285–298, 2005.

graphs represent asserting combinations of such elementary judgments; to make
it readable we use *relational graphs* as "rhetoric structure".

Definition 2. *A* relational graph *is a structure* (V, E, ν) *of two sets V and E
and a map $\nu : E \rightarrow \bigcup_{k=1,2,...} V^k$. The elements of V and E are called* vertices
and edges, *respectively. An edge e linked by ν with k vertices is called k-ary – in
symbols: $|e| := k$. Furthermore let $E^{(k)} := \{e \in E \mid |e| = k\}$ and $E^{(0)} := V$.* □

An example of a relational graph is depicted in Figure 1. There are two
vertices (the rectangles) and a 2-ary edge (the ellipse); the map ν is represented
by the lines (with the numbers) between them. Now, we are able to define concept
graphs of power context families (cf. [Wi02]):

Definition 3. *Let $\vec{\mathbb{K}} := (\mathbb{K}_0, \mathbb{K}_1, \mathbb{K}_2, ...)$ be a power context family. A structure
$\mathfrak{G} := (V, E, \nu, \kappa, \rho)$ is called a* concept graph *of $\vec{\mathbb{K}}$ if:*

- *(V, E, ν) is a relational graph,*
- *$\kappa : V \cup E \rightarrow \bigcup_{k=0,1,2,...} \mathfrak{B}(\mathbb{K}_k)$ such that $\kappa(E^{(k)}) \subseteq \mathfrak{B}(\mathbb{K}_k)$ for all $k \geq 0$,*
- *$\rho : V \rightarrow \mathfrak{P}(G_0) \setminus \{\emptyset\}$ is a map such that*
  - *$v \in V \Rightarrow \rho(v) \subseteq Ext(\kappa(v))$,*
  - *$e \in E$ with $\nu(e) = (v_1, ..., v_k) \Rightarrow \rho(e) := \rho(v_1) \times \cdots \times \rho(v_k) \subseteq Ext(\kappa(e))$.* □

A concept graph can be understood as a relational graph with entries of a
power context family. The concept graph in Figure 1 represents the expression:
"The woman Ruth skies with the man Peter." The underlying power context
family is given by Fig. 2.

Fig. 1. Dyadic concept graph

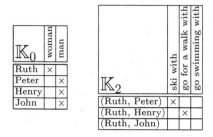

Fig. 2. Dyadic power context family

In order to take conditions or modalities in our considerations we recall the
basic notions of *Triadic Concept Analysis*. For a motivating discussion of the
following definitions of *triadic contexts* and *triadic concepts* see [LW95].

Definition 4. *A* triadic context *is a structure* $\mathbb{K} := (G, M, B, Y)$ *of three sets* G, M, B *and a ternary relation* Y *between them, i.e.* $Y \subseteq G \times M \times B$. *The elements of* G, M, *and* B *are called* objects, attributes, *and* conditions, *respectively.*
□

For introducing triadic concepts we need some kinds of derivation operators and, for defining them, it is useful to write K_1, K_2, K_3 instead of G, M, B. We define for $Z \subseteq K_i \times K_j$, $A_i \subseteq K_i$ and $\{i, j, k\} = \{1, 2, 3\}$ with $i < j$:

- $Z^{(k)} := \{a_k \in K_k \mid a_k, a_i, a_j \text{ are related by } Y \text{ for all } (a_i, a_j) \in Z\}$
- $A_j^{(i,k,A_i)} := A_i^{(j,k,A_j)} := \{a_k \in K_k \mid a_k, a_i, a_j \text{ are related by } Y \text{ for all } a_i \in A_i \text{ and } a_j \in A_j\}$
- $A_k^{(k)} := \{(a_i, a_j) \in K_i \times K_j \mid a_i, a_j, a_k \text{ are related by } Y \text{ for all } a_k \in A_k\}$

Thus, triadic concepts can be introduced as a natural generalization of dyadic concepts:

Definition 5. *A* triadic concept *of a triadic context* $\mathbb{K}$ *is defined as a triple* (A_1, A_2, A_3) *of subsets* $A_i \subseteq K_i$ $(i \in \{1, 2, 3\})$ *with* $A_k = (A_i \times A_j)^{(k)}$ *for* $\{i, j, k\} = \{1, 2, 3\}$ *with* $i < j$. *The sets* A_1, A_2 *and* A_3 *are called* extent, intent, *and* modus *of the concept* $\mathfrak{c} := (A_1, A_2, A_3)$ *and are denoted by* $Ext(\mathfrak{c})$, $Int(\mathfrak{c})$, *and* $Mod(\mathfrak{c})$. *The set of all triadic concepts of a triadic context* $\mathbb{K}$ *is denoted by* $\mathfrak{T}(\mathbb{K})$.
□

The mathematical structure theory of $\mathfrak{T}(\mathbb{K})$ is elaborated in [Wi95] and [Bi98]. Here, we only mention some simple properties of triadic concepts:

Suppose $\mathbb{K} := (K_1, K_2, K_3)$ is a triadic context. Then for $X_i \subseteq K_i$ and $X_k \subseteq K_k$ with $\{i, j, k\} = \{1, 2, 3\}$ we set

$$A_j := X_i^{(i,j,X_k)}$$
$$A_i := A_j^{(i,j,X_k)}$$
$$A_k := A_i^{(j,k,A_j)}$$

Then $\mathfrak{b}_{ik}(X_i, X_k) := (A_1, A_2, A_3)$ is a triadic concept. Notice that, in general, $\mathfrak{b}_{ik}(X_i, X_k) \neq \mathfrak{b}_{ki}(X_i, X_k)$, i.e. in particular, triadic concepts are not determined by their extent, intent or modus, but by two of these sets. Three special triadic concepts are always given by $\mathfrak{o}_1 := \mathfrak{b}_{23}(M, B) = \mathfrak{b}_{32}(M, B)$, $\mathfrak{o}_2 := \mathfrak{b}_{13}(G, B) = \mathfrak{b}_{31}(G, B)$, and $\mathfrak{o}_3 := \mathfrak{b}_{12}(G, M) = \mathfrak{b}_{21}(G, M)$.

Notice that a dyadic context $\mathbb{K} := (G, M, I)$ can be understood as a triadic context $\mathbb{K}_t := (G, M, \{b\}, Y)$ with only one condition b and $Y := I \times \{b\}$; then the map $(A_1, A_2) \mapsto (A_1, A_2, \{b\})$ is a bijection from $\mathfrak{B}(\mathbb{K}) \setminus \{(G, M)\}$ to $\mathfrak{T}(\mathbb{K}_t) \setminus \{\mathfrak{o}_3\}$. Thus, the triadic theory can be understood as a generalization of the dyadic theory.

Now, we introduce triadic power context families and triadic concept graphs as generalizations of dyadic power context families and concept graphs, respectively (cf. [Wi98], [GW00] and [SW03]):

Definition 6. *A sequence* $\vec{\mathbb{K}} := (\mathbb{K}_0, \mathbb{K}_1, \mathbb{K}_2, \ldots)$ *of triadic contexts* $\mathbb{K}_k := (G_k, M_k, B, Y_k)$ *with* $G_k \subseteq (G_0)^k$ *is called* triadic power context family. *The triadic concepts of the contexts* $\mathbb{K}_k$ ($k \geq 1$) *are called* (triadic) relation concepts *since their extents are mathematical relations on the set* G_0.   □

$\mathbb{K}_0$	in winter			in summer		
	woman	man	young	woman	man	young
Ruth	×			×		
Peter		×	×		×	×
Henry		×			×	
John		×			×	

$\mathbb{K}_2$	in winter					in summer				
	ski with	go for a walk with	go swimming with	play football with	play chess with	ski with	go for a walk with	go swimming with	play football with	play chess with
(Ruth, Peter)	×			×					×	
(Ruth, Henry)		×		×		×	×	×		
(Ruth, John)	×					×			×	

Fig. 3. Triadic power context family $\vec{\mathbb{K}} := (\mathbb{K}_0, \mathbb{K}_2)$

Definition 7. *Let* $\vec{\mathbb{K}} := (\mathbb{K}_0, \mathbb{K}_1, \mathbb{K}_2, \ldots)$ *be a triadic power context family. A structure* $\mathfrak{G} := (V, E, \nu, \kappa, \rho)$ *is called a* (triadic) concept graph *of* $\vec{\mathbb{K}}$ *if:*

- (V, E, ν) *is a relational graph,*
- $\kappa : V \cup E \to \bigcup_{k=0,1,2,\ldots} \mathfrak{T}(\mathbb{K}_k)$ *such that* $\kappa(E^{(k)}) \subseteq \mathfrak{T}(\mathbb{K}_k)$ *for all* $k \geq 0$,
- $\rho : V \to \mathfrak{P}(G_0) \setminus \{\emptyset\}$ *is a map such that*
  - $v \in V \Rightarrow \rho(v) \subseteq Ext(\kappa(v))(= (Int(\kappa(v)) \times Mod(\kappa(v)))^{(1)})$,
  - $e \in E$ *with* $\nu(e) = (v_1, \ldots, v_k) \Rightarrow \rho(e) := \rho(v_1) \times \cdots \times \rho(v_k) \subseteq Ext(\kappa(e))$ $(= (Int(\kappa(e)) \times Mod(\kappa(e)))^{(1)})$.

□

Remark: Concept graphs represent combinations of concepts which are linked by common objects. More exactly: a relation concept $\mathfrak{c}$ can only be linked (by ν and κ) with concepts $\mathfrak{c}_1, \ldots, \mathfrak{c}_k$ of $\mathbb{K}_0$, if there is an object in $Ext(\mathfrak{c})$ with components in the extents of $\mathfrak{c}_1, \ldots, \mathfrak{c}_k$, which is *mentioned* in the concept graph.

An example of a concept graph of the triadic power context family $\vec{\mathbb{K}} := (\mathbb{K}_0, \mathbb{K}_2)$ of Fig. 3 is given in Fig. 4. Here, the triadic concepts are WOMAN:= $\mathfrak{b}_{23}^0(\{woman\}, B)$, MAN:= $\mathfrak{b}_{23}^0(\{man\}, B)$ and SKI IN WINTER WITH := $\mathfrak{b}_{23}^2$ ($\{ski\ with\}, \{in\ winter\}$).

Fig. 4. Triadic concept graph

2   Conceptual Contents

Now, we want to discuss the question: Which information is given by a triadic concept graph? Obviously, there are the "triadic concept instances" $(g, \mathfrak{c})$ of objects g and triadic concepts $\mathfrak{c}$ coded directly in the concept graph. But moreover, we assume that there is some "background information" given by "object implications" and "concept implications" (coded in the underling power context family). Here triadic concepts are identified by intent-modus-pairs. This approach is introduced in [Wi02] for the dyadic case and in the following section we generalize this (in the abovementioned sense) to the triadic case.

2.1   Triadic Implications

In the following sections we identify triadic concepts with their intent-modus-pairs. First, we consider implications between triadic concepts: Let $\mathfrak{c}, \mathfrak{d}$ be concepts of a given triadic context $\mathbb{K} := (G, M, B, Y)$, then

$$\mathfrak{c} \to \mathfrak{d} \text{ in } \mathbb{K} : \Longleftrightarrow Ext(\mathfrak{c}) \subseteq Ext(\mathfrak{d}).$$

In the triadic context $\mathbb{K}_2$ of Fig. 3 we obtain for example the implications:

$$\mathfrak{b}_{23}(\{\text{ski with}\}, \{\text{in winter}\}) \to \mathfrak{b}_{23}(\{\text{play football with}\}, \{\text{in summer}\})$$

and

$$\mathfrak{b}_{23}(\{\text{play chess with}\}, \{\text{in winter}\}) \to \mathfrak{b}_{23}(\{\text{play football with}\}, \{\text{in summer}\}).$$

We compare this kind of implication with the *triadic implications* introduced in [Bi98]; there K. Biedermann suggests for sets $R, S \subseteq M$ and $C \subseteq B$ the notion:

$$(R \to S)_C : \Longleftrightarrow (R \times C)^{(1)} \subseteq (S \times C)^{(1)}.$$

It is easy to check the following two propositions:

Proposition 1. *Let* $\mathbb{K} := (G, M, B, Y)$ *be a triadic context and* $R, S \subseteq M$, $C \subseteq B$. *Then*

$$
\begin{aligned}
(R \to S)_C &\Longleftrightarrow \mathfrak{b}_{23}(R, C) \to \mathfrak{b}_{23}(S, C) \\
&\Longleftrightarrow \mathfrak{b}_{32}(R, C) \to \mathfrak{b}_{32}(S, C) \\
&\Longleftrightarrow \mathfrak{b}_{23}(R, C) \to \mathfrak{b}_{32}(S, C) \\
&\Longleftrightarrow \mathfrak{b}_{32}(R, C) \to \mathfrak{b}_{23}(S, C).
\end{aligned}
$$

$\square$

Proposition 2. *Let* $\mathbb{K} := (G, M, B, Y)$ *be a triadic context with the concepts* (A_1, A_2, A_3) *and* (D_1, D_2, D_3). *If* $A_3 \subseteq D_3$, *then*

$$(A_1, A_2, A_3) \to (D_1, D_2, D_3) \Rightarrow (A_2 \to D_2)_{D_3}.$$

If $D_3 \subseteq A_3$, *then*

$$(A_2 \to D_2)_{A_3} \Rightarrow (A_1, A_2, A_3) \to (D_1, D_2, D_3).$$

$\square$

However, for our purposes we need not only implications between concepts, but between sets of concepts and between sets of objects:

Definition 8. *Let* $\mathbb{K} := (G, M, B, Y)$ *be a triadic context and* $D, F \subseteq \mathfrak{T}(\mathbb{K})$, *then* $D \to F$ *is a* concept implication *in* $\mathbb{K}$ *if* $\bigcap_{\mathfrak{d} \in D} Ext(\mathfrak{d}) \subseteq \bigcap_{\mathfrak{f} \in F} Ext(\mathfrak{f})$. *Furthermore, for* $A, C \subseteq G : A \to C$ *is an* object implication *in* $\mathbb{K}$ *if* $A^{(1)} \subseteq C^{(1)}$. $\square$

For the "Basic Theorem of Conceptual Contents" (see section 3) the following lemma is essential because it reduces these "triadic object and concept implications" to object and attribute implications of a dyadic context:

Lemma 1. *For a triadic context* $\mathbb{K} := (G, M, B, Y)$ *we consider the dyadic context* $\mathbb{K}_d := (G, P, I)$ *wih* $P := \mathfrak{T}(\mathbb{K})$ *and* $gI\mathfrak{c} : \Longleftrightarrow g \in Ext(\mathfrak{c})$. *Then, for* $D, F \subseteq \mathfrak{T}(\mathbb{K})$, $A, C \subseteq G$ *it holds* $D \to F$ *in* $\mathbb{K}$ *if and only if* $D^I \subseteq F^I$, *and* $A \to C$ *in* $\mathbb{K}$ *if and only if* $A^I \subseteq C^I$.

Proof: Let $D, F \subseteq \mathfrak{T}(\mathbb{K})$, then $D \to F \iff \bigcap_{\mathfrak{d} \in D} Ext(\mathfrak{d}) \subseteq \bigcap_{\mathfrak{f} \in F} Ext(\mathfrak{f}) \iff D^I \subseteq F^I$.

Let $A, C \subseteq G$, let $A \to C, \mathfrak{a} \in A^I$ and $(m, b) \in Int(\mathfrak{a}) \times Mod(\mathfrak{a})$, this implies $\forall a \in A : (a, m, b) \in Y$, hence $(m, b) \in A^{(1)}$. With $A^{(1)} \subseteq C^{(1)}$ we obtain: $\forall c \in C : (c, m, b) \in Y$, and hence $\mathfrak{a} \in C^I$. This implies $A \to C \Rightarrow A^I \subseteq C^I$.

Conversely, let $A^I \subseteq C^I, (m, b) \in A^{(1)}$, and $\mathfrak{c} := \mathfrak{b}_{23}(m, b)$, this implies $\forall a \in A : a \in Ext(\mathfrak{c})$, hence $\forall c \in C : c \in Ext(\mathfrak{c})$, finally $(m, b) \in C^{(1)}$ This implies $A^I \subseteq C^I \Rightarrow A \to C$. $\square$

Lemma 2. *Suppose* $\mathbb{K} := (G, M, B, Y)$ *is a triadic context. Then the concept lattice of* $\mathbb{K}^{(1)} := (G, M \times B, Y^{(1)})$ *with* $(g, (m, b)) \in Y^{(1)} : \Longleftrightarrow (g, m, b) \in Y$ *is isomorphic to the concept lattice of the context* $\mathbb{K}_d$ *(of Lemma 1).*

Proof: We consider the maps $\gamma_d : G \to \mathfrak{B}(\mathbb{K}_d)$ and $\mu_d \circ \mathfrak{b}_{23} : M \times B \to \mathfrak{B}(\mathbb{K}_d)$. Obviously, $(g, (m, b)) \in Y^{(1)} \iff (g, m, b) \in Y \iff \gamma_d g \leq \mu_d \mathfrak{b}_{23}(b, m)$. With the Basic Theorem on Concept Lattices (cf. [GW99]) it remains to show that $\mu_d \circ \mathfrak{b}_{23}(M \times B)$ is $\bigwedge$-dense in $\mathfrak{B}(\mathbb{K}_d)$, i.e. for $(A, C) \in \mathfrak{B}(\mathbb{K}_d)$:

$$\bigcap \{Ext(\mathfrak{b}_{23}(m, b)) \mid \forall a \in A : (a, m, b) \in Y\} \subseteq A.$$

Let $g \notin A$, then $\exists c \in C : g \notin Ext(\mathfrak{c})$, then $\exists (m, b) \in Int(\mathfrak{c}) \times Mod(\mathfrak{c}) : (g, m, b) \notin Y$, and hence $g \notin \bigcap \{Ext(\mathfrak{b}_{23}(m, b)) \mid \forall a \in A : (a, m, b) \in Y\}$. $\square$

Finally, we remark that for subsets $A, F \subseteq \mathfrak{T}(\mathbb{K})$ the implication $A \to F$ holds if and only if $\{(m, b) \mid \exists \mathfrak{a} \in A : (m, b) \in Int(\mathfrak{a}) \times Mod(\mathfrak{a})\}^{(1)} \subseteq \{(m, b) \mid \exists \mathfrak{f} \in F : (m, b) \in Int(\mathfrak{f}) \times Mod(\mathfrak{f})\}^{(1)}$.

2.2   Conceptual Contents

The implications of Definition 8 and Lemma 1 give rise to the closure system $\mathcal{C}(\mathbb{K})$ on $\mathbb{S}^{imp}(\mathbb{K}_d) := \{(g, \mathfrak{c}) \in G \times \mathfrak{T}(\mathbb{K}) \mid g \in Ext(\mathfrak{c})\}(= I)$ consisting of all subsets $X \subseteq \mathbb{S}^{imp}(\mathbb{K}_d)$ with the following property:

$$A \times B \subseteq X, \; A \to C, \; B \to D \Rightarrow C \times D \subseteq X.$$

Now, we are able to define the *conceptual content* of triadic concept graphs (cf. [Wi04]):

Definition 9. *The k-ary conceptual content (or $\mathbb{K}_k$-conceptual content) $C_k(\mathfrak{G})$ of a concept graph $\mathfrak{G} := (V, E, \nu, \kappa, \rho)$ of a triadic power context family $\vec{\mathbb{K}} := (\mathbb{K}_0, \mathbb{K}_1, \mathbb{K}_2, \ldots)$ is defined as the closure of the set*

$$\{(g, \mathfrak{c}) \mid \exists u \in E^{(k)} : g \in \rho(u), \ \mathfrak{c} = \kappa(u)\}$$

with respect to the closure system $\mathcal{C}(\mathbb{K}_k)$ ($k = 1, 2, \ldots$). The 0-ary conceptual content (or $\mathbb{K}_0$-conceptual content) $C_0(\mathfrak{G})$ is defined as the closure of

$$\{(g, \mathfrak{c}) \mid \exists u \in V : g \in \rho(u), \ \mathfrak{c} = \kappa(u)\} \cup$$

$$\{(g_i, \mathfrak{o}_2) \mid \exists k \ \exists (g_1, \ldots, g_k, \mathfrak{c}) \in C_k(\mathfrak{G}) : g_i \in \{g_1, \ldots, g_k\}\}$$

(with $\mathfrak{o}_2 := (G_0, (G_0 \times B)^{(2)}, B)$) with respect to the closure system $\mathcal{C}(\mathbb{K}_0)$. Then the disjoint union

$$C(\mathfrak{G}) := C_0(\mathfrak{G}) \dot{\cup} C_1(\mathfrak{G}) \dot{\cup} C_2(\mathfrak{G}) \dot{\cup} \cdots$$

is called the $(\vec{\mathbb{K}}\text{-})$conceptual content of the concept graph $\mathfrak{G}$. □

The conceptual content of the triadic concept graph $\mathfrak{G}$ of Fig. 4 is $C(\mathfrak{G}) := C_0(\mathfrak{G}) \cup C_2(\mathfrak{G})$ with

$$C_0(\mathfrak{G}) = \{\text{Ruth}\} \times \{\text{WOMAN}, \mathfrak{o}_2, \mathfrak{o}_3\} \cup \{\text{Peter}\} \times \{\text{MAN}, \mathfrak{o}_2, \mathfrak{o}_3\} \text{ and}$$

$$C_2(\mathfrak{G}) = \{(\text{Ruth, Peter}), (\text{Ruth, John})\} \times \{\text{SKI IN WINTER WITH, PLAY FOOTBALL IN SUMMER WITH}, \mathfrak{o}_2, \mathfrak{o}_3\}.$$

The conceptual contents of $\mathbb{K}$ are exactly those elements $C := (C_0, C_1, C_2, \ldots)$ of the closure system $\mathcal{C}(\vec{\mathbb{K}})$ for which $(g_1, \ldots, g_k, \mathfrak{c}) \in C_k$ implies $(g_i, \mathfrak{o}_2) \in C_0$ ($k = 1, 2, \ldots;$ $i \in \{1, 2, \ldots, k\}$). Notice, that there is no informtion carried by the instances $(g_i, \mathfrak{o}_2)$. Furthermore, notice that the conceptual content is based on the elementary judgments $(g, \mathfrak{c})$ and implications in the contexts $\mathbb{K}_k$ – ignoring the relations between the contexts of the power context family.

Conceptual contents give rise to an information order on the class of concept graphs of a given power context family:

Definition 10. *For triadic concept graphs, $\mathfrak{G}_1$ is defined to be* less informative *(more general)* than $\mathfrak{G}_2$ *(in symbols: $\mathfrak{G}_1 \lesssim \mathfrak{G}_2$) if $C_k(\mathfrak{G}_1) \subseteq C_k(\mathfrak{G}_2)$ for $k = 0, 1, 2, \ldots;$ $\mathfrak{G}_1$ and $\mathfrak{G}_2$ are called* equivalent *(in symbols: $\mathfrak{G}_1 \sim \mathfrak{G}_2$) if $\mathfrak{G}_1 \lesssim \mathfrak{G}_2$ and $\mathfrak{G}_2 \lesssim \mathfrak{G}_1$ (i.e., $C_k(\mathfrak{G}_1) = C_k(\mathfrak{G}_2)$ for $k = 0, 1, 2, \ldots$). The set of all equivalence classes of concept graphs of a triadic power context family $\vec{\mathbb{K}}$ together with the order induced by the quasi-order $\lesssim$ is an ordered set denoted by $\vec{\Gamma}(\vec{\mathbb{K}})$.* □

Proposition 3. *$\widetilde{\vec{\Gamma}}(\vec{\mathbb{K}})$ is always a complete lattice. For a set $\{\mathfrak{G}_i \mid i \in I\}$ of triadic concept graphs $\mathfrak{G}_i := (V_i, E_i, \nu_i, \kappa_i, \rho_i)$ of $\vec{\mathbb{K}}$ the supremum of $[\mathfrak{G}_i]$ ($i \in I$) is given by $[\bigsqcup_{i \in I} \mathfrak{G}_i]$, where $\bigsqcup_{i \in I} \mathfrak{G}_i := (\bigcup_{i \in I} V_i, \bigcup_{i \in I} E_i, \bigcup_{i \in I} \nu_i, \bigcup_{i \in I} \kappa_i, \bigcup_{i \in I} \rho_i)$.*
The infimum of $[\mathfrak{G}_i]$ ($i \in I$) is $[\mathfrak{G}]$ where $\mathfrak{G} := (V, E, \nu, \kappa, \rho)$ is given by $E^{(k)} := \bigcap_{i \in I} C_k(\mathfrak{G}_i)$, $\nu(g_1, \ldots, g_k, \mathfrak{c}) := ((g_1, \mathfrak{o}_2), \ldots, (g_k, \mathfrak{o}_2))$, $\kappa(g, \mathfrak{c}) := \mathfrak{c}$ and $\rho(g, \mathfrak{c}) := \{g\}$. □

3   The Basic Theorems on Conceptual Contents

In section 1 triadic concept graphs are introduced as a generalization of dyadic concept graphs. The conceptual contents of these graphs are based on triadic object and concept implications. But in Lemma 1 these implications turn out to "be dyadic"; thus it is not surprising that results of the dyadic theory carry over to the triadic theory. In this section we discuss this phenomenon by the Basic Theorems on Conceptual Contents.

3.1   Information Contexts of Lattices

The Basic Theorem (which we discuss in the next section) allows to represent the $\mathbb{K}_k$-conceptual contents independent of triadic concept graphs – as extents of a (dyadic) "information context" $\mathbb{K}^{inf}(\mathbb{K}_k)$ with $\widetilde{\Gamma}((\mathbb{K}_k)) \cong \underline{\mathfrak{B}}(\mathbb{K}^{inf}(\mathbb{K}_k))$. We will deduce the Basic Theorem from a more general result about "information contexts" of lattices (see Proposition 4) which we take over from [Wi03]. First, we need the following definitions:

Definition 11. *An* implicational lattice context *is defined as a formal context* (K, L, R) *formed by complete lattices* K *and* L *satisfying*

- $a \leq_K c$ *in* K, $d \leq_L b$ *in* L *and* $(c, d) \in R$ *imply* $(a, b) \in R$,
- $A \times B \subseteq R$ *implies* $(\bigvee_K A, \bigwedge_L B) \in R$. $\qquad\square$

Definition 12. *Let* $(L, L, \leq_L)$ *be an implicational lattice context. The* implicational lattice structure *of* $(L, L, \leq_L)$ *is defined as the ordered structure* $\underline{\mathbb{S}}^{imp}(L) := (\mathbb{S}^{imp}(L), \leq, \widetilde{\bigvee})$ *with* $\mathbb{S}^{imp}(L) := \{(a, b) \in L^2 \mid 0 <_L a \leq_L b\}$ *where the order* $\leq$ *and the partial supremum* $\widetilde{\bigvee}$ *are fixed by*

- $(a, b) \leq (c, d) :\Leftrightarrow a \leq_L c$ *and* $d \leq_L b$,
- $\widetilde{\bigvee}(A \times B) := (\bigvee_L A, \bigwedge_L B)$ *for* $A \times B \subseteq \mathbb{S}^{imp}(L)$.

In this structure, the implicational closure $C^{imp}(X)$ *of a subset* $X \subseteq \mathbb{S}^{imp}(L)$ *is the smallest order ideal of* $(\mathbb{S}^{imp}(L), \leq)$ *containig* X *closed under* $\widetilde{\bigvee}$. *The set of all implicational closures of* $\underline{\mathbb{S}}^{imp}(L)$ *ordered by set inclusion is denoted by* $\underline{\mathcal{C}}(\mathbb{S}^{imp}(L))$. $\qquad\square$

Definition 13. *Let* $\mathfrak{S}(L)$ *be the set of all subsets* S *of* $L \setminus \{0\}$ *for which* $S \cup \{0\}$ *is a complete sublattice of the intervall* $[0, \bigvee S]$ *of* L. *Then for an implicational lattice context* $(L, L, \leq_L)$, *the corresponding* conceptual information context *shall be defined as the formal context*

$$\mathbb{K}^{inf}(L) := (\mathbb{S}^{imp}(L), \mathfrak{S}(L), \hat{\Delta})$$

with $(x, y)\hat{\Delta}S :\Leftrightarrow [x, y] \cap S \neq \emptyset$. $\qquad\square$

Proposition 4. *For an implicational lattice context* $(L, L, \leq_L)$, *the extents of the corresponding conceptual information context* $\mathbb{K}^{inf}(L)$ *are exactly the implicational closures of the implicational lattice structure* $\underline{\mathbb{S}}^{imp}(L)$.

For the proof of this proposition see [Wi03], but notice that it is not sufficient to take only the convex elements of $\mathfrak{S}(L)$. $\qquad\square$

3.2   The Basic Theorems

In this section we will introduce information contexts corresponding to triadic contexts and formulate the Basic Theorems on Conceptual Contents of triadic concept graphs.

Definition 14. *Let* $\mathbb{K} := (G, M, B, Y)$ *be a triadic context and* $\mathbb{K}_d := (G, P, I)$ *where* $P := \mathfrak{T}(\mathbb{K})$ *and* $gIp :\Leftrightarrow g \in Ext(p)$. *The information context corresponding to* $\mathbb{K}$ *is defined as the formal context*

$$\mathbb{K}^{inf}(\mathbb{K}) := (\mathbb{S}^{imp}(\mathbb{K}_d), \mathfrak{S}(\underline{\mathfrak{B}}(\mathbb{K}_d)), \bar{\Delta})$$

with $\mathbb{S}^{imp}(\mathbb{K}_d) := I$ *and* $(g, p)\bar{\Delta}S :\Leftrightarrow [\gamma g, \mu p] \cap S \neq \emptyset$. *The implicational context structure of a triadic context* $\mathbb{K} := (G, M, B, Y)$ *is then defined as the ordered structure* $\underline{\mathbb{S}}^{imp}(\mathbb{K}_d) := (\mathbb{S}^{imp}(\mathbb{K}_d), \leq, \widetilde{\bigvee})$, *where the order* $\leq$ *and the partial supremum* $\widetilde{\bigvee}$ *are fixed by*

- $(g, p) \leq (h, q) :\Leftrightarrow \gamma g \leq \gamma h$ *and* $\mu q \leq \mu p$,
- $\widetilde{\bigvee}(A \times B) := \{(g, p) \mid g \in A^{II}, p \in B^{II}\}$ *for* $A \times B \subseteq \mathbb{S}^{imp}(\mathbb{K}_d)$.

In this structure, the implicational closure $C^{imp}(X)$ *of a subset* $X \subseteq \mathbb{S}^{imp}(\mathbb{K}_d)$ *is the smallest order ideal of* $(\mathbb{S}^{imp}(\mathbb{K}_d), \leq)$ *containing* X *such that* $A \times B \subseteq C^{imp}(X)$ *implies* $\widetilde{\bigvee}(A \times B) \subseteq C^{imp}(X)$. *The set of all implicational closures of* $\mathbb{S}^{imp}(\mathbb{K}_d)$ *ordered by set inclusion is denoted by* $\underline{\mathcal{C}}(\mathbb{S}^{imp}(\mathbb{K}_d))$. □

Now, we are able to formulate the main result about the conceptual contents of triadic contexts $\mathbb{K} := (G, M, B, Y)$ (which are understood as the conceptual contents of the triadic concept graphs of the power conext family $\vec{\mathbb{K}} := (\mathbb{K})$), which is quite similar to its dyadic analogue in [Wi03].

Theorem 1. (Basic Theorem on $\mathbb{K}$-Conceptual Contents of Triadic Contexts $\mathbb{K}$) *For a triadic context* $\mathbb{K} := (G, M, B, Y)$ *with* $\emptyset^{(1)(1)} = \emptyset$, *the extents of the corresponding conceptual informaion context* $\mathbb{K}^{inf}(\mathbb{K})$ *are exactly the implicational closures of the implicational context structure* $\underline{\mathbb{S}}^{imp}(\mathbb{K}_d)$; *those implicational closures are exactly the conceptual contents of* $\mathbb{K}$.

Remark to the proof: Since the proof is quite similar to its dyadic analogue in [Wi03] here only some hints are given: In [Wi03] the set of all concept instances of a dyadic context $\mathbb{K} := (G, M, I)$ is defined to be $\mathbb{S}^{imp}(\mathbb{K}) := \{(g, \mathfrak{b}) \in G \times \mathfrak{B}(\mathbb{K}) \mid g \in Ext(\mathfrak{b})\}$, hence a set of object-concept-pairs. Thus, to use the result of Proposition 4 about $\mathbb{K}^{inf}(\mathfrak{B}(\mathbb{K}))$ for the proof of the dyadic Basic Theorem one needs suitable maps ordering the relevant sets of concept-concept-pairs to sets of object-concept-pairs. In the triadic case one has object-attribute-pairs (cf. Lemma 1 and Definition 14) instead of object-concept-pairs. But, the treatment of the attributes here works dually to the treatment of the objects in [Wi03]. □

The Basic Theorem shall be extended to the general case of limited triadic power context families. Let $\vec{\mathbb{K}} := (\mathbb{K}_0, \mathbb{K}_1, \ldots, \mathbb{K}_n)$ be a triadic power context family with $\mathbb{K}_k := (G_k, M_k, B, Y_k)$ $(k = 0, 1, 2, \ldots n)$. The *conceptual information context* corresponding to $\vec{\mathbb{K}}$ is defined as the formal context

$$\mathbb{K}^{inf}(\vec{\mathbb{K}}) := \mathbb{K}^{inf}(\mathbb{K}_0) + \mathbb{K}^{inf}(\mathbb{K}_1) + \cdots + \mathbb{K}^{inf}(\mathbb{K}_n).$$

An extent U of $\mathbb{K}^{inf}(\vec{\mathbb{K}})$ is said to be *rooted* if $((g_1, \ldots, g_k), \mathfrak{c}) \in U$ always implies $(g_i, \mathfrak{o}_2) \in U$ for $i = 1, \ldots, k$. Now, we are able to formulate the desired theorem:

Theorem 2. (Basic Theorem on $\vec{\mathbb{K}}$-Conceptual Contents of Triadic Power Context Families $\vec{\mathbb{K}}$) *For a (limited) triadic power context family $\vec{\mathbb{K}}$ with $\emptyset^{(1)_k(1)_k} = \emptyset$ for $k = 0, 1, 2, \ldots,$ the conceptual contents of the concept graphs of $\vec{\mathbb{K}}$ are exactly the rooted extents of the corresponding conceptual information context $\mathbb{K}^{inf}(\vec{\mathbb{K}})$.*

Proof: The conceptual content of a concept graph $\mathfrak{G}$ of $\vec{\mathbb{K}} := (\mathbb{K}_0, \mathbb{K}_1, \ldots, \mathbb{K}_n)$ is the disjoint union $C(\mathfrak{G}) := C_0(\mathfrak{G}) \dot\cup C_1(\mathfrak{G}) \dot\cup \cdots \dot\cup C_n(\mathfrak{G})$ where $C_k(\mathfrak{G})$ is an extent of $\mathbb{K}^{inf}(\mathbb{K}_k)$ $(k = 0, \ldots, n)$. Therefore $C(\mathfrak{G})$ is an extent of $\mathbb{K}^{inf}(\vec{\mathbb{K}})$; this extent is rooted as a direct consequence of the definition of $C_0(\mathfrak{G})$. Conversely, let U be a rooted extent of $\mathbb{K}^{inf}(\vec{\mathbb{K}})$. Then $U_k := U \cap C_k(\mathfrak{G})$ is an extent of $\mathbb{K}^{inf}(\mathbb{K}_k)$ for each $k = 0, 1, \ldots, n$ and hence an implicational closure of $\underline{\mathbb{S}}^{imp}(\mathbb{K}_{kd})$ by the Basic Theorem on $\mathbb{K}$-Conceptual Contents. This suggests the following construction of a triadic concept graph $(V, E, \nu, \kappa, \rho)$ of $\vec{\mathbb{K}}$:

$E^{(k)} := U_k$, $\nu((g_1, \ldots, g_k), \mathfrak{c}) := ((g_1, \mathfrak{o}_2), \ldots, (g_k, \mathfrak{o}_2))$,
$\kappa(g, \mathfrak{c}) := \mathfrak{c}$ and $\rho(g, \mathfrak{c}) := \{g\}$. □

Theorem 1 and Theorem 2 can be understood as generalizations of the dyadic Basic Theorems if one consider instead of a dyadic power context family $\vec{\mathbb{K}} := (\mathbb{K}_0, \mathbb{K}_1, \mathbb{K}_2, \ldots)$ the triadic $\vec{\mathbb{K}}_t := (\mathbb{K}_{t0}, \mathbb{K}_{t1}, \mathbb{K}_{t2}, \ldots)$ (see Definition 5 – Definition 6).

4   Concept Graphs with Subdivision

In this section we will discuss a semantic approach to concept graphs with subdivision. The meaning of the subdivision is to indicate in a larger concept graph which parts of it "belong" to the same modus; and moreover to identify this modus. First, we have to "extend" triadic power context families and relational graphs (cf. [SW03]):

Definition 15. *A triadic power context family with conceptual objects is defined as a triadic power context family $\vec{\mathbb{K}} := (\mathbb{K}_0, \mathbb{K}_1, \mathbb{K}_2, \ldots)$ with a partial map ξ from the concept set G_0 of the context $\mathbb{K}_0$ into the set $\mathfrak{T}(\mathbb{K}_0)$ of the triadic concepts of $\mathbb{K}_0$. The partial map ξ is called* subdivision map *of $\vec{\mathbb{K}}$ and the elements of $\mathrm{dom}\xi$ are called* conceptual objects. □

Definition 16. *A relational graph with subdivision is a structure (V, E, ν, σ) where (V, E, ν) is a relational graph and $\sigma : V \to \mathfrak{P}(V \cup E)$ is a map such that $v \notin \sigma^n(v)$ for all $v \in V$ and $n \in \mathbb{N}$ (where $\sigma(X) := \bigcup_{w \in X \cap V} \sigma(w)$).* □

Definition 17. *Let $\vec{\mathbb{K}} := (\mathbb{K}_0, \mathbb{K}_1, \mathbb{K}_2, \ldots)$ be a triadic power context family with subdivision map ξ. Then a structure $\mathfrak{G} := (V, E, \nu, \sigma, \kappa, \rho)$ is called a* triadic concept graph with subdivision *of $\vec{\mathbb{K}}$ if (V, E, ν, σ) is a relational graph with subdivision, $(V, E, \nu, \kappa, \rho)$ is a triadic concept graph of $\vec{\mathbb{K}}$ and if, additionally,*

- *$\sigma(w) \neq \emptyset \Rightarrow |\rho(w)| = 1$ and $\rho(w) \subseteq dom\xi$,*
- *$v \in \sigma(w) \cap V \Rightarrow \rho(v) \subseteq (Int(\kappa(v)) \times Mod(\xi(\rho(w))))^{(1)}$,*
- *$e \in \sigma(w) \cap E$ with $\nu(e) := (v_1, \ldots, v_k) \Rightarrow \rho(e) := \rho(v_1) \times \cdots \times \rho(v_k) \subseteq (Int(\kappa(e)) \times Mod(\xi(\rho(w))))^{(1)}$.* □

The underlying power context family of the concept graph with subdivision depicted in Fig. 5 is the triadic power context family $\vec{\mathbb{K}}$ of Fig. 3 extended by the conceptual objects "winter" and "summer", the attribute "season" and the "self-referential" information, that the winter (summer) is a season in winter (summer). The subdivision map ξ is given by "winter" $\mapsto$ ({winter}, {season}, {in winter}) =: "SEASON IN WINTER" and "summer" $\mapsto$ ({summer}, {season}, {in summer}) =: "SEASON IN SUMMER".

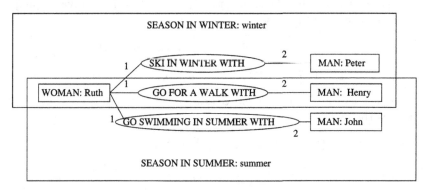

Fig. 5. Triadic concept graph with subdivision

Notice, that in the suggested definition of concept graphs with subdivision the "modus of the subdivision" is determined by the conceptual objects. This idea is taken over from the approach in [Wi98]. Of course, other approaches are conceivable. In our example the modi determined (via subdivision map) by the conceptual objects are {in winter}, resp. {in summer}. More examples of concept graphs with subdivision can be found in [Wi98], [GW00] and [SW03].

One can define conceptual contents of concept graphs with subdivision as a generalization of contents of ordinary triadic concept graphs. To do this we define for a context $\mathbb{K}_k$ of a triadic power context family with subdivision map ξ object and concept implications:

Definition 18. *Let $\mathbb{K}_k$ be a context of a triadic power context family $\vec{\mathbb{K}} := (\mathbb{K}_0, \mathbb{K}_1, \mathbb{K}_2, \ldots)$ with subdivision map ξ. For $A, C \subseteq G_k$ and $D, F \subseteq P_k^{\xi} :=$*

$\{(\mathfrak{c}, \mathfrak{d}) \mid \mathfrak{c} \in \mathfrak{T}(\mathbb{K}_k), \mathfrak{c} = \mathfrak{d}, \text{ or } \mathfrak{d} \in im \; \xi\}$ *we say that* $A \rightarrow C$ *is an* object impli-*cation in* $\mathbb{K}_k$ *if* $A^{(1)k} \subseteq C^{(1)k}$ *and* $D \rightarrow F$ *is a* concept implication *in* $\mathbb{K}_k$ *(with respect to* ξ*) if*

$$\bigcap_{(\mathfrak{d}_1, \mathfrak{d}_2) \in D} (Int(\mathfrak{d}_1) \times Mod(\mathfrak{d}_2))^{(1)} \subseteq \bigcap_{(\mathfrak{f}_1, \mathfrak{f}_2) \in F} (Int(\mathfrak{f}_1) \times Mod(\mathfrak{f}_2))^{(1)}.$$

□

Lemma 3. *Let* $\mathbb{K}_k$ *be a triadic context of a triadic power context family* $\vec{\mathbb{K}} := (\mathbb{K}_0, \mathbb{K}_1, \mathbb{K}_2, \ldots)$ *with subdivision map* ξ*. We define the dyadic context* $\mathbb{K}_{kd}^{\xi} := (G_k, P_k^{\xi}, I_k^{\xi})$ *with* $g I_k^{\xi}(\mathfrak{c}, \mathfrak{d}) :\Leftrightarrow g \in (Int(\mathfrak{c}) \times Mod(\mathfrak{d}))^{(1)}$*. Then for* $A, C \subseteq G_k$ *the implication* $A \rightarrow C$ *is equivalent to* $C \subseteq A^{I_k^{\xi} I_k^{\xi}}$*, and for* $B, D \subseteq P_k^{\xi}$ *the implication* $B \rightarrow D$ *is equivalent to* $D \subseteq B^{I_k^{\xi} I_k^{\xi}}$*.*

Proof: The proof is almost equal to the proof of Lemma 1.   □

So we get a closure system $\mathcal{C}^{\xi}(\mathbb{K}_k)$ on the set $\mathbb{S}^{imp}(\mathbb{K}_{kd}^{\xi}) := I_k^{\xi}$. Then the *k-ary conceptual content* $C_k^{\xi}(\mathfrak{G})$ of a concept graph $\mathfrak{G} := (V, E, \nu, \sigma, \kappa, \rho)$ of a triadic power context family $\vec{\mathbb{K}} := (\mathbb{K}_0, \mathbb{K}_1, \mathbb{K}_2, \ldots)$ with subdivision map ξ is defined as the closure of the set

$$\{(g, \mathfrak{c}, \mathfrak{c}) \mid \exists u \in E^{(k)} : \; g \in \rho(u), \; \mathfrak{c} = \kappa(u)\} \cup$$

$$\{(g, \mathfrak{c}, \mathfrak{d}) \mid \exists w \in V. \exists u \in E^{(k)} \cap \sigma(w) : g \in \rho(u), \; \mathfrak{c} = \kappa(u), \; \mathfrak{d} = \xi(\rho(w))\}$$

with respect to the closure system $\mathcal{C}^{\xi}(\mathbb{K}_k)$ $(k = 1, 2, \ldots)$. Moreover, $C_0^{\xi}(\mathfrak{G})$ is defined as the closure of

$$\{(g, \mathfrak{c}, \mathfrak{c}) \mid \exists u \in V : \; g \in \rho(u), \; \mathfrak{c} = \kappa(u)\} \cup$$

$$\{(g, \mathfrak{c}, \mathfrak{d}) \mid \exists w \in V. \exists u \in V \cap \sigma(w) : g \in \rho(u), \; \mathfrak{c} = \kappa(u), \; \mathfrak{d} = \xi(\rho(w))\} \cup$$

$$\{(g_i, \mathfrak{o}_2, \mathfrak{o}_2) \mid \exists k \; \exists (g_1, \ldots, g_k, \mathfrak{c}, \mathfrak{d}) \in C_k^{\xi}(\mathfrak{G}) : g_i \in \{g_1, \ldots, g_k\}\}$$

with respect to the closure system $\mathcal{C}^{\xi}(\mathbb{K}_0)$ and $C^{\xi}(\mathfrak{G}) := C_0^{\xi}(\mathfrak{G}) \dot{\cup} C_1^{\xi}(\mathfrak{G}) \dot{\cup} \cdots$.

The *information context* corresponding to $\mathbb{K}_k$ (with repect to ξ) is defined as the formal context

$$\mathbb{K}^{inf}(\mathbb{K}_k^{\xi}) := (\mathbb{S}^{imp}(\mathbb{K}_{kd}^{\xi}), \mathfrak{S}(\mathfrak{B}(\mathbb{K}_{kd}^{\xi})), \bar{\Delta})$$

with $(g, p) \bar{\Delta} S :\Leftrightarrow [\gamma g, \mu p] \cap S \neq \emptyset$. Moreover, we set

$$\mathbb{K}^{inf}(\vec{\mathbb{K}}^{\xi}) := \mathbb{K}^{inf}(\mathbb{K}_0^{\xi}) + \mathbb{K}^{inf}(\mathbb{K}_1^{\xi}) + \cdots + \mathbb{K}^{inf}(\mathbb{K}_n^{\xi})$$

for a limited power context family with subdivision map ξ. An extent U of $\mathbb{K}^{inf}(\vec{\mathbb{K}}^{\xi})$ is said to be *rooted with respect to* ξ if $((g_1, \ldots, g_k), \mathfrak{c}, \mathfrak{d}) \in U$ always implies $(g_i, \mathfrak{o}_2, \mathfrak{o}_2) \in U$ for $i = 1, \ldots, k$ and if for every $U_k := U \cap \mathcal{C}^{\xi}(\mathbb{K}_k)$ $(k \geq 0)$ there is a generating subset $\underline{U}_k$ of U_k with: $(g, \mathfrak{c}, \mathfrak{d}) \in \underline{U}_k \Rightarrow (g, \mathfrak{c}, \mathfrak{c}) \in \underline{U}_k$. Now, we obtain the following theorem:

Theorem 3. *For a (limited) triadic power context family $\vec{\mathbb{K}}$ with subdivision map ξ with $\emptyset^{(1)_k(1)_k} = \emptyset$ for $k = 0, 1, 2, \ldots$, the conceptual contents of the concept graphs with subdivision of $\vec{\mathbb{K}}$ are exactly the extents of the corresponding conceptual information context $\mathbb{K}^{inf}(\vec{\mathbb{K}}^\xi)$ which are rooted with respect to ξ.*

Proof: The conceptual content of a concept graph $\mathfrak{G}$ of $\vec{\mathbb{K}} := (\mathbb{K}_0, \mathbb{K}_1, \ldots, \mathbb{K}_n)$ is the disjoint union $C^\xi(\mathfrak{G}) := C_0^\xi(\mathfrak{G}) \cup C_1^\xi(\mathfrak{G}) \cup \cdots \cup C_n^\xi(\mathfrak{G})$ where $C_k^\xi(\mathfrak{G})$ is an extent of $\mathbb{K}^{inf}(\mathbb{K}_k^\xi)$ $(k = 0, \ldots, n)$ with a generating subset $\underline{C}_k^\xi(\mathfrak{G})$ such that $[(g, \mathfrak{c}, \mathfrak{d}) \in \underline{C}_k^\xi(\mathfrak{G}) \Rightarrow (g, \mathfrak{c}, \mathfrak{c}) \in \underline{C}_k^\xi(\mathfrak{G})]$ by the Basic Theorem on $\mathbb{K}$-Conceptual Contents and the definition of $C_k^\xi(\mathfrak{G})$. Therefore $C^\xi(\mathfrak{G})$ is an extent of $\mathbb{K}^{inf}(\vec{\mathbb{K}})$; this extent is rooted with respect to ξ as a direct consequence of the definition of $C_0^\xi(\mathfrak{G})$. Conversely, let U be a rooted extent of $\mathbb{K}^{inf}(\vec{\mathbb{K}})$. Then $U_k := U \cap C_k^\xi(\mathfrak{G})$ is an extent of $\mathbb{K}^{inf}(\mathbb{K}_k^\xi)$ for each $k = 0, 1, \ldots, n$ with a generating subset $\underline{U}_k$ with $[(g, \mathfrak{c}, \mathfrak{d}) \in \underline{U}_k \Rightarrow (g, \mathfrak{c}, \mathfrak{c}) \in \underline{U}_k]$ and hence an implicational closure of $\underline{\mathbb{S}}^{imp}(\mathbb{K}_{kd}^\xi)$ by the Basic Theorem on $\mathbb{K}$-Conceptual Contents. This suggests the following construction of a triadic concept graph with subdivision $(V, E, \nu, \sigma, \kappa, \rho)$ of $\vec{\mathbb{K}}$:

$E^{(k)} := \underline{U}_k$, $\nu((g_1, \ldots, g_k), \mathfrak{c}, \mathfrak{d}) := ((g_1, \mathfrak{o}_2, \mathfrak{o}_2), \ldots, (g_k, \mathfrak{o}_2, \mathfrak{o}_2))$,

$\sigma(h, \mathfrak{o}_2, \mathfrak{o}_2) := \{(g, \mathfrak{c}, \mathfrak{d}) \in V \cup E \mid \mathfrak{c} \neq \mathfrak{d} \text{ and } \xi(h) = \mathfrak{d}\}$,

$\sigma(g, \mathfrak{c}, \mathfrak{d}) := \emptyset$ if $\mathfrak{c} \neq \mathfrak{o}_2$ or $\mathfrak{d} \neq \mathfrak{o}_2$,

$\kappa(g, \mathfrak{c}, \mathfrak{d}) := \mathfrak{c}$ and $\rho(g, \mathfrak{c}, \mathfrak{d}) := \{g\}$. $\qquad\qquad\square$

References

[Bi98]   K. Biedermann: *A foundation of the theory of trilattices.* Dissertation, TU Darmstadt 1998. Shaker Verlag, Aachen 1998.

[GW99]   B. Ganter, R. Wille: *Formal Concept Analysis: mathematical foundations.* Springer, Heidelberg 1999.

[GW00]   B. Groh, R. Wille: Lattices of triadic concept graphs. In: B. Ganter, G. W. Mineau (eds.): *Conceptual structures: logical, linguistic and computational issues.* LNAI **1867**. Springer, Heidelberg 2000, 332-341.

[LW95]   F. Lehmann, R. Wille: A triadic approach to formal concept analysis. In: G. Ellis, R. Levinson, W. Rich, J. F. Sowa (eds.): *Conceptual structures: applications, implementations, and theory.* LNAI **954**. Springer, Heidelberg 1995, 32–43.

[SW03]   L. Schoolmann, R. Wille: Concept graphs with subdivision: a semantic approach. In: A. de Moor, W. Lex, B. Ganter (eds.): *Conceptual structures for knowledge creation and comunication.* LNAI **2746**. Heidelberg 2003, 271-281.

[Wi95]   R. Wille: The basic theorem of Triadic Concept Analysis. In: *Order 12* (1995), 149-158.

[Wi98]   R. Wille: Triadic concept graphs. In: M. L. Mugnier, M. Chein (eds.): *Conceptual structures: theory, tools and applications.* LNAI **1453**. Springer, Heidelberg 1998, 194-208.

[Wi00]   R. Wille: Contextual Logic summary. In: G. Stume (eds.): *Working with Conceptual Structures. Contributions to ICCS 2000.* Shaker, Aachen 2000, 265-276.

[Wi02]   R. Wille: Existential concept graphs of power context families. In: U. Priss, D. Corbett, G. Angelova (eds.): *Conceptual structures: integration and interfaces*. LNAI **2393**. Springer, Heidelberg 2002, 382–395.

[Wi03]   R. Wille: Conceptual contents as information - basics for Contextual Judgment Logic. In: A. de Moor, W. Lex, B. Ganter (eds.):*Conceptual structures for knowledge creation and comunication*. LNAI **2746**. Springer, Heidelberg 2003, 1-15.

[Wi04]   R. Wille: Implicational concept graphs. In: H. Pfeiffer, K. E. Wolff, H. Delugach (eds.): *Conceptual Structures at work*. LNAI **3127**. Springer, Heidelberg 2004.

Alpha Galois Lattices: An Overview

Véronique Ventos[1] and Henry Soldano[2]

[1] LRI, UMR-CNRS 8623, Université Paris-Sud, 91405 Orsay, France
ventos@lri.fr
[2] L.I.P.N, UMR-CNRS 7030, Université Paris-Nord,
93430 Villetaneuse, France
soldano@lipn.univ-paris13.fr

Abstract. What we propose here is to reduce the size of Galois lattices still conserving their formal structure and exhaustivity. For that purpose we use a preliminary partition of the instance set, representing the association of a "type" to each instance. By redefining the notion of *extent* of a term in order to cope, to a certain degree (denoted as α), with this partition, we define a particular family of Galois lattices denoted as *Alpha Galois lattices*. We also discuss the related implication rules defined as inclusion of such α-extents and show that Iceberg concept lattices are Alpha Galois lattices where the partition is reduced to one single class.

1 Introduction

Galois lattices (or concept lattices) are well-defined and exhaustive representations of the concepts embedded in a data set since they allow us to obtain every subset of instances distinguishable according to the chosen attributes. However, when dealing with real-world data sets the size of such a lattice can be too large to be handled. Various techniques have been proposed to reduce the size of concept lattices by eliminating part of the nodes (e.g. [7]). In particular, Iceberg concept lattices [14, 17] represent the topmost part of a concept lattice w.r.t. a global criterion of frequency: only nodes with an *extent* cardinality satisfying a threshold according to the whole data set are kept. In this paper, we present more flexible Galois lattices in which the number of nodes is controlled according to a local criterion of frequency linked to a prior partition of the set of instances.

The partition is a set of *basic classes* which are clusters of instances sharing the same basic type. For instance, in real data concerning the electronic catalog of computer products C/Net (http://www.cnet.com), there are 59 different basic types (e.g. *Laptops*, *HardDrives*, *NetworkStorage*) for 2274 instances. Basic classes are then used in order to add a local criterion of frequency to the notion of *extent* as follows: an instance i now belongs to $ext_\alpha(T)$, the α-extent of a subset T of the set of attributes, when it belongs to $ext(T)$, the extent of T, (i.e. i has every of T's properties), and when at least α % of the instances of the basic class of i also belong to $ext(T)$. This new notion of α-extent is used in the

B. Ganter and R. Godin (Eds.): ICFCA 2005, LNAI 3403, pp. 299–314, 2005.

Galois connection related to the family of *Alpha Galois lattices*. Alpha Galois lattices were first introduced in [12] as a part of the system ZooM.

In comparison with concept lattices, Alpha Galois lattices are mainly characterized by the following properties:

- For the same set of attributes, the same set of individuals, and for any value of α, the Alpha Galois lattice G_α is coarser than the concept lattice G, i.e. the set of nodes of G_α is a subset of the set of nodes of the concept lattice G.
- G_0 exactly is G, and G_{100} also is a concept lattice built from a set of instances that each represents one basic class.
- The values of α define a total order on *Alpha Galois lattices* where the *Alpha Galois lattice* induced by $ext_{\alpha 1}$ is coarser than the *Alpha Galois lattice* induced by $ext_{\alpha 2}$ if $\alpha 1 \geq \alpha 2$.
- When all individuals belong to a single basic class, the corresponding Alpha Galois lattice is an Iceberg concept lattice where $\frac{\alpha}{100} = minsupp$.
- A property (i.e. an attribute) can belong to an *intent* of an Alpha Galois lattice G_α even if it is not globally frequent. For instance, in G_{90} the "support" property will appear since in the *HardDrives* basic class, 92 % of the instances of *HardDrives* were sold with support. Actually, this property is not globally frequent (13 products out of 2274, i.e. 0.5 %) and so would not apppear in the corresponding Iceberg concept lattice with $minsupp = 0.9$
- The inclusion of α-*extent* corresponds to particular implication rules, representing some kind of approximation of usual implication rules, that depends on the selected partition of the instances.

The general framework of Galois lattices is given in section 2. In section 3, we present Alpha Galois lattices illlustrated with a simple example. Section 4 presents experimental results on the C/net data set and discusses the ability of such a representation to deal with exceptional data (α near 0 or near 100). Section 5 first discusses Iceberg Alpha Galois lattices together with α-implication rules, and then briefly addresses theoretical issues as the nature of the objects of a formal context which concept lattice is isomorphic to an Alpha Galois lattice. Finally, related work and future work are discussed in section 6.

2   Preliminaries and Definitions

Detailed definitions, results and proofs regarding Galois connections and lattices may be found in [1, 2]. Other results concerning Galois lattices in the field of Formal Concept Analysis can be found in [4]. However we need a more general presentation than the one in [4] as our main goal is to construct Galois lattices where the notion of *extent* is not the usual one. In the rest of the paper we denote as Galois lattice the formal structure that we define hereunder and we will denote as concept lattice the Galois lattice as presented in [4].We consider in our presentation that the reader is familiar with the definitions of *ordered set* and *lattice*. We also recall that a mapping w from an ordered set M to M

is called a closure operator iff for any pair (x, y) of elements of M we have a) $x \leq w(x)$ (extensity), b) if $x \leq y$ then $w(x) \leq w(y)$ (monotonicity), and c) $w(x) = w(w(x))$ (idempotency). An element of M such that $x = w(x)$ is called a *closed element* of M w.r.t. w.

Definition 1 (Galois Connection). *Let m1: $P \to Q$ and m2: $Q \to P$ be maps between two ordered sets $(P, \leq_P)$ and $(Q, \leq_Q)$. Such a pair of maps is called a Galois connection if for all p, p1, p2 in P and for all q, q1, q2 in Q:*
  C1- $p1 \leq_P p2 \Rightarrow m1(p2) \leq_Q m1(p1)$
  C2- $q1 \leq_Q q2 \Rightarrow m2(q2) \leq_P m2(q1)$
  C3- $p \leq_P m2(m1(p))$ and $q \leq_Q m1(m2(q))$

The following simple example will be used in order to illustrate the different notions presented in section 2 and in section 3.

Example 1. The two ordered sets are $(\mathcal{L}, \preceq)$ and $(\mathcal{P}(I), \subseteq)$. $\mathcal{L}$ is a language a term of which is a subset of a set of attributes $\mathcal{A} = \{t1, t2, t3, a3, a4, a5, a6, a7, a8\}$. Here $c1 \preceq c2$ means that $c1 \subseteq c2$. I is a set of individuals $= \{i1, i2, i3, i4, i5, i6, i7, i8\}$. Let *int* and *ext* be the two maps int: $\mathcal{P}(I) \to \mathcal{L}$ and ext: $\mathcal{L} \to \mathcal{P}(I)$ such that $int(e1)$ is the subset of attributes common to all the individuals in $e1$ and $ext(c1)$ is the subset of individuals of I which have all the attributes of $c1$. Example 1 is fully described in Figure 1 where each line i represents the *intent* $int(\{i\})$ of an individual of I and each column j represents the *extent* $ext(\{j\})$ of an attribute of $\mathcal{A}$.

Together with $\mathcal{L}$ and $\mathcal{P}(I)$, *int* and *ext* define a Galois connection.

	t1	t2	t3	a3	a4	a5	a6	a7	a8
i1	1			1	1		1		1
i2	1			1		1	1		
i3		1			1		1		1
i4		1			1		1	1	
i5		1		1			1		1
i6			1	1			1		1
i7			1	1			1		1
i8			1	1		1	1		1

Fig. 1. Example 1. $Tab(i, j) = 1$ if the jth attribute belongs to the ith individual

Definition 2 (Galois Lattices). *Let m1: $P \to Q$ and m2: $Q \to P$ be maps between two lattices $(P, \leq_P)$ and $(Q, \leq_Q)$, such that $(m1, m2)$ is a Galois connection.*
Let G = { (p, q) with p an element of P and q an element of Q such that $p = m2(q)$ and $q = m1(p)$}

Let $\leq$ be defined by: $(p1,q1) \leq (p2,q2)$ iff $q1 \leq_Q q2$.
$(G, \leq)$ is a lattice called a Galois lattice. When necessary it will be denoted as $G(P, m1, Q, m2)$.
Example: *In example 1, we have $G = \{(c,e) \mid c \in \mathcal{L}, e \in \mathcal{P}(\mathcal{I}), e = ext(c) \text{ and } c = int(e)\}$. Then $(G, \leq)$ is a Galois lattice where $\leq$ is defined by: $(c,e) \leq (c_1, e_1)$ iff $e \subseteq e_1$ (which is equivalent to $c \supseteq c_1$). The Galois lattice corresponding to example 1 is presented in Figure 2.*

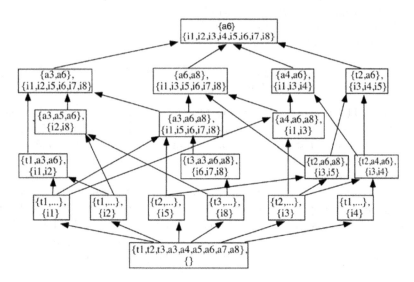

Fig. 2. The Galois Lattice corresponding to example 1

In a Galois connection $m1 \circ m2$ and $m2 \circ m1$ are closure operators on $(P, \leq_P)$ and $(Q, \leq_Q)$. As a consequence, a node of a Galois lattice is a pair of closed elements of P and Q.

Example: *In example 1, $ext(\{a4\}) = \{i1, i3, i4\}$, $int(\{i1, i3, i4\}) = \{a4, a6\}$. The term $\{a4, a6\}$ is therefore a closed term as $int(ext(\{a4\})) = \{a4, a6\}$*

Furthermore the functions $m1$ and $m2$ define equivalence relations on the lattices P and Q as follows:

Definition 3 (Equivalence Relations on P and Q). *Let $\equiv_P$ and $\equiv_Q$ denote the equivalence relations defined on P and Q by the mappings $m1$ and $m2$, i.e. let $p1, p2$ be elements of P and $q1, q2$ be elements of Q:*
$p1 \equiv_P p2$ iff $m1(p1) = m1(p2)$, and $q1 \equiv_Q q2$ iff $m2(q1) = m2(q2)$

Lemma 1. *Let p be an element of P, and q be an element of Q, then $m2(m1(p))$ is the greatest element of the equivalence class of $\equiv_P$ containing p and $m1(m2(q))$ is the greatest element of the equivalence class of $\equiv_Q$ containing q.*

So, a characteristic property of Galois lattices is that each node (p, q) is a pair of representatives of their respective equivalence classes.

In our previous example, we used the language $\mathcal{L}$, defined as the powerset $\mathcal{P}(\mathcal{A})$ of a set of attributes A, as the first lattice, and the powerset $\mathcal{P}(\mathcal{I})$ of a set of individuals I as the second lattice. Such a Galois lattice is known as a *concept lattice*[4]. In concept lattices, a node (c, e) is a concept, c is the *intent* and e is the *extent* of the concept. The relationship between I and A is expressed as the *formal context* (I, A, R) where $R \subseteq I \times A$ is the binary relation such that iRa if an only if the individual i has the attribute a. We have then $int(e) = \{a \in A \mid \forall i \in e, iRa\}$ and $ext(c) = \{i \in I \mid \forall a \in c, iRa\}$. The Galois lattice presented in Figure 2 is then the concept lattice defined by the formal context of Figure 1.

Concept lattices are interesting both from a practical point of view, as they express in a rigorous way the two sides of a concept, and from a theoretical point of view, as any complete lattice is isomorphic to a concept lattice [4].

3   Alpha Galois Lattices

In what follows we consider, with no loss of generality, $\mathcal{L} = \mathcal{P}(A)$ and we start with the concept lattice $G(\mathcal{L}, ext, \mathcal{P}(I), int)$ as previously examplified. Then we will discuss a variation on ext whose purpose is to obtain an equivalence relation $\equiv'_{\mathcal{L}}$ coarser than the original one (see definition 9) thus resulting in larger equivalence classes on $\mathcal{L}$ and so on less nodes in the corresponding Galois lattice.

The new ext function relies on the association of a predefined type to each individual of I. The corresponding clusters of instances, which form a partition of I are denoted as *basic classes*. The first idea is then to gather such clusters rather than individuals (see [12]). For instance, let us assume that the attributes $t1, t2, t3$ express the types of the individuals of example 1. These types corresponds to three basic classes $BC1$, $BC2$, $BC3$ whose descriptions are the following:

$BC1=\{i1,i2\}$, $int(BC1)= \{t1,a3,a6\}$; $BC2=\{i3,i4,i5\}$, $int(BC2)= \{t2,a6\}$; $BC3=\{i6,i7,i8\}$, $int(BC3)= \{t3,a3,a6,a8\}$.

Let us consider the concept lattice built on a set of individuals $\{bc1,bc2,bc3\}$, that we call the *prototypes* of their respective basic classes, and that are such that, for any index i, $int(BCi) = int(\{bci\})$. This concept lattice is represented in Figure 3 as a particular case of an Alpha Galois lattice, and is much smaller than the original concept lattice.

Now, we propose an intermediate approach where the entities gathered can be other subsets of I than either individuals or whole basic classes. This leads to the definition of Alpha Galois lattices.

3.1   Alpha Definitions

Definition 4 (Alpha Satisfaction). *Let α belong to [0,100]. Let $e=\{i_1,\ldots,i_n\}$ be a set of individuals and T be a term of $\mathcal{L}$. Then,*

$$e \; \alpha - satisfies \; T \; (e \; sat_\alpha \; T) \; iff \mid ext(T) \cap e \mid \; \geq \frac{|e|.\alpha}{100}$$

Since the Alpha satisfaction is defined according to a set of individuals and to a term of the language $\mathcal{L}$, we can use it to check whether at least α % of a basic class satisfies a term of $\mathcal{L}$ and add this constraint to isa, the classical membership relation between individuals and terms. In what follows i isa T means $i \in ext(T)$. We call this notion (membership relation plus Alpha satisfaction of the basic class) the *Alpha membership relation*.

Definition 5 (Alpha Membership Relation). *Let I be a set of individuals and $\mathcal{BC}$ be a partition of I into a set of basic classes. Let $BCl : I \to \mathcal{BC}$ be such that $BCl(i)$ is the basic class to which belongs i, and let T be a term of $\mathcal{L}$, then:*
i isa_α T *iff i isa T and $BCl(i)$ sat_α T*

Example (Example 1). *Let $T=\{a6,a8\}$, $ext(T) = \{i1,i3,i5,i6,i7,i8\}$. $BC1$ sat_{50} T since $i1$ isa T and $\mid BC1 \mid$ $=2$. As a result $i1$ isa_{50} T. $BC2$ sat_{60} T since $\mid ext(T) \cap BC2 \mid \geq \frac{|BC2|.60}{100}$. So we have $i3$ and $i5$ isa_{60} T. Finally $BC3$ sat_{100} T since 100 % of the individuals of $BC3$ belong to the extent of T. So we have $i6$, $i7$, and $i8$ isa_{100} T.*

Finally, we use the *Alpha membership relation* to define the notion of *extent* used in Alpha Galois Lattices.

Definition 6 (Alpha Extent of a Term). *The α-extent of T in I w.r.t. the set $\mathcal{BC}$ of basic classes is the following set:*

$$ext_\alpha(T) = \{i \in I \mid i \ isa_\alpha \ T\}$$

Example (Example 1) *: Let $T=\{a6,a8\}$, then $ext_0(T)= ext(T) = \{i1, i3,i5, i6,i7,i8\}$, $ext_{60}(T)= \{i3,i5,i6,i7,i8\}$ and $ext_{100}(T)= \{i6,i7,i8\}$.*

The following proposition about the new Galois connection needs the definition of E_α, a subset of $\mathcal{P}(\mathcal{I})$ whose elements are made of sufficiently large parts of basic classes.

Proposition 1. *Let E_α be the following subset of $\mathcal{P}(\mathcal{I})$:*
$E_\alpha = \{e \in \mathcal{P}(I) \mid \forall i \in e \mid e \cap BCl(i) \mid \geq \frac{|BCl(i)|.\alpha}{100}\}$.
Then int and ext_α define a Galois connection on $\mathcal{L}$ and E_α.

Proof: *The proof relies on theorem 1 given in the next section and is presented as the proof of a corollary.*

We can therefore define Galois lattices from this new Galois connection and we called them *Alpha Galois lattices*.

Definition 7 (Alpha Galois Lattices). *The Galois lattice $G(\mathcal{L}, ext_\alpha, E_\alpha, int)$ corresponding to the Galois connection defined above is called an Alpha Galois lattice and is denoted as G_α.*

When α is equal to 0, $E_\alpha = \mathcal{P}(I)$ and $ext_\alpha = ext$. Therefore, the Alpha Galois lattice is the concept lattice corresponding to the same attributes and instances. When α is equal to 100, the nodes of the Galois lattice are only whole basic classes gathered. As a consequence the Alpha Galois lattice is the concept lattice obtained by considering as instances the *prototypes* of the basic classes.

The Alpha Galois lattice G_{100} of Example 1 is represented in Figure 3. Figure 4 presents the topmost part of G_{60}. Note that *intents* of the nodes of G_{100} are also intents of nodes of G_{60} that in turn are all intents of nodes of the original concept lattice G_0 (see Figure 2).

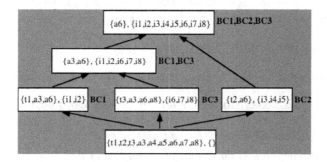

Fig. 3. When $\alpha = 100$ the Alpha Galois lattice G_{100} of example 1 is much smaller than the original concept lattice presented in Figure 2

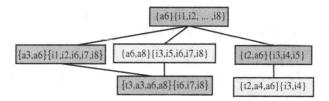

Fig. 4. $\alpha = 60$: The topmost part of G_{60} of example 1. New nodes, w.r.t. G_{100} are the lighter ones

Moreover, there exists a total order on Alpha Galois lattices defined in the next section.

3.2   Alpha Galois Lattice Order

In [5] the authors give a formal view to the extension of formal concept analysis to more sophisticated languages of terms and use the notion of *projection* as a way to obtain smaller lattices by reducing the language. [12] independently uses the same notion of projection with a similar scope and also introduce *extensional* projections to modify the *ext* function. We recall hereunder the notion of projection:

Definition 8 (Projection). *Proj is a projection of an ordered set (M,≤) iff for any pair (x,y) of elements of M:*
$x \geq Proj(x)$ *(minimality),*
if $x \leq y$ then $Proj(x) \leq Proj(y)$ (monotonicity),
$Proj(x) = Proj(Proj(x))$ *(idempotency).*

Applying first the mapping *ext* and then an extensional projection yields an equivalence relation $\equiv'_{\mathcal{L}}$ which is coarser than the original one, thus resulting in larger equivalence classes on $\mathcal{L}$ [12].

Definition 9. *Let $\equiv^1_{\mathcal{L}}$ be the equivalence relation defined on $\mathcal{L}$ by the mapping ext^1, and let $\equiv^2_{\mathcal{L}}$ be the equivalence relation defined on $\mathcal{L}$ by the mapping ext^2, then, $\equiv^2_{\mathcal{L}}$ is said coarser than $\equiv^1_{\mathcal{L}}$ iff for any pair (c1,c2) of elements of $\mathcal{L}$ we have:*
if $ext^1(c1) = ext^1(c2)$ then $ext^2(c1) = ext^2(c2)$.

The following theorem [12] has a corollary that proves the proposition 1:

Theorem 1 (An Extensional Order on Galois Connections).
Let int and ext define a Galois connection on $\mathcal{L}$ and E, and let proj be a projection of E. Let $E^1 = proj(E)$ and $ext^1 = proj \circ ext$. Then:

1) int, ext^1 define a Galois connection on $\mathcal{L}$ and E^1.
2) The Galois lattice $G^1(\mathcal{L}, ext^1, E^1, int)$ has the following property: for any node $g^1 = (c, e^1)$ in G^1 there exists a node $g = (c, e)$ in $G(\mathcal{L}, ext, E, int)$, with the same intent c, such that $e^1 = proj(e)$.
3) $\equiv^1_{\mathcal{L}}$ is coarser than $\equiv_{\mathcal{L}}$.

We will say then that G^1 is coarser than (or nested in) G and write $G^1 = proj(G)$. Let (c,e) be a node of G, then $proj(c,e) = (int \circ proj(e), proj(e))$ is the projected node in G^1.

Corollary 1. *Let $G(\mathcal{L}, ext, P(I), int)$ be a Galois lattice. Let $\alpha \in [0, 100]$ and for $e \in \mathcal{P}(\mathcal{I})$, let :*

$$- proj_\alpha(e) = e - \{i \mid i \in e \text{ and } \mid e \cap BCl(i) \mid < \frac{|BCl(i)|.\alpha}{100}\}$$
$$- ext_\alpha = proj_\alpha \circ ext \text{ and } E_\alpha = proj_\alpha(\mathcal{P}(\mathcal{I}))$$

Then:
- int, ext_α define a Galois connection on $\mathcal{L}$ and E_α and $G(\mathcal{L}, ext_\alpha, E_\alpha, int)$ is a Galois lattice coarser than G.

proof: *In order to prove this corollary, we simply have to show that $proj_\alpha$ is a projection: -$proj_\alpha(e)$ is included in e since we remove elements of e, so $proj_\alpha$ is minimal. - If e is included in e', every element of e removed when applying $proj_\alpha$ on e' will also be removed when applying $proj_\alpha$ on e, so $proj_\alpha$ is monotonic. - finally, $proj_\alpha$ is idempotent since no more element of $proj_\alpha(e)$ can be removed by applying again $proj_\alpha$.*

Furthermore, we can order the *alpha extents* according to the value of α: For every pair $(\alpha1, \alpha2)$ such that $\alpha1 \leq \alpha2$, $ext_{\alpha2} = proj_\alpha \circ ext_{\alpha1}$ with $\alpha = \alpha2$. As a consequence, the value of α defines a total order on Alpha Galois lattices:

Proposition 2 (A Total Order on Alpha Galois Lattices). *Let us denote as $\equiv_\alpha$ the equivalence relation on $\mathcal{L}$ associated to ext_α. Then for every pair ($\alpha1$, $\alpha2$) such that $\alpha1 \leq \alpha2$, $\equiv^{\alpha2}_{\mathcal{L}}$ is coarser than $\equiv^{\alpha1}_{\mathcal{L}}$.*

proof: $proj_\alpha$ *is a projection for every value of* α *belonging to [0,100].*
$ext_{\alpha 1} = proj_\alpha \circ ext$ *with* $\alpha = \alpha 1$ *and* $ext_{\alpha 2} = proj_\alpha \circ ext_{\alpha 1}$ *with* $\alpha = \alpha 2$. *According to 3) of Theorem 1,* $\equiv_{\mathcal{L}}^{\alpha 2}$ *is then coarser than* $\equiv_{\mathcal{L}}^{\alpha 1}$.

Example: $\equiv_{\mathcal{L}}^{100}$ *is coarser than* $\equiv_{\mathcal{L}}^{60}$ *which is in turn coarser than* $\equiv_{\mathcal{L}}^{0}$ *that is the equivalence relation* $\equiv_{\mathcal{L}}$ *of the concept lattice.*

The previous proposition is the basis to make successive refinements in Alpha Galois lattices (see section 4).

There is also a partial order associated to the initial partition $\mathcal{BC}$ of I in basic classes. Let us suppose that we substract some basic classes from I, and so from $\mathcal{BC}$, thus obtaining a reduced instance set I' together with a reduced partition $\mathcal{BC}'$. It is then easy to show (proof omitted here) that there is a projection $proj$ such that the corresponding E'_α simply rewrites as $proj(E_\alpha)$. As a consequence we have the following property where we denote as $G_\alpha^{\mathcal{B}}$ the Alpha Galois lattice built from the partition $\mathcal{B}$.

Proposition 3 (A Partial Order on Alpha Galois Lattices). *Let* $\mathcal{BC}'$ *be a subset of the set of basic classes* $\mathcal{BC}$, *then the Alpha Galois lattice* $G_\alpha^{\mathcal{BC}'}$ *is coarser than the Alpha Galois lattice* $G_\alpha^{\mathcal{BC}}$.

An interesting case is the one of the partition $\{I\}$ in which we consider only one single class, i.e. the case in which all individuals share the same type. The corresponding Alpha Galois lattice is the topmost part of the concept lattice defined by the same language $\mathcal{L}$ and the same set I of individuals. The lattice then only contains nodes whose extents have a size greater than $\frac{\alpha}{100}|I|$ (plus the bottom node whose extent is empty). This structure has been previously investigated and is denoted as an *Iceberg* (or *frequent*) *concept lattice* [14, 17] where $\frac{\alpha}{100}$ corresponds to the value of the support threshold *minsupp*.

Note that because of Proposition 3, the Iceberg lattice of any basic class BCi of a partition $\mathcal{BC}$ is always coarser than the Alpha Galois lattice corresponding to $\mathcal{BC}$.

4   Experiments

The program ALPHA that computes Alpha Galois lattices relies on a straightforward top-down procedure in which nodes are generated as follows: a current node intent c is specialized by adding a new attribute a, then $int \circ ext_\alpha$ is applied to $c \cup \{a\}$ in order to obtain a closed term; the corresponding node has then to be compared to previous nodes in order to avoid duplicates.

We have experimented with ALPHA on a real dataset composed of 2274 computer products extracted from the C/Net catalog. Each product is described using a subset of 234 attributes. There are 59 types of products and each product is labelled by one and only one type.

In our first experiment we have built G_{100} using the whole data set (so practically restricted to 59 prototypical instances). Then we smoothly lowered the value of α and recomputed the corresponding G_α lattice. As we can see here-under the number of nodes (and so the CPU time) exponentially grows from 211 concepts to 165369 as α varies from 100 to 91. This means that it is here impossible to have a complete view of the data at the level of instances ($\alpha=0$) and that even relaxation of the basic class constraint (starting with $\alpha=100$) has to be limited:

Alpha	100	98	96	94	92	91
Nodes	211	664	8198	44021	107734	165369

Our second experiment concerns the part of G_{100} between the node whose extent contains the 3 basic classes (*Laptop* (252 instances, 39 attributes involved), *Hard-drive*(45 instances, 22 attributes), *Network-storage*(4 instances, 16 attributes)) and the *Bottom* node.

The new G_{100} contains now 5 nodes (to be compared to the maximum number of $2^3 = 8$ nodes). Here computation of G_α is performed for a set of values $\alpha \in [0, 100]$ together with the corresponding Iceberg lattices (see Figure 5). We

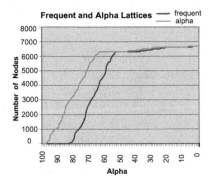

Fig. 5. Number of nodes vs Alpha values for Iceberg lattices and Alpha Galois lattices

are first interested in what happens with high values of α. Starting from G_{100}, new nodes appear as α slowly decreases. For instance at $\alpha = 99$, a new node appears under the G_{100} node standing for the basic class *Laptop*. The intent of the new node now contains the attribute "network-card". This is due to the fact that most instances of the class *Laptop* do possess a network card. So by relaxing the basic class constraint we get rid of the few, exceptional, instances of *Laptop* found in the catalog and that were hiding this "default" property of *Laptop* in G_{100}. In the same way most *hard-drives* are sold with "support". So at $\alpha = 92$, a new node representing *hard-drives* with "support" appears. Note that in this case, the attribute "support" is infrequent when considering all the instances ("support" appears in 13 products out of 301) and so would not be considered in a *Iceberg concept lattice*, whereas it is frequent within the *hard-drive* class (13 products out of 15) and so comes out in the Alpha Galois lattice G_{90}. As a summary, by slowly decreasing α from 100 we have a more accurate view of our data by revealing properties that are relevant to at least some basic classes.

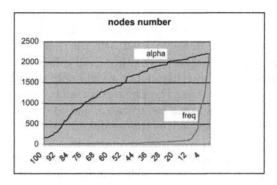

Fig. 6. Number of nodes vs Alpha values for Iceberg lattices and Alpha Galois lattices

Now as α slowly grows from 0 to small values (say 10), some instances, whose behavior is *exceptional* within their basic class w.r.t. some term t of $\mathcal{L}$, will disappear from the corresponding α-extent. These instances are exceptional as they belong to the extent of the term t whereas very few instances of the same basic class do belong to this extent. As a result some properties that are very infrequent within some basic class will no longer be allowed to discriminate concepts. For example, only few *Laptops* have the property "Digital-Signal-Protocol", and so when $\alpha = 6$, nodes whose intent contains the "Digital Signal Protocol" property no more include instances of *Laptop* in their extent. As a result terms including "Digital-Signal-Protocol" become equivalent whenever their extent only differed because of Laptop instances, thus resulting on a smaller (and so simpler) lattice. However a closer look to Figure 5 shows there can be a large number of nodes even for high values of α. In this particular example this is due to the fact that one basic class, namely *Laptop*, has a huge Iceberg lattice that invades the Alpha Galois lattice (data not shown). An experiment with 24 basic classes and 1187 objects (some large basic classes are removed thus resulting in a more homogenous class size distribution) shows that the size of Alpha Galois lattices can be really different from the one of Iceberg lattices (see also Figure 6) :

Alpha Values	100	80	50	30	0
Alpha Nodes	158	842	1493	1900	2202
Frequent Nodes	2	18	18	50	2202

5   Issues Related to the Alpha View of Data

5.1   Combining Global and Local Frequency Constraints: Frequent Alpha Lattices

On one hand, in Iceberg concept lattices we apply a global frequency constraint to the concept lattice: all nodes whose extents are small enough are eliminated (i.e. sent to the bottom node). When the threshold is high this unfortunately tends to eliminate many intents that, though globally infrequent, are frequent in some basic classes. On the other hand in Alpha Galois lattices we apply a local frequency constraint in such a way that intents frequent in at least one basic class appear in the lattice. However, a side effect is that an Alpha Galois lattice may still be very large, especially when using small values of *alpha*. Our proposal here is to combine these two constraints : we will only consider nodes whose α-*extent* is large enough. Applying the global constraint allows to eliminate nodes that are locally frequent on some basic classes, and so would be interesting, but still represent few instances and so can be discarded when we want a simpler view of the data.

The result of such a filter is again a Galois lattice. More precisely, for any real number f with $0 \leq f \leq 1$, consider the function $proj^f$ on E_α such that $proj^f(e) = e$ whenever $\frac{|e|}{|I|} \geq f$ and $proj^f(e) = \emptyset$ otherwise. $proj^f$ clearly is a projection, and therefore $G_\alpha^f = proj^f(G_\alpha)$ is a Galois lattice coarser than G_α. More precisely it corresponds to the topmost part of G_α plus a bottom node.

We denote G_α^f as the *Iceberg Alpha lattice* associated to the instance set I, the partition $\mathcal{BC}$ of I, the Alpha value α and the global frequency threshold f. The corresponding α-implication rules (see next section) have a support greater than f. Note that we will speak here of an α-support since the support is computed using α-extents.

5.2   Alpha Implication Rules

Association rules, as usually defined in data mining, are implications whose truth values are observed on a set of instances I. Each association rule has a *support* value, i.e. the frequency of its antecedent part within the instance set I, together with a *confidence* value. When its confidence value is 1, an association rule is called an *implication rule*. When considering concept lattices, the partial order induced on terms by the Galois connection can be related to a set of implication rules. More precisely $ext_I(T_1) \subseteq ext_I(T_2)$ means that the implication $T_1 \rightarrow T_2$ holds for all instances of I. In such rules, T_1 will be denoted as the left part and T_2 as the right part. In Iceberg concept lattices, the extent of a term is redefined as empty whenever the term is infrequent in I, i.e., when its original extent contains less than $minsupp * |I|$ instances of I. As a consequence the corresponding implication rules all have a support greater than $minsupp$.

Association rules are efficiently constructed in two steps, first constructing the Iceberg concept lattice corresponding to the instance set I. The intents of the concepts of an Iceberg concept lattice are usually denoted as *closed frequent itemsets*. Association rules are then built using closed frequent itemsets [11, 18]. The basic idea is that, as mentioned before, a node in the concept lattice corresponds to an equivalence class of terms, all sharing the same extent. In particular, the intent of the node, i.e., the unique greatest term, has the same extent as all the smallest terms (also called *generators*). We obtain then for each node several implication rules whose left part are these generators, and whose right part is the intent of the node. Part of the set of all these rules extracted from the concept lattice produces the non-redundant Guigues-Duquenne basis of implication rules [6]. For sake of clarity, the left part of each rule is substracted from the right part. For instance, let the node be $(\{a, b, c\}, \{i1, i2, i3\})$ and suppose that the generators of the corresponding equivalence class are $\{\{a\}, \{b\}\}$ (this means that $ext_I(\{a\}) = ext_I(\{b\}) = \{i1, i2, i3\}$). We obtain then the implication rules $\{\{a\} \rightarrow \{a, b, c\}, \{b\} \rightarrow \{a, b, c\}\}$ that are rewritten as $\{\{a\} \rightarrow \{b, c\}, \{b\} \rightarrow \{a, c\}\}$.

Now, in Alpha Galois lattices, whenever $ext_\alpha(T_1) \subseteq ext_\alpha(T_2)$ we will say that the α-implication $T_1 \rightarrow_\alpha T_2$ holds on the pair $(I, \mathcal{BC})$. Because they are derived from a Galois lattice, α-implication are transitive, monotonic and additive:

–If $T_1 \rightarrow_\alpha T_2$ and $T_2 \rightarrow_\alpha T_3$, then $T_1 \rightarrow_\alpha T_3$

–If $T_1 \rightarrow_\alpha T_2$, and $T_1 \subseteq T$, then $T \rightarrow_\alpha T_2$

–If $T_1 \rightarrow_\alpha T_2$ and $T_3 \rightarrow_\alpha T_4$, then $T_1 \cup T_3 \rightarrow_\alpha T_2 \cup T_4$. Furthermore we have the *modus ponens* as an inference rule:

If $i \; isa_\alpha \; T_1$ and $T_1 \rightarrow_\alpha T_2$, then $i \; isa_\alpha \; T_2$

The Guigues-Duquenne basis of implication rules has been extended to rules with a minimal support *minsupp*. Also the Luxenburger basis of association rules [10] summarizes rules whose confidence is greater or equal to a minimal confidence level *minconf* and has also been extended to rules with a minimal support. Both extended bases are computed using the closed terms of the corresponding Iceberg lattice [11, 13]. Hereunder we adapt definitions of support and confidence to Alpha rules by changing *extents* to α-*extents*:

Definition 10. *An α-association rule is a pair of terms T_1 and T_2, denoted as $T_1 \rightarrow_\alpha T_2$.*

The support and confidence of an α-association rule $r = T_1 \rightarrow_\alpha T_2$ are defined as follows :

$$\alpha\text{-}supp(r) = \frac{|ext_\alpha(T_1 \cup T_2)|}{|I|}$$

$$\alpha\text{-}conf(r) = \frac{|ext_\alpha(T_1 \cup T_2)|}{|ext_\alpha(T_1)|}$$

The α-association rule $r = T_1 \rightarrow_\alpha T_2$ holds on the pair $(I, \mathcal{BC})$ whenever $\alpha\text{-}supp(r) \geq minsupp$ and $\alpha\text{-}conf(r) \geq minconf$.

Note that when we consider the implication rules derived from a Galois lattice, the right part T_2 of the rule is an *intent* and the left part T_1 is smaller than T_2. As a consequence we have $T_1 \cup T_2 = T_2$ and the α-support rewrite as $\frac{|ext_\alpha(T_2)|}{|I|}$. This means that the set of rules whose α-support is greater than *minsupp* is obtained from the nodes of the Iceberg Alpha lattice $G_\alpha^{minsupp}$. The adaptation of the methods proposed in [11, 13] to compute these bases, starting from the Iceberg lattice (or equivalently from the set of closed terms), is straightforward (basically we simply have to compute α-extents rather than extents when adapting existing algorithms).

We would now emphasize by an example the meaning and usefulness of such rules to handle exceptions when individuals are labelled with basic classes as proposed in this paper. For this purpose, let us suppose that we have divided animals (i.e. individuals) into basic classes as *mammals, birds, insects* and that we search for general rules in the data. An intuitive rule is the following : an animal that *flies* should have *wings*. This rule holds for birds (unflying birds, as ostriches, do not contradict the rule) as well as for insects. The rule should also hold for mammals, that generally do not fly, but is falsified by a flying squirrel. The Alpha approach benefits here from the fact that very few mammals fly (in other words the antecedent part of the rule is infrequent within the basic class to which belong the individual that falsifies the rule). When using α-extents, the flying-squirrel is removed from the antecedent part of the rule. Here, a small value of α is sufficient to obtain an α-*implication* rule expressing that flying animals have wings. Of course greater values of α, namely close to 100, also preclude falsifying the rule. However in the latter case α-implication rules express something different: they apply to individual whenever the antecedent part is common to *most* individuals of the same basic class. In our example, only *birds* would be concerned with such a rule, as most of them fly, but not *insects*.

5.3   Theoretical Issues

A first question concerns what happens if we allow the basic classes to over-lap. A natural modification of definition 5 consists then to require that at least one of the basic classes to which belong the instance α-satisfies the term. Alpha membership is then defined as follows: : i isa_α T iff i isa T and there exists a basic class BC such that $i \in BC$ and BC sat_α T. By accordingly modifying the mapping $proj_\alpha$ (for each individual i in e there must be at least one basic class BC such that $i \in BC$ and $\mid e \cap BC \mid \geq \frac{|BC|.\alpha}{100}$) we again obtain an extensional projection, and so a Galois connection and a Galois lattice. The partial and total orders mentionned in section 3.2 are also preserved. A second question concerns the relationship between Alpha Galois lattices and formal concept analysis. To obtain a *representation* formal context [5] for an Alpha Galois lattice G_α, i.e. a formal context whose concept lattice is isomorphic to an Alpha Galois lattice, we consider as objects particular subsets of the basic classes. More precisely, for each basic class BCi we consider the smallest elements of $proj_\alpha(\mathcal{P}(BCi))$ strictly greater than $\emptyset$. We denote as I_α the set of all these subsets. For instance, when considering our example 1, we obtain $I_{60} = \{\{i1, i2\}, \{i3, i4\}, \{i4, i5\}, \{i3, i5\}, \{i6, i7\}, \{i7, i8\}, \{i6, i8\}\}$. The incidence relation R_α between the set I_α of objects and the set A of attributes is then defined as follows : $oR_\alpha a$ iff $o \subseteq ext(\{a\})$. Let us denote as ext_{I_α} and int_{I_α} the mappings of this formal context. In example 1 we have $ext_{60}(\{a8\}) = \{i3, i5, i6, i7, i8\}$ and $ext_{60}(\{a4\}) = \{i3, i4\}$ and so $ext_{60}(\{a8, a4\}) = proj_{60}(\{i3\}) = \emptyset$. We also have $ext_{I_{60}}(\{a8\}) = \{\{i3, i5\}, \{i6, i7\}, \{i7, i8\}, \{i6, i8\}\}$ and $ext_{I_{60}}(\{a4\}) = \{\{i3, i4\}\}$ and so $ext_{I_{60}}(\{a8, a4\}) = ext_{I_{60}}(\{a8\}) \cap ext_{I_{60}}(\{a4\}) = \emptyset$. Note that I_0 is then made of the singletons of I and I_{100} is the set of *prototypes* of the basic classes. We refer to I_α as the set of the $\alpha - prototypes$ of $\mathcal{BC}$. Clearly, we have for any α-prototype o, $int_{I_\alpha}(\{o\}) = int(o)$ and more generally $int_{I_\alpha}(\{o1, o2, ..., on\}) = int(o1 \cup o2 \cup ... \cup on)$.

6   Related Work and Conclusion

Recent work in Knowledge Representation and Machine Learning investigates Galois connections and lattices based on languages of terms more sophisticated than those used in concept lattices, so modifying the notion of intent of a concept [4, 3, 9, 5]. We have shown here that by restricting the notion of extent of a term with respect to a given partition of the instance set I, we also modify the lattice of extents which is no longer $\mathcal{P}(I)$ and we obtain a new family of Galois lattices. As mentioned above Iceberg concept lattices [17, 14] formally are Alpha Galois lattices in which all individuals belong to the same basic class. Besides, the implication rules related to Alpha-Galois lattices simply correspond to inclusion of α-extents, and such α-implication can be extracted from the Alpha-Galois lattices in the same way as implication rules are extracted from Iceberg concept lattices. Note that α-implication rules inherit from the Galois lattice structure properties (as transitiviy) unusual when dealing with "approximate" rules. About the construction of Alpha Galois lattices, it should be interesting to adapt efficient algorithms (e.g. [8]). Furthermore, as a consequence of property 3, another way [16] to build Alpha Galois lattices is to first build the iceberg lattices corresponding to each basic class and then combine them using a *subposition* operator as previously proposed by [15] to efficiently build concept lattices Note that this is the basis of the *basic class incrementality* of Alpha Galois lattices. We have also seen in 5.3 that the objects of a *representation* formal context for an Alpha Galois lattice are the minimal subsets of the basic classes that satisfy a cardinality constraint (we call them the α-prototypes of each basic class). As a conclusion there is still much work to experiment and to investigate theoretical issues and practical use of Alpha Galois lattices and corresponding α-implication rules. However they represent a flexible tool to investigate data and handle exceptions that are relative to a preliminary view of the data.

Acknowledgments. Many thanks to Nathalie Pernelle for its valuable contribution to the work presented here, and to Philippe Dague for his reading of an earlier draft of this paper.

References

1. M. Barbu and M. Montjardet. *Ordre et Classification, Algèbre et Combinatoire 2.* Hachette Université, 1970.
2. Garrett Birkhoff. *Lattice Theory.* American Mathematical Society Colloquium Publications, Rhode Island, 1973.
3. J-G. Ganascia. TDIS: an algebraic formalization. In *Int. Joint Conf. on Art. Int.*, volume 2, pages 1008–1013, 1993.
4. B. Ganter and R. Wille. *Formal Concept Analysis: Logical Foundations.* Springer Verlag, 1999.
5. Bernhard Ganter and Sergei O. Kuznetsov. Pattern structures and their projections. *ICCS-01, LNCS*, 2120:129–142, 2001.
6. J.L. Guigues and V. Duquenne. Famille non redondante d'implications informatives résultant d'un tableau de données binaires. *Mathématiques et Sciences humaines*, 95:5–18, 1986.

7. Joachim Hereth, Gerd Stumme, Rudolf Wille, and Uta Wille. Conceptual knowledge discovery and data analysis. In *Ganter, B., Mineau, G. (eds.): Conceptual Structures: Logical, Linguistic and Computational Issues*, pages 421–437. LNAI 1867, Springer, Berlin-Heidelberg, New-York, 2000.

8. S. Kuznetsov and S. Obiedkov. Comparing performance of algorithms for generating concept lattices. *J. of Experimental and Theoretical Art. Int.*, 2/3(14):189–216, 2002.

9. M. Liquiere and J. Sallantin. Structural machine learning with Galois lattice and graphs. In *ICML98, Morgan Kaufmann*, 1998.

10. M. Luxenburger. Implications partielles dans un contexte. *Mathématiques, Informatique et Sciences Humaines*, 29(113):35–55, 1991.

11. N. Pasquier, Y. Bastide, R. Taouil, and L. Lakhal. Efficient mining of association rules using closed itemset lattices. *Information Systems*, 24(1):25–46, 1999.

12. N. Pernelle, M-C. Rousset, H. Soldano, and V. Ventos. Zoom: a nested Galois lattices-based system for conceptual clustering. *J. of Experimental and Theoretical Artificial Intelligence*, 2/3(14):157–187, 2002.

13. Gerd Stumme, Rafik Taouil, Yves Bastide, Nicolas Pasquier, and Lotfi Lakhal. Intelligent structuring and reducing of association rules with formal concept analysis. *Lecture Notes in Computer Science*, 2174:335–349, 2001.

14. Gerd Stumme, Rafik Taouil, Yves Bastide, Nicolas Pasquier, and Lotfi Lakhal. Computing iceberg concept lattices with titanic. *Data and Knowledge Engineering*, 42(2):189–222, 2002.

15. P. Valtchev, R. Missaoui, and P. Lebrun. A partition-based approach towards building Galois (concept) lattices. *Discrete Mathematics*, 256(3):801–829, 2002.

16. V. Ventos and H. Soldano. Les treillis de Galois alpha ou De l?influence d?une partition a priori des donneés. *Revue d'Intelligence Artificielle*, to appear, 2005.

17. K. Waiyamai and L. Lakhal. Knowledge discovery from very large databases using frequent concept lattices. In *11th Eur. Conf. on Machine Learning, ECML'2000*, pages 437–445, 2000.

18. M. J. Zaki. Generating non-redundant association rules. *Intl. Conf. on Knowledge Discovery and DataMining (KDD 2000)*, 2000.

A Finite State Model for On-Line Analytical Processing in Triadic Contexts

Gerd Stumme

Chair of Knowledge & Data Engineering, Department of Mathematics and Computer Science,
University of Kassel, Wilhelmshöher Allee 73, D–34121 Kassel, Germany
http://www.kde.cs.uni-kassel.de

Abstract. About ten years ago, triadic contexts were presented by Lehmann and Wille as an extension of Formal Concept Analysis. However, they have rarely been used up to now, which may be due to the rather complex structure of the resulting diagrams. In this paper, we go one step back and discuss how traditional line diagrams of standard (dyadic) concept lattices can be used for exploring and navigating triadic data.

Our approach is inspired by the *slice & dice* paradigm of On-Line-Analytical Processing (OLAP). We recall the basic ideas of OLAP, and show how they may be transferred to triadic contexts. For modeling the navigation patterns a user might follow, we use the formalisms of finite state machines. In order to present the benefits of our model, we show how it can be used for navigating the IT Baseline Protection Manual of the German Federal Office for Information Security.

1 Introduction

Concept lattices have proven their high potential for visualizing and exploring datasets in many applications during the last 25 years. This success of Formal Concept Analysis incited researchers to extend it to other types of knowledge representation. Among them are for instances logical extensions, relational data, and power context families. One of these extensions are *triadic contexts*, which were introduced ten years ago by Fritz Lehmann and Rudolf Wille in [14]. They defined a *triadic formal context* as a quadruple $\mathbb{K} := (G, M, B, Y)$ where G, M, and B are sets, and Y is a ternary relation between G, M, and B, i.e., $Y \subseteq G \times M \times B$. The elements of G, M, and B are called *(formal) objects, attributes*, and *conditions*, resp, and $(g, m, b) \in Y$ is read "object g has attribute m under condition b. A *triadic concept* of $\mathbb{K}$ is a triple (A_1, A_2, A_3) with $A_1 \subseteq G$, $A_2 \subseteq M$, and $A_3 \subseteq B$ where $A_1 \times A_2 \times A_3 \subseteq Y$ such that none of its three components can be enlarged without violating this condition.

Lehmann and Wille present an extension of the theory of ordered sets and (concept) lattices to the triadic case, and discuss structural properties. This approach initiated research on the theory of *concept trilattices*, which was followed by several researchers (e. g., [1, 2, 3, 4, 5, 6, 8, 10, 11, 15, 16, 17, 18, 20, 21, 22]). Already in the first paper on this topic, Lehmann and Wille elaborated also a visualization of concept trilattices in *triadic diagrams*. But even though there are applications where the natural representation of the data are triadic contexts, the visualization by triadic diagrams never made it into

B. Ganter and R. Godin (Eds.): ICFCA 2005, LNAI 3403, pp. 315–328, 2005.

practice, and there exist only few visualizations of rather small concept trilattices. This is probably due to the complex structure of the diagrams. In this paper, we go one step back and discuss how traditional line diagrams of dyadic concept lattices can be used for exploring and navigating triadic data.

The idea of deriving dyadic contexts from the triadic one is not new. Lehmann and Wille present, for instance, in [14] the derived dyadic context $\mathbb{K}^{(1)} := (G, M \times B, Y^{(1)})$ with $(g, (m, b)) \in Y^{(1)} : \iff (g, m, b) \in Y$ (marked by 'W' below), and its two symmetric variations. In [8], the set B is used to define two modal operators (marked by '∃' and '∀' below). We will use these derivation modes later, but will set them in a common navigation framework.

Our approach for navigating triadic data is inspired by the *slice & dice* paradigm of On-Line-Analytical Processing (OLAP). We present the basic ideas of OLAP in the next section, and show how they may be transferred to triadic contexts. For modeling the navigation patterns a user might follow, we use the formalisms of finite state machine (see Section 3). In order to present the benefits of our model, we show in Section 4 how it can be used for navigating the IT Baseline Protection Manual of the German Federal Office for Information Security. As this model is only a first step to a comprehensive navigation environment for triadic (and possibly other) data, many interesting research questions remain open. They conclude the article.

2   On-Line Analytical Processing and Triadic Contexts

The expression *On-Line Analytical Processing (OLAP)* has been coined by E. F. Codd et al in [7], and stands for the analysis of multi-dimensional data. We will first give a short introduction in the main features of OLAP as far as they are needed in this paper, before informally outlining how we adapt them to triadic contexts.

OLAP relies on the metaphor of a (high-dimensional) cube containing data. One might for instance want to structure sales facts along the dimensions region, product and time. These dimensions span a three-dimensional cube as shown in Fig. 1. The cube is composed of cells, one for each combination of a region, a product, and a day. The cell contains a numerical value indicating how many items of that product have been purchased in the specific region on the given day.

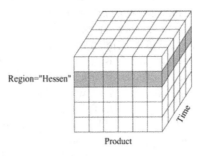

Region="Hessen"

Time

Product

Fig. 1. A data cube

The analyst may ask queries like 'Give me the sales facts for all products over all days in Hessen' or 'Give me the total number of items sold of product X in Hessen within the whole time period'. Both queries reduce the dimensionality of the answer: the first query returns a two-dimensional answer (the 'slice' which is indicated in Fig. 1), while the second query returns a one-dimensional answer. This reduction of the dimensionality is known as *slicing* in OLAP. The second query applies additionally an *aggregation function* as it sums up numbers of items. There are usually predefined hierarchies on the dimensions along which the aggregation takes place. For instance, days may sum up to months, and months to years. In this paper, however, we won't make use of these hierarchies.[1] An additional feature of OLAP is *dicing*[2] which rotates the data cube. This is particularly useful if the results are presented by spreadsheets, as it allows to permute rows and columns.

The association between OLAP data cubes and triadic contexts is now straight forward. The latter can in fact be considered as an OLAP data cube with three dimensions $(G, M,$ and $B)$, and the content of the cells represents the membership function of the relation Y.

As in OLAP, we want to be able to dice. This means that we want to allow to use any of the three sets as the set of objects at some point in time, depending on the task on hand. Therefore, we will not fix the roles of the three sets in advance. Instead, we consider a triadic context as symmetric structure, where all three sets are of equal importance. The decision which of the sets is considered as object set, attribute set, and condition set, resp., is made later by the user. For easier handling, we will therefore denote the triadic context $\mathbb{K} := (G, M, B, Y)$ alternatively by $\mathbb{K} := (K_1, K_2, K_3, Y)$ in the sequel. We consider all (triadic) contexts in this paper to be finite.

We will usually not work on the full triadic context (K_1, K_2, K_3, Y). Instead, we allow to focus on subsets of interest. Hence, for each of the three dimensions, we allow to restrict the set K_i to any subset X_i we are currently interested in. Thus, the current triadic context is just the sub-context $(X_1, X_2, X_3, Y \cap (X_1 \times X_2 \times X_3))$. This reduction is inspired by the slicing operation in OLAP.

Dicing is modeled by a permutation on the set $\{1, 2, 3\}$, i.e., by an element σ of the full symmetric group S_3. Such a permutation indicates that currently $X_{\sigma(1)}$ is considered as set of objects, $X_{\sigma(2)}$ as set of objects, and $X_{\sigma(3)}$ as set of conditions.

The aggregation mode is determined by one of the four options '∃', '∀', '𝕆', and '𝕎'.

– In the first case, we consider the concept lattice of the dyadic context

$$\mathbb{K}^{\sigma, \exists}_{X_1, X_2, X_3} := (X_{\sigma(1)}, X_{\sigma(2)}, I)$$

with $(x_{\sigma(1)}, x_{\sigma(2)}) \in I$ if and only if there exists $x_{\sigma(3)} \in X_{\sigma(3)}$ with $(x_1, x_2, x_3) \in Y$.
– In the second case, we consider the concept lattice of the dyadic context

$$\mathbb{K}^{\sigma, \forall}_{X_1, X_2, X_3} := (X_{\sigma(1)}, X_{\sigma(2)}, I)$$

with $(x_{\sigma(1)}, x_{\sigma(2)}) \in I$ if and only if for all $x_{\sigma(3)} \in X_{\sigma(3)}$ holds $(x_1, x_2, x_3) \in Y$.

---

[1] For a combination of these hierarchies with Formal Concept Analysis, see [19].

[2] Here the terminology is diverging in the literature. In some papers this is called *pivoting*, while 'dicing' is used for 'slicing' with resulting slices of dimension 3 or higher.

- In the third case, we consider the concept lattice of the dyadic context

$$\mathbb{K}^{\sigma,\mathfrak{D}}_{X_1,X_2,X_3} := (X_{\sigma(1)} \times X_{\sigma(3)}, X_{\sigma(2)}, I)$$

with $((x_{\sigma(1)}, x_{\sigma(3)}), x_{\sigma(2)}) \in I$ if and only if $(x_1, x_2, x_3) \in Y$.
- In the fourth case, we consider the concept lattice of the dyadic context

$$\mathbb{K}^{\sigma,\mathsf{W}}_{X_1,X_2,X_3} := (X_{\sigma(1)}, X_{\sigma(2)} \times X_{\sigma(3)}, I)$$

with $(x_{\sigma(1)}), (x_{\sigma(2)}, x_{\sigma(3)})) \in I$ if and only if $(x_1, x_2, x_3) \in Y$.

Concluding, the binary context (and its concept lattice) that we consider at a given moment depends on the following selections:

- the choice of three subsets $X_1 \subseteq K_1$, $X_2 \subseteq K_2$, and $X_3 \subseteq K_3$,
- a permutation $\sigma \in S_3$, and
- the choice of the aggregation mode $q \in \{\exists, \forall, \mathfrak{D}, \mathsf{W}\}$.

Up to now, we have discussed how single concept lattices can be derived from a triadic context. In order to support navigation, however, we need a mechanism which allows us to come from one concept lattice to the next. This will be discussed in the next section. We make use of the model of a finite state machine. Single concept lattices will correspond to states, while the navigation steps are captured by state transitions.

3   The Finite State Model

As said above, there are many binary contexts that can be derived from a triadic one. In this section, we discuss how the navigation between them may go on. We model this by a finite state machine.

We recall that a finite state machine is a model of computation consisting of a set of states, a start state, an input alphabet, and a transition function that maps pairs of input symbols and current states to a next state. Thus, it is a tuple $A = (E, S, \delta, s_0)$ where E is a finite set, the *input alphabet*, S is a finite set, the *set of states*, δ is the transition function, i. e., a mapping from $E \times S$ to S, and $s_0 \in S$ is the start state. In our approach, the contexts are considered as the states, and the navigation through the set of contexts is modeled by the transition function.

Next, we give the formal definition of our finite state machine. Readers unfamiliar with mathematical notations might first go to Section 4 in order to get a feeling for the approach, before returning here.

As discussed in Section 2, the derivation of a binary context depends on a set of choices. The combination of these choices makes up the state:

Definition 1. *Let* $\mathbb{K} := (K_1, K_2, K_3, Y)$ *be a triadic context. A state is then a tuple*

$$s := (X_1, X_2, X_3, \sigma, q)$$

where $X_1 \subseteq K_1$, $X_2 \subseteq K_2$, *and* $X_3 \subseteq K_3$, $\sigma \in S_3$, *and* $q \in \{\exists, \forall, \mathfrak{D}, \mathsf{W}\}$. *The set of all states of a triadic context is denoted by* $S(\mathbb{K})$, *or simply by* S *if* $\mathbb{K}$ *is unambiguous from the context.*

For a given state $s := (X_1, X_2, X_3, \sigma, q)$, *we let* $\mathbb{K}(s) := \mathbb{K}^{\sigma,q}_{X_1,X_2,X_3}$, *and* $\underline{\mathfrak{B}}(s) := \underline{\mathfrak{B}}(\mathbb{K}(s))$.

We initialize our state machine with the *start state*

$$s_0 := (K_1, K_2, K_3, \mathrm{id}, \exists) \ .$$

This setting allows a first, global overview over the data, which may be refined later. The initial choice of the existential quantifier for q follows the observation that this is the most frequently used in applications for deriving a dyadic context.

The *input alphabet E* of the machine is given by

$$E := \{\texttt{slice}(i, A) \mid i \in \{1, 2, 3\}, A \subseteq K_i\}$$
$$\cup \{\texttt{dice}(\sigma) \mid \sigma \in S_3\}$$
$$\cup \{\texttt{mode}(\exists), \texttt{mode}(\forall), \texttt{mode}(\mathfrak{D}), \texttt{mode}(\mathsf{W})\} \ .$$

Last but not least we define the *transition function*. We split up the definition into three parts, corresponding to the three types of elements of the input alphabet listed above.

Definition 2. *Let $s := (X_1, X_2, X_3, \sigma, q)$ be a state, and $e \in E$. The transition function δ is defined as follows. If $e = \texttt{slice}(i, A)$ with $i \in \{1, 2, 3\}$ and $A \subseteq K_i$, then $\delta(s, e) := (X_1', X_2', X_3', \sigma, q)$ with $X_i' := A$ and $X_j' := X_j$, for $j \neq i$.*

For simplifying the navigation, one could restrict the set A to be a concept extent or intent of some suitable concept lattice. Experiments with the example discussed below, however, revealed the need for selecting arbitrary sets. Note that we also allow to enlarge sets again, as there is no constraint saying that A has to be a subset of X_i. This allows to extend sets again during the navigation process. In practice however, A often is a subset of X_i. We discuss some properties of this case before continuing the definition of the transition function.

Let $A \subseteq X_i$. If $\sigma(1) = i$ or $q = \mathfrak{D}$ and $\sigma(3) = i$, then the slice operation reduces the current set of objects. The resulting concept lattice is thus isomorphic to a $\vee$-sub-semilattice of the previous one. If $\sigma(2) = i$ or $q = \mathsf{W}$ and $\sigma(3) = i$, then the slice operation reduces the current set of attributes, and the resulting concept lattice is isomorphic to a $\wedge$-sub-semilattice of the previous one. In all these cases, the information presented in the lattice is thus reduced, just as a slicing operation in OLAP would do. In the two remaining cases, however, the analogy to OLAP fails. If $\sigma(3) = i$ and $q = \exists$, then the binary relation of the current dyadic context decreases; if $q = \forall$ then the binary relation increases. Simple examples show that there is no pre-determined relationship between the current and the following lattice. In both situations the concept lattice can either shrink or grow, depending on the constellation.

Definition 2 (contd.). *If $e = \texttt{dice}(\sigma')$ with $\sigma' \in S_3$, then $\delta(s, e) := (X_1, X_2, X_3, \sigma \circ \sigma', q)$.*

We may denote the elements $\sigma \in S_3$ by $(123), (132), \ldots, (321)$ where, e. g., (132) means that the role of the object set remains unchanged while the attribute and the condition sets interchange their roles. The transition $\texttt{dice}(1, 2, 3)$ doesn't do anything; and the transition $\texttt{dice}(2, 1, 3)$ interchanges the roles of objects and attributes. This means that the concept lattice is turned upside down.

Definition 2 (contd.). *If* $e = \text{mode}(q)$ *with* $q \in \{\exists, \forall, \mho, \mathsf{W}\}$ *then* $\delta(s, e) := (X_1, X_2, X_3, \sigma, q)$.

In practice it turns out that the '$\exists$' mode is the mostly used one. The change from either '$\exists$' or '$\forall$' to '$\mho$' or 'W' can be considered as drill-down, since the resulting concept lattice provides more detailed information. A change of mode in the other direction is a roll-up, as the information becomes more summarized.

The definition of our finite state machine is now complete. Next we introduce two additional shortcuts.

Definition 3. *We let* $\text{delG} := \text{slice}(\sigma^{-1}(1), \{\{x\} \in X_{\sigma^{-1}(1)} \mid \{x\}' \neq \emptyset\}$ *and* $\text{delM} := \text{slice}(\sigma^{-1}(2), \{x \in X_{\sigma^{-1}(2)} \mid x' \neq \emptyset\}$, *where the derivation* $\cdot'$ *is computed in the current dyadic context.*

These two derived operators serve the following purpose. After a `slice` operation, one usually also wants to prune the remaining sets to the relevant elements. For instance, if one reduced the set of objects, then there may be some attributes which do not relate to any of the remaining objects. In most cases, one may want to remove these attributes (which would all be attached to the bottom concept), as they do not provide any further insight. This removing of 'superfluous' attributes is performed by `delM`, while `delG` removes all objects which are not covered by at least one attribute any more.

As known from basics about finite state machines, we can now extend the transition function such that it applies not only to single symbols of the input alphabet, but also to words. We denote the set of words over the input alphabet E by E^*, which includes the empty word λ. The transition function δ is then recursively extended to $\delta^*: E^* \times S \to S$ by $\delta^*(\lambda, s) := s$, and $\delta^*((w, e), s) := \delta^*(w, \delta(e, s))$, for $w \in E^*$ and $e \in E$.

An element w of E^* can naturally be considered as a program (which is executed from right to left). Its semantics is given by the function $[\![w]\!]: S \to S$ which is given by $[\![w]\!](s) := \delta^*(w, s)$. Specifically, one can determine the current state of the system by storing all previous interactions of the user as $w \in E^*$. The current state is then just $[\![w]\!](s_0)$. In an implementation of the framework, w may be shown as *navigation history* to the user, and supports an 'undo' function or 'back' button.

4   Navigation Within the Triadic Information System

In this section, we show by an example, how our framework supports navigation in a real world dataset. The IT Baseline Security Manual [9] of the German Federal Office for Information Security provides a description of the threat scenario that is globally assumed, standard security measures for typical IT systems, and detailed descriptions of safeguards to assist with their implementation.[3]

The core data of the manual can be considered as a triadic context. We consider the the possible threats as objects, the IT components as attributes, and the safeguards as

---

[3] The online version of the manual can be found at http://www.bsi.de/gshb/ .

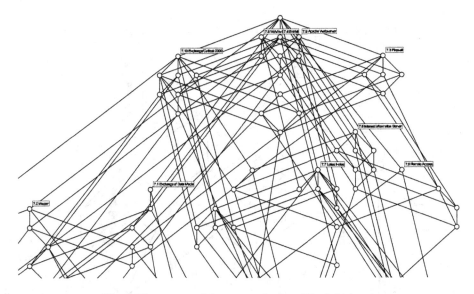

Fig. 2. The top part of the concept lattice of the initial state

conditions.[4] For presentation purposes, we restrict the example as follows. $K_1 := \{$Exchange of Data Media, ..., Exchange/Outlook 2000$\}$ is the set of all ten data transmission systems listed in Section 7 of [9], K_2 is the set of all threats against at least one of the data transmission systems, and K_3 is set of all safeguards against at least one of these threats for at least one of the data transmission systems.[5] The top part of the concept lattice of the start state is shown in Fig. 2. The names of the threats (which play the role of objects at the moment) are omitted due to representation issues. From this initial state, we will perform a series of navigation steps to illustrate the different features of the model.

First we want to reduce the components to those which are currently used at our research group. We perform thus the operations

op_1: slice(1, {7.3 Firewall,   7.4 E-Mail,   7.5 WWW-Server,   7.6 Remote Access, 7.9 Apache Webserver})

op_2: delG .

The resulting concept lattice is shown in Fig. 3. The concept in the middle of the diagram indicates for instance that the two threats 'T 3.38 Errors in configuration and operation' and 'T 4.39 Software conception errors' are the threats which are directed against all the three components Firewall, WWW-Server, and Apache Webserver. The two threats 'T 5.2

---

[4] Other assignments have been done in [8, 18, 22]. But as our approach considers all sets equivalently, this assignment influences the start state only. Any initial arrangement can be reached from any other by one dice operation.

[5] This restricted scenario can also be reached by three consecutive slice operations within the larger system that comprises all components, threads and safeguards discussed in the manual.

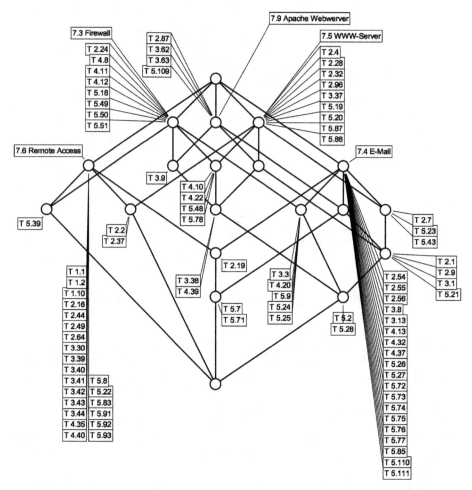

Fig. 3. After slicing the set of components to Firewall, E-Mail, WWW-Server, Remote Access, Apache Webserver

Manipulation of data and software' and 'T 5.28 Denial of services' are directed against these components, but they additionally concern the E-Mail system, as the lower right concept indicates. The complete list of threats can be found at the website mentioned above.

A major set of threats are deliberate attacks to the system. We now want to study which of these attacks are potentially dangerous to the data transmission systems of our research group. We perform thus the operation

op_3: slice(2, $X_2 \cap$ {'T 5.1 Manipulation or destruction of IT equipment or accessories',..., 'T 5.111 Misuse of active content of E-Mails'})

(where X_2 is the current set of objects) and obtain the line diagram in Fig. 4. The diagram shows that there are many rather specific threats, as they are related to only one component each.

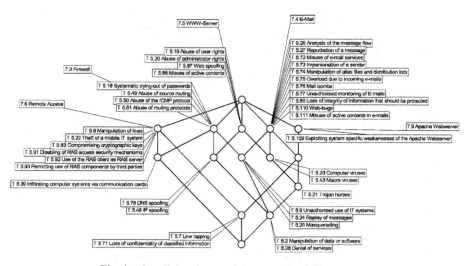

Fig. 4. After slicing the set of threats to the deliberate acts

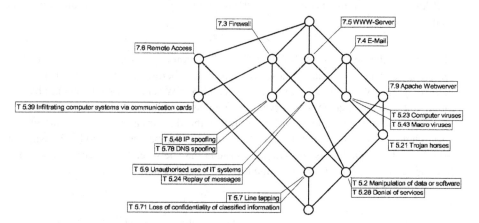

Fig. 5. After slicing away all threats which are against single components

In order to analyze in more depth deliberate acts against combinations of components, we prune away all deliberate acts against single components. The operation

$$op_4\colon \texttt{slice}(2, \{x \in \tilde{X}_2 | \text{card}(\{x\}') \neq 1\})$$

(where $\tilde{X}_2$ is the current set of objects) yields the concept lattice in Fig. 5. In the lattice, we can for instance discover that there are four threats which endanger at the same time firewalls and WWW servers: IP spoofing, DNS spoofing, manipulation of data or software, and denial of services.

For studying which other components are threatened by denial of service attacks, we could make use of the same diagram. Another option — which supports the more natural way of reading from top to bottom — is to turn the diagram upside down first. As

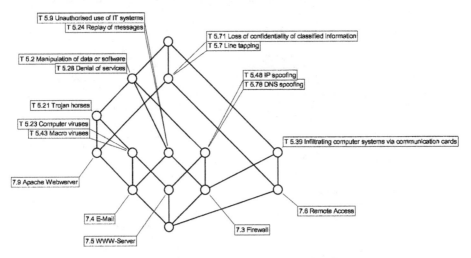

Fig. 6. After dicing (exchange of objects and attributes)

well-known in FCA, this is obtained by interchanging the roles of objects and attributes on the context level:

$$op_5: \texttt{dice}(2,1,3)$$

The result of the operation is shown in Fig. 6. The components threatened by denial of service attacks are now exactly those which are listed below 'T 5.28 Denial of services' in the diagram. We see that not only firewalls and WWW servers are endangered by this threat, but also the Apache web server and email systems.

What can now be done in order to protect the data transmission systems of our research group against deliberate acts? This question can be approached by focusing on the relation between threats and safeguards rather than between threats and components. In other words, we have to interchange the role of components and safeguards:

$$op_6: \texttt{dice}(3,2,1)$$

The result is shown in Fig. 6. The encoding of the safeguards can be found online in [9]. The lattice is rather complex — which indicates that there is no easy solution for protecting our IT environment.

For getting a better insight, we focus (first) on those safeguards which are related to hard- and software:

$$op_7: \texttt{slice}(3, \{x \in \tilde{X}_3 | x = \text{"S 4. ...")}\})$$

(where $\tilde{X}_3$ is the current set of objects). This yields the concept lattice in Fig. 8. It shows for instance that there is no hard- or software related safeguard against IP spoofing. This threat has to be countered by other means. On the other side we discover that even with only two safeguards, 'S 4.95 Minimal operating system' and 'S 4.34 Using encryption, checksums or digital signatures', many of the listed threats can be countered.

Figure 8 doesn't show in detail if a safeguard is designed against a threat for all relevant components or just for specific ones. One could expect that the choice of a safeguard is independent of the component. By drill-down, we can analyze this hypothesis. Figure 9 shows the result of

$$op_8: \mathtt{mode}(\mathfrak{D}) \ ,$$

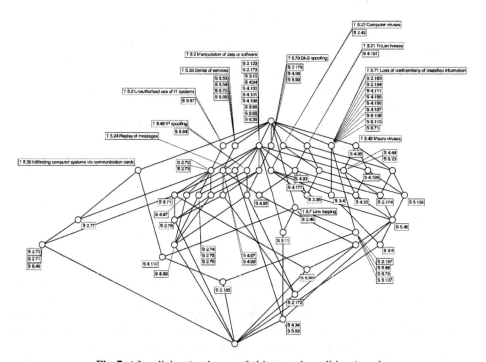

Fig. 7. After dicing (exchange of objects and conditions) again

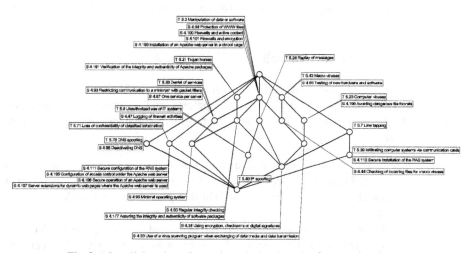

Fig. 8. After slicing the safeguards to the hard- and software related ones

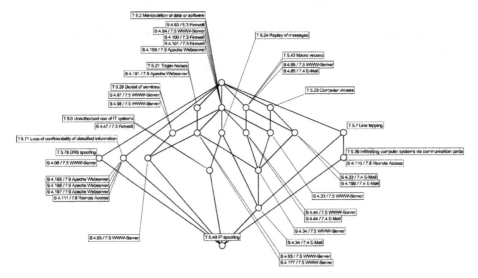

Fig. 9. After drill-down to the components

i. e., the concept lattice

$$\mathfrak{B}(\llbracket op_8, \ldots, op_1 \rrbracket(s_0)) \ .$$

By comparing Figs. 8 and 9, we discover that this hypothesis is indeed almost true, since the concept lattices are quite similar. However, there are some differences. For instance, safeguard 'S 4.33 Use of a virus scanning program while exchanging of data media and data transmission' is adequate against macro viruses and computer viruses for email systems, while it protects web servers against computer viruses and manipulation of data or software. At least, this is what the data provided by the IT baseline manual indicate. One may now discuss if this is adequately modeling the situation. We, however, will now leave this example, and return to a more general discussion on future research.

5   Conclusion and Outlook

In this paper, we integrated bits and pieces which were available for analyzing triadic contexts by line diagrams of derived dyadic contexts, and set them into a navigation framework. As this is only a first step, many interesting research questions remain open. They include:

-   Further development of automated drawing routines is important, as there are so many potential concept lattices which can be derived from a triadic context that one cannot create them all beforehand. (In this paper, we have made the layout manually).
-   The framework should be implemented and evaluated in more scenarios. This includes a concretization of the means of interaction by which the user can perform the slice & dice operations.

- How can techniques like conceptual scaling or iceberg lattices be exploited for more sophisticated navigation means?
- Are there other aggregation modes? For instance, one might want to integrate power scales [13] or relational scales [12].
- In OLAP, the dimensions carry an additional structure: Each dimension goes along with a hierarchy along which aggregation is performed. On a small scale, such hierarchies also exist in our IT example, by means of clustering each of the sets of components, threats, and safeguards into different sections of the IT Baseline Manual. How can these hierarchies be incorporated into the model?
- Relational databases are an important application domain which provides a lot of interesting applications. From the perspective of Formal Concept Analysis, many-valued contexts and multi-contexts are data structures closely related to relational databases. How can the framework presented here be extended to incorporate them as well?

References

1. K. Biedermann. How triadic diagrams represent conceptual structures. In D. Lukose, H. S. Delugach, M. Keeler, L. Searle, and J. F. Sowa, editors, *Conceptual Structures: Fulfilling Peirce's Dream*, number 1257 in LNAI, pages 304–317, Heidelberg, 1997. Springer.
2. K. Biedermann. Triadic Galois connections. In K. Denecke and O. Lüders, editors, *General algebra and applications in discrete mathematics*, pages 23–33, Aachen, 1997. Shaker Verlag.
3. K. Biedermann. Completion of triordered sets and trilattices. In D. Dorninger, G. Eigenthaler, H.K. Kaiser, H. Kautschitsch, W. More, and W.B. Müller, editors, *Contributions to General Algebra*, number 10, pages 61–78, Klagenfurt, 1998. Johannes Heyn Verlag.
4. K. Biedermann. *A foundation of the theory of trilattices*. Dissertation, TU Darmstadt, Aachen, 1998.
5. K. Biedermann. Powerset trilattices. In M. Mugnier and M. Chein, editors, *Conceptual Structures: Theory, Tools and Applications*, volume 1453 of *Lecture Notes in Computer Science*. Springer, 1998.
6. K. Biedermann. An equational theory for trilattices. *Algebra Universalis*, 42:253–268, 1999.
7. E. F. Codd, S. B. Codd, and C. T. Salley. Providing OLAP (On-Line Analytical Processing) to User-Analysis: An IT Mandate, 1993. White paper.
8. F. Dau and R. Wille. On the modal unterstanding of triadic contexts. In R. Decker and W. Gaul, editors, *Classification and Information Processing at the Turn of the Millenium*, Proc. Gesellschaft für Klassifikation, 2001.
9. German Federal Office for Information Security. IT Baseline Protection Manual. http://www.bsi.de/gshb/, October 2003.
10. B. Ganter and S. A. Obiedkov. Implications in triadic contexts. In *Conceptual Structures at Work: 12th International Conference on Conceptual Structures*, volume 3127 of *Lecture Notes in Computer Science*. Springer, 2004.
11. B. Groh and R. Wille. Lattices of triadic concept graphs. In B. Ganter and G. W. Mineau, editors, *Conceptual Structures: Logical, Linguistic, and Computational Issues*, volume 1867 of *Lecture Notes in Computer Science*. Springer, 2000.
12. J. Hereth. Relational scaling and databases. In U. Priss, D. Corbett, and G. Angelova, editors, *Conceptual Structures: Integration and Interfaces, 10th International Conference on Conceptual Structures, ICCS 2002, Borovets, Bulgaria, July 15-19, 2002, Proceedings*, volume 2393 of *Lecture Notes in Computer Science*, pages 62–76. Springer, 2002.

13. J. Hereth and G. Stumme. Reverse pivoting in conceptual information systems. In H. S. Delugach and G. Stumme, editors, *Conceptual Structures: Broadening the Base*, volume 2120 of *Lecture Notes in Computer Science*, pages 202–215. Springer, 2001.

14. F. Lehmann and R. Wille. A triadic approach to formal concept analysis. In G. Ellis, R. Levinson, W. Rich, and J. F. Sowa, editors, *Conceptual Structures: Applications, Implementation and Theory*, volume 954 of *Lecture Notes in Computer Science*. Springer, 1995.

15. S. Prediger. Nested concept graphs and triadic power context families. In B. Ganter and G. W. Mineau, editors, *Conceptual Structures: Logical, Linguistic, and Computational Issues. Proc. ICCS '00*, number 1867 in LNAI, pages 249–262, Heidelberg, 2000. Springer.

16. U. Priss. A triadic model of information flow. In G.W. Mineau, editor, *Conceptual Structures: Extracting and Representing Semantics*, pages 159–170, Quebec, Canada, 2001. Dept. of Computer Science, University Laval.

17. L. Schoolmann and R. Wille. Concept graphs with subdivision: A semantic approach. In Aldo de Moor, Wilfried Lex, and Bernhard Ganter, editors, *Conceptual Structures for Knowledge Creation and Communication*, volume 2746 of *Lecture Notes in Computer Science*, pages 271–281. Springer, 2003.

18. H. Söll. Begriffliche Analyse triadischer Daten: Das IT-Grundschutzhandbuch des Bundesamts für Sicherheit in der Informationstechnik. Diploma thesis, FB Mathematik, TU Darmstadt, Darmstadt, April 1998.

19. G. Stumme. On-line analytical processing with conceptual information systems. In K. Tanaka and S. Ghandeharizadeh, editors, *Proc. 5th Intl. Conf. on Foundations of Data Organization (FODO'98)*, pages 117–126, November 12-13, 1998.

20. R. Wille. The basic theorem of triadic concept analysis. *Order*, 12:149–158, 1995.

21. R. Wille. Triadic Concept Graphs. In M.-L. Mugnier and M. Chein, editors, *Conceptual structures: theory, tools and application*, number 1453 in LNAI, pages 194–208, Berlin-Heidelberg-New York, 1998. Springer.

22. R. Wille and M. Zickwolff. Grundlagen einer triadischen Begriffsanalyse. In G. Stumme and R. Wille, editors, *Begriffliche Wissensverarbeitung. Methoden und Anwendungen*, pages 125–150, Berlin-Heidelberg, 2000. Springer-Verlag.

Complete Subalgebras of Semiconcept Algebras and Protoconcept Algebras

Björn Vormbrock

Technische Universität Darmstadt, Fachbereich Mathematik
Schloßgartenstr. 7, D-64289 Darmstadt
vormbrock@mathematik.tu-darmstadt.de

Abstract. In order to define a negation on formal concepts in Formal Concept Analysis, the more general notions of semiconcepts and protoconcepts were introduced. The theory of the resulting protoconcept and semiconcept algebras is developed in Boolean Concept Logic as a part of Contextual Logic. In this paper it is shown that each complete subalgebra of a semiconcept algebra is itself the semiconcept algebra of an appropriate context. An analogous result holds for the complete subalgebras of protoconcept algebras. These contexts can be obtained from the original context through partitions of the object and the attribute set satisfying certain conditions. Characterizations of the complete subalgebras of semiconcept and protoconcept algebras in terms of contexts, in terms of subsets, and through closed subrelations are given.

1 Introduction

Formal Concept Analysis developed as a mathematical theory of concepts is used succesfully in the area of knowledge representation and knowledge processing. The advantage of this approach lies in the close relation to conceptual human thinking. Yet, while negations of concepts are common in human language and human thinking (e.g. non-smoker, non-profit, non-fiction, NGO), negated concepts generally cannot be represented in concept lattices. Boolean Concept Logic is a theory that extends Formal Concept Analysis by introducing negations of concepts. It is therefore a part of Contextual Logic whose aim is to mathematize the philosophical logic with its doctrines of concepts, judgments, and conclusions (for a brief introduction to Contextual Logic we refer to [Wi00b] or [DK03], an overview over existing theories is given in [KV03]. For more detailed work on Concept Logic see [VW03], [Vo03], [Vo04], recent work on Contextual Judgment Logic can be found in [Da03], [Wi01]).

In Boolean Concept Logic, the negation of a concept is modeled by taking set complements. Since formal concepts consist of two sets, the extent and the intent, we distinguish two kinds of negation: Firstly, the operation $\neg(A, B) := (G \backslash A, (G \backslash A)')$ on the extent side, which will also be called "negation". Secondly, we define an operation $\lnot(A, B) := ((M \backslash B)', M \backslash B)$ on the intent side, which will be called "opposition" of a formal concept (A, B) in a context $\mathbb{K}$. This distinction

B. Ganter and R. Godin (Eds.): ICFCA 2005, LNAI 3403, pp. 329–343, 2005.

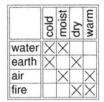

	cold	moist	dry	warm
water	✕	✕		
earth	✕		✕	
air		✕		✕
fire			✕	✕

Fig. 1. The four elements and their attributes in greek philosophy

can already be found in ancient philosophical logic (c.f. [Wi00a]). The negation ¬ corresponds to those negations we think of when using words like "non-smoker": In an appropiate context (e.g. with all humans as objects and with at least the attribute "smokes") the word "non-smoker" refers to all objects that are not in the extent of the concept generated by "smokes". The opposition ⌐ is best illustrated in a context where the attribute set consists of pairs of attributes that mutually exclude each other (i.e. for every such pair $\{m_1, m_2\}$ we have $(g, m_1) \in I \Leftrightarrow (g, m_2) \notin I$ for every object g). The context given in Figure 1, for example, has as attributes the dychotomic pairs "dry" ↔ "moist" and "cold" ↔ "warm". The opposition ⌐($\{water\}, \{moist, cold\}$) of the concept generated by water yields the concept ($\{fire\}, \{dry, warm\}$), which is the opposite of this element in ancient greek philosophy (cp. [Wi00a]).

As in general the set complement of an extent (intent) is not an extent (intent) itself, the notion of formal concept is generalized to semiconcepts:

Definition 1. *A semiconcept of a formal context* $\mathbb{K} := (G, M, I)$ *is a pair* (A, B) *with* $A \subseteq G$ *and* $B \subseteq M$ *such that* $A' = B$ *or* $B' = A$. *We denote the set of all semiconcepts of a context* $\mathbb{K}$ *by* $\mathfrak{H}(\mathbb{K})$ *and define on* $\mathfrak{H}(\mathbb{K})$ *operations* ⊓, ⊔, ⌐, ¬, ⊤ *and* ⊥ *by:*

$$(A_1, B_1) \sqcap (A_2, B_2) := (A_1 \cap A_2, (A_1 \cap A_2)')$$
$$(A_1, B_1) \sqcup (A_2, B_2) := ((B_1 \cap B_2)', B_1 \cap B_2)$$
$$\neg(A, B) := (G \setminus A, (G \setminus A)')$$
$$\lnot(A, B) := ((M \setminus B)', M \setminus B)$$
$$\top := (G, \emptyset)$$
$$\bot := (\emptyset, M)$$

The set of all semiconcepts of a context $\mathbb{K}$ *together with these operations is called the* semiconcept algebra *of* $\mathbb{K}$ *and denoted by* $\underline{\mathfrak{H}}(\mathbb{K})$.

The operations are called "meet" (⊓), "join" (⊔), "negation" (¬), "opposition" (⌐), "all" (⊤) and "nothing" (⊥).

In [KV03], an example is given that illustrates the higher expressiveness resulting from these additional operations and its possible benefits in applications of Formal Concept Analysis.

The aim of this paper is to contribute to the development of the algebraic theory of semiconcept algebras and of the more general protoconcept algebras, since a mathematical theory of these algebras is a prerequisite for their application in knowledge representation and knowledge processing. We therefore focus on the mathematical theory of the complete subalgebras. However, subalgebras may be of practical interest for their own sake: Since semiconcept and protoconcept algebras are more complex than concept lattices, a major problem is that with increasing number of objects and attributes in a context, the number of semi- and protoconcepts grows much faster than the number of concepts. In cases where not all the information of the context is needed in all instances, a restriction to appropriate subalgebras may be helpful. In this paper, a characterization of the subalgebras of semiconcept algebras is presented that allows to describe the subalgebras without calculating the semiconcepts of a context.

In the next section, basic definitions and results on semiconcept and protoconcept algebras are provided. We assume that the reader is familiar with the basic notions of Formal Concept Analysis (a textbook on Formal Concept Analysis is [GW99]). Section 3 contains the characterization of the complete subalgebras of semiconcept algebras through partitions of the object and the attribute set. This yields a criterion to decide whether a given subset of a semiconcept algebra is a complete subalgebra. In Formal Concept Analysis the sublattices of concept lattices correspond to closed subrelations of the incidence relation. In Section 4, it is shown which closed subrelations correspond to complete subalgebras of semiconcept algebras. The results of Section 5 are extended to complete subalgebras of protoconcepts in Section 5. Finally, some perspectives for further research are discussed in the last section.

2 Semiconcept and Protoconcept Algebras

In order to develop Boolean Concept Logic it is useful to introduce protoconcepts as a generalization of semiconcepts:

Definition 2. *A* protoconcept *of a formal context* $\mathbb{K} := (G, M, I)$ *is a pair* (A, B) *with* $A \subseteq G$ *and* $B \subseteq M$ *such that* $A' = B''$ *or, equivalently,* $B' = A''$. *We denote the set of all protoconcepts of a context* $\mathbb{K}$ *by* $\mathfrak{P}(\mathbb{K})$. *On* $\mathfrak{P}(\mathbb{K})$ *the operations* $\sqcap$, $\sqcup$, $\lnot$, $\neg$, $\top$ *and* $\bot$ *are defined as in Definition 1. The set of all protoconcepts of a context* $\mathbb{K}$ *together with these operations is called the* protoconcept algebra *of* $\mathbb{K}$ *and denoted by* $\underline{\mathfrak{P}}(\mathbb{K})$.

Note that the result of any operation on protoconcepts is a semiconcept. Thus, for a protoconcept algebra $\underline{\mathfrak{P}}(\mathbb{K})$ of a context $\mathbb{K}$ and a set of protoconcepts $P \subseteq \mathfrak{P}(\mathbb{K}) \setminus \mathfrak{H}(\mathbb{K})$, the set $\underline{\mathfrak{P}}(\mathbb{K}) \setminus P$ is a subalgebra of $\underline{\mathfrak{P}}(\mathbb{K})$. In particular, $\underline{\mathfrak{H}}(\mathbb{K})$ is a subalgebra of $\underline{\mathfrak{P}}(\mathbb{K})$. In [Wi00a], a description of the equational class generated by protoconcept algebras, the double Boolean algebras, is given:

Theorem 1. *A basis for all equations which are valid in all protoconcept algebras is given by the following equations which are the equational axioms of the so-called double Boolean algebras:*

$1a)$ $(x \sqcap x) \sqcap y = x \sqcap y$ $\qquad$ $1b)$ $(x \sqcup x) \sqcup y = x \sqcup y$

$2a)$ $x \sqcap y = y \sqcap x$ $\qquad$ $2b)$ $x \sqcup y = y \sqcup x$

$3a)$ $x \sqcap (y \sqcap z) = (x \sqcap y) \sqcap z$ $\qquad$ $3b)$ $x \sqcup (y \sqcup z) = (x \sqcup y) \sqcup z$

$4a)$ $x \sqcap (x \sqcup y) = x \sqcap x$ $\qquad$ $4b)$ $x \sqcup (x \sqcap y) = x \sqcup x$

$5a)$ $x \sqcap (x \sqcupdot y) = x \sqcap x$ $\qquad$ $5b)$ $x \sqcup (x \sqcapdot y) = x \sqcup x$

$6a)$ $x \sqcap (y \sqcupdot z) = (x \sqcap y)\,(x \sqcap z)$ $\qquad$ $6b)$ $x \sqcup (y \sqcapdot z) = (x \sqcup y)\sqcapdot(x \sqcup z)$

$7a)$ $\neg\neg(x \sqcap y) = x \sqcap y$ $\qquad$ $7b)$ $\lrcorner\lrcorner(x \sqcup y) = x \sqcup y$

$8a)$ $\neg(x \sqcap x) = \neg x$ $\qquad$ $8b)$ $\lrcorner(x \sqcup x) = \lrcorner x$

$9a)$ $x \sqcap \neg x = \bot$ $\qquad$ $9b)$ $x \sqcup \lrcorner x = \top$

$10a)$ $\neg \bot = \top \sqcap \top$ $\qquad$ $10b)$ $\lrcorner \top = \bot \sqcup \bot$

$11a)$ $\neg \top = \bot$ $\qquad$ $11b)$ $\lrcorner \bot = \top$

$$12) \quad (x \sqcap x) \sqcup (x \sqcap x) = (x \sqcup x) \sqcap (x \sqcup x)$$

with the operations $\sqcupdot$ and $\sqcapdot$ defined as $x \sqcupdot y := \neg(\neg x \sqcap \neg y)$ and $x \sqcapdot y := \lrcorner(\lrcorner x \sqcup \lrcorner y)$.

Semiconcept algebras even satisfy the stronger condition

$$x = x \sqcap x \text{ or } x = x \sqcup x.$$

A double Boolean algebra satisfying this condition is called *pure*. For a double Boolean algebra $\underline{D}$ and $x \in \underline{D}$ we define $x_{\sqcap} := x \sqcap x$, $x_{\sqcup} := x \sqcup x$, $D_{\sqcap} := \{x \in D \mid x = x_{\sqcap}\}$, $D_{\sqcup} := \{x \in D \mid x = x_{\sqcup}\}$ and the *pure subalgebra* $\underline{D}_p$ of $\underline{D}$ as the subalgebra over the set $D_p := D_{\sqcap} \cup D_{\sqcup}$. For a context $\mathbb{K}$, the semiconcepts of type (A, A') for $A \subseteq G$, i.e. those satisfying $x = x_{\sqcap}$, are called $\sqcap$-*semiconcepts*. Dually, the semiconcepts of type (B', B) for $B \subseteq M$, i.e. those satisfying $x = x_{\sqcup}$, are called $\sqcup$-*semiconcepts*.

Note that for a double Boolean algebra $\underline{D}$ the set $D_{\sqcap}$ together with the operations $\sqcap, \sqcupdot, \bot, \top := \neg\bot$ forms a Boolean algebra. Dually, the set $D_{\sqcup}$ together with the operations $\sqcapdot, \sqcup, \bot := \lrcorner\top, \top$ forms a Boolean algebra.

On double Boolean algebras a *quasi-order* $\sqsubseteq$ is defined by

$$x \sqsubseteq y :\Leftrightarrow x \sqcap y = x_{\sqcap} \text{ and } x \sqcup y = y_{\sqcup}.$$

For protoconcept algebras this definition yields an order, and $(A_1, B_1) \sqsubseteq (A_2, B_2)$ is equivalent to $A_1 \subseteq A_2$ and $B_2 \subseteq B_1$.

Example 1. Figure 2 depicts a context and its protoconcept algebra. The elements represented by filled circles are formal concepts. The circles with the upper half filled represent $\sqcup$-semiconcepts, those with the lower half filled represent $\sqcap$-semiconcepts.

Basic Theorems on semiconcept algebras and on protoconcept algebras were established in [VW03]. In order to quote them here we have to introduce the notions of contextual, fully contextual and complete double Boolean algebras: A

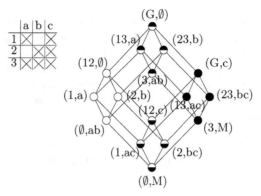

Fig. 2. A context and its protoconcept algebra

double Boolean algebra $\underline{D}$ is called *contextual* if its quasiorder $\sqsubseteq$ is antisymmetric, i.e. the relation $\sqsubseteq$ is an order on $\underline{D}$. A contextual double Boolean algebra $\underline{D}$ is said to be *fully contextual* if, in addition, for each $x \in \underline{D}_\sqcap$ and $y \in \underline{D}_\sqcup$ with $x_\sqcup = y_\sqcap$ there is a unique $z \in \underline{D}$ with $z_\sqcap = x$ and $z_\sqcup = y$. The double Boolean algebra $\underline{D}$ is called complete if and only if its Boolean algebras $\underline{D}_\sqcap$ and $\underline{D}_\sqcup$ are complete.

Theorem 2 (The Basic Theorem on Semiconcept Algebras). *For a context $\mathbb{K} := (G, M, I)$, the semiconcept algebra $\underline{\mathfrak{H}}(\mathbb{K})$ is a complete pure double Boolean algebra whose Boolean algebras $\underline{\mathfrak{H}}_\sqcap(\mathbb{K})$ and $\underline{\mathfrak{H}}_\sqcup(\mathbb{K})$ are atomic. The (arbitrary) meet and join of $\underline{\mathfrak{H}}(\mathbb{K})$ are given by*

$$\prod_{t \in T}(A_t, B_t) = (\bigcap_{t \in T} A_t, (\bigcap_{t \in T} A_t)') \quad and \quad \bigsqcup_{t \in T}(A_t, B_t) = ((\bigcap_{t \in T} B_t)', \bigcap_{t \in T} B_t).$$

In general, a complete pure double Boolean algebra $\underline{D}$ whose Boolean algebras $\underline{D}_\sqcap$ and $\underline{D}_\sqcup$ are atomic, is isomorphic to $\underline{\mathfrak{H}}(\mathbb{K})$ if and only if there exist a bijection $\tilde{\gamma}$ from G onto the set $\mathcal{A}(\underline{D}_\sqcap)$ of all atoms of $\underline{D}_\sqcap$ and a bijection $\tilde{\mu}$ from M onto the set $\mathcal{C}(\underline{D}_\sqcup)$ of all coatoms of $\underline{D}_\sqcup$ such that $gIm \Leftrightarrow \tilde{\gamma}(g) \sqsubseteq \tilde{\mu}(m)$ for all $g \in G$ and $m \in M$. In particular, for any complete pure double Boolean algebra $\underline{D}$ whose Boolean algebras are atomic, we get $\underline{D} \cong \underline{\mathfrak{H}}(\mathcal{A}(\underline{D}_\sqcap), \mathcal{C}(\underline{D}_\sqcup), \sqsubseteq)$, i.e., the semiconcept algebras are up to isomorphism the complete pure double Boolean algebras $\underline{D}$ whose Boolean algebras $\underline{D}_\sqcap$ and $\underline{D}_\sqcup$ are atomic.

Theorem 3 (The Basic Theorem on Protoconcept Algebras). *For a context $\mathbb{K} := (G, M, I)$, the protoconcept algebra $\underline{\mathfrak{P}}(\mathbb{K})$ of $\mathbb{K}$ is a complete fully contextual double Boolean algebra whose Boolean algebras $\underline{\mathfrak{H}}_\sqcap(\mathbb{K})$ and $\underline{\mathfrak{H}}_\sqcup(\mathbb{K})$ are atomic. The (arbitrary) meet and join of $\underline{\mathfrak{P}}(\mathbb{K})$ are given by*

$$\prod_{t \in T}(A_t, B_t) = (\bigcap_{t \in T} A_t, (\bigcap_{t \in T} A_t)') \quad and \quad \bigsqcup_{t \in T}(A_t, B_t) = ((\bigcap_{t \in T} B_t)', \bigcap_{t \in T} B_t).$$

In general, a complete fully contextual double Boolean algebra $\underline{D}$ whose Boolean algebras $\underline{D}_\sqcap$ and $\underline{D}_\sqcup$ are atomic, is isomorphic to $\underline{\mathfrak{P}}(\mathbb{K})$ if and only if there *exist a bijection $\tilde{\gamma}$ from G onto the set $\mathcal{A}(\underline{D}_\sqcap)$ of all atoms of $\underline{D}_\sqcap$ and a bijection $\tilde{\mu}$ from M onto the set $\mathcal{C}(\underline{D}_\sqcup)$ of all coatoms of $\underline{D}_\sqcup$ such that $gIm \Leftrightarrow \tilde{\gamma}(g) \sqsubseteq \tilde{\mu}(m)$ for all $g \in G$ and $m \in M$. In particular, for any complete fully contextual double Boolean algebra $\underline{D}$ whose Boolean algebras are atomic, we get $\underline{D} \cong \mathfrak{P}(\mathcal{A}(\underline{D}_\sqcap), \mathcal{C}(\underline{D}_\sqcup), \sqsubseteq)$, i.e., the protoconcept algebras are up to isomorphism the complete fully contextual double Boolean algebras $\underline{D}$ whose Boolean algebras $\underline{D}_\sqcap$ and $\underline{D}_\sqcup$ are atomic.*

Both basic theorems are powerful tools for the investigation of double Boolean algebras and will be used several times in proofs about the structure of subalgebras.

3   Complete Subalgebras of Semiconcept Algebras

First, we show that every complete subalgebra of a semiconcept algebra is a semiconcept algebra itself (Proposition 1). This is done by using the Basic Theorem and the following lemma:

Lemma 1. *Every complete subalgebra S of a complete atomistic Boolean algebra $\mathcal{B}$ is atomistic.*

Proof: For every atom a of $\mathcal{B}$ we set $a' := \bigwedge([a) \cap S)$. For two atoms a_1, a_2 of $\mathcal{B}$ we obtain $a_1' = a_2'$ or a_1' and a_2' are incomparable: If we assume $a_1' > a_2'$, then $a_1' \wedge \neg a_2' \in S$ and $0 < a_1' \wedge \neg a_2' < a_1'$. Since $\mathcal{B}$ is atomistic, it follows from $a_1 \leq a_1' = (a_1' \wedge \neg a_2') \vee a_2'$ that $a_1 \leq a_1' \wedge \neg a_2'$ or $a_1 \leq a_2'$, which yields a contradiction to the construction of a_1'.

We obtain that, for every atom a of $\mathcal{B}$, the element a' is an atom of S: If an element $x \in S$ satisfying $0 < x < a'$ existed, then we would find an atom b of $\mathcal{B}$ such that $b \leq x < a'$ which would yield $b' < a'$. Since $\mathcal{B}$ is atomic, we find for every $x \in S$ an atom a of $\mathcal{B}$ with $a \leq x$, which implies $a' \leq x$. Hence S is atomic, which is equivalent to atomistic in complete Boolean algebras (cp. [GP81]). $\quad\square$

Proposition 1. *Every complete subalgebra $\underline{S}$ of a semiconcept algebra $\mathfrak{H}(\mathbb{K})$ is a semiconcept algebra itself.*

Proof: Obviously every subalgebra of a pure double Boolean algebra is pure. The preceeding lemma yields that $\underline{S}_\sqcap$ and $\underline{S}_\sqcup$ are atomic, hence we obtain from Theorem 2 $\underline{S} \cong \mathfrak{H}(\mathcal{A}(\underline{S}_\sqcap), \mathcal{C}(\underline{S}_\sqcup), \sqsubseteq)$. $\quad\square$

Our goal is to derive for each subalgebra S of $\mathfrak{H}(\mathbb{K})$ a context $\mathbb{K}^S$ with $S \cong \mathfrak{H}(\mathbb{K}^S)$. Proposition 2 establishes a link between certain partitions of G and M and the subalgebras of $\mathfrak{H}(\mathbb{K})$. After that, Theorem 4 gives a bijection between the set of these partitions and the subalgebras. Finally, Corollary 1 yields a criterion to decide whether a given subset of a semiconcept algebra is a complete subalgebra.

Proposition 2. *Let* $\mathbb{K} := (G, M, I)$ *be a context. If* $\underline{D}$ *is a complete subalgebra of* $\mathfrak{H}(\mathbb{K})$ *then:*

1. $\Gamma := \{A \subseteq G \mid (A, A') \in \mathcal{A}(\underline{D_{\sqcap}})\}$ *is a partition of* G.
2. $\Theta := \{B \subseteq M \mid (B', B) \in \mathcal{C}(\underline{D_{\sqcup}})\}$ *is a partition of* M.
3. *For every* $A \in \Gamma$ *the set* A' *is a union of some classes of* Θ.
4. *For every* $B \in \Theta$ *the set* B' *is a union of some classes of* Γ.
5. $\underline{D} \cong \mathfrak{H}(\Gamma, \Theta, I_{\Gamma,\Theta})$ *with* $A\, I_{\Gamma,\Theta}\, B :\Leftrightarrow B \subseteq A'$

<u>Proof:</u> 1) As $(A_1, A_1') \sqcap (A_2, A_2') = (\emptyset, M) = \bot$ holds for atoms (A_1, A_1'), (A_2, A_2'), we have $A_1 \cap A_2 = \emptyset$ for all $A_1, A_2 \in \Gamma$. Moreover, $\bigsqcup\{(A, A') \mid (A, A') \in \mathcal{A}(\underline{D_{\sqcap}})\} = \top = (G, G')$, thus $\bigcup_{A \in \Gamma} A = G$.
2) follows dually.
3) Suppose there is a class $A \in \Gamma$ and a class $B \in \Theta$ such that $\emptyset \neq A' \cap B \neq B$. Then $(A, A') \in \underline{D}$ and $(B', B) \in \underline{D}$ implies $(A, A') \sqcup (B', B) = ((A' \cap B)', A' \cap B) \in \underline{D}$. From $B \neq A' \cap B \neq \emptyset$ we obtain $(B', B) \sqsubset ((A' \cap B)', A' \cap B) \sqsubset \top$, hence (B', B) was no coatom of $\underline{D_{\sqcup}}$. Since Θ is a partition of M, it follows that A' is a union of classes of Θ.
    4) follows dually.
    5) follows immediately from the Basic Theorem on semiconcept algebras. $\square$

The previous proposition suggests, that pairs of partitions of the object and attribute set which are linked as described in 3. and 4. of Proposition 2 are very close to complete subalgebras of semiconcept algebras. This leads to the following definition.

Definition 3. *Let* $\mathbb{K} := (G, M, I)$ *be a context. A pair* (Γ, Θ) *consisting of a partition* Γ *of* G *and a partition* Θ *of* M *is called a* subalgebra generating pair *iff for all* $A \in \Gamma$ *and* $B \in \Theta$

1. $B \subseteq A'$ *or* $B \cap A' = \emptyset$
2. $A \subseteq B'$ *or* $A \cap B' = \emptyset$.

For a subset A of a set B and a partition Ψ of B we say that A *is saturated by* Ψ iff A is a union of classes of Ψ. Note that if, for a context $\mathbb{K} := (G, M, I)$, the pair (Γ, Θ) is a subalgebra generating pair and A is saturated by Γ then A' is saturated by Θ. Dually, if B is a subset of M saturated by Θ then B' is saturated by Γ.

We introduce an order on the subalgebra generating pairs of a given context $\mathbb{K}$ by:

$$(\Gamma_1, \Theta_1) \leq (\Gamma_2, \Theta_2) :\Leftrightarrow \Gamma_1 \leq \Gamma_2 \text{ and } \Theta_1 \leq \Theta_2,$$

where $P_1 \leq P_2$ for partitions P_1, P_2 of a set if and only if P_1 is a refinement of P_2.

Now we have the means to prove the first main result of this paper:

Theorem 4 (Complete subalgebras of semiconcept algebras).
For a context $\mathbb{K} := (G, M, I)$ and partitions Γ and Θ on G and M, respectively, we define

$$\mathfrak{H}_{(\Gamma,\Theta)}(\mathbb{K}) := \{(A, B) \in \mathfrak{H}(\mathbb{K}) \mid A \text{ is saturated by } \Gamma \text{ and } B \text{ is saturated by } \Theta\}.$$

Then the map

$$\phi : (\Gamma, \Theta) \mapsto \underline{\mathfrak{H}}_{(\Gamma,\Theta)}(\mathbb{K})$$

is a dual isomorphism between the ordered set of all subalgebra generating pairs of $\mathbb{K}$ and the lattice of all complete subalgebras of $\underline{\mathfrak{H}}(\mathbb{K})$.

Proof: Let (Γ, Θ) be a subalgebra generating pair. Then $\top = (G, \emptyset) \in \mathfrak{H}_{(\Gamma,\Theta)}(\mathbb{K})$, $\bot = (\emptyset, M) \in \mathfrak{H}_{(\Gamma,\Theta)}(\mathbb{K})$. Moreover, for $(A, B) \in \mathfrak{H}_{(\Gamma,\Theta)}(\mathbb{K})$ we have $(A, B)_\sqcap = (A, A') \in \mathfrak{H}_{(\Gamma,\Theta)}(\mathbb{K})$, since A is saturated by Γ and therefore A' is saturated by Θ. Likewise we obtain $(A, B)_\sqcup \in \mathfrak{H}_{(\Gamma,\Theta)}(\mathbb{K})$. It is easy to check that $(A_i, B_i) \in \mathfrak{H}_{(\Gamma,\Theta)}(\mathbb{K})$ for $i \in \mathcal{I}$ also yields that $\neg(A_i, B_i), \lnot(A_i, B_i), \bigsqcap_{i \in \mathcal{I}}(A_i, B_i)$ and $\bigsqcup_{i \in \mathcal{I}}(A_i, B_i)$ are contained in $\mathfrak{H}_{(\Gamma,\Theta)}(\mathbb{K})$. Thus every subalgebra generating pair yields a complete subalgebra of $\underline{\mathfrak{H}}(\mathbb{K})$. Different subalgebra generating pairs yield obviously different subalgebras and the previous proposition shows that we find a subalgebra generating pair for every complete subalgebra of $\underline{\mathfrak{H}}(\mathbb{K})$, hence the map ϕ is bijective.

Now suppose $\underline{\mathfrak{H}}_{\Gamma_1,\Theta_1}(\mathbb{K}) \subseteq \underline{\mathfrak{H}}_{\Gamma_2,\Theta_2}(\mathbb{K})$. Then the extents of the atoms of the Boolean algebra $\underline{\mathfrak{H}}_{\sqcap(\Gamma_1,\Theta_1)}(\mathbb{K})$ are unions of extents of atoms of $\underline{\mathfrak{H}}_{\sqcap(\Gamma_2,\Theta_2)}(\mathbb{K})$, hence $\Gamma_2 \le \Gamma_1$. Likewise we obtain $\Theta_2 \le \Theta_1$ and thus $\underline{\mathfrak{H}}_{(\Gamma_1,\Theta_1)}(\mathbb{K}) \le \underline{\mathfrak{H}}_{(\Gamma_2,\Theta_2)}(\mathbb{K})$ implies $(\Gamma_1, \Theta_1) \ge (\Gamma_2, \Theta_2)$. Conversely, if $(\Gamma_1, \Theta_1) \ge (\Gamma_2, \Theta_2)$, then every $A \subseteq G$ saturated by Γ_2 is also saturated by Γ_1 and every $B \subseteq M$ saturated by Θ_2 is also saturated by Θ_1, which implies $\underline{\mathfrak{H}}_{\Gamma_1,\Theta_1}(\mathbb{K}) \subseteq \underline{\mathfrak{H}}_{\Gamma_2,\Theta_2}(\mathbb{K})$. Thus we have

$$\underline{\mathfrak{H}}_{\Gamma_1,\Theta_1}(\mathbb{K}) \subseteq \underline{\mathfrak{H}}_{\Gamma_2,\Theta_2}(\mathbb{K}) \Leftrightarrow (\Gamma_1, \Theta_1) \ge (\Gamma_2, \Theta_2).$$

and therefore ϕ is a dual isomorphism. □

The preceeding theorem allows to identify subalgebras of semiconcept algebras through their contexts. In addition, a description in terms of subsets of a given semiconcept algebra may be derived.

Corollary 1. *Let $\underline{\mathfrak{H}}(\mathbb{K})$ be the semiconcept algebra of a context $\mathbb{K} := (G, M, I)$. A subset $\mathfrak{U} \subseteq \mathfrak{H}(\mathbb{K})$ is a complete subalgebra of $\underline{\mathfrak{H}}(\mathbb{K})$ if and only if it fulfills the following conditions:*

1. *$\mathfrak{U}_\sqcap := \mathfrak{U} \cap \underline{\mathfrak{H}}(\mathbb{K})_\sqcap$ is a complete subalgebra of the Boolean algebra $\underline{\mathfrak{H}}(\mathbb{K})_\sqcap$.*
2. *$\mathfrak{U}_\sqcup := \mathfrak{U} \cap \underline{\mathfrak{H}}(\mathbb{K})_\sqcup$ is a complete subalgebra of the Boolean algebra $\underline{\mathfrak{H}}(\mathbb{K})_\sqcup$.*
3. *For every atom $\mathfrak{a}$ of $\underline{\mathfrak{U}}_\sqcap$ and for every coatom $\mathfrak{c}$ of $\underline{\mathfrak{U}}_\sqcup$ holds either*

      a) $\mathfrak{a} \sqsubseteq \mathfrak{c}$ or
      b) $\mathfrak{a} \sqcup \mathfrak{c} = \top$ and $\mathfrak{a} \sqcap \mathfrak{c} = \bot$.

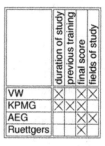

	duration of study	previous training	final score	fields of study
VW		X	X	X
KPMG	X	X	X	
AEG			X	X
Ruettgers			X	

Fig. 3. A context

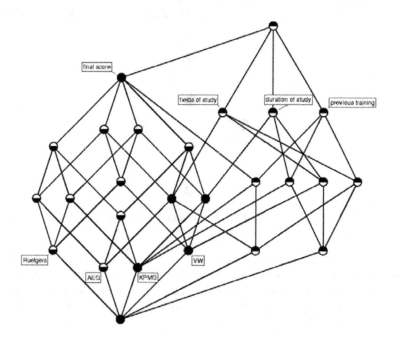

Fig. 4. The semiconcept algebra of the context in Figure 3

Proof: Let $\mathfrak{U}$ be a subset of $\mathfrak{H}(\mathbb{K})$ satisfying the conditions 1 to 3. We show that $\Gamma_{\mathfrak{U}} := \{A \subseteq G \mid (A, A') \text{ is an atom of } \underline{\mathfrak{U}}_{\sqcap}\}$ and $\Theta_{\mathfrak{U}} := \{B \subseteq M \mid (B', B) \text{ is a coatom of } \underline{\mathfrak{U}}_{\sqcup}\}$ form a subalgebra generating pair of $\mathbb{K}$. Conditions 1) and 2) yield that $\Gamma_{\mathfrak{U}}$ and $\Theta_{\mathfrak{U}}$ are partitions of G and M, respectively. Now, let $A \in \Gamma_{\mathfrak{U}}$ and let $B \in \Theta_{\mathfrak{U}}$. Then (A, A') is an atom of $\underline{\mathfrak{U}}_{\sqcap}$ and (B', B) is a coatom of $\underline{\mathfrak{U}}_{\sqcup}$. Condition 3) yields that either $(A, A') \sqsubseteq (B', B)$ which implies $B \subseteq A'$, or that $(A, A') \sqcup (B', B) = \top$, and thus $B \cap A' = \emptyset$. Analogously we obtain that B' is saturated by $\Gamma_{\mathfrak{U}}$, hence $(\Gamma_{\mathfrak{U}}, \Theta_{\mathfrak{U}})$ is a subalgebra generating pair. Since the classes of $\Gamma_{\mathfrak{U}}$ and $\Theta_{\mathfrak{U}}$ correspond to the atoms of $\underline{\mathfrak{U}}_{\sqcap}$ and the coatoms of $\underline{\mathfrak{U}}_{\sqcup}$, respectively,

Table 1. Subalgebra generating pairs

Nr.	Partitions
1	$\Gamma = \{\{\text{VW, KPMG, AEG, Rütgers}\}\}$ $\Theta = \{\{\text{duration of study, previous training, fields of study}\}, \{\text{final score}\}\}$
2	$\Gamma = \{\{\text{Rütgers}\}, \{\text{VW, AEG, KPMG}\}\}$ $\Theta = \{\{\text{duration of study, previous training, fields of study}\}, \{\text{final score}\}\}$
3	$\Gamma = \{\{\text{AEG, Rütgers}\}, \{\text{VW, KPMG}\}\}$ $\Theta = \{\{\text{previous training, fields of study}\}, \{\text{duration of study}\}, \{\text{final score}\}\}$
4	$\Gamma = \{\{\text{VW, Rütgers}\}, \{\text{AEG, KPMG}\}\}$ $\Theta = \{\{\text{duration of study, previous training, fields of study}\}, \{\text{final score}\}\}$
5	$\Gamma = \{\{\text{Rütgers}\}, \{\text{VW, AEG}\}, \{\text{KPMG}\}\}$ $\Theta = \{\{\text{duration of study, previous training}\}, \{\text{fields of study}\}, \{\text{final score}\}\}$
6	$\Gamma = \{\{\text{Rütgers}\}, \{\text{VW}\}, \{\text{AEG}\}, \{\text{KPMG}\}\}$ $\Theta = \{\{\text{duration of study}\}, \{\text{previous training}\}, \{\text{fields of study}\}, \{\text{final score}\}\}$

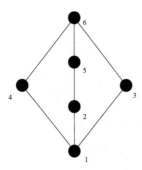

Fig. 5. The lattice of subalgebras of the semiconcept algebra shown in Figure 4

the Basic Theorem on Semiconcept Algebras then yields $\mathfrak{H}(\Gamma_{\mathfrak{U}}, \Theta_{\mathfrak{U}}, I_{\Gamma_{\mathfrak{U}}, \Theta_{\mathfrak{U}}}) \cong \mathfrak{U}$. Conversely, if $\mathfrak{U}$ is a subalgebra of $\mathfrak{H}(\mathbb{K})$ we obtain immediately that 1) and 2) must be satisfied. Let $\mathfrak{a}$ be an atom of $\mathfrak{U}_{\sqcap}$ and let $\mathfrak{b}$ be a coatom of $\mathfrak{U}_{\sqcup}$. From $\mathfrak{b} \sqsubseteq \mathfrak{a} \sqcup \mathfrak{b} \in \mathfrak{U}$ it follows that $\mathfrak{a} \sqcup \mathfrak{b} = \mathfrak{b}$ (and thus $\mathfrak{a} \sqsubseteq \mathfrak{b}$) or $\mathfrak{a} \sqcup \mathfrak{b} = \top$. Dually, $\mathfrak{a} \sqsubseteq \mathfrak{a} \sqcap \mathfrak{b} \in \mathfrak{U}_{\sqcap}$ yields $\mathfrak{a} \sqsubseteq \mathfrak{b}$ or $\mathfrak{a} \sqcap \mathfrak{b} = \bot$.   □

Example 2. For every context $\mathbb{K} := (G, M, I)$ the partitions corresponding to the equivalence relations $(g, h) \in \Gamma :\Leftrightarrow g' = h'$ and $(m, n) \in \Theta :\Leftrightarrow m' = n'$ form a subalgebra generating pair. The context $(\Gamma, \Theta, I_{(\Gamma, \Theta)})$ is usually called the clarified context of $\mathbb{K}$.

Example 3. Figure 3 shows a subcontext of a context used in [GZ90], and Figure 4 the respective semiconcept algebra. Fig. 5 depicts its lattice of subalgebras. The numbers correspond to subalgebra generating pairs as shown in Table 1.

The relation between subalgebra generating pairs and closed subrelations is illustrated in the next section, while the generalization of Theorem 4 to proto-concept algebras is given in Section 5.

4   Complete Subalgebras and Closed Subrelations

In Formal Concept Analysis, closed subrelations of incidence relations correspond to complete sublattices of the respective concept lattice. In this section we briefly recall the definition of closed subrelations and two propositions from [GW99]. Then it is shown how every subalgebra generating pair of a context $\mathbb{K}$ yields a closed subrelation of $\mathbb{K}$. Moreover, the closed subrelations obtained in this way are also characterized.

Definition 4. *A relation $J \subseteq I$ is called a closed relation of the context (G, M, I) if every concept of the context (G, M, J) is also a concept of (G, M, I).*

Note that if J is a closed subrelation of a context $\mathbb{K} := (G, M, I)$ then $\mathfrak{B}(G, M, J)$ is a complete sublattice of $\mathfrak{B}(\mathbb{K})$. Conversely, for every complete sublattice $\underline{V}$ of $\mathfrak{B}(\mathbb{K})$ we find a closed subrelation J of $\mathbb{K}$ such that $\mathfrak{B}(G, M, J) \cong \underline{V}$ (cf. [GW99]).

Proposition 3 ([GW99]). *A subrelation $J \subseteq I$ is closed if and only if*

$$X^{JJ} \supseteq X^{JI}$$

for each subset $X \subseteq G$ and for each subset $X \subseteq M$.

Proposition 4 ([GW99]). *The closed relations of a context (G, M, I) are precisely those subrelations $J \subseteq I$ which satisfy the following condition:*

(C) $(g, m) \in I \setminus J$ *implies* $(h, m) \notin I$ *for some* $h \in G$ *with* $g^J \subseteq h^J$ *as well as* $(g, n) \notin I$ *for some* $n \in M$ *with* $m^J \subseteq n^J$.

For a given subalgebra generating pair of a context $\mathbb{K}$, a closed subrelation of $\mathbb{K}$ can be constructed such that the concept lattice contained in the subalgebra is isomorphic to the sublattice obtained from the closed subrelation:

Proposition 5. *Let $\mathbb{K} := (G, M, I)$ be a context and let (Γ, Θ) be a subalgebra generating pair of $\mathbb{K}$. The relation $J_{(\Gamma,\Theta)} := \bigcup\{A \times B \mid A \in \Gamma, B \in \Theta$ and $A\, I_{(\Gamma,\Theta)}\, B\}$ is a closed subrelation of $\mathbb{K}$ and $\mathfrak{B}(G, M, J) \cong \mathfrak{B}(\Gamma, \Theta, I_{(\Gamma,\Theta)})$.*

Proof: In the following we set $J := J_{(\Gamma,\Theta)}$. First we show that J is a subrelation of I: From $(g, m) \in J \Leftrightarrow [g]_\Gamma\, I_{\Gamma,\Theta}\, [m]_\Theta \Leftrightarrow \forall h \in [g]_\Gamma. \forall n \in [m]_\Theta : hIn$ we obtain $(g, m) \in I$.

We use Proposition 3 in order to prove that J is closed: For $X \subseteq G$ we find that X^J is saturated by Θ since $X^J = (\pi_\Gamma(X))^J$ where π_Γ denotes the canonical projection $G \to \Gamma$. For any class $[m]_\Theta \in X^J$ we have $[m]_\Theta^J = [m]_\Theta^I$, hence $X^{JI} = X^{JJ}$. Analogously, it is shown $Y^{JI} = Y^{JJ}$ for $Y \subseteq M$ and thus J is a closed subrelation of I. The context $(\Gamma, \Theta, I_{(\Gamma,\Theta)})$ is obtained from (G, M, J) by clarification, which does not affect the concept lattice, hence $\mathfrak{B}(G, M, J) \cong \mathfrak{B}(\Gamma, \Theta, I_{(\Gamma,\Theta)})$.   □

The closed subrelations of a context obtained in this way from subalgebra generating pairs can be described as those satisfying an even stronger condition than Condition (C) given in Proposition 4:

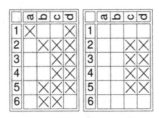

Fig. 6. a) A context b) A closed subrelation of the context

Proposition 6. *The closed relations of a context* $\mathbb{K} := (G, M, I)$ *obtained from subalgebra generating pairs of* $\mathbb{K}$ *satisfy the condition:*

> **(C')** $(g, m) \in I \setminus J$ *implies* $(h, m) \notin I$ *for some* $h \in G$ *with* $g^J = h^J$
> *as well as* $(g, n) \notin I$ *for some* $n \in M$ *with* $m^J = n^J$.

Conversely, for every subrelation $J \subseteq I$ *of* $\mathbb{K}$ *satisfying (C') there exists a sub-algebra generating pair* (Γ_J, Θ_J) *such that* $J = J_{(\Gamma_J, \Theta_J)}$.

<u>Proof:</u> Let (Γ, Θ) be a subalgebra generating pair of $\mathbb{K}$, let $J_{(\Gamma,\Theta)}$ be the corresponding closed subrelation as above, and let $(g, m) \in I \setminus J_{(\Gamma,\Theta)}$. It follows that $([g]_\Gamma, [m]_\Theta) \notin I_{\Gamma,\Theta}$ and thus $m \notin [g]^I_\Gamma$. Therefore there is an $h \in [g]_\Gamma$ with $(h, m) \notin I$. Likewise we find an $n \in [m]_\Theta$ with $(g, n) \notin I$.

Conversely, let J be a subrelation of I satisfying (C'). Proposition 4 immediately yields that J is a closed subrelation. By setting $\Gamma_J := \bigcup_{g \in G} A_g$ with $A_g := \{h \in G \mid g' = h'\}$ and $\Theta_J := \bigcup_{m \in M} B_m$ with $B_m := \{n \in M \mid m' = n'\}$ a subalgebra generating pair is obtained: Let $A \in \Gamma_J$. In order to prove that A^I is saturated by Θ_J we show $A^I = A^J$. From $J \subseteq I$ we obtain $A^J \subseteq A^I$. If $m \in A^I \setminus A^J$ then there exists a $g \in A$ such that $(g, m) \in I \setminus J$. Condition (C') then yields an $h \in A$ with $(h, m) \notin I$ which is a contradiction to $m \in A^I$. Thus we have $A^I = A^J$. Since obviously A^J is saturated by Θ_J, A^I is saturated as well. Dually we find that for every $B \in \Theta_J$ its derivation B^I is saturated by Γ_J.

In order to show $J = J_{(\Gamma_J, \Theta_J)}$, we first assume gJm. Then $m \in [g]^J_{\Gamma_J} = [g]^I_{\Gamma_J} \Leftrightarrow [m]_{\Theta_J} \subseteq [g]^I_{\Gamma_J} \Leftrightarrow [g]_{\Gamma_J} I_{(\Gamma_J, \Theta_J)} [m]_{\Theta_J}$ and thus $(g, m) \in J_{(\Gamma_J, \Theta_J)}$. Secondly, in the case $(g, m) \notin I$, it is clear that $([g]_{\Gamma_J}, [m]_{\Theta_J}) \notin I_{(\Gamma_J, \Theta_J)}$. If $(g, m) \in I \setminus J$ then condition (C') yields $m \in [g]^I_{\Gamma_J}$, hence $([g]_{\Gamma_J}, [m]_{\Theta_J}) \notin I_{(\Gamma_J, \Theta_J)}$. Therefore we obtain $J = J_{(\Gamma_J, \Theta_J)}$. $\square$

The correspondence between subalgebra generating pairs and closed subrelations is not one-to-one. If we take for example the context presented in Fig. 6a) then $\Gamma_1 := \{\{1, 6\}, \{2, 3\}, \{4, 5\}\}$ and $\Theta := \{\{a, b\}, \{c, d\}\}$ form a subalgebra generating pair. But the subalgebra generating pair $\Gamma_2 := \{\{1, 6\}, \{2, 3, 4, 5\}\}, \Theta$ is assigned to the corresponding closed subrelation J presented in Fig. 6b). In general, for a subalgebra generating pair (Γ_J, Θ_J) obtained from a closed subrelation J as in the proof of Proposition 6, the context $(\Gamma_J, \Theta_J, I_{(\Gamma_J, \Theta_J)})$ is clarified, while J may be obtained also from subalgebra generating pairs (Γ, Θ) where $(\Gamma, \Theta, I_{(\Gamma,\Theta)})$ is not clarified. Yet, not every combination of equal lines and rows of (G, M, J) will yield a subalgebra generating pair of (G, M, I): In our example there is no subalgebra generating pair with $\Gamma_3 := \{\{1,6\},\{2,5\},\{3,4\}\}$ as partition of G, although the lines 2, 3, 4 and 5 are identical in (G,M,J). If such a subalgebra generating pair would exist, then the semiconcepts $(25, bcd)$ and $(34, cd)$ would be elements of the generated subalgebra. But $((25, bcd) \sqcup \lnot(34, cd)) \sqcap (25, bcd) = (6, bc)$ in contradiction to $\{1, 6\} \in \Gamma_3$.

5   Complete Subalgebras of Protoconcept Algebras

The subalgebras of protoconcept algebras cannot be uniquely described through subalgebra generating pairs. Yet, for a given subalgebra generating pair they cannot vary too much. Remember that if $\mathbb{K} := (G, M, I)$ is a context and $V \subseteq \mathfrak{P}(\mathbb{K}) \setminus \mathfrak{H}(\mathbb{K})$ then $\mathfrak{P}(\mathbb{K}) \setminus V$ yields a complete subalgebra of $\mathfrak{P}(\mathbb{K})$. In particular, $\mathfrak{H}(\mathbb{K})$ is a complete subalgebra. This result can immediately be extended to: If $\mathbb{K} := (G, M, I)$ is a context and $\underline{D}$ a complete subalgebra of $\mathfrak{P}(\mathbb{K})$ then the pure subalgebra $\underline{D}_p \leq \underline{D}$ is a complete subalgebra of $\mathfrak{P}(\mathbb{K})$ and of $\mathfrak{H}(\mathbb{K})$. This can be used to categorize the subalgebras of $\mathfrak{P}(\mathbb{K})$ through the contained semiconcept algebras.

Proposition 7. *Let $\mathbb{K} := (G, M, I)$ be a context and (Γ, Θ) a subalgebra generating pair of $\mathbb{K}$. We set*

$$\mathfrak{P}_{(\Gamma,\Theta)}(\mathbb{K}) := \{(A, B) \in \mathfrak{P}(\mathbb{K}) \mid A \text{ is saturated by } \Gamma \text{ and } B \text{ is saturated by } \Theta\}.$$

Then $\mathfrak{P}_{(\Gamma,\Theta)}(\mathbb{K})$ is a complete subalgebra of $\mathfrak{P}(\mathbb{K})$ with

$$\underline{\mathfrak{P}}_{(\Gamma,\Theta)}(\mathbb{K}) \cong \underline{\mathfrak{P}}(\Gamma, \Theta, I_{(\Gamma,\Theta)})$$

<u>Proof:</u> From Theorem 4 it follows that $\mathfrak{P}_{(\Gamma,\Theta)}(\mathbb{K}) \cap \mathfrak{H}(\mathbb{K})$ is a complete pure subalgebra of $\mathfrak{P}(\mathbb{K})$. Moreover, for $(A, B) \in \mathfrak{P}_{(\Gamma,\Theta)}(\mathbb{K}) \setminus \mathfrak{H}(\mathbb{K})$, we have $(A, B)_\sqcap$, $(A, B)_\sqcup \in \mathfrak{H}_{(\Gamma,\Theta)}(\mathbb{K}) \subseteq \mathfrak{P}_{(\Gamma,\Theta)}(\mathbb{K})$. Since $\prod_{i \in I}(A_i, B_i) = \prod_{i \in I}(A_i, B_i)_\sqcap \in \mathfrak{H}_{(\Gamma,\Theta)}(\mathbb{K})$ and $\bigsqcup_{i \in I}(A_i, B_i) = \bigsqcup_{i \in I}(A_i, B_i)_\sqcup \in \mathfrak{H}_{(\Gamma,\Theta)}(\mathbb{K})$ we have that $\underline{\mathfrak{P}}_{(\Gamma,\Theta)}(\mathbb{K})$ is a complete subalgebra of $\mathfrak{P}(\mathbb{K})$. As $\mathfrak{P}(\mathbb{K})$ is fully contextual, $\mathfrak{P}_{(\Gamma,\Theta)}(\mathbb{K})$ is fully contextual, too and Theorem 3 yields $\underline{\mathfrak{P}}_{(\Gamma,\Theta)}(\mathbb{K}) \cong \underline{\mathfrak{P}}(\Gamma, \Theta, I_{\Gamma,\Theta})$.   □

Now we have the means to describe the subalgebras of protoconcept algebras similar to Theorem 4:

Theorem 5 (Complete subalgebras of protoconcept algebras). *Let $\mathbb{K} :=$ (G, M, I) be a context and let $\underline{\mathfrak{U}}$ be a complete subalgebra of $\mathfrak{P}(\mathbb{K})$. Then there is a subalgebra generating pair (Γ, Θ) of $\mathbb{K}$ with $\mathfrak{U}_p = \mathfrak{H}_{(\Gamma,\Theta)}(\mathbb{K})$. Moreover, $\underline{\mathfrak{H}}_{(\Gamma,\Theta)}(\mathbb{K}) \leq \underline{\mathfrak{U}} \leq \underline{\mathfrak{P}}_{(\Gamma,\Theta)}(\mathbb{K})$ holds in the lattice of subalgebras and the interval $[\underline{\mathfrak{H}}_{(\Gamma,\Theta)}(\mathbb{K}), \underline{\mathfrak{P}}_{(\Gamma,\Theta)}(\mathbb{K})]$ is isomorphic to the powerset lattice of the set $\mathfrak{P}_{(\Gamma,\Theta)}(\mathbb{K}) \setminus$ $\mathfrak{H}_{(\Gamma,\Theta)}$.*

<u>Proof:</u> Since $\mathfrak{U}_p$ is a complete pure subalgebra of $\mathfrak{P}(\mathbb{K})$, it is also a complete subalgebra of $\mathfrak{H}(\mathbb{K})$. Theorem 4 yields the subalgebra generating pair (Γ, Θ)

with $\underline{\mathfrak{U}}_p = \underline{\mathfrak{H}}_{(\Gamma,\Theta)}(\mathbb{K})$. Suppose $\mathfrak{U} \not\subseteq \mathfrak{P}_{(\Gamma,\Theta)}(\mathbb{K})$. Then there exists a protoconcept (A, B) in $\mathfrak{U}$ for which A is not saturated by Γ or B is not saturated by Θ.

1) If A is not saturated by Γ then there is a $C \in \Gamma$ such that $\emptyset \neq C \cap A \neq A$. Since $(C, C') \in \mathfrak{H}_{(\Gamma,\Theta)}(\mathbb{K}) \subseteq \mathfrak{U}$ it follows that $(C \cap A, (C \cap A)') \in \mathfrak{U}_p$, thus (C, C') is not an atom of $\underline{\mathfrak{U}}_p$ and $\underline{\mathfrak{U}}_p \neq \underline{\mathfrak{H}}_{(\Gamma,\Theta)}(\mathbb{K})$.

2) Dually we obtain $\underline{\mathfrak{U}}_p \neq \underline{\mathfrak{H}}_{(\Gamma,\Theta)}(\mathbb{K})$ in the case that B is not saturated by Θ.

Thus we have $\underline{\mathfrak{H}}_{(\Gamma,\Theta)}(\mathbb{K}) \leq \underline{\mathfrak{U}} \leq \underline{\mathfrak{P}}_{(\Gamma,\Theta)}(\mathbb{K})$. If $P \subseteq \mathfrak{P}_{(\Gamma,\Theta)}(\mathbb{K}) \setminus \mathfrak{H}_{(\Gamma,\Theta)}(\mathbb{K})$ then $\mathfrak{P}_{(\Gamma,\Theta)}(\mathbb{K}) \setminus P$ is a subalgebra since the elements of P cannot be generated from the other elements of $\mathfrak{P}_{(\Gamma,\Theta)}(\mathbb{K}) \setminus P$ and of course $\mathfrak{H}_{(\Gamma,\Theta)}(\mathbb{K}) \subseteq (\mathfrak{P}_{(\Gamma,\Theta)}(\mathbb{K}) \setminus P) \subseteq \mathfrak{P}_{(\Gamma,\Theta)}(\mathbb{K})$. □

Remark 1. Note that Corollary 1 also yields a criterion to decide whether a given subset $\mathfrak{U}$ of a protoconcept algebra $\mathfrak{P}(\mathbb{K})$ is a complete subalgebra. Since $\mathfrak{U} \cap \underline{\mathfrak{H}}(\mathbb{K})$ must be a complete subalgebra of $\underline{\mathfrak{H}}(\mathbb{K})$ it has to satisfy the conditions 1 to 3 of Corollary 1. It can be easily seen that $\mathfrak{U}$ is a subalgebra if, in addition, for every $\mathfrak{p} \in \mathfrak{U} \setminus \mathfrak{H}(\mathbb{K})$ the semiconcepts $\mathfrak{p}_\sqcap$ and $\mathfrak{p}_\sqcup$ are contained in $\mathfrak{U}$. Equivalent, but closer to the preceeding theorem is the following condition:

4)   For every $\mathfrak{p} \in \mathfrak{U} \setminus \underline{\mathfrak{H}}(\mathbb{K})$:

   $a)$ $\mathfrak{p} \sqcap \mathfrak{a} = \mathfrak{a}$ or $\mathfrak{p} \sqcap \mathfrak{a} = \bot$ for every atom $\mathfrak{a}$ of $\mathfrak{U}_\sqcap$

   $b)$ $\mathfrak{p} \sqcup \mathfrak{c} = \mathfrak{c}$ or $\mathfrak{p} \sqcup \mathfrak{c} = \top$ for every coatom $\mathfrak{c}$ of $\mathfrak{U}_\sqcup$

In Theorem 4 and Theorem 5 of this paper we have thus shown how the complete subalgebras of semiconcept algebras and of protoconcept algebras can be characterized through partitions of the object and attribute sets of their contexts. Moreover, their relation to the theory of closed subrelations has been illustrated.

6   Further Research

This paper contributes to the development of an algebraic structure theory of double Boolean algebras. After having found results on subalgebras and on congruence relations (cf. [Vo03]), a next step is the investigation of homomorphisms between double Boolean algebras.

A better understanding of the free double Boolean algebras is crucial for the development of a Boolean Concept Logic. A first step was made in [Vo04]. This approach will be further elaborated.

References

[Da03]   F. Dau: The Logic System of Concept Graphs with Negation. LNAI **2892**. Springer, Heidelberg 2003.

[DK03] F. Dau, J. Klinger: From Formal Concept Analysis to Contextual Logic. 2003
        <http://www.dr-dau.net/Papers/ICFCA2003.pdf >

[GW99] B. Ganter, R. Wille: Formal Concept Analysis: Mathematical Foundations.
        Springer, Heidelberg 1999.

[GZ90] B. Ganter, M. Zickwolff: Nach welchen Kriterien wählen Firmen Hochschul-
        absolventen aus? - Auswertung einer Befragung. FB4-Preprint 1343, TH-
        Darmstadt 1990.

[GP81] H.-P. Gumm, W. Poguntke: Boolesche Algebra. B.I.-Wissenschaftsverlag,
        Mannheim-Wien-Zürich 1981.

[KV03] J. Klinger, B. Vormbrock: Contextual Boolean Logic: How did it develop? In:
        B. Ganter, A. de Moor (eds.): Using Conceptual Structures. Contributions to
        ICCS 2003. Shaker, Aachen 2003, 143–156.

[Vo03] B. Vormbrock: Congruence Relations on Double Boolean Algebras. FB4-
        Preprint 2287, TU Darmstadt 2003.

[Vo04] B. Vormbrock: A First Step Towards Protoconcept Exploration. In: P. Eklund
        (ed.): Concept Lattices. LNAI **2961**. Springer Heidelberg 2004, 208–221.

[VW03] B. Vormbrock, R. Wille: Semiconcept and Protoconcept Algebras: The Basic
        Theorems. FB4-Preprint 2309, TU Darmstadt 2003.

[Wi00a] R. Wille: Boolean Concept Logic. In: B. Ganter, G. W. Mineau (eds.): Con-
        ceptual structures: logical, linguistic, and computational issues. LNAI **1867**.
        Springer, Heidelberg 2000, 317–331.

[Wi00b] R. Wille: Contextual Logic Summary. In: G. Stumme (ed.): Working with Con-
        ceptual Structures. Contributions to ICCS 2000. Shaker, Aachen 2000, 265–
        276.

[Wi01] R. Wille: Boolean Judgment Logic. In: H. Delugach, G. Stumme (eds.): Con-
        ceptual structures: broadening the base. LNAI **2120**. Springer, Heidelberg
        2001, 115–128.

Coherence Networks of Concept Lattices: The Basic Theorem

Sascha Karl Dörflein and Rudolf Wille

Technische Universität Darmstadt, Fachbereich Mathematik,
Schloßgartenstr. 7, D–64289 Darmstadt
wille@mathematik.tu-darmstadt.de

Abstract. For representing different views and their connections, networks of formal contexts are considered which are coded by so-called *multicontexts*. The coincidences between the network contexts of a multicontext give rise to a *coherence network of concept lattices*. It is the aim of this paper to state and to prove the *Basic Theorem on Coherence Networks of Concept Lattices* as an extension of the Basic Theorem on Concept Lattices.

Contents

1   Introduction

Formal contexts are the basic mathematical structures of *Formal Concept Analysis*. They allow to speak mathematically about *formal objets, formal attributes*, and the binary relation which indicates when a formal object *has a* formal attribute. This foundational setting was created for mathematizing concepts and concept hierarchies by so-called *formal concepts* and *concept lattices* of formal contexts (see [Wi82], [GW99]). Since data tables can be mathematized by formal contexts, Formal Concept Analysis has found extensive applications, in particular in data analysis and knowledge processing (cf. [GWW87],[Wi92],[WZ94], [Wi97], [SW00],[Wi00],[Wi02],[Ek04]).

Quite often, the object-attribute-relation is considered under different views so that a representation by a single formal context is not sufficient. This has led to generalizations of the notion of formal context. One approach is based on the mathematization of the ternary relationship that an object has an attribute under a certain condition. This approach, based on *triadic contexts* and derived

B. Ganter and R. Godin (Eds.): ICFCA 2005, LNAI 3403, pp. 344–359, 2005.

triadic concepts, has been elaborated under the heading *Triadic Concept Analysis* (see [Wi95],[LW95], [Bi98],[WZ00],[DW00]).

The approach we discuss in this paper maintains the vivid dyadic setting, but enables to represent different views and their connections by a network of formal contexts coded in a so-called *multicontext* [Wi96]. In contrast to triadic contexts, multicontexts are composed by formal contexts which may have different object sets and different attribute sets. Those context sets might have some elements in common, where elements could even be objects in one formal context and attributes in another. The coincidences between the context sets of a multicontext give rise to a *coherence network of concept lattices* which are derived from the formal contexts of the multicontext. It is the aim of this paper to state and to prove the so-called *Basic Theorem on Coherence Networks of Concept Lattices* [Dö99] as an extension of the Basic Theorem on Concept Lattices [Wi82] which is stated as follows (cf. [GW99]):

Basic Theorem on Concept Lattices. *Let* $\mathbb{K} := (G, M, I)$ *be a formal context. Then the set* $\mathfrak{B}(\mathbb{K})$ *of all formal concepts of* $\mathbb{K}$ *ordered by the subconcept-superconcept-relation is a complete lattice, called the* concept lattice *of* (G, M, I), *for which infima and suprema can be described in the following way:*

$$\bigwedge_{t \in T}(A_t, B_t) = (\bigcap_{t \in T} A_t, (\bigcup_{t \in T} B_t)^{II}),$$

$$\bigvee_{t \in T}(A_t, B_t) = ((\bigcup_{t \in T} A_t)^{II}, \bigcap_{t \in T} B_t).$$

In general, a complete lattice L is isomorphic to $\mathfrak{B}(\mathbb{K})$ *if and only if there exist mappings* $\tilde{\gamma} : G \longrightarrow L$ *and* $\tilde{\mu} : M \longrightarrow L$ *such that* $\tilde{\gamma}G$ *is* supremum-dense *in* L *(i.e.* $L = \{\bigvee X \mid X \subseteq \tilde{\gamma}G\}$*),* $\tilde{\mu}M$ *is* infimum-dense *in* L *(i.e.* $L = \{\bigwedge X \mid X \subseteq \tilde{\mu}M\}$*), and* $gIm \iff \tilde{\gamma}g \leq \tilde{\mu}m$ *for* $g \in G$ *and* $m \in M$; *in particular,* $L \cong \mathfrak{B}(L, L, \leq)$.

2   Multicontexts and Coherence Mappings

First we generalize the notion of a formal context to that of a multicontext which can be viewed as a network of formal contexts.

Definition 1. *A* multicontext of signature *$\sigma : P \to I^2$, where I and P are nonempty sets, is defined as a pair (S_I, R_P) consisting of a family $S_I := (S_i)_{i \in I}$ of sets and a family $R_P := (R_p)_{p \in P}$ of binary relations with $R_p \subseteq S_i \times S_j$ if $\sigma p = (i, j)$. A multicontext (S_I, R_P) of signature $\sigma : P \to I^2$ can be understood as a network of formal contexts $\mathbb{K}_p := (S_i, S_j, R_p)$ with $\sigma p = (i, j)$.*

The common elements which emerge in different formal contexts induce a coherence between the concept lattices of those formal contexts. Even with identical sets of objects and sets of attributes, the relation of distinct formal contexts could vary from each other so that one could hardly see any coherence between them. Therefore the coherence within a network of formal contexts has to be considered not only by the common elements, but also by the conceptual conformity of the formal contexts. The following definition provides mappings which respect these conceptual conformities ($X \mapsto X^q$ denotes the derivation operators of $\mathbb{K}_q$).

Definition 2. *Let (S_I, R_P) be a multicontext, let $\mathbb{K}_p := (S_i, S_j, R_p)$ and $\mathbb{K}_q := (S_k, S_l, R_q)$ be formal contexts of (S_I, R_P), and let (A, B) be a formal concept of $\mathbb{K}_p$. Then there are four coherence mappings from $\underline{\mathfrak{B}}(\mathbb{K}_p)$ to $\underline{\mathfrak{B}}(\mathbb{K}_q)$ defined by*

$$\lambda_{pq}(A, B) := ((A \cap S_k)^{qq}, (A \cap S_k)^q),$$
$$\varrho_{pq}(A, B) := ((B \cap S_l)^q, (B \cap S_l)^{qq}),$$
$$\varphi_{pq}(A, B) := ((B \cap S_k)^{qq}, (B \cap S_k)^q),$$
$$\psi_{pq}(A, B) := ((A \cap S_l)^q, (A \cap S_l)^{qq}).$$

The concept lattices of the formal contexts of (S_I, R_P) together with all coherence mappings form a coherence network of concept lattices, *which shall be denoted by $\mathfrak{N}(S_I, R_P)$.*

The mapping λ_{pq} elucidates the conformity of the extents of the formal contexts $\mathbb{K}_p$ and $\mathbb{K}_q$; the mapping ϱ_{pq} does the same for the intents. φ_{pq} delineates the similarity of image extents and domain intents; ψ_{pq} works vice versa. Starting from a concept lattice of special interest, the coherence mappings yield a partial conceptual information about the image lattice that depend on the concepts of the domain lattice. The richer the images of the coherence mappings, the more related the corresponding concept lattices are. Clearly, the mappings λ_{pq} and ϱ_{pq} are isotone and the mappings φ_{pq} and ψ_{pq} are antitone. Concatenations of coherence mappings yield the following inequalities:

$$\lambda_{pq}\lambda_{qp}\lambda_{pq}(A, B) \geq \lambda_{pq}(A, B),$$
$$\varrho_{pq}\varrho_{qp}\varrho_{pq}(A, B) \leq \varrho_{pq}(A, B),$$
$$\varphi_{pq}\psi_{qp}\varphi_{pq}(A, B) \geq \varphi_{pq}(A, B),$$
$$\psi_{pq}\varphi_{qp}\psi_{pq}(A, B) \leq \psi_{pq}(A, B).$$

If the considered context sets coincide completely, we obtain even simpler inequalities:

$$\lambda_{qp}\lambda_{pq}(A, B) \geq (A, B) \quad \text{if } S_i = S_k,$$
$$\rho_{qp}\rho_{pq}(A, B) \leq (A, B) \quad \text{if } S_j = S_l,$$
$$\psi_{qp}\varphi_{pq}(A, B) \leq (A, B) \quad \text{if } S_j = S_k,$$
$$\varphi_{qp}\psi_{pq}(A, B) \geq (A, B) \quad \text{if } S_i = S_l.$$

The listed inequalities are immediate consequences of the slightly stronger conditions stated in the following proposition as valid in all multicontexts:

Proposition 1. *Let (S_I, R_P) be a multicontext and let $(A, B) \in \mathfrak{B}(\mathbb{K}_p)$ with $p \in P$. Then the following equations are valid:*

$$\lambda_{pq}(A, B) = \lambda_{pq}((A, B) \wedge \lambda_{qp}\lambda_{pq}(A, B)),$$
$$\varrho_{pq}(A, B) = \varrho_{pq}((A, B) \vee \varrho_{qp}\varrho_{pq}(A, B)),$$
$$\varphi_{pq}(A, B) = \varphi_{pq}((A, B) \vee \psi_{qp}\varphi_{pq}(A, B)),$$
$$\psi_{pq}(A, B) = \psi_{pq}((A, B) \wedge \varphi_{qp}\psi_{pq}(A, B)).$$

Furthermore, $\lambda_{pq}((A \cap S_k)^{pp}, (A \cap S_k)^p) = \lambda_{pq}(A, B)$ and

$$((A \cap S_k)^{pp}, (A \cap S_k)^p) \leq \lambda_{qp}\lambda_{pq}((A \cap S_k)^{pp}, (A \cap S_k)^p).$$

Proof: The first equality may be verified as follows: The isotony of λ_{pq} yields the inequality $\lambda_{pq}(A, B) \geq \lambda_{pq}((A, B) \wedge \lambda_{qp}\lambda_{pq}(A, B))$. The dual inequality becomes clear by $(A \cap S_k)^{qq} = (A \cap ((A \cap S_k) \cap S_i) \cap S_k)^{qq} \subseteq (A \cap ((A \cap S_k)^{qq} \cap S_i)^{pp} \cap S_k)^{qq}$. The other inequalities follow analogously.

Since $\lambda_{pq}(A, B)$ is defined as the formal concept $((A \cap S_k)^{qq}, (A \cap S_k)^q)$ and $(A \cap S_k)^{pp} \cap S_k = A \cap S_k$, it follows that $\lambda_{pq}((A \cap S_k)^{pp}, (A \cap S_k)^p) = \lambda_{pq}(A, B)$. Furthermore, $(A \cap S_k)^{pp} = ((A \cap S_k) \cap S_i)^{pp} \subseteq ((A \cap S_k)^{qq} \cap S_i)^{pp}$ yields $((A \cap S_k)^{pp}, (A \cap S_k)^p) \leq \lambda_{qp}\lambda_{pq}((A \cap S_k)^{pp}, (A \cap S_k)^p)$.

For $\sigma p = (i, j)$, $\mathbb{K}_{\overline{p}} := (S_j, S_i, R_p^{-1})$ is the dual context of $\mathbb{K}_p$ and the map $(A, B) \mapsto (B, A)$ is an antiisomorphism from $\mathfrak{B}(\mathbb{K}_p)$ onto $\mathfrak{B}(\mathbb{K}_{\overline{p}})$. Using this duality, all cohehrence mappings can be derived from one type of coherence mappings:

Proposition 2. *The mapping $\varphi_{p\overline{p}}$ $(= \psi_{p\overline{p}})$ is an antiisomorphism from $\mathfrak{B}(\mathbb{K}_p)$ onto $\mathfrak{B}(\mathbb{K}_{\overline{p}})$ and $\varrho_{pq} = \varphi_{\overline{q}q}\lambda_{\overline{p}q}\varphi_{p\overline{p}}$, $\varphi_{pq} = \lambda_{\overline{p}q}\varphi_{p\overline{p}}$, $\psi_{pq} = \varphi_{\overline{q}q}\lambda_{p\overline{q}}$.*

3   An Example

An interesting example of a multicontext is given by the data tables in the "IT-Grundschutzhandbuch 1996" [IT96] issued by the "Bundesamt für Sicherheit in der Informationstechnik", an institution of the Federal Republic of Germany. This example is a contextual section from the area of threats and countermeasures. Threats and countermeasures are related if the measure is suitable for averting the threat, for instance, a fire-extinguisher and a fire form a pair of this relation. The multicontext of the "IT-Grundschutzhandbuch" portrays these measures and threats that affect certain technical appliances. Since the same threats threaten various appliances, there is a great deal of conformity within the multicontext.

We take just two data tables from the handbook to illustrate the coherence mappings. The two derived formal contexts $\mathbb{K}_1$ and $\mathbb{K}_2$ take into account measures and threats which affect data medium archives (Figure 1) and rooms for technical infrastructure (Figure 2). The concept lattices are represented by line diagrams. In this example the mappings λ_{12}, ϱ_{12}, λ_{21} and ϱ_{21} have non trivial images. The effect of the coherence mappings are represented in the Figures 3

	fire	water	changing humidity/temperature	dust/dirt	missing rules	room access	building access	theft	vandalism
burning instruction	×								
fire–extinguisher	×								
considering fire while alloting rooms	×								
security doors	×					×	×	×	×
closed doors/windows		×		×		×	×	×	×
supervising system	×	×				×	×	×	×
locked doors					×	×	×	×	×
avoiding of water pipes		×							
airconditioning			×						
key management					×	×	×	×	×
company of visitors					×	×	×	×	×
access control					×	×	×	×	×
supervision walks	×	×			×	×	×	×	×
smoking prohibition	×				×				

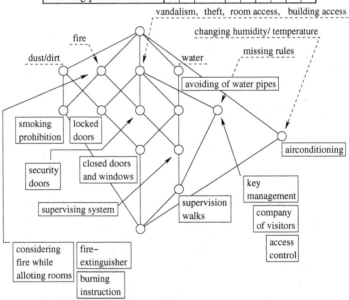

Fig. 1. data medium archives

	fire	water	changing humidity/temperature	missing rules	room access	manipulation/destruction	building access	theft	vandalism
partition of electric circuit	×								
burning instruction	×								
fire–extinguisher	×								
considering fire while alloting rooms	×								
security doors	×				×	×	×	×	×
closed doors/windows		×			×	×	×	×	×
supervising system	×	×				×	×	×	×
locked doors					×	×	×	×	×
avoiding of water pipes		×							
fuses	×								
emergency switch off	×	×							
airconditioning			×						
remote control	×	×					×		
key management				×	×	×	×	×	×
company of visitors	×			×	×	×	×	×	×
access control	×			×	×	×	×	×	×
supervision walks	×				×	×	×	×	×
smoking prohibition	×								

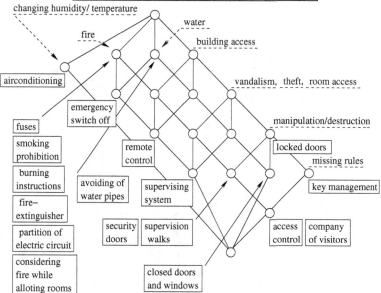

Fig. 2. room for technical infrastructure

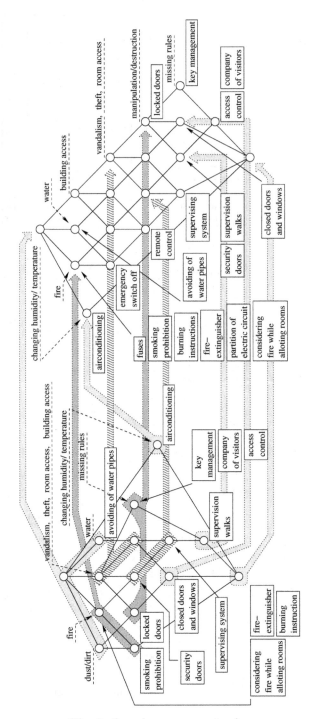

Fig. 3. the coherence mapping λ_{12}

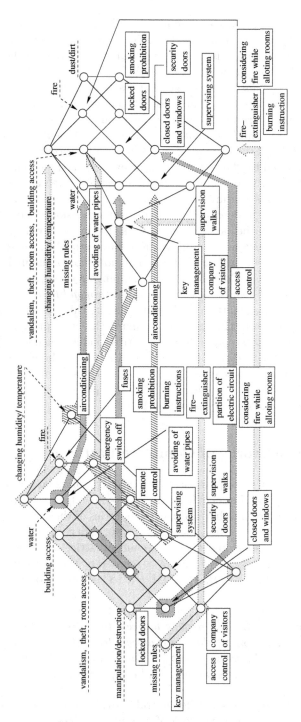

Fig. 4. the coherence mapping λ_{21}

and 4. For instance, in Figure 3 all concepts which have the same image under λ_{12} are surrounded by the same area. The wide arrows show the corresponding image. The following figure is done in the same way. What coherence is elucidated by the coherence mappings λ_{12} and λ_{21}?

For λ_{12} the measures *"closed doors and windows"* and *"airconditioning"* that avert by themselves a certain combination of dangers concerning data medium archives, play the same part in a room for technical infrastructure. Whereas *"supervision walks"* and *"smoking prohibition"* lose this particular importance. For *"smoking prohibition"* the reason is immediately recognizable. In $\mathbb{K}_2$ it is among others only related to *"fire"*, while in $\mathbb{K}_1$ it has a special meaning because of the threat *"dust/dirt"*. Concerning *"supervision walks"*, we can see that in a room for technical infrastructure, this countermeasure is no longer a means against water. Therefore, it does not form its own extension. The fact that neighbours in $\mathfrak{B}(\mathbb{K}_1)$ have the same image in $\mathfrak{B}(\mathbb{K}_2)$ is not surprising. For two concepts with disjointed extensions this is more interesting. In our example, the threat *"manipulation/destruction"* of $\mathbb{K}_2$ that does not occur in $\mathbb{K}_1$ is the reason for this phenomenon. Considering the mappings λ_{21}, we see, alike λ_{12}, that the measures *"closed doors and windows"* and *"airconditioning"* maintain their significance in the image lattice, and again some neighbours have the same image under λ_{21}. More interesting is that λ_{21} maps the concept generated by the object *"avoiding of water pipes"* on the correspondig concept in $\mathfrak{B}(\mathbb{K}_1)$. Secondly, there are two bigger domain areas. The reason for the chain of three concepts are objects that do not emerge in $\mathfrak{B}(\mathbb{K}_1)$. For the domain area of six concepts, we perceive that the extension of the concept, generated by *"access control"* and *"security doors"*, is a coatom in $\mathfrak{B}(\mathbb{K}_1)$.

4   Coherence Networks of Complete Lattices

The coherence networks of concept lattices consist of complete lattices and connecting coherence mappings. For clarifying the order-theoretic structure of those networks, we have to determine characterizing properties of their structural constituents. As preparation for defining coherence networks of complete lattices, we first abstract the dual relationship between the concept lattices of the formal contexts $\mathbb{K}_p := (S_i, S_j, R_p)$ and $\mathbb{K}_{\overline{p}} := (S_j, S_i, R_p^{-1})$, caused by the antiisomorphisms $\delta_{p\overline{p}} : (A, B) \mapsto (B, A)$ and $\delta_{\overline{p}p} : (B, A) \mapsto (A, B)$, to the purely order-theoretic level:

Definition 3. *In a family L_P of complete lattices, the lattices L_p and $L_{\overline{p}}$ with $p, \overline{p} \in P$ are* dual companions *if there are designated antiisomorphisms $\delta_{p\overline{p}} : L_p \to L_{\overline{p}}$ and $\delta_{\overline{p}p} : L_{\overline{p}} \to L_p$ with $\delta_{\overline{p}p} = \delta_{p\overline{p}}^{-1}$. The family L_P is called* dyadic *if each of its lattices has one (and only one) dual companion in L_P.*

Now, we are ready to introduce the central abstract notion of this paper, namely the notion of a coherence network of complete lattices:

Definition 4. *A* coherence network of complete lattices *is defined as a family* $L_P := (L_p)_{p \in P}$ *of complete lattices together with isotone mappings* $\widehat{\lambda}_{pq} : L_p \to L_q$ *and* $\widehat{\varrho}_{pq} : L_p \to L_q$ *satisfying*

$$\widehat{\lambda}_{pq} x = \widehat{\lambda}_{pq}(x \wedge \widehat{\lambda}_{qp}\widehat{\lambda}_{pq} x) \ and$$
$$\widehat{\varrho}_{pq} x = \widehat{\varrho}_{pq}(x \vee \widehat{\varrho}_{qp}\widehat{\varrho}_{pq} x) \ for \ all \ \ x \in L_p,$$

and antitone mappings $\widehat{\varphi}_{pq} : L_p \to L_q$ *and* $\widehat{\psi}_{pq} : L_p \to L_q$ *satisfying*

$$\widehat{\varphi}_{pq} x = \widehat{\varphi}_{pq}(x \vee \widehat{\psi}_{qp}\widehat{\varphi}_{pq} x) \ and$$
$$\widehat{\psi}_{pq} x = \widehat{\psi}_{pq}(x \wedge \widehat{\varphi}_{qp}\widehat{\psi}_{pq} x) \ for \ all \ x \in L_p.$$

Furthermore, $\widehat{\lambda}_{pp}$ *and* $\widehat{\varrho}_{pp}$ *are the identity on* L_p *and, for each pair* $(L_p, L_{\overline{p}})$ *of dual companions in* L_P, $\widehat{\varphi}_{p\overline{p}}$ $(= \widehat{\psi}_{p\overline{p}})$ *and* $\widehat{\varphi}_{\overline{p}p}$ $(= \widehat{\psi}_{\overline{p}p})$ *are the corresponding designated antiisomorphism. For such a coherence network we use the denotations* $\hat{m}_{pq} := (\widehat{\lambda}_{pq}, \widehat{\varrho}_{pq}, \widehat{\varphi}_{pq}, \widehat{\psi}_{pq})$, $\hat{m}_{P^2} := ((\widehat{\lambda}_{pq}, \widehat{\varrho}_{pq}, \widehat{\varphi}_{pq}, \widehat{\psi}_{pq}) | (p, q) \in P^2)$, *and* $\mathfrak{N}_P := (L_P, \hat{m}_{P^2})$. *A coherence network of complete lattices is said to be* dyadic *if its family* $L_P := (L_p \mid p \in P)$ *of complete lattices is dyadic.*

We will see that a coherence network $\mathfrak{N}(S_I, R_P) := ((\underline{\mathfrak{B}}(\mathbb{K}_p) \mid p \in P), m_{P^2})$ of concept lattices with $m_{P^2} := ((\lambda_{pq}, \varrho_{pq}, \varphi_{pq}, \psi_{pq}) \mid (p, q) \in P^2)$ is a coherence network of complete lattices as defined in Definition 4. For finding out how well coherence networks of concept lattices can be order-theoretically characterized, we need an adequate notion of *isomorphism* for coherence networks of complete lattices as introduced by the next definition:

Definition 5. *Two coherence networks of complete lattices* $\mathfrak{N}_P^1 := (L_P^1, \hat{m}_{P^2}^1)$ *and* $\mathfrak{N}_P^2 := (L_P^2, \hat{m}_{P^2}^2)$ *are said to be* isomorphic *if there is a family of isomorphisms* $\iota_P := (\iota_p)_{p \in P}$ (*called a* network isomorphism) *so that, for all* $(p, q) \in P^2$, *the following diagram schema commutes:*

$$
\begin{array}{ccc}
& \hat{m}_{pq}^1 & \\
L_p^1 & \longrightarrow & L_q^1 \\
\iota_p \downarrow & & \downarrow \iota_q \\
L_p^2 & \longrightarrow & L_q^2 \\
& \hat{m}_{pq}^2 &
\end{array}
$$

i.e. $\iota_q\widehat{\lambda}_{pq}^1 = \widehat{\lambda}_{pq}^2\iota_p, \quad \iota_q\widehat{\varrho}_{pq}^1 = \widehat{\varrho}_{pq}^2\iota_p, \quad \iota_q\widehat{\varphi}_{pq}^1 = \widehat{\varphi}_{pq}^2\iota_p, \quad \iota_q\widehat{\psi}_{pq}^1 = \widehat{\psi}_{pq}^2\iota_p.$

Proposition 3. *Let* $(L_p, L_{\overline{p}})$ *and* $(L_q, L_{\overline{q}})$ *be pairs of complete lattices which are pairs of dual companions with the designated antiautomorphisms* $\delta_{p\overline{p}} : L_p \to L_{\overline{p}}$, $\delta_{\overline{p}p} : L_{\overline{p}} \to L_p$, *and* $\delta_{q\overline{q}} : L_q \to L_{\overline{q}}$, $\delta_{\overline{q}q} : L_{\overline{q}} \to L_q$, *respectively; furthermore, let* $\widehat{\lambda}_{\overline{p}\overline{q}} : L_{\overline{p}} \to L_{\overline{q}}$, $\widehat{\lambda}_{\overline{p}q} : L_{\overline{p}} \to L_q$, *and* $\widehat{\lambda}_{p\overline{q}} : L_p \to L_{\overline{q}}$ *be isotone mappings satisfying*

$$\widehat{\lambda}_{\overline{p}\overline{q}} x = \widehat{\lambda}_{\overline{p}\overline{q}}(x \wedge \widehat{\lambda}_{\overline{q}\overline{p}}\widehat{\lambda}_{\overline{p}\overline{q}} x), \quad \widehat{\lambda}_{\overline{p}q} x = \widehat{\lambda}_{\overline{p}q}(x \wedge \widehat{\lambda}_{q\overline{p}}\widehat{\lambda}_{\overline{p}q} x), \quad \widehat{\lambda}_{p\overline{q}} x = \widehat{\lambda}_{p\overline{q}}(x \wedge \widehat{\lambda}_{\overline{q}p}\widehat{\lambda}_{p\overline{q}} x).$$

Then, if we define $\hat{\varrho}_{pq} := \delta_{\overline{q}q}\hat{\lambda}_{\overline{pq}}\delta_{p\overline{p}}$, $\hat{\varrho}_{qp} := \delta_{\overline{p}p}\hat{\lambda}_{\overline{qp}}\delta_{q\overline{q}}$, $\hat{\varphi}_{pq} = \hat{\lambda}_{\overline{pq}}\delta_{p\overline{p}}$, $\hat{\varphi}_{qp} = \hat{\lambda}_{\overline{qp}}\delta_{q\overline{q}}$, $\hat{\psi}_{pq} = \delta_{\overline{q}q}\hat{\lambda}_{p\overline{q}}$, $\hat{\psi}_{qp} = \delta_{\overline{p}p}\hat{\lambda}_{q\overline{p}}$, we obtain an isotone mapping $\hat{\varrho}_{pq} : L_p \to L_q$ satisfying

$$\hat{\varrho}_{pq}x = \hat{\varrho}_{pq}(x \vee \hat{\varrho}_{qp}\hat{\varrho}_{pq}x)$$

and antitone mappings $\hat{\varphi}_{pq} : L_p \to L_q$ and $\hat{\psi}_{pq} : L_p \to L_q$ satisfying

$$\hat{\varphi}_{pq}x = \hat{\varphi}_{pq}(x \vee \hat{\psi}_{qp}\hat{\varphi}_{pq}x) \quad \text{and} \quad \hat{\psi}_{pq}x = \hat{\psi}_{pq}(x \wedge \hat{\varphi}_{qp}\hat{\psi}_{pq}x).$$

Proof: The asserted equations can be proved as follows: $\hat{\varrho}_{pq}(x \vee \hat{\varrho}_{qp}\hat{\varrho}_{pq}x) =$
$\delta_{\overline{q}q}\hat{\lambda}_{\overline{pq}}\delta_{p\overline{p}}(x \vee \delta_{\overline{p}p}\hat{\lambda}_{\overline{qp}}\delta_{q\overline{q}}\delta_{\overline{q}q}\hat{\lambda}_{\overline{pq}}\delta_{p\overline{p}}x) = \delta_{\overline{q}q}\hat{\lambda}_{\overline{pq}}\delta_{p\overline{p}}(\delta_{\overline{p}p}\delta_{p\overline{p}}x \vee \delta_{\overline{p}p}\hat{\lambda}_{\overline{qp}}\hat{\lambda}_{\overline{pq}}\delta_{p\overline{p}}x) =$
$\delta_{\overline{q}q}\hat{\lambda}_{\overline{pq}}\delta_{p\overline{p}}\delta_{\overline{p}p}(\delta_{p\overline{p}}x \wedge \hat{\lambda}_{\overline{qp}}\hat{\lambda}_{\overline{pq}}\delta_{p\overline{p}}x) = \delta_{\overline{q}q}\hat{\lambda}_{\overline{pq}}\delta_{p\overline{p}}x = \hat{\varrho}_{pq}x$,
$\hat{\varphi}_{pq}(x \vee \hat{\psi}_{qp}\hat{\varphi}_{pq}x) = \hat{\lambda}_{\overline{pq}}\delta_{p\overline{p}}(x \vee \delta_{\overline{p}p}\hat{\lambda}_{q\overline{p}}\hat{\lambda}_{\overline{pq}}\delta_{p\overline{p}}x) = \hat{\lambda}_{\overline{pq}}\delta_{p\overline{p}}(\delta_{\overline{p}p}\delta_{p\overline{p}}x \vee \delta_{\overline{p}p}\hat{\lambda}_{q\overline{p}}\hat{\lambda}_{\overline{pq}}\delta_{p\overline{p}}x)$
$= \hat{\lambda}_{\overline{pq}}\delta_{p\overline{p}}\delta_{\overline{p}p}(\delta_{p\overline{p}}x \wedge \hat{\lambda}_{q\overline{p}}\hat{\lambda}_{\overline{pq}}\delta_{p\overline{p}}x) = \hat{\lambda}_{\overline{pq}}\delta_{p\overline{p}}x = \hat{\varphi}_{pq}x$, and
$\hat{\psi}_{pq}(x \wedge \hat{\varphi}_{qp}\hat{\psi}_{pq}x) = \delta_{\overline{q}q}\hat{\lambda}_{p\overline{q}}(x \wedge \hat{\lambda}_{\overline{qp}}\delta_{q\overline{q}}\delta_{\overline{q}q}\hat{\lambda}_{p\overline{q}}x) = \delta_{\overline{q}q}\hat{\lambda}_{p\overline{q}}x = \hat{\psi}_{pq}x$.

5   The Basic Theorem

In preparation of proving the Basic Theorem on Coherence Networks of Concept Lattices, we offer a construction method for formal contexts which correspond to complete lattices and connecting coherence mappings, by the following proposition (improving Proposition 1 in [Wi96]):

Proposition 4. Let $\hat{\lambda}_{pq} : L_p \to L_q$ and $\hat{\lambda}_{qp} : L_q \to L_p$ be isotone mappings between complete lattices which satisfy

$$\hat{\lambda}_{pq}x = \hat{\lambda}_{pq}(x \wedge \hat{\lambda}_{qp}\hat{\lambda}_{pq}x) \quad \text{and} \quad \hat{\lambda}_{qp}y = \hat{\lambda}_{qp}(y \wedge \hat{\lambda}_{pq}\hat{\lambda}_{qp}y);$$

furthermore, let $\hat{G}_p := \{(x,y) \in \hat{\lambda}_{pq} \cup \hat{\lambda}_{qp}^{-1} \mid y \leq_q \hat{\lambda}_{pq}\hat{\lambda}_{qp}y\}$ and $\hat{G}_q := \{(x,y) \in \hat{\lambda}_{pq} \cup \hat{\lambda}_{qp}^{-1} \mid x \leq_p \hat{\lambda}_{qp}\hat{\lambda}_{pq}x\}$. Then, for the contexts $\hat{\mathbb{K}}_p := (\hat{G}_p, L_p, \sqsubseteq_p)$ and $\hat{\mathbb{K}}_q := (\hat{G}_q, L_q, \sqsubseteq_q)$ with $(x,y) \sqsubseteq_p a :\Leftrightarrow x \leq_p a$ and $(x,y) \sqsubseteq_q b :\Leftrightarrow y \leq_q b$, there are order isomorphisms $\iota_p : \underline{\mathfrak{B}}(\hat{\mathbb{K}}_p) \to L_p$ and $\iota_q : \underline{\mathfrak{B}}(\hat{\mathbb{K}}_q) \to L_q$ with $\iota_q\hat{\lambda}_{pq} = \hat{\lambda}_{pq}\iota_p$ and $\iota_p\hat{\lambda}_{qp} = \hat{\lambda}_{qp}\iota_q$. Additionally, we have $\hat{\lambda}_{pq} \bigvee \pi_1 A = \bigvee \pi_2(A \cap \hat{G}_q)$ for all $(A,B) \in \underline{\mathfrak{B}}(\hat{\mathbb{K}}_p)$.

Proof: In $\hat{\mathbb{K}}_p$, an attribute a has the derivation

$$a^p = \{(x,y) \in \hat{\lambda}_{pq} \cup \hat{\lambda}_{qp}^{-1} \mid x \leq_p a \text{ and } y \leq_q \hat{\lambda}_{pq}\hat{\lambda}_{qp}y\};$$

furthermore, since $(x,y) \in \hat{\lambda}_{pq}$ implies $y = \hat{\lambda}_{pq}x = \hat{\lambda}_{pq}(x \wedge \hat{\lambda}_{qp}\hat{\lambda}_{pq}x) = \hat{\lambda}_{pq}(x \wedge \hat{\lambda}_{qp}y) \leq_q \hat{\lambda}_{pq}\hat{\lambda}_{qp}y$ and therefore $(a, \hat{\lambda}_{pq}a) \in a^p$, we have $a^{pp} = [a]_p := \{x \in L_p \mid x \geq_p a\}$. Because of $\bigcap_{t \in T}[a_t]_p = [\bigvee_{t \in T} a_t]_p$, all intents of $\hat{\mathbb{K}}_p$ are of the form $[a]_p$ with $a \in L_p$. Thus, an order isomorphism $\iota_p : \underline{\mathfrak{B}}(\hat{\mathbb{K}}_p) \to L_p$ can be defined by

$\iota_p(a^p, [a]_p) := a$ for $a \in L_p$. Analogously, an order isomorphism $\iota_q : \underline{\mathfrak{B}}(\widehat{\mathbb{K}}_q) \to L_q$ can be defined by $\iota_q(b^q, [b]_q) := b$ for $b \in L_q$.

For proving the asserted equalities, we use $(x, y) \sqsubseteq_p a$ implies $y \leq_q \widehat{\lambda}_{pq}a$ for $(x, y) \in \widehat{G}_p$ and $a \in L_p$; this implication can be verified as follows: for $(x, y) \in \widehat{\lambda}_{pq}$, since $\widehat{\lambda}_{pq}$ is isotone, $x \leq_p a$ and $\widehat{\lambda}_{pq}x = y$ imply $y \leq_q \widehat{\lambda}_{pq}a$ and, for $(x, y) \in \widehat{\lambda}_{qp}^{-1}$, using the special assumption of the definition of $\widehat{G}_p$, $x \leq_p a$ and $\widehat{\lambda}_{qp}y = x$ with $y \leq_q \widehat{\lambda}_{pq}\widehat{\lambda}_{qp}y$ yield

$$y = y \wedge \widehat{\lambda}_{pq}\widehat{\lambda}_{qp}y = y \wedge \widehat{\lambda}_{pq}x \leq_q \widehat{\lambda}_{pq}a.$$

Now, for $\underline{a} := a \wedge \widehat{\lambda}_{qp}\widehat{\lambda}_{pq}a$, since $\widehat{\lambda}_{pq}a = \widehat{\lambda}_{pq}\underline{a}$ and $(\underline{a}, \widehat{\lambda}_{pq}\underline{a}) \in (a^p \cap \widehat{G}_q)$ and therefore $(a^p \cap \widehat{G}_q)^q = [\widehat{\lambda}_{pq}a)_q$, we can conclude $\iota_q \lambda_{pq}(a^p, [a]_p) = \iota_q((a^p \cap \widehat{G}_q)^{qq}, (a^p \cap \widehat{G}_q)^q) = \iota_q((\widehat{\lambda}_{pq}a)^q, [\widehat{\lambda}_{pq}a)_q) = \widehat{\lambda}_{pq}a = \widehat{\lambda}_{pq}\iota_p(a^p, [a]_p)$ and so $\iota_q \lambda_{pq} = \widehat{\lambda}_{pq}\iota_p$; analogously, we obtain $\iota_p \lambda_{qp} = \widehat{\lambda}_{qp}\iota_q$.

Finally, for $(A, B) \in \underline{\mathfrak{B}}(\widehat{\mathbb{K}}_p)$ with $A = a^p$, we conclude $\widehat{\lambda}_{pq} \bigvee \pi_1 A = \widehat{\lambda}_{pq}a = \widehat{\lambda}_{pq}\underline{a} \in \pi_2(A \cap \widehat{G}_q)$ and, because of $(x, y) \sqsubseteq_p a \Rightarrow y \leq_q \widehat{\lambda}_{pq}a$, even $\widehat{\lambda}_{pq}\underline{a} = \bigvee \pi_2(A \cap \widehat{G}_q)$; hence $\widehat{\lambda}_{pq} \bigvee \pi_1 A = \bigvee \pi_2(A \cap \widehat{G}_q)$

Now we are able to state and to prove the desired order-theoretic characterization of the coherence networks of concept lattices corresponding to multicontexts:

Basic Theorem of Coherence Networks of Concept Lattices.
For a multicontext (S_I, R_P), the corresponding network $(\underline{\mathfrak{B}}((\mathbb{K}_p) \mid p \in P), m_{P^2})$ is a coherence network of complete lattices. In general, a coherence network $\mathfrak{N}_P := (L_P, \hat{m}_{P^2})$ of complete lattices is isomorphic to the coherence network $\mathfrak{N}(S_I, R_P) := (\underline{\mathfrak{B}}((\mathbb{K}_p) \mid p \in P), m_{P^2})$ of the concept lattices of a multicontext (S_I, R_P) with signature σ if and only if, for every formal context $\mathbb{K}_p$ of (S_I, R_P) with $\sigma p = (i, j)$, there exist mappings $\tilde{\gamma}_p : S_i \to L_p$ and $\tilde{\mu}_p : S_j \to L_p$ such that $\tilde{\gamma}_p S_i$ is supremum-dense in L_p, $\tilde{\mu}_p S_j$ is infimum-dense in L_p, $(s_i, s_j) \in R_p \Leftrightarrow \tilde{\gamma}_p s_i \leq_p \tilde{\mu}_p s_j$ for all $s_i \in S_i$ and $s_j \in S_j$, and, for all $(p, q) \in P^2$ with $\sigma p = (i, j)$, $\sigma q = (k, l)$, and $(A, B) \in \underline{\mathfrak{B}}(\mathbb{K}_p)$, the coherence mappings of $\mathfrak{N}_P$ satisfy

$$\bigvee \tilde{\gamma}_q(A \cap S_k) = \widehat{\lambda}_{pq} \bigvee \tilde{\gamma}_p A, \quad \bigwedge \tilde{\mu}_q(B \cap S_l) = \widehat{\varrho}_{pq} \bigwedge \tilde{\mu}_p B,$$
$$\bigvee \tilde{\gamma}_q(B \cap S_k) = \widehat{\varphi}_{pq} \bigwedge \tilde{\mu}_p B, \quad \bigwedge \tilde{\mu}_q(A \cap S_l) = \widehat{\psi}_{pq} \bigvee \tilde{\gamma}_p A;$$

in particular, a dyadic coherence network $\mathfrak{N}_P := (L_P, \hat{m}_{P^2})$ of complete lattices is isomorhic to the corresponding coherence network $((\underline{\mathfrak{B}}(\widehat{\mathbb{K}}_p) \mid p \in P), m_{P^2})$ of concept lattices with a strict linear order $<$ on P,

$$\overline{G}_p := \dot{\bigcup}_{p<q \in P}\{(x, y, 1) \mid (x, y) \in \widehat{\lambda}_{pq} \cup \widehat{\lambda}_{qp}^{-1} \text{ and } y \leq_q \widehat{\lambda}_{pq}\widehat{\lambda}_{qp}y\}$$
$$\dot{\cup} \ \dot{\bigcup}_{p>r \in P}\{(y, x, 2) \mid (y, x) \in \widehat{\lambda}_{rp} \cup \widehat{\lambda}_{pr}^{-1} \text{ and } x \leq_p \widehat{\lambda}_{rp}\widehat{\lambda}_{pr}x\},$$

$(x, y, 1)\bar{\sqsubseteq}_p a :\Leftrightarrow x \leq_p a$, $(y, x, 2)\bar{\sqsubseteq}_p a :\Leftrightarrow x \leq_p a$, and $\widehat{\mathbb{K}}_p := (\overline{G}_p, L_p, \bar{\sqsubseteq}_p)$; consequently, the coherence networks of concept lattices are up to isomorphism the coherence networks of complete lattices.

Proof: By Proposition 1, the corresponding network $(\underline{\mathfrak{B}}((\mathbb{K}_p) \mid p \in P), m_{P2})$ of a multicontext (S_I, R_P) satisfies the defining equations of coherence networks of complete lattices and, for $\lambda_{pq} : \underline{\mathfrak{B}}(\mathbb{K}_p) \to \underline{\mathfrak{B}}(\mathbb{K}_q)$ and $(A, B) \in \underline{\mathfrak{B}}(\mathbb{K}_p)$, $\lambda_{pq}((A \cap S_k)^{pp}, (A \cap S_k)^p) = \lambda_{pq}(A, B)$ and satifies the inequality

$$((A \cap S_k)^{pp}, (A \cap S_k)^p) \le \widehat{\lambda}_{qp}\widehat{\lambda}_{pq}((A \cap S_k)^{pp}, (A \cap S_k)^p);$$

analogous concepts exist for the other coherence mappings. Furthermore, λ_{pp} and ϱ_{pp} are the identity on $\underline{\mathfrak{B}}(\mathbb{K}_p)$ and, for each pair $(\underline{\mathfrak{B}}(\mathbb{K}_p), \underline{\mathfrak{B}}(\mathbb{K}_{\bar{p}}))$ of dual companions, $\varphi_{p\bar{p}} (= \psi_{p\bar{p}})$ and $\varphi_{\bar{p}p} (= \psi_{\bar{p}p})$ are the corresponding designated antiisomorphism by Proposition 2. Altogether implies that $(\underline{\mathfrak{B}}((\mathbb{K}_p) \mid p \in P), m_{P2})$ is a coherence network of complete lattices.

Now, we show that the coherence networks $\mathfrak{N}(S_I, R_P)$ and $\mathfrak{N}_P$ are isomorphic if we have mappings $\tilde{\gamma}_p$ and $\tilde{\mu}_p$ $(p \in P)$ satisfying the conditions of the theorem. By the Basic Theorem on Concept Lattices, the mappings $\tilde{\gamma}_p$ and $\tilde{\mu}_p$ can be used to define an isomorphism ι_p from $\underline{\mathfrak{B}}(\mathbb{K}_p)$ onto L_p by

$$\iota_p(A, B) := \bigvee_{s_i \in A} \tilde{\gamma}_p s_i \left(= \bigwedge_{s_j \in B} \tilde{\mu}_p s_j\right).$$

We prove that $\iota_q \lambda_{pq}(A, B) = \widehat{\lambda}_{pq}\iota_p(A, B)$ is the consequence of $\bigvee \tilde{\gamma}_q(A \cap S_k) = \widehat{\lambda}_{pq} \bigvee \tilde{\gamma}_p A$ as follows:

$$\iota_q \lambda_{pq}(A, B) = \iota_q((A \cap S_k)^{qq}, (A \cap S_k)^q) = \bigvee \tilde{\gamma}_q (A \cap S_k)^{qq}$$
$$= \bigvee \tilde{\gamma}_q(A \cap S_k) = \widehat{\lambda}_{pq} \bigvee \tilde{\gamma}_p A$$
$$= \widehat{\lambda}_{pq}\iota_p(A, B).$$

For the mappings $\widehat{\varrho}_{pq}$, $\widehat{\varphi}_{pq}$, and $\widehat{\psi}_{pq}$, we similarly obtain

$$\iota_q \varrho_{pq}(A, B) = \iota_q((B \cap S_l)^q, (B \cap S_l)^{qq}) = \bigwedge \tilde{\mu}_q(B \cap S_l)$$
$$= \widehat{\varrho}_{pq} \bigwedge \tilde{\mu}_p B = \widehat{\varrho}_{pq}\iota_p(A, B),$$

$$\iota_q \varphi_{pq}(A, B) = \iota_q((B \cap S_k)^{qq}, (B \cap S_k)^q) = \bigvee \tilde{\gamma}_q(B \cap S_k)$$
$$= \widehat{\varphi}_{pq} \bigwedge \tilde{\mu}_p B = \widehat{\varphi}_{pq}\iota_p(A, B),$$

$$\iota_q \psi_{pq}(A, B) = \iota_q((A \cap S_l)^q, (A \cap S_l)^{qq}) = \bigwedge \tilde{\mu}_q(A \cap S_l)$$
$$= \widehat{\psi}_{pq} \bigvee \tilde{\gamma}_p A = \widehat{\psi}_{pq}\iota_p(A, B).$$

Hence, we can conclude that $\iota_P := (\iota_p)_{p \in P}$ is a network isomorphism.

Conversely, we assume that there is a network isomorphism ι_P from $\mathfrak{N}(S_I, R_P)$ onto $\mathfrak{N}_P$. Then the desired mappings $\tilde{\gamma}_p$ and $\tilde{\mu}_p$ $(p \in P)$ can be defined by $\tilde{\gamma}_p :=$

$\iota_p \circ \gamma_p$ and $\tilde{\mu}_p := \iota_p \circ \mu_p$, where $\gamma_p g := (\{g\}^{pp}, \{g\}^p)$ and $\mu_p m := (\{m\}^p, \{m\}^{pp})$. The mappings $\tilde{\gamma}_p$ and $\tilde{\mu}_p$ obviously satisfy the conditions of the Basic Theorem on Concept Lattices for $\mathfrak{B}(\mathbb{K}_p)$ and L_p. Finally, we obtain

$$\bigvee \tilde{\gamma}_q(A \cap S_k) = \bigvee \iota_q \gamma_q(A \cap S_k) = \iota_q \bigvee \gamma_q(A \cap S_k)$$
$$= \iota_q((A \cap S_k)^{qq}, (A \cap S_k)^q) = \iota_q \lambda_{pq}(A, B)$$
$$= \widehat{\lambda}_{pq}\iota_p(A, B) = \widehat{\lambda}_{pq}\iota_p \bigvee \gamma_p A$$
$$= \widehat{\lambda}_{pq} \bigvee \tilde{\gamma}_p A,$$

$$\bigwedge \tilde{\mu}_q(B \cap S_l) = \bigwedge \iota_q \mu_q(B \cap S_l) = \iota_q \bigwedge \mu_q(B \cap S_l)$$
$$= \iota_q((B \cap S_l)^q, (B \cap S_l)^{qq}) = \iota_q \varrho_{pq}(A, B)$$
$$= \widehat{\varrho}_{pq}\iota_p(A, B) = \widehat{\varrho}_{pq}\iota_p \bigwedge \mu_p B$$
$$= \widehat{\varrho}_{pq} \bigwedge \tilde{\mu}_p B,$$

$$\bigvee \tilde{\gamma}_q(B \cap S_k) = \bigvee \iota_q \gamma_q(B \cap S_k) = \iota_q \bigvee \gamma_q(B \cap S_k)$$
$$= \iota_q((B \cap S_k)^{qq}, (B \cap S_k)^q) = \iota_q \varphi_{pq}(A, B)$$
$$= \widehat{\varphi}_{pq}\iota_p(A, B) = \widehat{\varphi}_{pq}\iota_p \bigwedge \mu_p B$$
$$= \widehat{\varphi}_{pq} \bigwedge \tilde{\mu}_p B,$$

$$\bigwedge \tilde{\mu}_q(A \cap S_l) = \bigwedge \iota_q \mu_q(A \cap S_l) = \iota_q \bigwedge \mu_q(A \cap S_l)$$
$$= \iota_q((A \cap S_l)^q, (A \cap S_l)^{qq}) = \iota_q \psi_{pq}(A, B)$$
$$= \widehat{\psi}_{pq}\iota_p(A, B) = \widehat{\psi}_{pq}\iota_p \bigvee \gamma_p A$$
$$= \widehat{\psi}_{pq} \bigvee \tilde{\gamma}_p A.$$

Thus, all conditions of the theorem are satisfied.

For a given dyadic coherence network $\mathfrak{N}_P := (L_P, \hat{m}_{P^2})$ of complete lattices and the corresponding coherence network $((\mathfrak{B}(\widehat{\mathbb{K}}_p) \mid p \in P), m_{P^2})$ of concept lattices, we define mappings $\tilde{\gamma}_p : \overline{G}_p \to L_p$ and $\tilde{\mu}_p := L_p \to L_p$ by $\tilde{\gamma}_p(x, y, 1) := \tilde{\gamma}_p(y, x, 2) := x$ and $\tilde{\mu}_p(a) := a$. The mappings $\tilde{\gamma}_p$ and $\tilde{\mu}_p$ obviously satisfy the conditions of the Basic Theorem on Concept Lattices for $\mathfrak{B}(\widehat{\mathbb{K}}_p)$ and L_p; therefore, there are isomorphisms ι_p from $\mathfrak{B}(\widehat{\mathbb{K}}_p)$ onto L_p ($p \in P$) given by

$$\iota_p(A, B) := \bigvee_{g \in A} \tilde{\gamma}_p g \ \left(= \bigwedge_{b \in B} \tilde{\mu}_p b\right).$$

Finally, for $(A, B) \in \underline{\mathfrak{B}}(\widehat{\mathbb{K}}_p)$, we conclude with Proposition 4:

$$\widehat{\lambda}_{pq} \bigvee \tilde{\gamma}_p A = \widehat{\lambda}_{pq} \bigvee \pi_1 A = \bigvee \pi_2(A \cap \widehat{G}_q) = \bigvee \tilde{\gamma}_q(A \cap \widehat{G}_q).$$

For the mappings $\widehat{\varphi}_{pq}$, and $\widehat{\psi}_{pq}$, we now obtain by using Proposition 3 and 4:

$$\widehat{\varrho}_{pq} \bigwedge \tilde{\mu}_p B = \delta_{\overline{q}q} \widehat{\lambda}_{\overline{p}q} \delta_{p\overline{p}} \bigwedge \tilde{\mu}_p B = \delta_{\overline{q}q} \widehat{\lambda}_{\overline{p}q} \bigvee \tilde{\gamma}_{\overline{p}} B = \delta_{\overline{q}q} \bigvee \tilde{\gamma}_{\overline{q}}(B \cap \widehat{G}_{\overline{q}}) = \bigwedge \tilde{\mu}_q(B \cap \widehat{G}_q),$$
$$\widehat{\varphi}_{pq} \bigwedge \tilde{\mu}_p B = \widehat{\lambda}_{\overline{p}q} \delta_{p\overline{p}} \bigwedge \tilde{\mu}_p B = \widehat{\lambda}_{\overline{p}q} \bigvee \tilde{\gamma}_{\overline{p}} B = \bigvee \tilde{\gamma}_q(B \cap \widehat{G}_q),$$
$$\widehat{\psi}_{pq} \bigvee \tilde{\gamma}_p A = \delta_{\overline{q}q} \widehat{\lambda}_{p\overline{q}} \bigvee \tilde{\gamma}_p A = \delta_{\overline{q}q} \bigvee \tilde{\gamma}_{\overline{q}}(A \cap \widehat{G}_{\overline{q}}) = \bigwedge \tilde{\mu}_q(A \cap \widehat{G}_q).$$

Thus, with the already proved part of the Basic Theorem, we obtain that the (abstract) coherence network $\mathfrak{N}_P$ is isomorph to the (concrete) coherence network $((\underline{\mathfrak{B}}(\widehat{\mathbb{K}}_p) \mid p \in P), m_{P^2})$. Since an arbitrary coherence network of complete lattices can obviously be extended to a dyadic coherence network, it follows that the coherence networks of concept lattices are up to isomorphism the coherence networks of complete lattices.

6   Further Research

The Basic Theorem on Coherence Networks of Concept Lattice should be a useful foundation for futher research and applications of the conceptual structures and relationships in multicontexts. Some work has already been done and should be enriched by using and integrating the Basic Theorem, namely the analysis of formal methods for *aggregating components of a multicontext* [Wi96], the development of a theory of *many-valued multicontexts* [Ga96], and the *mathematical structure theory* of multicontexts and coherence networks of complete lattices [Dö99]. Mainly, there are two direction which should stimulate further research: concrete applications in *conceptual data analysis and knowledge processing*, and a dyadic development of *Contextual Logic on the modal level* which might even contribute to the relational database theory.

References

[Bi98]   K. Biedermann: *A foundation of the theory of trilattices.* Dissertation, TU Darmstadt. Shaker Verlag, Aachen 1998.

[DW00]   F. Dau, R. Wille: On the modal understanding of triadic contexts. In: R. Decker, W. Gaul (eds.): *Classification and information processing at the turn of the millennium.* Spinger-Verlag, Heidelberg 2000, 83–94.

[Dö99]   S. K. Dörflein: *Coherence networks of concept lattices.* Dissertation, TU Darmstadt. Shaker Verlag, Aachen 1999.

[Ek04]   P. Eklund (ed.): *Concept lattices.* LNAI **2961**, Springer-Verlag, Heidelberg 2004.

[GW99]   B. Ganter, R. Wille: *Formal Concept Analysis: mathematical foundations.* Springer-Verlag, Heidelberg 1999.

[GWW87]   B. Ganter, R. Wille, K. E. Wolf (Hrsg.): *Beiträge zur Begriffsanalyse.* B.I.-Wissenschaftsverlag, Mannheim 1987.

[Ga96]   P. Gast: *Begriffliche Strukturen mehrwertiger Multikontexte.* Diplomarbeit,
           TH Darmstadt 1996.
[IT96]   *IT-Grundschutzhandbuch 1996: Maßnahmenempfehlung für den mittleren
           Schutzbedarf,* Band 3. Bundesamt für Sicherheit in der Informationstechnik,
           Berlin 1996.
[LW95]   F. Lehmann, R. Wille: A triadic approach to Formal Concept Analysis. In:
           G. Ellis, R. Levinson, W. Rich, J. F. Sowa (eds.): *Conceptual structures:
           applications, implementation and theory.* LNAI **954**. Springer-Verlag, Hei-
           delberg 1995, 32–43.
[SW00]   G. Stumme, R. Wille (Hrsg.): *Begriffliche Wissensverarbeitung. Methoden
           und Anwendungen.* Springer-Verlag, Heidelberg 2000.
[Wi82]   R. Wille: Restructuring lattice theory: an approach based on hierarchies of
           concepts. In: I. Rival (ed.): *Ordered sets.* Reidel, Dordrecht-Boston 1982,
           445–470.
[Wi92]   R. Wille: Concept lattices and conceptual knowledge systems. *Computers
           & Mathematics with Applications* **23** (1992), 493–515.
[Wi95]   R. Wille: The basic theorem of Triadic Concept Analysis. *Order* **12** (1995),
           149–158.
[Wi96]   R. Wille: Conceptual structures of multicontexts. In: P. W. Eklund, G. El-
           lis, G. Mann (eds.): *Conceptual structures: knowledge representation as
           interlingua.* LNAI **1115**. Springer-Verlag, Heidelberg 1996, 23–39.
[Wi97]   R. Wille: Conceptual landscapes of knowledge: a pragmatic paradigm for
           knowledge processing. In: G. Mineau, A. Fall (eds.): *Proceedings of the
           International Symposium on knowledge representation, use, and storage
           efficiency.* Simon Fraser University, Vancouver 1997, 2–13; also in: W. Gaul,
           H. Locarek-Junge (eds.): *Classification in the information age.* Springer-
           Verlag, Heidelberg 1999, 344–356.
[Wi00]   R. Wille: Begriffliche Wissensverarbeitung: Theorie und Praxis. *Informatik
           Spektrum* **23** (2000), 357–369; gekürzte Version in: B. Schmitz (Hrsg.):
           Thema Forschung: Information, Wissen, Kompetenz. Heft 2/2000, TU
           Darmstadt, 128–140.
[Wi02]   R. Wille: Begriffliche Wissensverarbeitung in der Wirtschaft. *Information
           - Wissenschaft und Praxis* (Organ der Deutschen Gesellschaft für Informa-
           tionswissenschaft und Informationspraxis e.V.) **53** (2002), 149–160.
[WZ94]   R. Wille, M. Zickwolff (Hrsg.): *Begriffliche Wissensverarbeitung. Grund-
           fragen und Aufgaben.* B.I.-Wissenschaftsverlag, Mannheim 1994.
[WZ00]   R. Wille, M. Zickwolff: Grundlegung einer Triadischen Begriffsanalyse. In:
           G. Stumme, R. Wille (Hrsg.): *Begriffliche Wissensverarbeitung. Methoden
           und Anwendungen.* Springer-Verlag, Heidelberg 2000, 125–150.

Turing Machine Representation in Temporal Concept Analysis

Karl Erich Wolff[1] and Wendsomde Yameogo[2]

[1] Mathematics and Science Faculty,
Darmstadt University of Applied Sciences
Schoefferstr. 3, D-64295 Darmstadt, Germany
karl.erich.wolff@t-online.de
http://www.fbmn.fh-darmstadt.de/home/wolff
[2] Independent Researcher
wendsomde@yameogo.com
http://www.yameogo.com

Abstract. The purpose of this paper is to investigate the connection between the theory of computation and Temporal Concept Analysis, the temporal branch of Formal Concept Analysis.

The main idea is to represent for each possible input of a given algorithm the uniquely determined sequence of computation steps as a life track of an object in some conceptually described state space. For that purpose we introduce for a given Turing machine a Conceptual Time System with Actual Objects and a Time Relation (CTSOT) which yields the state automaton of a Turing machine as well as its configuration automaton. The conceptual role of the instructions of a Turing machine is understood as a set of background implications of the derived context of a Turing CTSOT.

1 Introduction: Computing as a Temporal Activity

Computing is usually understood as a temporal activity, starting with a certain input which is transformed in several steps into the output of the computation. Usually, the common mathematical representations of computations do not make explicit its temporal aspects. For example, while the notion of *time* is employed in nearly all informal descriptions of computations, it is usually not specified in mathematical representations of computations. Similarly the notion of *state* is often used only informally and not as a mathematical term in some specified temporal theory. It is well-known that the notion of *state* is chosen as a primitive notion in the definition of an automaton, but automata theory does not have an explicit time representation [Arb70, Eil74, Mal74]. The notions of *initial state*, *final state* and *transition* emphasize its temporal interpretation as well as the notion of a *successful path* leading from an initial to a final state.

Clearly, the notion of *state* is also used in many other sciences, for example in classical mechanics where the trajectories of particles seem to be quite similar to the paths in automata theory. This similarity has been made explicit in Temporal

B. Ganter and R. Godin (Eds.): ICFCA 2005, LNAI 3403, pp. 360–374, 2005.
© Springer-Verlag Berlin Heidelberg 2005

Concept Analysis by the introduction of the notion of *state* in a Conceptual Time System (CTS) [Wol00a] and the notion of the *life track of an object* in a Conceptual Time System with Actual Objects and a Time Relation (CTSOT) [Wol02b].

It was shown in the Map Reconstruction Theorem in [Wol02b] that each automaton is isomorphic to an automaton within a suitable CTSOT such that the paths of the given automaton are mapped onto the life tracks of objects of the CTSOT.

In the present paper we show that any Turing machine [Tur36, Loe76] can be represented by a CTSOT together with a specified set of *background implications* interpreting the instructions of the given Turing machine. That temporal representation preserves the state automaton of the given Turing machine as well as its tape automaton (Theorem 1 in section 6).

2   Turing Machines

Alan M. Turing [Tur36, Tur36a] and simultaneously E.L. Post [Pos36] investigated "computing machines" using simple mathematical models of a computer. Turing wrote in the first section of his paper [Tur36]:

Quote 1:
We may compare a man in the process of computing a real number to a machine which is only capable of a finite number of conditions q_1, q_2, ..., q_R which will be called "m-configurations". The machine is supplied with a "tape", (the analogue of paper) running through it, and divided into sections (called "squares") each capable of bearing a "symbol". At any moment there is just one square, say the r-th, bearing the symbol $S(r)$ which is "in the machine". We may call this square the "scanned square". The symbol on the scanned square may be called the "scanned symbol". The "scanned symbol" is the only one of which the machine is, so to speak, "directly aware".

These "computing machines" are now well-known under the name "Turing machines". There are many slightly different formal definitions of Turing machines. We assume that the reader is familiar with the main ideas and standard notions.

2.1   Definition of a Turing Machines

We first recall an often used definition of a Turing machine [Loe76, HU79]).

Definition 1. *"Turing Machine"*
A Turing machine is a tuple

$$\mathfrak{T}_0 := (Q, q_0, \Gamma, B, \Sigma, \delta)$$

*where Q is a finite set, called the set of **states**, $q_0 \in Q$ is called the **initial state**, Γ is a finite set, called the set of **tape symbols**, B is an element of Γ,*

*called the **blank**; Σ is a subset of $\Gamma \setminus \{B\}$, called the set of **input symbols**;*
δ *is a partial mapping, called the **next move function***

$$\delta : Q \times \Gamma \longrightarrow Q \times \Gamma \times \{L, O, R\}$$

*where $\{L, O, R\}$ is a set of three elements called **left**, **zero**, and **right**, respectively.*

For the purpose of a mathematically clear temporal representation of Turing machines we now introduce the notion of a "temporal Turing machine".

2.2   Temporal Turing Machines

In the following definitions we introduce a "temporal Turing machine" as a Turing machine together with a "temporal extension" which describes for each element of a given set $\mathfrak{I}$ of "input words" its "sequence of computing steps" in form of a data table as indicated in Table 1 and Table 2. For that purpose the "tape" and its "squares" are described by the group $(\mathbb{Z}, +)$ of integers; for defining explicitly for each input word p the "time t after starting with input p" we use the ordered monoid $(\mathbb{N}_0, +, \leq)$ of nonnegative integers; furthermore we introduce for each "current input" (p, t) the current state $\sigma(p, t)$ of the temporal Turing machine, the current tape content $\tau_{(p,t)}$, and the current head position $h(p, t)$. We also introduce an encoding mapping α and a decoding mapping ω. That is done in the following definition.

Definition 2. *"Temporal Extension of a Turing Machine"*
Let $\mathfrak{T}_0 := (Q, q_0, \Gamma, B, \Sigma, \delta)$ be a Turing machine. The tuple

$$\mathfrak{E} := (\mathfrak{I}, (\mathbb{Z}, +), (\mathbb{N}_0, +, \leq), \alpha, \sigma, \tau, h, \omega)$$

is called a temporal extension of $\mathfrak{T}_0$ if the following conditions $(1) - (8)$ hold:

(1) $\mathfrak{I}$ *is a subset of Σ^*, called the **input set**;*

(2) $(\mathbb{Z}, +)$ *is the group of integers; the elements of $\mathbb{Z}$ are interpreted as **squares** on the **tape**; α is a mapping, called the **encoding***

$$\alpha : \mathfrak{I} \longrightarrow \Gamma^{\mathbb{Z}}, \ p \mapsto \alpha(p);$$

$\alpha(p)$ *is called the **encoded stream** of the **input** p; the elements in $\Gamma^{\mathbb{Z}}$ are called **streams**;*

(3) $(\mathbb{N}_0, +, \leq)$ *is the monoid of non-negative integers; the elements of $\mathbb{N}_0$ are interpreted as **time granules**; the partial mappings σ, τ, h, and ω together with their domains are defined as follows: σ, τ, and h have a common domain $D_{\sigma,\tau,h}$, the domain of ω is denoted by D_ω;*
$$D_{\sigma,\tau,h} := \bigcup \{D_{\sigma,\tau,h}^{(p)} | p \in \mathfrak{I}\};$$
$$D_\omega := \bigcup \{D_\omega^{(p)} | p \in \mathfrak{I}\};$$
for each $p \in \mathfrak{I}$ the sets $D_{\sigma,\tau,h}^{(p)}$ and $D_\omega^{(p)}$ are defined inductively in (8):

(4) σ is a partial mapping, called the **state mapping**

$$\sigma : \mathfrak{I} \times \mathbb{N}_0 \longrightarrow Q, \ (p, t) \mapsto \sigma(p, t);$$

the state $\sigma(p, t)$ is called the **state of the Turing machine at** (p, t) **or at time** t **after starting with input** p;

(5) τ is a partial mapping, called the **tape content**

$$\tau : \mathfrak{I} \times \mathbb{N}_0 \longrightarrow \Gamma^{\mathbb{Z}}, \ (p, t) \mapsto \tau(p, t);$$

the stream $\tau_{(p,t)} := \tau(p, t)$ is called the **tape content at** (p, t);

(6) h is a partial mapping, called the **head position**

$$h : \mathfrak{I} \times \mathbb{N}_0 \longrightarrow \mathbb{Z}, \ (p, t) \mapsto h(p, t);$$

the integer $h(p, t)$ is called the **head position at** (p, t) and $v(p, t) := \tau_{(p,t)}(h(p, t))$ the **head value (or the scanned symbol) at** (p, t);

(7) ω is a partial mapping, called the **decoding**

$$\omega : \Gamma^{\mathbb{Z}} \longrightarrow \Sigma^*, \ s \mapsto \omega(s).$$

(8.0) For any input $p \in \mathfrak{I}$:
  - $D_{\sigma,\tau,h}^{(p,0)} := \{(p, 0)\};$
  - $\sigma(p, 0) := q_0,$
  - $\tau_{(p,0)} := \alpha(p),$
  - $h(p, 0) := 0;$

(8.1) For any input $p \in \mathfrak{I}$ and for any time $t \in \mathbb{N}_0$:

let $(p, t) \in D^- :\Leftrightarrow (p, t) \in D_{\sigma,\tau,h}^{(p,t)}$ and $(\sigma(p, t), v(p, t)) \notin D_\delta;$

let $(p, t) \in D^+ :\Leftrightarrow (p, t) \in D_{\sigma,\tau,h}^{(p,t)}$ and $(\sigma(p, t), v(p, t)) \in D_\delta;$

STOP Condition:

$(8.1.1)$ If $(p, t) \in D^-$, then let

  - $D_\omega^{(p)} := \{\tau_{(p,t)}\}$ and
  - $D_{\sigma,\tau,h}^{(p)} := D_{\sigma,\tau,h}^{(p,t)}$;

Next Step Condition:

$(8.1.2)$ If $(p, t) \in D^+$ and $\delta(\sigma(p, t), v(p, t)) =: (q', a', \Delta)$ then

  - $D_{\sigma,\tau,h}^{(p,t+1)} := D_{\sigma,\tau,h}^{(p,t)} \cup \{(p, t + 1)\}$;

  - $\sigma(p, t + 1) := q',$
  - $\tau_{(p,t+1)}(z) := \tau_{(p,t)}(z) \qquad$ if $z \in \mathbb{Z} \setminus \{h(p, t)\},$
  - $\tau_{(p,t+1)}(z) := a' \qquad\qquad$ if $z = h(p, t);$
  - $h(p, t + 1) := h(p, t) - 1 \qquad$ if $\Delta = \mathrm{L},$
  - $h(p, t + 1) := h(p, t) \qquad\quad$ if $\Delta = \mathrm{O},$
  - $h(p, t + 1) := h(p, t) + 1 \qquad$ if $\Delta = \mathrm{R}.$

Definition 3. *"Temporal Turing Machine"*
A temporal Turing machine is a pair $\mathfrak{T} := (\mathfrak{T}_0, \mathfrak{E})$ where $\mathfrak{T}_0$ is a Turing machine and $\mathfrak{E}$ a temporal extension of $\mathfrak{T}_0$.

Lemma 1. *"Computing Time Lemma"*
Let $\mathfrak{T} := (\mathfrak{T}_0, \mathfrak{E})$ be a temporal Turing machine and $p \in \mathfrak{J}$.

(i) *If $s, t \in \mathbb{N}_0$, $(p, t) \in D^-$, $s < t$ and $(p, s) \in D_{\sigma, \tau, h}$,*
    then $(p, s) \in D^+$ and $(p, s + 1) \in D_{\sigma, \tau, h}$.
(ii) *There exist at most one $t \in \mathbb{N}_0$ such that $(p, t) \in D^-$.*

Proof. (i) Let $p \in \mathfrak{J}$, $s, t \in \mathbb{N}_0$, $(p, t) \in D^-$, $s < t$ and $(p, s) \in D_{\sigma, \tau, h}$, hence $(p, s) \in D_{\sigma, \tau, h}^{(p, s)}$. Assume that $(p, s) \in D^-$, then $D_{\sigma, \tau, h}^{(p)} = D_{\sigma, \tau, h}^{(p, s)}$ by *(8.1.1)*, and since $s < t$ we obtain that $(p, t) \notin D_{\sigma, \tau, h}^{(p)}$ contradicting $(p, t) \in D^-$. Hence $(p, s) \in D^+$ which implies $(p, s + 1) \in D_{\sigma, \tau, h}^{(p, s+1)}$ by *(8.1.2)*.
(ii) Assume that there exist $s, t \in \mathbb{N}_0, s < t$, $(p, s) \in D^-$, $(p, t) \in D^-$, then we get a contradiction by (i).

Definition 4. *"Computing Time and Output"*
Let $\mathfrak{T} := (\mathfrak{T}_0, \mathfrak{E})$ be a temporal Turing machine. We call the set
$\mathfrak{D}_\mathfrak{T} := \{p | p \in \mathfrak{J} \text{ and } \exists_t (p, t) \in D^-\}$ *the set of **accepted words** of $\mathfrak{T}$.*
*For any $p \in \mathfrak{D}_\mathfrak{T}$ there is exactly one $t \in \mathbb{N}_0$ satisfying $(p, t) \in D^-$ by Lemma 1. That non-negative integer is called the **computing time needed by the Turing machine to process** p and is denoted by t_p. By (8.1.1, 8.0) $(p, t_p) \in D_{\sigma, \tau, h}$ and $\tau(p, t_p) \in D_\omega$. Hence $\omega(\tau_{(p, t_p)}) \in \Sigma^*$ and this word is called the **output of** p. The mapping $f_\mathfrak{T} : \mathfrak{D}_\mathfrak{T} \longrightarrow \Sigma^*$ where $f_\mathfrak{T}(p) := \omega(\tau_{(p, t_p)})$ is called the **word function** of $\mathfrak{T}$.*

In the next section we introduce for a temporal Turing machine two automata.

3   Automata of a Temporal Turing Machine

For any temporal Turing machine $\mathfrak{T} := (\mathfrak{T}_0, \mathfrak{E})$ we define two automata, the **state automaton** of the Turing machine $\mathfrak{T}_0$ and the **configuration automaton** of the temporal Turing machine $\mathfrak{T}$. First, we recall the definition of an automaton.

Definition 5. *"Automaton"*
*A tuple $\mathfrak{A} := (S, S^{(i)}, S^{(f)}, A, T)$ is called an **automaton** if S is a set (of **states**), $S^{(i)}$ and $S^{(f)}$ are subsets of S, called the set of **initial** and **final** states, respectively. A is a set, called the set of **actions** (or **labels**) and $T \subseteq S \times A \times S$ is called the set of **transitions** of $\mathfrak{A}$.*

Definition 6. *"State Automaton"*
Let $\mathfrak{T} := (\mathfrak{T}_0, \mathfrak{E})$ be a temporal Turing machine. Then the tuple
$\mathfrak{A}_s := (S_s, S_s^{(i)}, S_s^{(f)}, A_s, T_s)$ *where*

$S_{\boldsymbol{s}} := \sigma D_{\sigma,\tau,h}$

$S_{\boldsymbol{s}}^{(i)} := \{\sigma(p,0)|p \in \mathfrak{I}\}$

$S_{\boldsymbol{s}}^{(f)} := \{\sigma(p,t_p)|p \in \mathfrak{D}_{\mathfrak{I}}\}$

$A_{\boldsymbol{s}} := \{v(p,t)|(p,t) \in D_{\sigma,\tau,h}\}$

$T_{\boldsymbol{s}} := \{(\sigma(p,t), v(p,t), \sigma(p,t+1))|(p,t) \in D^+\}$

*is an automaton, called the **state automaton** of $\mathfrak{T}$.*

The state automaton $\mathfrak{A}_{\boldsymbol{s}}$ represents the transitions within the set Q of states of the Turing machine $\mathfrak{T}_0$. Replacing in the previous definition σ by $\sigma \times \tau \times h$ we obtain the following definition of a "configuration automaton" which represents the transitions between the "configurations" $(\sigma \times \tau \times h)(p,t)$ where $(p,t) \in D_{\sigma,\tau,h}$.

Definition 7. "*Configuration Automaton*"
Let $\mathfrak{T} := (\mathfrak{T}_0, \mathfrak{E})$ be a temporal Turing machine. Then the tuple
$\mathfrak{A}_{\boldsymbol{c}} := (S_{\boldsymbol{c}}, S_{\boldsymbol{c}}^{(i)}, S_{\boldsymbol{c}}^{(f)}, A_{\boldsymbol{c}}, T_{\boldsymbol{c}})$ *where*
$S_{\boldsymbol{c}} := (\sigma \times \tau \times h)D_{\sigma,\tau,h}$
$S_{\boldsymbol{c}}^{(i)} := \{(\sigma \times \tau \times h)(p,0)|p \in \mathfrak{I}\}$
$S_{\boldsymbol{c}}^{(f)} := \{(\sigma \times \tau \times h)(p,t_p)|p \in \mathfrak{D}_{\mathfrak{I}}\}$
$A_{\boldsymbol{c}} := \{v(p,t)|(p,t) \in D_{\sigma,\tau,h}\}$
$T_{\boldsymbol{c}} := \{((\sigma \times \tau \times h)(p,t), v(p,t), (\sigma \times \tau \times h)(p,t+1))|(p,t) \in D^+$
*is an automaton, called the **configuration automaton** of $\mathfrak{T}$.*

The following "6-shaped Path Lemma" describes the behavior of a temporal Turing machine in the configuration automaton. Each input is processed either in a "6-shaped path"(see a) or in a finite (b_1) or infinite (b_2) path without repetitions.

Lemma 2. "*6-shaped Path Lemma*"
Let $\mathfrak{T} := (\mathfrak{T}_0, \mathfrak{E})$ be a temporal Turing machine, $p \in \mathfrak{I}$,
$D^{(p)} := \{(p,t)|(p,t) \in D_{\sigma,\tau,h}\}$.

(a) *If $(\sigma \times \tau \times h)|_{D^{(p)}}$ is not injective, then $p \notin \mathfrak{D}_{\mathfrak{I}}$ and there exist exactly one pair $(r_p, s_p) \in \mathbb{N}_0$ such that $r_p < s_p$ and $(\sigma \times \tau \times h)(p,r_p) = (\sigma \times \tau \times h)(p,s_p)$ and r_p and s_p are minimal with respect to this condition.*
Let $D^{[p]} := \{(p,t)|0 \le t < s_p\}$; then
$(\sigma \times \tau \times h)|_{D^{[p]}}$ is injective and
$(\sigma \times \tau \times h)D^{[p]} = (\sigma \times \tau \times h)D^{(p)}$;
(b_1) *if $(\sigma \times \tau \times h)|_{D^{(p)}}$ is injective and $p \in \mathfrak{D}_{\mathfrak{I}}$, then $D^{(p)} = \{(p,t)|0 \le t \le t_p\}$;*
(b_2) *if $(\sigma \times \tau \times h)|_{D^{(p)}}$ is injective and $p \notin \mathfrak{D}_{\mathfrak{I}}$, then $(\sigma \times \tau \times h)D^{(p)}$ is infinite.*

Proof. (a) Let $p \in \mathfrak{I}$. We assume that $(\sigma \times \tau \times h)|_{D^{(p)}}$ is not injective. Then there exist $r, s \in \mathbb{N}_0, r < s$ such that $(\sigma \times \tau \times h)(p,r) = (\sigma \times \tau \times h)(p,s)$; then there exists exactly one pair $(r_p, s_p) \in \mathbb{N}_0 \times \mathbb{N}_0$ satisfying $(\sigma \times \tau \times h)(p,s_p) = (\sigma \times \tau \times h)(p,r_p)$ and $r_p < s_p$ such that r_p and s_p are minimal with respect to that condition.
Since $v(p,r) = \tau_{(p,r)}(h(p,r)) = \tau_{(p,s)}(h(p,s)) = v(p,s)$
we get $\{(p,r),(p,s)\} \subseteq D^+$ by 8.1 and Lemma 1.

Hence the sequence $((\sigma \times \tau \times h)(p,t)|r_p \leq t \leq s_p)$ is a cycle of length $c_p := s_p - r_p$ in the directed graph of the configuration automaton $\mathfrak{A}_C$. Therefore $(p,t) \in D^+$ for all $t \in \mathbb{N}_0$. Hence $p \notin \mathfrak{D}_{\mathfrak{T}}$ and $(\sigma \times \tau \times h)|_{D^{[p]}}$ is injective and $(\sigma \times \tau \times h)D^{[p]} = (\sigma \times \tau \times h)D^{(p)}$.

(b_1) Let $(\sigma \times \tau \times h)|_{D^{(p)}}$ be injective and $p \in \mathfrak{D}_{\mathfrak{T}}$. By Lemma 1 $D^{(p)} = \{(p,t)|0 \leq t \leq t_p\}$ and $(\sigma \times \tau \times h)D^{(p)}$ is the set of vertices of a finite path in the directed graph of the configuration automaton $\mathfrak{A}_C$.

(b_2) Let $(\sigma \times \tau \times h)|_{D^{(p)}}$ be injective and $p \notin \mathfrak{D}_{\mathfrak{T}}$, then $(\sigma \times \tau \times h)D^{(p)}$ is the set of vertices of an infinite path in the directed graph of the configuration automaton $\mathfrak{A}_C$.

Clearly, we could define several other automata of a temporal Turing machine, for example a "state value automaton" using the mapping $\sigma \times v$, but we will discuss mainly the state and the configuration automaton in this paper. From the following representation of temporal Turing machines in Temporal Concept Analysis we will understand that these automata can be constructed easily from a suitable conceptual time system.

That is prepared in the following section using a small example.

4   Example: A Simple Temporal Turing Machine

We construct a temporal Turing machine which computes for any given word w over the alphabet $\Sigma := \{a, b\}$ its first symbol f(w) [Loe76], p.43. For achieving this, we start with the following simple idea for such a computation:

Start with the first symbol and preserve it, then delete all the following symbols up to the last one.

We distinguish three stages of such a computation for an input word w:
$S_0 :=$ checking the first position of the word w,
$S_1 :=$ checking the second or a later position of the word w,
$S_2 :=$ reaching the end of the word w.

Now, we construct a Turing machine $\mathfrak{T}_0 := (Q, q_0, \Gamma, B, \Sigma, \delta)$ for this computation. According to the three stages we choose $Q := \{q_0, q_1, q_2\}$, take q_0 as the initial state, $\Sigma := \{a, b\}$, $\Gamma := \{a, b, B\}$ and define
$\delta : Q \times \Gamma \longrightarrow Q \times \Gamma \times \{L, O, R\}$ by:
(1) $\delta(q_0, a) := (q_1, a, R)$,
(2) $\delta(q_0, b) := (q_1, b, R)$,
(3) $\delta(q_0, B) := (q_0, B, O)$,
(4) $\delta(q_1, a) := (q_1, B, R)$,
(5) $\delta(q_1, b) := (q_1, B, R)$,
(6) $\delta(q_1, B) := (q_2, B, O)$.

To visualize a temporal extension of this Turing machine $\mathfrak{T}_0$ we represent the computation of the word ab in Table 1:

Table 1. A temporal representation of a computation

input	time	tape	head position	TM-state	head value	change	output
ab	0	$...Ba\underline{b}B...$	0	q_0	a	(1)	
ab	1	$...Ba\underline{b}B...$	1	q_1	b	(5)	
ab	2	$...BaB\underline{B}...$	2	q_1	B	(6)	
ab	3	$...BaB\underline{B}...$	2	q_2	B	STOP	a

The input word ab is encoded at time granule 0 onto the tape as a stream
$...BabB...$ where the first symbol is at square 0, the following symbols are placed
on the following squares, and the other squares are filled with blanks which is
indicated by dots in Figure 1. For time granule 0 the head position is at square
0 and the state of the Turing machine is chosen to be q_0. Then the partial
mapping δ tells us the changes depending on the current state and the current
head value. In this case, instruction (1) in the definition of δ tells us, that the
Turing machine changes from state q_0 to q_1, and the head position from 0 to
1, since the head moves to the right. That determines the next row of Table 1
uniquely, if we repeat the input and increase the time value by 1. At time granule
3 the Turing machine is in the state q_2 and stops since none of the instructions of
δ is applicable. From the current tape stream the result is obtained by applying
the map ω which is chosen to "omit all blanks". In the following we restrict
ourselves to the set $\mathfrak{I}_{\leq 2}$ of all words of $\{a, b\}^*$ of length smaller or equal 2. The
corresponding temporal extension $\mathfrak{E}_{\leq 2} := (\mathfrak{I}_{\leq 2}, (\mathbb{Z}, +), (\mathbb{N}_0, +, \leq), \alpha, \sigma, \tau, h, \omega)$
will be used as our leading example.

The tabular representation of the computation steps for each input word
p is the basis for the construction of Conceptual Time Systems with actual
Objects and a Time Relation (CTSOT) for a given temporal Turing machine.
This construction is explained in the next sections.

5   Temporal Concept Analysis

We assume that the reader is familiar with the basic notions in Formal Concept
Analysis [GaWi99]. In the following we also need the standard notions in Temporal Concept Analysis [Wol02a, Wol02b]. The most important ones are recalled
in the next subsection.

5.1   Conceptual Time Systems with Actual Objects and a Time Relation

The idea of a simple temporal data table which records for each time granule g
its "temporal meaning and the events happening at that time granule" in the
row of g is mathematically described in the following definition.

Definition 8. *"Conceptual Time System"*
Let $\boldsymbol{T} := ((G, M_T, W_T, I_T), (\boldsymbol{S}_m | m \in M_T))$ and $\boldsymbol{C} := ((G, M_C, W_C, I_C), (\boldsymbol{S}_m | m \in M_C))$ be scaled many-valued contexts on the same object set G. Then the pair

(T, C) is called a **conceptual time system (CTS)** on the set G of **time granules**. T is called the **time part** and C the **event part** of (T, C). The set of object concepts of the derived context $K_T | K_C$ of (T, C) is called the **situation space**, the set of object concepts of K_C is called the **state space of** (T, C). The elements of the situation space are called **situations**, those of the state space are called **states**.

To describe persons or particles or other objects which are commonly understood to be at each time granule in exactly one state we use the following definition.

Definition 9. "CTSOT"
"conceptual time systems with actual objects and a time relation"
Let P be a set (of **persons**, or **objects**) and G a set (of **time granules**) and $\Pi \subseteq P \times G$. Let (T, C) be a conceptual time system on Π, and $\hat{R} \subseteq \Pi \times \Pi$. Then the tuple $(P, G, \Pi, T, C, \hat{R})$ is called a **conceptual time system** (on $\Pi \subseteq P \times G$) **with actual objects and a time relation**, in short a **CTSOT**. For each object $p \in P$ the set $p^\Pi := \{g \in G | (p, g) \in \Pi\}$ is called the **time of p in Π**. Then the set $\boldsymbol{R}_p := \{(g, h) | ((p, g), (p, h)) \in R\}$ is called the set of **R-transitions of p**.

In this paper we will take the input words of a temporal Turing machine as the objects of a suitable CTSOT. Then the "computation way" of an input can be represented as a "life track of an object" in the sense of the following definition.

Definition 10. "life track of an object"
Let $(P, G, \Pi, T, C, \hat{R})$ be a CTSOT, and $p \in P$. Then for any mapping $f : \{p\} \times p^\Pi \to X$ (into some set X) the set $f = \{((p, g), f(p, g)) | g \in p^\Pi\}$ is called the **f-life track of p in X**.

The two most useful examples for such mappings are the object concept mappings γ and γ_C of the derived contexts $K_T | K_C$ and K_C of the conceptual time system $(T, C, \hat{R})$ on Π, each of them restricted to the set $\{p\} \times p^\Pi$ of actual objects. They are called the **life track of p in the situation space** and the **life track of p in the state space** respectively.

In the following section we represent temporal Turing machines by CTSOTs.

6   CTSOT Representations of Temporal Turing Machines

We show that any temporal Turing machine can be represented by a CTSOT. For the purpose of a flexible representation we even construct a class of "Turing CTSOTs" of the given temporal Turing machine such that each Turing CTSOT gives a specific insight into the behavior of the given Turing machine. For example, we construct Turing CTSOTs which yield automata isomorphic to the configuration automaton of the given temporal Turing machine.

For that purpose we have to choose the scales such that they preserve "all the information" of the values of their many-valued attributes. The most simple, but often too rigid choice is a nominal scale, but for the purpose of separating different values by different object concepts it is sufficient to choose "object clarified" scales which are introduced in the following.

6.1   Object Clarified Scales

First we recall the well-known definition that a formal context (G,M,I) is called **object clarified** if $\forall_{g,h \in G}$ $(g^{\uparrow} = h^{\uparrow} \Rightarrow g = h)$. In the following we use the notion $g^{I} := g^{\uparrow}$ to distinguish several formal contexts.

Lemma 3.
Let $((G, M, W, I), (\boldsymbol{S_m}|m \in M))$ be a scaled many-valued context where $\boldsymbol{S_m} = (G_m, N_m, I_m)$. Let $\mathbb{K} = (G, \{(m, n)|m \in M, n \in N_m\}, J)$ be its derived context where $g\ J\ (m, n) :\Leftrightarrow m(g)I_m n$. If $\boldsymbol{S_m}$ is object clarified for all $m \in M$, then

$$\forall_{g_1,g_2 \in G}(\ (\forall_{m \in M}\ m(g_1) = m(g_2)) \Leftrightarrow \gamma(g_1) = \gamma(g_2)\ ).$$

Proof. The implication "$\Rightarrow$" is trivial by definition of the derived context. To prove the converse, let $\gamma(g_1) = \gamma(g_2)$, hence
$\bigcup\{g_1^{J_m}|m \in M\} = g_1^{J} = g_2^{J} = \bigcup\{g_2^{J_m}|m \in M\}$. Hence $g_1^{J_m} = g_2^{J_m}$ for each $m \in M$ since $g_1^{J_m} = \{(m, n)|m(g_1)I_m n\}$.
Therefore $m(g_1)^{I_m} = m(g_2)^{I_m}$, which yields $m(g_1) = m(g_2)$ since $\boldsymbol{S_m}$ is object clarified.

6.2   The Definition of a Turing CTSOT

Let $\mathfrak{T}$ be a temporal Turing machine. To have a flexible notation for the intended construction of the class of "Turing CTSOTs" of $\mathfrak{T}$ we introduce a "set of relevant mappings" of $\mathfrak{T}$, defined by $M_{\mathfrak{T}} := \{time, \alpha', \sigma, \tau, h, \Omega, v, \sigma', v', \chi\}$ where
$time : D_{\sigma,\tau,h} \longrightarrow \mathbb{N}_0$, $time(p, t) := t$;
$\alpha' : D_{\sigma,\tau,h} \longrightarrow \Sigma^*$ where $\alpha'(p, t) := p$;
σ, τ, h, v are the previously defined mappings, and $\Omega, \sigma', v', \chi$ are partial mappings, described as mappings from $D_{\sigma,\tau,h}$ using the sign "/" as a "missing value":
$\Omega : D_{\sigma,\tau,h} \longrightarrow \Sigma^* \cup \{/\}$, $\Omega(p, t_p) := \omega(\tau(p, t_p))$ if $p \in D_{\mathfrak{T}}$, otherwise $\Omega(p, t) := /$,
$\sigma' : D_{\sigma,\tau,h} \longrightarrow \sigma D^+$, $\sigma'(p, t) := \sigma(p, t + 1)$, if $(p, t + 1) \in D^+$, otherwise $\sigma'(p, t) := /$,
$v' : D_{\sigma,\tau,h} \longrightarrow vD^+$, $v'(p, t) := v(p, t+1)$ if $(p, t+1) \in D^+$, otherwise $v'(p, t) := /$,
$\chi : D_{\sigma,\tau,h} \longrightarrow \{L, O, R\}$, $\chi(p, t) := \Delta$ (where Δ is the value $\in \{L, O, R\}$ satisfying $\delta(\sigma(p, t), v(p, t)) = (q', a', \Delta)$) if $(p, t + 1) \in D^+$, otherwise $\chi(p, t) := /$.
The mappings σ', v', χ will be used to represent the instructions of δ as implications in the derived context of a suitable CTSOT.

Table 2 shows a typical outline of a CTSOT of a temporal Turing machine where we have selected a subset $M_T := \{time\}$ as the set of attributes of the time part and a subset $M_C := \{\sigma, \tau, h, \Omega\}$ as the set of attributes of the event part.

Table 2. An outline of a CTSOT of a Turing machine

current word	time	$TM-state\ \sigma$	tape τ	head position h	output Ω
$(p,0)$	0	$\sigma(p,0)$	$\tau(p,0)$	$h(p,0)$	/
(p,t)	t	$\sigma(p,t)$	$\tau(p,t)$	$h(p,t)$	/
$(p,t+1)$	$t+1$	$\sigma(p,t+1)$	$\tau(p,t+1)$	$h(p,t+1)$	/
(p,t_p)	t_p	$\sigma(p,t_p)$	$\tau(p,t_p)$	$h(p,t_p)$	$\omega(\tau(p,t_p))$

Definition 11. *"Turing CTSOT"*
Let $\mathfrak{T}_0 := (Q, q_0, \Gamma, B, \Sigma, \delta)$ be a Turing machine and
$\mathfrak{E} := (\mathfrak{J}, (\mathbb{Z}, +), (\mathbb{N}_0, +, \leq), \alpha, \sigma, \tau, h, \omega)$ a temporal extension of $\mathfrak{T}_0$. Let $\mathfrak{T} := (\mathfrak{T}_0, \mathfrak{E})$ and $M_{\mathfrak{T}} := \{time, \alpha', \sigma, \tau, h, \Omega, v, \sigma', v', \chi\}$ its set of relevant mappings.
Let $\mathfrak{S} := (P, G, \Pi, \boldsymbol{T}, \boldsymbol{C}, \hat{R})$ be a CTSOT where $\boldsymbol{T} := ((\Pi, M_T, W_T, I_T), (\boldsymbol{S}_m | m \in M_T))$ and $\boldsymbol{C} := ((\Pi, M_C, W_C, I_C), (\boldsymbol{S}_m | m \in M_C))$.
Then $\mathfrak{S}$ is called a Turing CTSOT of $\mathfrak{T}$ if the following conditions hold:

(a) $P = \mathfrak{J}$, $G = \mathbb{N}_0$, $\Pi = D_{\sigma,\tau,h}$, $\hat{R} = \{((p,t), (p,t+1)) | (p,t) \in D^+\}$;
(b) $M_T \subseteq M_{\mathfrak{T}}$,
$W_T := \bigcup \{m D_{\sigma,\tau,h} | m \in M_T\}$,
$I_T := \{((p,t), m, m(p,t)) | (p,t) \in D_{\sigma,\tau,h}, m \in M_T\}$;
(c) $M_C \subseteq M_{\mathfrak{T}}$,
$W_C := \bigcup \{m D_{\sigma,\tau,h} | m \in M_C\}$,
$I_C := \{((p,t), m, m(p,t)) | (p,t) \in D_{\sigma,\tau,h}, m \in M_C\}$;
(d) All scales $\boldsymbol{S}_m$ ($m \in M_T \cup M_C$) are object clarified.

Definition 12. *"The State Automaton of a Turing CTSOT"*
Let $\mathfrak{T} := (\mathfrak{T}_0, \mathfrak{E})$ be a temporal Turing machine, and $\mathfrak{S}$ be a Turing CTSOT of $\mathfrak{T}$, $\mathbb{K} := (\mathbb{K}_T, \mathbb{K}_C)$ the derived context of $\mathfrak{S}$, γ_C the object concept mapping of $\mathbb{K}_C$.
Let $\mathfrak{A}_{\mathfrak{S}} := (S_{\mathfrak{S}}, S_{\mathfrak{S}}^{(i)}, S_{\mathfrak{S}}^{(f)}, A_{\mathfrak{S}}, T_{\mathfrak{S}})$, where
$S_{\mathfrak{S}} := \gamma_C D_{\sigma,\tau,h}$ is the set of states of $\mathfrak{S}$,
$S_{\mathfrak{S}}^{(i)} := \gamma_C \{(p,0) | p \in \mathfrak{J}\}$,
$S_{\mathfrak{S}}^{(f)} := \gamma_C \{(p,t_p) | p \in D_{\mathfrak{T}}\}$,
$A_{\mathfrak{S}} := v D_{\sigma,\tau,h}$,
$T_{\mathfrak{S}} := \{(\gamma_C(p,t), v(p,t), \gamma_C(p,t+1)) | (p,t) \in D^+\}$;
$\mathfrak{A}_{\mathfrak{S}}$ is an automaton, called the **state automaton of the Turing CTSOT** $\mathfrak{S}$.

The most interesting examples of Turing CTSOTs arise by the following choices for (M_T, M_C):

(1) $(\{time\}, (\{\sigma, \tau, h, \Omega\})$;
(2) $(\{time\}, \{\sigma, \tau, h\})$;
(3) $(\{time\}, \{\sigma\})$;
(4) $(\{time\}, \{\sigma, v, \sigma', v', \chi\})$.

Definition 13. *"Automata Isomorphisms"*

Let $\mathfrak{A}_j := (S_j, S_j^{(i)}, S_j^{(f)}, A_j, T_j)$, $(j \in \{1, 2\})$ be automata. A pair (ι, κ) is called an isomorphism from $\mathfrak{A}_1$ to $\mathfrak{A}_2$ if

(1) $\iota : S_1 \longrightarrow S_2$ is a bijection and
(2) $\kappa : A_1 \longrightarrow A_2$ is a bijection such that
(3) $\iota S_1^{(i)} = S_2^{(i)}, \iota S_1^{(f)} = S_2^{(f)}$ and
(4) $\forall s, t \in S_1 \; \forall a \in A_1 \; ((s, a, t) \in T_1 \Leftrightarrow (\iota(s), \kappa(a), \iota(t)) \in T_2)$.

Theorem 1. *"Turing CTSOT Theorem"*

Let $\mathfrak{T} := (\mathfrak{T}_0, \mathfrak{E})$ be a temporal Turing machine. Then

(1) the configuration automaton $\mathfrak{A}_C$ of $\mathfrak{T}$ is isomorphic to the state automaton $\mathfrak{A}_{\mathfrak{S}}$ of any CTSOT $\mathfrak{S}$ of $\mathfrak{T}$ with event set $M_C = \{\sigma, \tau, h\}$;
(2) the state automaton $\mathfrak{A}_{\mathfrak{S}}$ of $\mathfrak{T}$ is isomorphic to the state automaton $\mathfrak{A}_{\mathfrak{S}}$ of any CTSOT $\mathfrak{S}$ of $\mathfrak{T}$ with event set $M_C = \{\sigma\}$.

Proof. (1) Let $\mathfrak{T} := (\mathfrak{T}_0, \mathfrak{E})$ be a temporal Turing machine and $\mathfrak{S}$ be a CTSOT of $\mathfrak{T}$ with event set $M_C = \{\sigma, \tau, h\}$. To construct an isomorphism from $\mathfrak{A}_C$ to

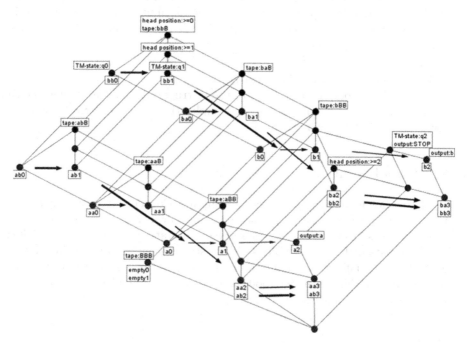

Fig. 1. A transition diagram of the concept lattice of the Turing CTSOT with event set $\{\sigma, \tau, h, \Omega\}$ of the temporal Turing machine $(\mathfrak{T}_0, \mathfrak{E}_{\leq 2})$ (from section 4) processing on the set $\mathfrak{I}_{\leq 2}$ of all words of length ≤ 2 over $\{a, b\}$. The head position has an ordinal scale, the tape attribute τ is scaled lexicographically using the order "$B < a < b$". The cycle of length 1 generated by the empty word is not drawn

$\mathfrak{A}_\mathfrak{S}$ we use Lemma 3 applied to the derived context $\mathbb{K}_C$ of the CTSOT $\mathfrak{S}$ which yields that $\sigma \times \tau \times h$ and the object concept mapping γ_C of $\mathbb{K}_C$ have the same kernel. Hence there exist exactly one bijection $\iota : (\sigma \times \tau \times h)D_{\sigma\tau h} \longrightarrow \gamma_C D_{\sigma,\tau,h}$ satisfying $\iota \circ (\sigma \times \tau \times h) = \gamma_C$.

We show that the pair (ι, κ), where κ is the identity on $A_\mathfrak{S} := vD_{\sigma,\tau,h}$ is an isomorphism from the configuration automaton $\mathfrak{A}_C$ onto the state automaton $\mathfrak{A}_\mathfrak{S}$. Clearly, (1) and (2) in Def. 14 are satisfied. Condition (3) and (4) hold since $\iota \circ (\sigma \times \tau \times h) = \gamma_C$.

(2) The proof for the event set $\{\sigma\}$ runs in the same way as for the event set $\{\sigma, \tau, h\}$ in (1).

Figure 1 shows a transition diagram of the concept lattice of the derived context of the event part of a Turing CTSOT for the temporal Turing machine in the example in section 4 with the input set $\mathfrak{I}_{\leq 2}$ and the event set $\{\sigma, \tau, h, \Omega\}$.

6.3   Instructions as Implications

To represent the instructions of δ of a given temporal Turing machine $\mathfrak{T}$ as valid implications of a formal context we construct a Turing CTSOT of $\mathfrak{T}$ with event set $\{\sigma, v, \sigma', v', \chi\}$ and take nominal scales for these event attributes such that each value appears also as a scale attribute. Then each instruction $((q, a), (q', a', \Delta))$ of δ can be represented by the implication

$$(\{(\sigma, q), (v, a)\}, \{(\sigma', q'), (v', a'), (\chi, \Delta)\})$$

which is valid in the derived context $\mathbb{K}_C$.

Hence the "program" consisting of the instructions of a Turing machine can be understood as a set of "background implications" in the derived context of the event part of a Turing CTSOT.

7   Conclusion

We have shown that the methods of Temporal Concept Analysis can be used to represent for a given Turing machine $\mathfrak{T}$ and for any input p the sequence of computation steps for p as a life track of p in the state space of a Turing CTSOT. The automata of Turing CTSOTs yield the state automaton of a Turing machine as well as its configuration automaton. The conceptual role of the instructions of a Turing machine is understood as a set of background implications of the derived context of a Turing CTSOT.

References

[Arb70]   Arbib, M.A.: *Theory of Abstract Automata*. Prentice Hall, Englewood Cliffs, N.J., 1970.

[Ber69]   Bertalanffy, L.v.; *General System Theory*. George Braziller, New York, 1969.

[Bir67]   Birkhoff, G.: *Lattice theory*. 3rd ed., Amer.Math.Soc., Providence 1967.

[But99]   Butterfield, J. (ed): *The Arguments of Time*. Oxford University Press, 1999.

[ButIsh99] Butterfield, J., C.J. Isham: On the Emergence of Time in Quantum Gravity. In Butterfield, J. (ed.): *The Arguments of Time*. Oxford University Press, 1999.

[Cast98]   Castellani, E.(ed.): *Interpreting Bodies: Classical and Quantum Objects in Modern Physics*. Princeton University Press 1998.

[Eil74]   Eilenberg, S.: *Automata, Languages, and Machines*. Vol. A. Academic Press 1974.

[GHR94]   Gabbay, D.M., I. Hodkinson, M. Reynolds: *Temporal Logic - Mathematical Foundations and Computational Aspects*. Vol.1, Clarendon Press Oxford 1994.

[GaWi99]   Ganter, B., R. Wille: *Formal Concept Analysis: mathematical foundations*. (translated from the German by Cornelia Franzke) Springer, Berlin-Heidelberg 1999.

[HU79]   Hopcroft,J.E., J.D. Ullman: *Introduction to Automata Theory, Languages, and Computation*. Addison-Wesley, 1979.

[Loe76]   Loeckx, J.: *Algorithmentheorie*. Springer, Berlin-Heidelberg 1976.

[Mal74]   Malcev, A.I.: *Algorithmen und rekursive Funktionen*. Vieweg-Verlag, Braunschweig, 1974.

[Pos36]   Post, E.L.: Finite combinatory processes - Formulation. J.Symbolic Logic **1** (1936)103-105.

[Tur36]   Turing, A.M.: On computable numbers, with an application to the Entscheidungsproblem. Proc. London Math. Soc.,2: 42, 230-265. A correction, ibid. 43, pp. 544-546, 1936.

[Tur36a]   Turing, A.M.: On computable numbers, with an application to the Entscheidungsproblem.
            http://www.abelard.org/turpap2/tp2-ie.asp#section-9

[vBe95]   van Benthem, J.: Temporal Logic. In: Gabbay, D.M., C.J. Hogger, J.A. Robinson: *Handbook of Logic in Artificial Intelligence and Logic Programming*. Vol. **4**, *Epistemic and Temporal Reasoning*. Clarendon Press, Oxford, 1995, 241-350.

[Wil82]   Wille, R.: Restructuring lattice theory: an approach based on hierarchies of concepts. In: Rival, I. (ed.): *Ordered Sets*. Reidel, Dordrecht-Boston 1982, 445-470.

[Wil97a]   Wille, R.: Introduction to Formal Concept Analysis. In: G. Negrini (ed.): *Modelli e modellizzazione. Models and modelling*. Consiglio Nazionale delle Ricerche, Istituto di Studi sulla Ricerca e Documentatione Scientifica, Roma 1997, 39-51.

[Wil97b]   Wille, R.: Conceptual graphs and Formal Concept Analysis. In: D. Lucose, H. Delugach, M. Keeler, L. Searle, J.F. Sowa (eds.): *Conceptual structures: Fulfilling Peirce's dream*. LNAI **1257**. Springer, Heidelberg 1997, 290-303.

[Wol94]   Wolff, K.E.: A first course in Formal Concept Analysis - How to understand line diagrams. In: Faulbaum, F. (ed.): SoftStat 93, *Advances in Statistical Software 4*, Gustav Fischer Verlag, Stuttgart 1994, 429-438.

[Wol00a]   Wolff, K.E.: Concepts, States, and Systems. In: Dubois, D.M. (ed.): *Computing Anticipatory Systems*. CASYS99 - Third International Conference, Liège, Belgium, 1999, American Institute of Physics, Conference Proceedings **517**, 2000, pp. 83-97.

[Wol00b]   Wolff, K.E.: Towards a Conceptual System Theory. In: B. Sanchez, N. Nada, A. Rashid, T. Arndt, M. Sanchez (eds.): *Proceedings of the World Multiconference on Systemics, Cybernetics and Informatics, SCI 2000, Vol. II: Information Systems Development, International Institute of Informatics and Systemics,* 2000, 124-132.

[Wol00c]   Wolff, K.E.: A Conceptual View of Knowledge Bases in Rough Set Theory. In: Ziarko, W., Yao, Y. (eds.): *Rough Sets and Current Trends in Computing.* Second International Conference, RSCTC 2000, Banff, Canada, October 16-19, 2000, Revised Papers, 220-228.

[Wol01]   Wolff, K.E.: Temporal Concept Analysis. In: E. Mephu Nguifo & al. (eds.): *ICCS-2001 International Workshop on Concept Lattices-Based Theory, Methods and Tools for Knowledge Discovery in Databases,* Stanford University, Palo Alto (CA), 91-107.

[Wol02a]   Wolff, K.E.: Transitions in Conceptual Time Systems. In: D.M.Dubois (ed.): *International Journal of Computing Anticipatory Systems,* vol. **11**, CHAOS 2002, p.398-412.

[Wol02b]   Wolff, K.E.: Interpretation of Automata in Temporal Concept Analysis. In: U. Priss, D. Corbett, G. Angelova (eds.): *Integration and Interfaces.* Tenth International Conference on Conceptual Structures. LNAI **2393**, Springer 2002, 341-353.

[WoYa03]   Wolff, K.E., W. Yameogo: Time Dimension, Objects, and Life Tracks - A Conceptual Analysis. In: A. de Moor, W. Lex, B. Ganter (eds.): *Conceptual structures for knowledge creation and communication.* LNAI **2746**. Springer, Heidelberg 2003, 188-200.

[Wol04a]   Wolff, K.E.: 'Particles' and 'Waves' as Understood by Temporal Concept Analysis. In: K.E. Wolff, H.D. Pfeiffer, H.S. Delugach (eds.): *Conceptual Structures at Work.* 12th International Conference on Conceptual Structures, ICCS 2004. Huntsville, AL, USA, July 2004. Proceedings. Springer Lecture Notes in Artificial Intelligence, LNAI 3127, Springer-Verlag, Berlin Heidelberg 2004, 126-141.

[Yam03]   Yameogo, W.: Time Conceptual Foundations of Programming. Master Thesis. Department of Computer Science at Darmstadt University of Applied Sciences, 2003.

Protoconceptual Contents and Implications

Joachim Hereth Correia[1] and Julia Klinger[2]

[1] Dresden University of Technology, Department of Mathematics,
Institute for Algebra, D-01062 Dresden
`heco@math.tu-dresden.de`
[2] Darmstadt University of Technology, Department of Mathematics,
Schloßgartenstr. 7, D-64289 Darmstadt
`jklinger@mathematik.tu-darmstadt.de`

Abstract. The development of a mathematical model for judgments understood as compositions of concepts and relations has been an important branch of research in recent years. It led to the definitions of *concept* and *protoconcept graphs* which are based on information contained in a *power context family*, where incidence relations between objects (or tuples of objects) and attributes are stored.

A theory of the information those graphs represent (called *conceptual content*) has been developed for concept graphs in [PW99] and [Wi03]. In [HK04], an extension of this theory to protoconcept graphs not considering object implications (as it is done for concept graphs) has been established. The first part of this paper concentrates on the investigation of the protoconceptual content of protoconcept graphs respecting both protoconceptual and object implications.

The second part compares the different structures of conceptual and protoconceptual contents of a given power context family, showing how more background information (using object implications and concepts instead of protoconcepts) reduces the number of possible contents.

The third and final part analyzes how the different approaches can be generalized. Here we will concentrate on the (generalized) conceptual content of a formal context.

In each part an *information context* will be defined, which provides an accessible representation of the lattice of (proto-)conceptual closures.

1 Introduction

Concept and protoconcept graphs are means to graphically represent relational information, i. e. that an object or a tuple of objects is in a certain relation (or is not for the case of protoconcept graphs). While the original reason for the development of the theory of concept and protoconcept graphs is the aspect of visualization and the support of rational communication (see [Wi97, Wi00b, DK03]), we will concentrate here on the aspect of (proto-)conceptual content.

Different (proto-)concept graphs may represent the same information, resulting in a twofold problem: on the one hand, the problem of finding in the set of graphs representing the same information one that is suited for the purpose

B. Ganter and R. Godin (Eds.): ICFCA 2005, LNAI 3403, pp. 375–390, 2005.
© Springer-Verlag Berlin Heidelberg 2005

of rational communication; and, on the other hand, analyzing the structure of representable information, i. e. of the (proto-)conceptual contents generated by the graphs.

As seen in [PW99] the conceptual contents of concept graphs form a complete lattice, and this remains true for the protoconceptual contents of protoconcept graphs (see also [HK04]). For the case of conceptual contents Wille extended in [Wi03] the notion of conceptual content and found a formal context having the conceptual contents as extents. We adapted this approach in [HK04] with a restricted notion of content – this restriction will be removed with the results in Section 3.

As concepts are special protoconcepts and concept graphs are special protoconcept graphs (up to isomorphism), we are interested in the relationship of the resulting structures. Starting from protoconceptual contents without object implications and gradually adding background information up to the case of conceptual content with object implications we find that the set of possible contents is reduced. We can identify the structure of contents of the case with more background information as substructure of the one with less background information. This will be shown in Section 4.

All approaches, those for concept graphs and also the ones in [HK04] and in this paper have much in common. In Section 5 an approach is presented, which abstracts from the concrete cases allowing us to adapt the information context for variants of implications for objects or (proto-)concepts. Section 6 ends the paper with a conclusion.

2   Basic Definitions

In this section, we shortly recall the definitions for protoconcepts, protoconcept graphs and conceptual contents. The adaption of the definition of conceptual contents to protoconceptual contents (respecting object implications) will be discussed in the following section. For a more detailed introduction to these topics we refer the reader to [Wi00, Wi03] and [HK04].

A *protoconcept* (cf. [Wi00]) of $\mathbb{K} := (G, M, I)$ is defined as a pair (A, B) with $A \subseteq G$ and $B \subseteq M$ such that $A^I = B^{II}$ (which is equivalent to $A^{II} = B^I$). The set $\mathfrak{P}(\mathbb{K})$ of all protoconcepts of $\mathbb{K}$ is structured by the *generalization order* $\sqsubseteq$, defined by

$$(A_1, B_1) \sqsubseteq (A_2, B_2) :\Leftrightarrow A_1 \subseteq A_2 \text{ and } B_1 \supseteq B_2,$$

and by the following operations:

$$
\begin{aligned}
(A_1, B_1) \sqcap (A_2, B_2) &:= (A_1 \cap A_2, (A_1 \cap A_2)^I) \\
(A_1, B_1) \sqcup (A_2, B_2) &:= ((B_1 \cap B_2)^I, B_1 \cap B_2) \\
\neg(A, B) := (G\backslash A, (G\backslash A)^I) \quad &\text{and} \quad \text{⌐}(A, B) := ((M \backslash B)^I, M \backslash B) \\
\top := (G, \emptyset) \quad &\phantom{\text{and}} \quad \bot := (\emptyset, M).
\end{aligned}
$$

The set $\mathfrak{P}(\mathbb{K})$ together with the operations $\sqcap, \sqcup, \neg, \text{⌐}, \top$ and $\bot$ is called the *algebra of protoconcepts* of $\mathbb{K}$ and denoted by $\underline{\mathfrak{P}}(\mathbb{K})$. The operations are called *meet, join, negation, opposition, all* and *nothing*. In this paper, we also need some operations derived from these basic ones. For $\mathfrak{p}, \mathfrak{q} \in \mathfrak{P}(\mathbb{K})$ we set

$$\mathfrak{p} \sqcup \mathfrak{q} := \neg(\neg \mathfrak{p} \sqcap \neg \mathfrak{q}) \text{ and } \mathfrak{p} \sqcap\!\!\!\sqcap \mathfrak{q} := {\dashv}(\neg \mathfrak{p} \sqcup \neg \mathfrak{q}),$$
$$\top := \neg\bot \text{ and } \bot := {\dashv}\top.$$

Semiconcepts are special protoconcepts, which fulfill the stronger condition $A^I = B$ or $B^I = A$. We define $\mathfrak{H}(\mathbb{K})$ to be the set of all semiconcepts of $\mathbb{K}$. For any protoconcept (A, B), we have that $(A, B)_\sqcap := (A, B) \sqcap (A, B) = (A, A^I)$ and $(A, B)_\sqcup := (A, B) \sqcup (A, B) = (B^I, B)$ are semiconcepts. To differentiate these two types of semiconcepts, we set $\mathfrak{H}_\sqcap(\mathbb{K})$ to be the semiconcepts of the form (A, A^I) (the $\sqcap$-semiconcepts), and $\mathfrak{H}_\sqcup(\mathbb{K})$ to be the semiconcepts of the form (B^I, B) (the $\sqcup$-semiconcepts). Note that whenever an operation is performed, we obtain a semiconcept. In particular, the result of the operations $\sqcap, \neg, \bot$ is a $\sqcap$-semiconcept, and any result of the operations $\sqcup, \dashv, \top$ is a $\sqcup$-semiconcept. Finally, *(formal) concepts* are those protoconcepts, that are $\sqcap$- and $\sqcup$-semiconcepts at once, i.e. with $A = B^I$ and $B = A^I$.

Since concept and protoconcept graphs store relational information, the notion of formal context was extended to a family of formal contexts in [Wi97]:

Definition 1. *A* power context family *$\overrightarrow{\mathbb{K}} := (\mathbb{K}_k)_{k=0,1,2,...}$ is a family of formal contexts $\mathbb{K}_k := (G_k, M_k, I_k)$ such that $G_k \subseteq G_0^k$ for $k \in \mathbb{N}$. The power context family is said to be of* limited *type $n \subset \mathbb{N}$ if $\overrightarrow{\mathbb{K}} := (\mathbb{K}_0, \mathbb{K}_1, \ldots, \mathbb{K}_n)$, otherwise it is called* unlimited.

Now, we can recall the definition of protoconcept graphs introduced in [Wi02]: The underlying structure is a so-called *relational graph*, which is a triple (V, E, ν) consisting of a set V of vertices, a set E of edges and a mapping $\nu \colon E \to \bigcup_{k=1,2,...} V^k$ which maps each edge to the ordered tuple of its adjacent vertices. If $\nu(e) = (v_1, \ldots, v_k)$, then we say that the *arity* of e is k. The vertices are said to have arity 0, i.e. we set $E^{(0)} := V$. Moreover, let $E^{(k)}$ $(k = 0, 1, \ldots)$ be the set of all k-ary edges.

Definition 2. *A* protoconcept graph *of a power context family $\overrightarrow{\mathbb{K}} := (\mathbb{K}_0, \mathbb{K}_1, \ldots)$ with $\mathbb{K}_k := (G_k, M_k, I_k)$ for $k = 0, 1, 2, \ldots$ is a structure $\mathfrak{G} := (V, E, \nu, \kappa, \varrho)$ for which*

- *(V, E, ν) is a relational graph,*
- *$\kappa \colon V \cup E \to \bigcup_{k=0,1,...} \mathfrak{P}(\mathbb{K}_k)$ is a mapping with $\kappa(u) \in \mathfrak{P}(\mathbb{K}_k)$ for all $u \in E^{(k)}$ $(k = 0, 1, \ldots)$,*
- *$\varrho \colon V \to \mathfrak{P}(G_0) \setminus \{\emptyset\}$ is a mapping with $\varrho^+(v) := \varrho(v) \cap \mathrm{Ext}(\kappa(v))$ and $\varrho^-(v) := \varrho(v) \setminus \varrho^+(v)$ satisfying for $\nu(e) = (v_1, \ldots, v_k)$, that $\varrho^+(v_j) \neq \emptyset$ for all $j = 1, \ldots, k$ or $\varrho^-(v_j) \neq \emptyset$ for all $j = 1, \ldots, k$, and that $\varrho^+(v_1) \times \cdots \times \varrho^+(v_k) \subseteq \mathrm{Ext}(\kappa(e))$ and $\varrho^-(v_1) \times \cdots \times \varrho^-(v_k) \subseteq (G_0)^k \setminus \mathrm{Ext}(\kappa(e))$.*

We canonically extend the mapping ϱ from vertices to edges: For $\nu(e) = (v_1, \ldots, v_k)$, let $\varrho(e) := \varrho^+(e) \cup \varrho^-(e)$ with $\varrho^+(e) := \varrho^+(v_1) \times \cdots \times \varrho^+(v_k)$ and $\varrho^-(e) := \varrho^-(v_1) \times \cdots \times \varrho^-(v_k)$.

Before we turn to the definition of conceptual content of protoconcept graphs, we recall some definitions for the case of concept graphs from [Wi03] (formally, concept graphs can be identified with protoconcept graphs, where $\kappa(u) \in \mathfrak{B}(\mathbb{K}_k)$ for all $u \in E^{(k)}$ and $\varrho^-(v) = \emptyset$ for all $v \in V$). The concept graph contains the *information units* $(g, \kappa(u))$ with $u \in E^{(k)}$ for $k = 0, 1, 2, \ldots$ and $g \in \varrho(u)$. For a context $\mathbb{K}_k$ of the underlying power context family, we define $\mathfrak{B}^{inst}(\mathbb{K}_k) := \{(g, \mathfrak{p}) \in G_k \times \mathfrak{B}(\mathbb{K}_k) \mid g \in \mathrm{Ext}(\mathfrak{p})\}$, the set of possible information units for $\mathbb{K}_k$. The material inferences coded in the underlying power context family (see [Wi02] and [Wi03]) are made explicit in the following way. Let $k = 0, 1, 2, \ldots$, $A, C \subseteq G_k$ and $\mathfrak{B}, \mathfrak{D} \subseteq \mathfrak{B}(\mathbb{K}_k)$. Then we say

$$\mathbb{K}_k \text{ satisfies } A \to C :\Longleftrightarrow A^{I_k} \subseteq C^{I_k} \tag{1}$$

and

$$\mathbb{K}_k \text{ satisfies } \mathfrak{B} \to \mathfrak{D} :\Longleftrightarrow \bigwedge \mathfrak{B} \leq \bigwedge \mathfrak{D} \text{ in } \mathfrak{B}(\mathbb{K}_k) \tag{2}$$

These implications give rise to a closure system $\mathcal{C}(\mathfrak{B}^{inst}(\mathbb{K}_k))$, consisting of all subsets Y which have the following property:

If $A \times \mathfrak{B} \subseteq Y$ and $\mathbb{K}_k$ satisfies both $A \to C$ and $\mathfrak{B} \to \mathfrak{D}$, then $C \times \mathfrak{D} \subseteq Y$ (3)

Based on this, the *k-ary conceptual content* $C_k(\mathfrak{G})$ of a concept graph $\mathfrak{G}$ is defined as the closure of $\{(g, \kappa(u)) \mid u \in E^{(k)} \text{ and } g \in \varrho(u)\}$, and the full *conceptual content* $C(\mathfrak{G})$ of the graph is the disjoint union of all k-ary conceptual contents for $k = 0, 1, 2, \ldots$.

3   Protoconceptual Content of Protoconcept Graphs

In this section, we discuss how the equations (1)–(3) can be interpreted for the case of protoconcept graphs, and investigate the structure of the resulting closure system. To simplify writing, we will use $\mathbb{K}$ in the following instead of $\mathbb{K}_k$.

The translation of the first definition poses no problem, we understand as *information units* of a protoconcept graph the pairs $(g, \kappa(u))$ and $(h, \neg\kappa(u))$ with $u \in E^{(k)}$, $g \in \varrho^+(u)$ and $h \in \varrho^-(u)$ for $k = 0, 1, 2, \ldots$. Analogously to $\mathfrak{B}^{inst}(\mathbb{K})$ we define $\mathfrak{P}^{inst}(\mathbb{K}) := \{(g, \mathfrak{p}) \mid g \in \mathrm{Ext}(\mathfrak{p}), \mathfrak{p} \in \mathfrak{P}(\mathbb{K})\}$. The object implications in Equation (1) are independent from the notion of concept and can therefore be accepted. However, in the Equation (2), we use an operation in the concept lattice, and have to look for an alternative in the protoconcept algebra. The solution proposed in [Wi02], which also has been applied in [HK04], is to replace $\bigwedge$ by $\bigsqcap$, i.e. we have (with $\mathfrak{B}, \mathfrak{D} \subseteq \mathfrak{P}(\mathbb{K})$):

$$\mathbb{K} \text{ satisfies } \mathfrak{B} \to \mathfrak{D} :\Longleftrightarrow \bigsqcap \mathfrak{B} \sqsubseteq \bigsqcap \mathfrak{D} \text{ in } \mathfrak{P}(\mathbb{K}) \tag{$2_\sqcap$}$$

For concepts, the operation $\sqcap$ in $\mathfrak{P}(\mathbb{K})$ yields the same result as $\wedge$ in the concept lattice $\underline{\mathfrak{B}}(\mathbb{K})$. However, the operation $\sqcap$ is defined on the extensional side and also the information units are defined by objects. This led us to consider the operation $\sqcap\!\!\!\sqcap$ as an alternative, which is (indirectly) defined on the intensional side $((A,B)\sqcap\!\!\!\sqcap(C,D) = ((B \cup D)', B \cup D))$:

$$\mathbb{K} \text{ satisfies } \mathfrak{B} \to \mathfrak{D} :\Longleftrightarrow \sqcap\!\!\!\sqcap \mathfrak{B} \sqsubseteq \sqcap\!\!\!\sqcap \mathfrak{D} \text{ in } \mathfrak{P}(\mathbb{K}) \qquad (2_{\sqcap\!\!\!\sqcap})$$

By comparison, it is weaker in the sense, that for $\mathfrak{B} \subseteq \mathfrak{P}(\mathbb{K})$ we have $\sqcap \mathfrak{B} \sqsubseteq \sqcap\!\!\!\sqcap \mathfrak{B}$,[1] or in other words $\mathfrak{B} \to \mathfrak{D}$ using $(2_{\sqcap\!\!\!\sqcap})$ implies $\mathfrak{B} \to \mathfrak{D}$ using $(2_{\sqcap})$ but not vice-versa. Using $(2_{\sqcap})$ we can therefore conclude more new information. For this reason, the latter approach is not useful if we can calculate $\sqcap$ for all sets of protoconcepts. However, if knowledge about I is not complete, the operation $\sqcap\!\!\!\sqcap$ might be interesting, as this operation is defined by the intensional side.

Now we turn to transforming Equation (3) for the case of protoconcepts. We have to consider the problem that with $A \times \mathfrak{B} \subseteq \mathfrak{P}^{inst}(\mathbb{K})$, $A \to C$ and $\mathfrak{B} \to \mathfrak{D}$ it might happen, that $C \times \mathfrak{D} \not\subseteq \mathfrak{P}^{inst}(\mathbb{K})$. For instance, let $g_1, g_2 \in G$ with $g_1^I \subseteq g_2^I$, then we have $g_1 \to g_2$, but with $\mathfrak{p} := (\{g_1\}, g_1^I)$ we get $\{g_1\} \times \{\mathfrak{p}\} \subseteq \mathfrak{P}^{inst}(\mathbb{K})$, $g_1 \to g_2$ and (trivially) $\mathfrak{p} \to \mathfrak{p}$, but $\{(g_2, \mathfrak{p})\} = \{g_2\} \times \{\mathfrak{p}\} \not\subseteq \mathfrak{P}^{inst}(\mathbb{K})$. This was the reason we dropped object implications in [HK04].

In this paper, we extend this approach as follows. We will consider the closure system of all subsets $Y \subseteq \mathfrak{P}^{inst}(\mathbb{K})$, which fulfill the following condition:

$$\text{If } A \times \mathfrak{B} \subseteq Y \text{ and if } \mathbb{K} \text{ satisfies both } A \to C \text{ and } \mathfrak{B} \to \mathfrak{D},$$
$$\text{then } (C \times \mathfrak{D}) \cap \mathfrak{P}^{inst}(\mathbb{K}) \subseteq Y \qquad (3_{\mathfrak{P}})$$

The two alternatives $(2_{\sqcap})$ and $(2_{\sqcap\!\!\!\sqcap})$ for the protoconcept inferences lead to different closure systems on $\mathfrak{P}^{inst}(\mathbb{K})$. To denote each alternative, we will use the symbols $(\sqcap)$ and $(\sqcap\!\!\!\sqcap)$. To treat both cases at once, we will use the symbol ∇. Any statement making use of ∇ stands for both variants, replacing ∇ by $\sqcap$ and $\sqcap\!\!\!\sqcap$.

For a protoconcept graph $\mathfrak{G}$ now we define the *k-ary protoconceptual content with respect to* ∇ denoted by $C_\nabla^{(k)}(\mathfrak{G})$ over the power context family $\overrightarrow{\mathbb{K}} := (\mathbb{K}_0, \mathbb{K}_1, \ldots)$ as the closure of $\{(g, \kappa(u)) \mid u \in E^{(k)}, g \in \varrho^+(u)\} \cup \{(g, \neg\kappa(u)) \mid u \in E^{(k)}, g \in \varrho^-(u)\} \subseteq \mathfrak{P}^{inst}(\mathbb{K}_k)$ with respect to $(3_{\mathfrak{P}})$ considering ∇. The *protoconceptual content* $C_\nabla(\mathfrak{G})$ of $\mathfrak{G}$ is defined as the disjoint union of the k-ary protoconceptual contents with respect to ∇ of $\mathfrak{G}$.

To investigate the structure of the closure system for the different variations we will first investigate the protoconceptual contents in more detail. In the following, we will use the notation known from the incidence relation of formal contexts, for instance, $\mathfrak{p}^Y$ means the set of all objects g such that $(g, \mathfrak{p})$ is an element of Y. If not otherwise mentioned, we assume that $Y \subseteq \mathfrak{P}^{inst}(\mathbb{K})$ is a protoconceptual content with respect to $(3_{\mathfrak{P}})$.

---

[1] We have $\mathrm{Ext}(\sqcap \mathfrak{B}) = \bigcap \mathrm{Ext}(\mathfrak{B}) \subseteq \bigcap \mathrm{Ext}(\mathfrak{B})^{II} = \bigcap \mathrm{Int}(\mathfrak{B})^I = (\bigcup \mathrm{Int}(\mathfrak{B}))^I = \mathrm{Ext}(\sqcap\!\!\!\sqcap \mathfrak{B})$ and $\mathrm{Int}(\sqcap \mathfrak{B}) = (\bigcap \mathrm{Ext}(\mathfrak{B}))^I \supseteq (\bigcap \mathrm{Ext}(\mathfrak{B})^{II})^I = (\bigcap \mathrm{Int}(\mathfrak{B})^I)^I = (\bigcup \mathrm{Int}(\mathfrak{B}))^{II} \supseteq \bigcup \mathrm{Int}(\mathfrak{B}) = \mathrm{Int}(\sqcap\!\!\!\sqcap \mathfrak{B})$ (Operations are considered element-wise, for instance $\mathrm{Ext}(\mathfrak{B})^{II} = \{\mathrm{Ext}(\mathfrak{b})^{II} \mid \mathfrak{b} \in \mathfrak{B}\}$). Considering the case of $\mathfrak{B}$ having only one element shows that left and right side are not equal in general.

In order to study the protoconceptual contents of a power context family, is is enough to concentrate on semiconcepts. Even more, we only need to consider either the $\sqcap$- or the $\sqcup$-semiconcepts:

Lemma 3. *Let* $\mathfrak{A} \subseteq \mathfrak{P}(\mathbb{K})$. *Then we have for the case* (∇):

$$\mathfrak{A} \to \nabla \mathfrak{A} \ and \ \nabla \mathfrak{A} \to \mathfrak{A}.$$

Remark 4. To unify the notation in the following, we will write $\mathfrak{p}_{\sqcap}$ to denote the $\sqcup$-semiconcept $\mathfrak{p}_{\sqcup}$.

Proof. For $\mathfrak{p} \in \mathfrak{H}_\nabla$ we have $\nabla\{\mathfrak{p}\} = \mathfrak{p}_\nabla = \mathfrak{p}$ and therefore $\nabla\nabla \mathfrak{A} = \nabla \mathfrak{A}$. With this, the assertion follows from (2_∇). $\qquad\square$

Corollary 5. *Let* $\mathfrak{A} \subseteq \mathfrak{P}(\mathbb{K})$. *Then we have for the case* (∇)

$$(\nabla \mathfrak{A})^Y = \bigcap_{\mathfrak{a} \in \mathfrak{A}} \mathfrak{a}^Y.$$

Proof. If $(g, \nabla \mathfrak{A}) \in Y$ $(\Leftrightarrow g \in (\nabla \mathfrak{A})^Y)$ then we have due to $\nabla \mathfrak{A} \to \mathfrak{A}$ that $\{g\} \times \mathfrak{A} \subseteq Y$, i.e. $(g, \mathfrak{a}) \in Y$ for all $\mathfrak{a} \in \mathfrak{A}$ and therefore $(\nabla \mathfrak{A})^Y \subseteq \bigcap_{\mathfrak{a} \in \mathfrak{A}} \mathfrak{a}^Y$. On the other hand, if $g \in \mathfrak{a}^Y$ for all $\mathfrak{a} \in \mathfrak{A}$, we have $\{g\} \times \mathfrak{A} \subseteq Y$ and with $\mathfrak{A} \to \nabla \mathfrak{A}$ that $(g, \nabla \mathfrak{A}) \in Y$. $\qquad\square$

Corollary 6. *Let* $\mathfrak{p} \in \mathfrak{P}(\mathbb{K})$. *For* $(\sqcap)$ *we have* $\mathfrak{p}^Y = \mathfrak{p}_\sqcap^Y = (\mathrm{Ext}(\mathfrak{p}), \mathrm{Ext}(\mathfrak{p})^I)^Y$ *and for the case* $(\sqcap)$ $\mathfrak{p}^Y = \mathfrak{p}_\sqcup^Y = (\mathrm{Int}(\mathfrak{p})^I, \mathrm{Int}(\mathfrak{p}))^Y$.

Proof. Using Corollary 5 with $\mathfrak{A} := \{\mathfrak{p}\}$ we get $(\nabla\{\mathfrak{p}\})^Y = \mathfrak{p}_\nabla^Y = \mathfrak{p}^Y$. $\qquad\square$

This shows that for a detailed investigation of the structure of the protoconceptual contents we do not have to consider the full set of protoconcepts but only the corresponding semiconcepts. The $\sqcap$- and $\sqcup$-semiconcepts each form a complete atomic Boolean subalgebra of $\mathfrak{P}(\mathbb{K})$, and every element of a complete atomic Boolean algebra can be described as meet of the coatoms above it. We will transfer this representation to our problem. However, while both $\mathfrak{H}_\sqcap(\mathbb{K})$ and $\mathfrak{H}_\sqcup(\mathbb{K})$ are Boolean substructures, they are not isomorphic in general. To allow a unified treatment, we fix some definitions:

Definition 7. *Let* $\mathbb{K} := (G, M, I)$ *be a formal context. We set*

$$\mathfrak{D}_\sqcap := \{(G \setminus \{g\}, (G \setminus \{g\})^I) \mid g \in G\} \subseteq \mathfrak{H}_\sqcap(\mathbb{K})$$
$$\mathfrak{D}_\sqcap := \{(m^I, \{m\}) \mid m \in M\} \subseteq \mathfrak{H}_\sqcup(\mathbb{K})$$

Lemma 8. *Let* $\mathfrak{p} \in \mathfrak{P}(G, M, I)$ *and* $\mathfrak{d} \in \mathfrak{D}_\nabla$. *Then there exists exactly one* $g_\mathfrak{d} \in G$ *such that* $\mathfrak{d} = (G \setminus \{g_\mathfrak{d}\}, (G \setminus \{g_\mathfrak{d}\})^I)$ *and* $\mathfrak{p} \to \mathfrak{d} \Leftrightarrow g_\mathfrak{d} \notin \mathrm{Ext}(\mathfrak{p})$. *In the case* $(\sqcap)$ *exists exactly one* $m_\mathfrak{d} \in M$, *such that* $\mathfrak{d} = (m_\mathfrak{d}^I, \{m_\mathfrak{d}\})$ *and* $\mathfrak{p} \to \mathfrak{d} \Leftrightarrow m_\mathfrak{d} \in \mathrm{Int}(\mathfrak{p})$.

Proof. For the case $(\sqcap)$ we have $\mathfrak{p} \to \mathfrak{d} \Leftrightarrow \bigsqcap\{\mathfrak{p}\} \sqsubseteq \bigsqcap\{\mathfrak{d}\} \Leftrightarrow \mathfrak{p}_\sqcap \sqsubseteq \mathfrak{d}_\sqcap \Leftrightarrow$ $\mathrm{Ext}(\mathfrak{p}) \subseteq \mathrm{Ext}(\mathfrak{d})$. By the definition of $\mathfrak{D}_\sqcap$ we know there is some $g_\mathfrak{d} \in G$ such that $\mathfrak{d} = (G \backslash \{g_\mathfrak{d}\}, (G \backslash \{g_\mathfrak{d}\})^I)$, so we have $\mathfrak{p} \to \mathfrak{d} \Leftrightarrow \mathrm{Ext}(\mathfrak{p}) \subseteq G \backslash \{g_\mathfrak{d}\} \Leftrightarrow g_\mathfrak{d} \notin \mathrm{Ext}(\mathfrak{p})$. The reasoning for the case $(\sqcap\!\!\!\sqcap)$ is analogous. $\qquad\square$

Now, we can represent $\mathfrak{p}^Y$ by means of the coatoms, hence the elements in $\mathfrak{D}_\nabla$:

Lemma 9. *We have for every $\mathfrak{p} \in \mathfrak{P}(\mathbb{K})$*

$$\mathfrak{p}^Y = \bigcap_{\substack{\mathfrak{d} \in \mathfrak{D}_\nabla \\ \mathfrak{p} \to \mathfrak{d}}} \mathfrak{d}^Y$$

Proof. Let $\mathfrak{c} := \bigvee\{\mathfrak{d} \in \mathfrak{D}_\nabla \mid \mathfrak{p} \to \mathfrak{d}\}$. For $\nabla = \sqcap$, we see by Lemma 8 that $\mathrm{Ext}(\mathfrak{c}) = \bigcap\{\mathrm{Ext}(\mathfrak{d}) \mid g_\mathfrak{d} \notin \mathrm{Ext}(\mathfrak{p})\} = \bigcap\{G \backslash \{g\} \mid g \notin \mathrm{Ext}(\mathfrak{p})\}\} = \mathrm{Ext}(\mathfrak{p})$. For $\nabla = \sqcap\!\!\!\sqcap$ we have analogously $\mathrm{Int}(\mathfrak{c}) = \bigcup\{\{m\} \mid m \in \mathrm{Int}(\mathfrak{p}) = \mathrm{Int}(\mathfrak{p})$. By Corollary 6 we get the assertion. $\qquad\square$

Lemma 10. *Let $\mathfrak{p} \in \mathfrak{P}(\mathbb{K})$ such that for some $g \in G$ we have $(g, \mathfrak{p}) \in Y$. Then*

$$(\mathfrak{p}^Y)^{II} \cap \mathrm{Ext}(\mathfrak{p}) = \mathfrak{p}^Y$$

Proof. We have $\mathfrak{p}^Y \times \{\mathfrak{p}\} \subseteq Y$. We can easily see, that $\mathfrak{p}^Y \to (\mathfrak{p}^Y)^{II}$ (as for any set $A \subseteq G$ we have $A \to A''$ because of $A' = A'''$). The implication $\mathfrak{p} \to \mathfrak{p}$ is obviously correct, and consequently $((\mathfrak{p}^Y)^{II} \times \{\mathfrak{p}\}) \cap \mathfrak{P}^{inst}(\mathbb{K}) = ((\mathfrak{p}^Y)^{II} \cap \mathrm{Ext}(\mathfrak{p})) \times \{\mathfrak{p}\} \subseteq Y$. Hence we obtain $(\mathfrak{p}^Y)^{II} \cap \mathrm{Ext}(\mathfrak{p}) \subseteq \mathfrak{p}^Y$, while $(\mathfrak{p}^Y)^{II} \cap \mathrm{Ext}(\mathfrak{p}) \supseteq \mathfrak{p}^Y$ is obvious. $\qquad\square$

Lemma 9 says, that the set $\mathfrak{p}^Y$ is defined by the sets $\mathfrak{d}^Y$ with $\mathfrak{d} \in \mathfrak{D}_\nabla$. Lemma 10 says that $\mathfrak{d}^Y$ can be calculated from the concept extent it generates (and is often equal). Combining these facts together, we get an easy way to represent any protoconceptual content $Y \subseteq \mathfrak{P}^{inst}(\mathbb{K})$. The following proposition shows how to transform an arbitrary mapping from $\mathfrak{D}_\nabla$ to concept extents into a protoconceptual content of $\mathbb{K}$.

Proposition 11. *Let $\mathbb{K}$ be a formal context and $\varphi : D_\nabla \to \mathrm{Ext}(\mathbb{K})$ be a mapping. Then the set*

$$Y^\varphi := \bigcup_{\mathfrak{p} \in \mathfrak{P}(\mathbb{K})} A_\mathfrak{p}^\varphi \times \{\mathfrak{p}\}$$

with

$$A_\mathfrak{p}^\varphi := \mathrm{Ext}(\mathfrak{p}) \cap \bigcap_{\substack{\mathfrak{d} \in \mathfrak{D}_\nabla \\ \mathfrak{p} \to \mathfrak{d}}} \varphi(\mathfrak{d})$$

is a protoconceptual content of $\mathbb{K}$.

Proof. As $A_{\mathfrak{p}}^{\varphi} \subseteq \text{Ext}(\mathfrak{p})$ it is easy to see that $Y^{\varphi} \subseteq \mathfrak{P}^{inst}(\mathbb{K})$. Now we will prove that Y^{φ} is closed under $(3_{\mathfrak{P}})$. Let $A, C \subseteq G$ and $\mathfrak{B}, \mathfrak{F} \subseteq \mathfrak{P}(\mathbb{K})$ such that $A \times \mathfrak{B} \subseteq Y^{\varphi}$ and the inferences $A \to C$ and $\mathfrak{B} \to \mathfrak{F}$ hold in $\mathbb{K}$.

Let $\mathfrak{f} \in \mathfrak{F}$ and $\mathfrak{d} \in \mathfrak{D}_{\nabla}$ with $\mathfrak{f} \to \mathfrak{d}$. For the case $(\sqcap)$ we know due to Lemma 8 that there exists some $g_{\mathfrak{d}} \in G$ such that $g_{\mathfrak{d}} \notin \text{Ext}(\mathfrak{f})$. Because of $\mathfrak{B} \to \mathfrak{F}$ we know that $\bigcap \mathfrak{B} \sqsubseteq \bigcap \mathfrak{F} \Leftrightarrow \bigcap\{\text{Ext}(\mathfrak{b}) \mid \mathfrak{b} \in \mathfrak{B}\} \subseteq \bigcap\{\text{Ext}(\mathfrak{f}) \mid \mathfrak{f} \in \mathfrak{F}\}$. Therefore there has to be some $\mathfrak{b} \in \mathfrak{B}$ with $g_{\mathfrak{d}} \notin \text{Ext}(\mathfrak{b}) \Leftrightarrow \mathfrak{b} \to \mathfrak{d}$. In the case $(\sqcap)$ we reason analogously that there has to be some $\mathfrak{b} \in \mathfrak{B}$ with $m_{\mathfrak{d}} \in \text{Int}(\mathfrak{b}) \Leftrightarrow \mathfrak{b} \to \mathfrak{d}$. This means that we have for each $\mathfrak{d} \in \mathfrak{D}_{\nabla}$ with $\mathfrak{f} \to \mathfrak{d}$ some $\mathfrak{b} \in \mathfrak{B}$ with $\mathfrak{b} \to \mathfrak{d}$, and we see $A \subseteq A_{\mathfrak{b}}^{\varphi} \subseteq \varphi(\mathfrak{d})$. Moreover, from $A \to C$ we know that $A^{I} \subseteq C^{I} \Rightarrow C \subseteq C^{II} \subseteq A^{II} \subseteq \varphi(\mathfrak{d})^{II} = \varphi(\mathfrak{d})$ (the latter equation holds because $\varphi(\mathfrak{d})$ is an extent of $\mathbb{K}$). Therefore $\text{Ext}(\mathfrak{f}) \cap C \subseteq A_{\mathfrak{f}}^{\varphi} \Leftrightarrow (C \times \{\mathfrak{f}\}) \cap \mathfrak{P}^{inst}(\mathbb{K}) \subseteq Y^{\varphi}$. This concludes the proof that we have $(C \times \mathfrak{F}) \cap \mathfrak{P}^{inst}(\mathbb{K}) \subseteq Y^{\varphi}$. $\square$

Moreover, every protoconceptual content may be obtained in this way:

Lemma 12. *Let* $Y \subseteq \mathfrak{P}^{inst}(\mathbb{K})$ *be a protoconceptual content of* $\mathbb{K}$ *with respect to* $(3_{\mathfrak{P}})$. *We define the mapping* $\varphi_Y : \mathfrak{D}_{\nabla} \to \text{Ext}(\mathbb{K})$ *by* $\varphi_Y(\mathfrak{d}) := (\mathfrak{d}^Y)^{II}$. *Then we have for the protoconceptual content* Y^{φ_Y} *as defined in Proposition 11*

$$Y = Y^{\varphi_Y}.$$

Proof. As we know that Y and Y^{φ_Y} are both protoconceptual contents, it suffices according to Lemma 9 to show that $\mathfrak{d}^Y = \mathfrak{d}^{Y^{\varphi_Y}}$ for all $\mathfrak{d} \in \mathfrak{D}_{\nabla}$. We have

$$\mathfrak{d}^{Y^{\varphi_Y}} = A_{\mathfrak{d}}^{\varphi} = \text{Ext}(\mathfrak{d}) \cap \varphi(\mathfrak{d}) = \text{Ext}(\mathfrak{d}) \cap (\mathfrak{d}^Y)^{II} = \mathfrak{d}^Y$$

according to Lemma 10. $\square$

This shows that any protoconceptual content can be represented by a corresponding mapping from $\mathfrak{D}_{\nabla}$ into the concept extents of the formal context. Instead of describing the protoconceptual contents directly, we will concentrate on a good description of those mappings. We notice, that if φ, ψ are two of those mappings, then $\varphi \cap \psi$ with $(\varphi \cap \psi)(\mathfrak{d}) = \varphi(\mathfrak{d}) \cap \psi(\mathfrak{d})$ is also a valid mapping because extents are closed under intersection. We aim at fixing a small set of mappings, such that every mapping can be represented as intersection of these mappings. The images of the $\mathfrak{d}$ are extents of the context which are always the intersection of the extents of the attribute concepts. This immediately leads to the following lemma:

Lemma 13. *Let* $\varphi : \mathfrak{D}_{\nabla} \to \text{Ext}(\mathbb{K})$ *be an arbitrary mapping. For any* $\mathfrak{d} \in \mathfrak{D}_{\nabla}$ *and any* $m \in M$ *we set* $\varphi_{\mathfrak{d}}^{m}(\mathfrak{e}) := m^I$ *if* $\mathfrak{e} = \mathfrak{d}$ *and* $\varphi_{\mathfrak{d}}^{m}(\mathfrak{e}) := G$ *otherwise. Then we have*

$$\varphi = \bigcap\{\varphi_{\mathfrak{d}}^{m} \mid (\mathfrak{d}, m) \in \mathfrak{D}_{\nabla} \times M, \varphi(\mathfrak{d}) \subseteq m^I\}.$$

Remark 14. The set $\mathfrak{D}_{\nabla} \times M$ in the above equation is not minimal. If m is a reducible attribute, its extent is the intersection of the attributes above it. In a doubly founded context, we have the notion of irreducible attributes (see [GW99,

p. 32] for details) and M can be replaced by the set of irreducible attributes. In particular, every finite context is doubly founded. Moreover, if for some $\mathfrak{d} \in \mathfrak{D}_\nabla$ there exist two (irreducible) attributes m, n such that $m^I, n^I \subseteq G \setminus \text{Ext}(\mathfrak{d})$, then we have $Y^{\varphi_\mathfrak{d}^m} = Y^{\varphi_\mathfrak{d}^n}$. For instance, for the case ($\sqcap$) this happens if for some $g \in G$ both $\emptyset$ and $\{g\}$ are extents of (irreducible) attributes. Therefore one of the mappings $\varphi_\mathfrak{d}^m, \varphi_\mathfrak{d}^n$ could be discarded. Finally, if for some $\mathfrak{d} \in \mathfrak{D}_\nabla$ and some (irreducible) attribute m we have $m^I = \text{Ext}(\mathfrak{d})$, then $Y^{\varphi_\mathfrak{d}^m} = \mathfrak{P}^{inst}(\mathbb{K})$, so the mapping $\varphi_\mathfrak{d}^m$ could be discarded, too. For our purpose of a simple description of the protoconceptual contents we will continue with the set $\mathfrak{D}_\nabla \times M$, but it could be replaced by smaller sets as mentioned in this remark.

Theorem 15 (Basic Theorem on Protoconceptual Contents of $\mathbb{K}$).
The protoconceptual contents of $\mathbb{K}$ are exactly the extents of the protoconceptual information context

$$\mathbb{K}_\nabla^{inf}(\mathbb{K}) := (\mathfrak{P}^{inst}(\mathbb{K}), \mathfrak{D}_\nabla \times M, \Delta)$$

with $((g, \mathfrak{p}), (\mathfrak{d}, m)) \in \Delta :\Leftrightarrow$ *If* $\mathfrak{p} \to \mathfrak{d}$ *then* $(g, m) \in I$.

Proof. Let $(\mathfrak{d}, m) \in \mathfrak{D}_\nabla \times M$ be an arbitrary attribute of $\mathbb{K}_\nabla^{inf}(\mathbb{K})$ and $(g, \mathfrak{p}) \in \mathfrak{P}^{inst}(\mathbb{K})$. We have $(g, \mathfrak{p}) \in Y^{\varphi_\mathfrak{d}^m} \Leftrightarrow (\mathfrak{p} \not\to \mathfrak{d}$ or $g \in m^I) \Leftrightarrow ($ If $\mathfrak{p} \to \mathfrak{d}$ then $(g, m) \in I) \Leftrightarrow (g, \mathfrak{p}) \in (\mathfrak{d}, m)^\Delta$. From this follows $(\mathfrak{d}, m)^\Delta = Y^{\varphi_\mathfrak{d}^m}$. And because the extents of the attribute concepts are closures in $\mathfrak{P}^{inst}(\mathbb{K})$ we can conclude that all extents of $\mathbb{K}_\nabla^{inf}(\mathbb{K})$ – being intersections of these closures – are closures too and therefore protoconceptual contents.

Now let $Y \in \mathcal{C}_\nabla(\mathfrak{P}^{inst}(\mathbb{K}))$ be a protoconceptual content. In view of Lemma 12 and 13 we get that $Y = \bigcap\{(\mathfrak{d}, m)^\Delta \mid (\mathfrak{d}, m) \in \mathfrak{D}_\nabla \times M$ and $(\mathfrak{d}^Y)^{II} \subseteq m^I\}$ is an extent of $\mathbb{K}_\nabla^{inf}(\mathbb{K})$. $\qquad\square$

Based on this theorem, we now have an approach to describe the protoconceptual contents of the protoconcept graphs of a given power context family. Its proof is similar to the corresponding ones in [Wi03] and [HK04] but will be shown in short form for the sake of completeness.

Let $\overrightarrow{\mathbb{K}} := (\mathbb{K}_0, \mathbb{K}_1, \dots, \mathbb{K}_n)$ be a power context family with $\mathbb{K}_k := (G_k, M_k, I_k)$ $(k = 0, 1, \dots, n)$. The *protoconceptual information context corresponding to* $\overrightarrow{\mathbb{K}}$ with respect to ∇ is defined as the formal context

$$\mathbb{K}_\nabla^{inf}(\overrightarrow{\mathbb{K}}) := \mathbb{K}_\nabla^{inf}(\mathbb{K}_0) + \mathbb{K}_\nabla^{inf}(\mathbb{K}_1) + \dots + \mathbb{K}_\nabla^{inf}(\mathbb{K}_n),$$

thus as the direct sum of the contexts $\mathbb{K}_\nabla^{inf}(\mathbb{K}_k)$ $(k = 0, \dots, n)$. An extent U of $\mathbb{K}_\nabla^{inf}(\overrightarrow{\mathbb{K}})$ is said to be *rooted* if for $k = 1, \dots, n$ we have $((g_1, \dots, g_k), \mathfrak{b}_k) \in U$ implies $(g_j, \top_\nabla^{(0)}) \in U$ for all $j = 1, \dots, k$ with $\top_\sqcap^{(0)} := (G_0, G_0^{I_0})$ and $\top_\sqcap^{(0)} := (G_0, \emptyset)$. Rooted extents are needed in order to identify the graphs with certain extents of the context $\mathbb{K}_\nabla^{inf}(\overrightarrow{\mathbb{K}})$. (An extent which is not rooted would correspond to a graph which has an edge but is missing at least one of the adjacent vertices.) Now we are able to formulate the desired theorem:

Theorem 16 (Basic Theorem on Protoconceptual Contents of $\overrightarrow{\mathbb{K}}$).

For a power context family $\overrightarrow{\mathbb{K}}$ of limited type n the protoconceptual contents of the protoconcept graphs of $\overrightarrow{\mathbb{K}}$ are exactly the rooted extents of the corresponding conceptual information context $\mathbb{K}_{\triangledown}^{inf}(\overrightarrow{\mathbb{K}})$.

Proof. From Theorem 15 follows that for any protoconcept graph $\mathfrak{G}$ and all $k = 0, 1, \ldots, n$ the closure $C_{\triangledown}^{(k)}(\mathfrak{G})$ is an extent of $\mathbb{K}_{\triangledown}^{inf}(\mathbb{K}_k)$. Therefore, the set $C_{\triangledown}(\mathfrak{G}) = C_{\triangledown}^{(0)}(\mathfrak{G}) \dot{\cup} C_{\triangledown}^{(1)}(\mathfrak{G}) \dot{\cup} \ldots \dot{\cup} C_{\triangledown}^{(n)}(\mathfrak{G})$ is an extent of $\mathbb{K}_{\triangledown}^{inf}(\overrightarrow{\mathbb{K}}) = \mathbb{K}_{\triangledown}^{inf}(\mathbb{K}_0) + \mathbb{K}_{\triangledown}^{inf}(\mathbb{K}_1) + \ldots + \mathbb{K}_{\triangledown}^{inf}(\mathbb{K}_n)$, which is by definition of $C_{\triangledown}(\mathfrak{G})$ rooted. Now, let U be a rooted extent of $\mathbb{K}_{\triangledown}^{inf}(\overrightarrow{\mathbb{K}})$. For all $k = 0, 1, \ldots, n$ the set $U_k := U \cap G_k$ is an extent of $\mathbb{K}_{\triangledown}^{inf}(\mathbb{K}_k)$, i.e. a protoconceptual content of $\mathbb{K}_k$. We define a protoconcept graph $\mathfrak{G} := (V, E, \nu, \kappa, \varrho)$ with $V := U_0$ and $E := \bigcup_{k=1,\ldots,n} U_k$, and the mappings are defined by $\nu\left((g_1, \ldots, g_k), \mathfrak{p}_k\right) := \left((g_1, \top_{\triangledown}^{(0)}), \ldots, (g_k, \top_{\triangledown}^{(0)})\right)$ (well-defined because U is rooted), $\kappa(g, \mathfrak{p}_0) := \mathfrak{p}_0$, $\kappa((g_1, \ldots, g_k), \mathfrak{p}_k) := \mathfrak{p}_k$ and $\varrho(g, \mathfrak{p}_0) := \{g\}$ (we have $\varrho^+ = \varrho$). It is easy to verify that this protoconcept graph has U as protoconceptual content. $\square$

4   Comparison to Former Approaches

After the discussion of the protoconceptual content of protoconcept graphs respecting object implications, we will now compare this result to previous work on protoconceptual and conceptual contents. The approaches have been basically the same: First, an information context $\mathbb{K}^{inf}(\mathbb{K})$ is constructed such that the extents of this context are exactly the (proto-)conceptual contents (with regard to the respective implications), then the information context $\mathbb{K}^{inf}(\overrightarrow{\mathbb{K}})$ is defined as direct sum of the information contexts of each single formal context. Because the second step is virtually the same (with only minor adaptions), we will concentrate on investigating how the definitions of the information context in the first step differ.

Influence of Object Implications

The problem solved in [HK04] was very similar to the one approached in the last section, the difference being that we did not consider object implications (the problem which finally led to the approach taken in this paper) and we restricted ourselves to protoconcept implications of the form ($\sqcap$). The information context for protoconceptual contents without object implications of a formal context $\mathbb{K} := (G, M, I)$ looks as follows:

$$\mathbb{K}_{old}^{inf}(\mathbb{K}) := (\mathfrak{P}^{inst}(\mathbb{K}), M^{irr}(\mathbb{K}), \in)$$

where $I(h) := \{\mathfrak{p} \in \mathfrak{P}(\mathbb{K}) \mid \mathfrak{p} \sqsubset (G \setminus \{h\}, (G \setminus \{h\})^I)\}$ for $h \in G$ and $M^{irr}(\mathbb{K}) := M_{\perp}^{irr}(\mathbb{K}) \cup M_{co}^{irr}(\mathbb{K})$ with

$$M_{\perp}^{irr}(\mathbb{K}) := \{\mathfrak{P}^{inst}(\mathbb{K}) \setminus (\{g\} \times \mathfrak{P}(\mathbb{K})) \mid g \in G\} \text{ and}$$
$$M_{co}^{irr}(\mathbb{K}) := \{\mathfrak{P}^{inst}(\mathbb{K}) \setminus (\{g\} \times I(h)) \mid g, h \in G, g \neq h\}.$$

The main difference in the two approaches is the direction of the mapping. While the present one uses a mapping assigning sets of objects (namely the attribute extents) to protoconcepts, the approach taken in [HK04] assigned sets of protoconcepts (actually closures with respect to the protoconcept implications) to each object, otherwise the constructions are similar.

However, as the two problems are so close, there should be a possibility to give a variant of our information context to match the case of protoconcept implications without object implications. A closer look at our approach shows that the extents we use for the objects sets are actually the closures generated by the object implications. Thus, if we remove the object implications gained by (2_∇), the closure system on the object set changes. Even without any special implications, we have still the trivial ones, that any set of objects implies itself or a subset. Therefore, if we consider any arbitrary set as closure (i. e. the closure system is just $\mathfrak{P}(G)$), we can follow the same steps. For the information context of protoconceptual contents with object implications we used the fact, that every extent can be described by the intersection of attribute extents. Likewise, we can describe an arbitrary set of objects by an intersection of sets of the form $G \setminus \{g\}$ for $g \in G$. This leads to the following context:

$$\mathbb{K}_{\not\nearrow}^{inf}(\mathbb{K}) := (\mathfrak{P}^{inst}(\mathbb{K}), \mathfrak{D}_\sqcap \times G, \Delta^-)$$

with $((g, \mathfrak{p}), (\mathfrak{d}, h)) \in \Delta^- :\Leftrightarrow$ If $\mathfrak{p} \to \mathfrak{d}$ then $g \neq h$.

Lemma 17. *The structure $\mathfrak{B}(\mathbb{K}_\sqcap^{inf}(\mathbb{K}))$ is isomorphic to a complete $\bigwedge$-sub-semilattice of $\mathfrak{B}(\mathbb{K}_{\not\nearrow}^{inf}(\mathbb{K}))$.*

Proof. Let $(\mathfrak{d}, m) \in \mathfrak{D}_\sqcap \times M$. It is easy see that $(\mathfrak{d}, m)^\Delta = \{(\mathfrak{d}, g) \mid g \in m^I\}^{\Delta^-}$. Therefore, all attribute extents of $\mathbb{K}_\sqcap^{inf}(\mathbb{K})$ are extents of $\mathbb{K}_{\not\nearrow}^{inf}(\mathbb{K})$. The set $\mathfrak{P}^{inst}(\mathbb{K})$ is in both lattices the extent of the top concept. As the extent of the meet of concepts is the intersection on the extents of the concepts and the extent of the top concept of $\mathfrak{B}(\mathbb{K}_{\not\nearrow}^{inf}(\mathbb{K}))$ is the same as of the one in $\mathfrak{B}(\mathbb{K}_\sqcap^{inf}(\mathbb{K}))$ (for the infimum of the empty set), the assertion follows.   □

We see that adding information about object implications is reducing the set of possible protoconceptual contents.

Protoconceptual and Conceptual Contents

In this part, we will compare our result with the result for conceptual contents of concept graphs, i. e. using object implications on the object side, but only concepts on the conceptual side.

In [Wi03] the information context is defined as

$$\mathbb{K}^{inf}(\mathbb{K}) := (\mathfrak{B}^{inst}(\mathbb{K}), \mathfrak{S}^{con}(\mathfrak{B}(\mathbb{K})), \bar{\Delta})$$

with $(g, \mathfrak{b})\bar{\Delta}S \iff [\gamma g, \mathfrak{b}] \cap S \neq \emptyset$, where $\mathfrak{S}^{con}(\mathfrak{B}(\mathbb{K}))$ is the set of all convex subsets S of $\mathfrak{B}(\mathbb{K}) \setminus \{(\emptyset, \emptyset^{II})\}$ for which $S \cup \{(\emptyset, \emptyset^{II})\}$ is a complete sublattice of the interval $[(\emptyset, \emptyset^{II}), \bigvee S]$ of $\underline{\mathfrak{B}}(\mathbb{K})$. The corresponding theorem is however restricted to formal contexts with $M^I = \emptyset$.

To compare more easily the conceptual contents of the approach in [Wi03] with the results in this paper, we provide a version more similar to ours. In both approaches object implications are considered (and equally defined). The approach taken with the protoconcepts can easily be transferred to the concept case, considering the similarity of the operations ∇ and $\wedge$ for the (proto-) conceptual implications and for the description of each (proto-)concept of the algebra. Therefore, the argumentation done in Section 3 can be transferred to the case of concepts. We have to define the set of describing concepts analogous to $\mathfrak{D}_\nabla$. As can be seen easily, the elements of $\mathfrak{D}_\nabla$ are the $\wedge$-irreducible elements in the algebra $\mathfrak{H}_\nabla$. For the case of concepts, we will use the (irreducible) attribute concepts (cf. Remark 14). The analog of the translation from mappings to conceptual contents is defined in the following way: Let $\mathfrak{D}$ be the set of (irreducible) attribute concepts and $\varphi : \mathfrak{D} \to \mathrm{Ext}(\mathbb{K})$ be a mapping from those concepts to the set of extents of the formal context. Then we define

$$Y^\varphi := \bigcup_{\mathfrak{c} \in \mathfrak{B}(\mathbb{K})} A_\mathfrak{c}^\varphi \times \{\mathfrak{c}\}$$

$$\text{with } A_\mathfrak{c}^\varphi := \mathrm{Ext}(\mathfrak{c}) \cap \bigcap_{\substack{\mathfrak{d} \in \mathfrak{D} \\ \mathfrak{c} \leq \mathfrak{d}}} \varphi(\mathfrak{d}).$$

Please note, that for the case of single concepts we have $\mathfrak{c} \to \mathfrak{d} \Leftrightarrow \mathfrak{c} \leq \mathfrak{d}$. Instead of using the attribute concepts from $\mathfrak{D}$, we substitute them by the attributes themselves for the information context. If $m_\mathfrak{d}$ generates $\mathfrak{d}$, then we have $\mathfrak{c} \leq \mathfrak{d} \Leftrightarrow m \in \mathrm{Int}(\mathfrak{c})$. This leads to the following information context:

$$\mathbb{K}_\mathfrak{B}^{inf}(\mathbb{K}) := (\mathfrak{B}^{inst}(\mathbb{K}), M \times M, \Delta_\mathfrak{B})$$

with $((g, \mathfrak{c}), (d, m)) \in \Delta_\mathfrak{B} :\Leftrightarrow$ If $d \in \mathrm{Int}(\mathfrak{c})$ then $(g, m) \in I$.

As the information contexts for protoconceptual and conceptual contents have different object sets, it is helpful to establish a closer relation between the two. As $\sqcap$ restricted to the set of concepts is equal to $\wedge$, we obtain the following equalities

Lemma 18. *For the case ($\sqcap$) we have that*

- *For any protoconceptual content X the set $X \cap \mathfrak{B}^{inst}(\mathbb{K})$ is a conceptual content,*
- *For any conceptual content Y we have $Y^{\triangle\triangle} \cap \mathfrak{B}^{inst}(\mathbb{K}) = Y$*

Proof. Obviously, we have $Y := X \cap \mathfrak{B}^{inst}(\mathbb{K}) \subseteq \mathfrak{B}^{inst}(\mathbb{K})$, therefore we have to show that Y is closed under Equation (3). Let $A, C \subseteq G$ and $\mathfrak{B}, \mathfrak{D} \subseteq \mathfrak{B}(\mathbb{K})$ with $A \to C$ and $\mathfrak{B} \to \mathfrak{D}$. The latter is meant in the sense of concept implications, however due to $\wedge \mathfrak{B} \leq \wedge \mathfrak{D} \Leftrightarrow (\bigcap\{\mathrm{Ext}(\mathfrak{b}) \mid \mathfrak{b} \in \mathfrak{B}\} \subseteq \bigcap\{\mathrm{Ext}(\mathfrak{d}) \mid \mathfrak{d} \in \mathfrak{D}\}) \Leftrightarrow \sqcap \mathfrak{B} \sqsubseteq \sqcap \mathfrak{D}$ it is also a protoconceptual implication, and as object implications are the same we get $C \times \mathfrak{D} \subseteq X$. Moreover, $C \times \mathfrak{D} \subseteq \mathfrak{B}^{inst}(\mathbb{K})$ is obvious, proving the first assertion.

Now let $Y \in \mathcal{C}(\mathfrak{B}^{inst}(\mathbb{K}))$ be an arbitrary conceptual content. We know that $Y^{\Delta\Delta}$ is a protoconceptual content and therefore $Z := Y^{\Delta\Delta} \cap \mathfrak{B}^{inst}(\mathbb{K})$ is a conceptual content and we have $Y \subseteq Z$. Let $\mathfrak{B} \subseteq \mathfrak{B}(\mathbb{K})$ and $\mathfrak{D} \subseteq \mathfrak{P}(\mathbb{K})$. Then we have for a protoconceptual implication $\mathfrak{B} \to \mathfrak{D} \Leftrightarrow \bigcap\{\mathrm{Ext}(\mathfrak{b}) \mid \mathfrak{b} \in \mathfrak{B}\} \subseteq \bigcap\{\mathrm{Ext}(\mathfrak{d}) \mid \mathfrak{d} \in \mathfrak{D}\} \Leftrightarrow \bigwedge \mathfrak{B} \sqsubseteq \bigcap \mathfrak{D}$ (the latter because both are $\mathfrak{H}_\sqcap$-semiconcepts). Therefore the set

$$Y_{\mathfrak{P}} := \bigcup_{A \times \mathfrak{B} \subseteq Y} A^{II} \times \{\mathfrak{p} \in \mathfrak{P}(\mathbb{K}) \mid \bigwedge \mathfrak{B} \sqsubseteq \mathfrak{p}\}$$

is a protoconceptual content containing Y but no additional $(g, \mathfrak{c}) \in \mathfrak{B}^{inst}(\mathbb{K})$ (otherwise Y were not closed under conceptual implications) and therefore $Y_{\mathfrak{P}} \cap \mathfrak{B}^{inst}(\mathbb{K}) = Y$ from which the second assertion follows. □

For the case ($\sqcap\!\!\!\sqcap$) we have a different notion of implications, such that $\mathfrak{B} \to \bigwedge \mathfrak{B}$ only if $\bigcup\{\mathrm{Int}(\mathfrak{b}) \mid \mathfrak{b} \in \mathfrak{B}\}$ is a concept intent. While all protoconceptual implications between sets of concepts translate into conceptual implications, the reverse is not true for the case ($\sqcap\!\!\!\sqcap$). For this reason, we will concentrate on the case ($\sqcap$) and get the following result:

Lemma 19. $\underline{\mathfrak{B}}(\mathbb{K}_{\mathfrak{B}}^{inf}(\mathbb{K}))$ *is isomorphic to a* $\wedge$*-sub-semilattice of* $\underline{\mathfrak{B}}(\mathbb{K}_{\sqcap}^{inf}(\mathbb{K}))$.

Proof. It is easy to see that the mapping $(A, B) \mapsto (A^{\Delta\Delta}, A^\Delta)$ for (A, B) is a $\wedge$-preserving homomorphism from $\underline{\mathfrak{B}}(\mathbb{K}_{\mathfrak{B}}^{inf}(\mathbb{K}))$ to $\underline{\mathfrak{B}}(\mathbb{K}_{\sqcap}^{inf}(\mathbb{K}))$. By Lemma 18 we see that $A_1^{\Delta\Delta} = A_2^{\Delta\Delta} \Rightarrow A_1 = A_1^{\Delta\Delta} \cap \mathfrak{B}^{inst}(\mathbb{K}) = A_2^{\Delta\Delta} \cap \mathfrak{B}^{inst}(\mathbb{K}) = A_2$, i.e. the mapping is injective. (The mapping actually preserves infima of arbitrary non-empty sets.) □

5   A More General Approach

In the previous section we have inspected the different structures of closure systems resulting from various versions of implications on objects, concepts and protoconcepts. Now we will try to generalize the approach taken in all those cases. In each case we had implications on the objects (in the case of [HK04] we assume the trivial implications) and also implications on the (proto-)concepts. The latter implications were defined by operations on sets of (proto-)concepts. Here we will abstract from those concrete definitions and investigate how the closure systems of sets of instances vary with the closure systems on objects and concepts.

Let $\mathcal{L}_G \subseteq G \times G$ and $\mathcal{L}_{\mathfrak{A}} \subseteq \mathfrak{A} \times \mathfrak{A}$. In the following, we will refer to the elements of $\mathfrak{A}$ as concepts, but they are used as a primitive notion, i.e. there are no assumptions with regard to their extent and intent. Therefore, $\mathfrak{A}$ could be a set of concepts or protoconcepts or even other mathematical objects.

For a pair $(A, B) \in \mathcal{L}_G$ we write $A \to B$ (and analogous $\mathfrak{B} \to \mathfrak{C}$ for $(\mathfrak{B}, \mathfrak{C}) \in \mathcal{L}_{\mathfrak{A}}$). Both sets $\mathcal{L}_G$ and $\mathcal{L}_{\mathfrak{A}}$ are supposed to fulfill the so-called Armstrong axioms (we will use the notation $\mathcal{L}_G$ only):

Definition 20 (Armstrong Axioms).

Let $A, B, C \subseteq G$, then

1. $B \subseteq A \Rightarrow A \to B$
2. $A \to B \Rightarrow A \cup C \to B \cup C$
3. $A \to B$ *and* $B \to C \Rightarrow A \to C$

It can easily be seen that the implications considered in the previous concrete examples all comply with these axioms. Next, we need a base set $P \subseteq G \times \mathfrak{A}$ which corresponds to $\mathfrak{B}^{inst}(\mathbb{K})$ and $\mathfrak{P}^{inst}(\mathbb{K})$ in the concrete approaches. A *general closure* is defined as a subset $Y \subseteq P$ such that for $A, C \subseteq G$ and $\mathfrak{B}, \mathfrak{D} \subseteq \mathfrak{A}$ with $A \to C$ and $\mathfrak{B} \to \mathfrak{D}$ we have

$$\text{If } A \times \mathfrak{B} \subseteq Y \text{ and } A \to C, \mathfrak{B} \to \mathfrak{D} \text{ then } (C \times \mathfrak{D}) \cap P \subseteq Y \qquad (3_\mathfrak{A})$$

In the following $\mathcal{C}(A)$ and $\mathcal{C}(\mathfrak{B})$ will denote the closures of the sets $A \subseteq G$ and $\mathfrak{B} \subseteq \mathfrak{A}$ with regard to $\mathcal{L}_G$ and $\mathcal{L}_\mathfrak{A}$ and $\mathcal{C}(\mathcal{L}_G)$ and $\mathcal{C}(\mathcal{L}_\mathfrak{A})$ the set of these closures respectively. We see that for any closure Y we have $Y = P \cap \bigcup_{A \times \mathfrak{B} \subseteq Y} \mathcal{C}(A) \times \mathcal{C}(\mathfrak{B})$, i.e. any closure can be represented by the union of products of closures. Using the notations of formal concept analysis, we get:

Lemma 21. *Let $Y \subseteq P \subseteq G \times \mathfrak{A}$ be a general closure. Then*

$$Y = \bigcup\{\text{Ext}(\mathfrak{c}) \times \text{Int}(\mathfrak{c}) \mid \mathfrak{c} \in \mathfrak{B}(G, \mathfrak{A}, Y)\}.$$

Proof. By definition we have for any $\mathfrak{c} \in \mathfrak{B}(G, \mathfrak{A}, Y)$ that $\text{Ext}(\mathfrak{c}) \times \text{Int}(\mathfrak{c}) \subseteq Y$, therefore the union is contained in Y. On the other hand, we have $(g, \mathfrak{c}) \in (\{g\}^{YY}, g^Y) \in \mathfrak{B}(G, \mathfrak{A}, Y)$ and therefore Y is contained in the union. $\square$

However, not all closures necessarily appear as extent (or intent) of a concept of $\mathfrak{B}(G, \mathfrak{A}, Y)$.

We are looking for a shorter description. We will apply the following Lemma, which directly follows from the main theorem in formal concept analysis:

Lemma 22. *Let $\mathfrak{B}, \mathfrak{D} \subseteq \mathfrak{A}$. Then we have $(\mathfrak{B} \cup \mathfrak{D})^Y = \mathfrak{B}^Y \cap \mathfrak{D}^Y$.*

As we see, we have intersection on one side and union on the other. Let $\mathcal{D} \subseteq \mathcal{C}(\mathcal{L}_\mathfrak{A})$ be a set of concept closures such that for all $\mathfrak{B} \in \mathcal{C}(\mathcal{L}_\mathfrak{A})$ we have $\mathfrak{B} = \mathcal{C}(\bigcup\{\mathfrak{D} \in \mathcal{D} \mid \mathfrak{B} \to \mathfrak{D}\})$, i.e. every closure is generated by a union of elements from $\mathcal{D}$.

In view of the above lemma it suffices to know for any $\mathfrak{D} \in \mathcal{D}$ the set $\mathfrak{D}^Y$ in order to calculate for any subset $\mathfrak{C} \subseteq \mathfrak{A}$ the set $\mathfrak{C}^Y = \bigcap_{\mathfrak{c} \in \mathfrak{C}} \mathfrak{c}^Y = \bigcap_{\mathfrak{c} \in \mathfrak{C}} \bigcap_{\mathfrak{D} \in \mathcal{D}, \mathfrak{c} \to \mathfrak{D}} \mathfrak{D}^Y$. Hence we get the following result:

Lemma 23. *Let $\varphi : \mathcal{D} \to \mathcal{C}(\mathcal{L}_G)$ be an arbitrary mapping. Then*

$$Y^\varphi := P \cap \bigcup_{\mathfrak{c} \in \mathfrak{A}} A_\mathfrak{c}^\varphi \times \{\mathfrak{c}\} \text{ with } A_\mathfrak{c}^\varphi := \bigcap_{\substack{\mathfrak{D} \in \mathcal{D} \\ \mathfrak{c} \to \mathfrak{D}}} \varphi(\mathfrak{D})$$

is a general closure.

Proof. As all $A_{\mathfrak{c}}^{\varphi}$ and $\mathfrak{D} \in \mathcal{D}$ are closures, we have $A \subseteq A_{\mathfrak{c}}^{\varphi} \Leftrightarrow \mathcal{C}(A) \subseteq A_{\mathfrak{c}}^{\varphi}$ and $\mathfrak{B} \to \mathfrak{D} \Leftrightarrow \mathcal{C}(\mathfrak{B}) \to \mathfrak{D}$ and therefore

$$A \times \mathfrak{B} \subseteq Y^{\varphi} \Leftrightarrow A \subseteq \bigcap_{\mathfrak{c} \in \mathfrak{B}} A_{\mathfrak{c}}^{\varphi} \Leftrightarrow \mathcal{C}(A) \subseteq \bigcap_{\mathfrak{c} \in \mathcal{C}(\mathfrak{B})} A_{\mathfrak{c}}^{\varphi} \Rightarrow P \cap (\mathcal{C}(A) \times \mathcal{C}(\mathfrak{B})) \subseteq Y^{\varphi}. \quad\square$$

Further, for any closure Y it is easy to verify that for the mapping φ_Y defined by $\varphi_Y(\mathfrak{D}) := \mathcal{C}(\mathfrak{D}^Y)$ for $\mathfrak{D} \in \mathcal{D}$ we get $Y = Y^{\varphi_Y}$, i.e. the set of mappings from $\mathcal{D}$ to $\mathcal{C}(\mathcal{L}_G)$ generate all general closures. Like the closures, the mappings are closed under intersection, so we are looking for a subset of mappings generating all others via intersection.

As the images are closures from $\mathcal{C}(\mathcal{L}_G)$, we can describe them by the $\bigcap$-irreducible elements of $\mathcal{C}(\mathcal{L}_G)$. Let $\mathcal{M} \subseteq \mathcal{C}(\mathcal{L}_G)$ be the set of these irreducible closures, then we can generate all mappings from the mappings of the form $\varphi_{\mathfrak{E}}^M$ with $(\mathfrak{E}, M) \in \mathcal{D} \times \mathcal{M}$ defined by $\varphi_{\mathfrak{E}}^M(\mathfrak{D}) := M$ if $\mathfrak{D} = \mathfrak{E}$ and $\varphi_{\mathfrak{E}}^M = G$ otherwise. Therefore, this yields the following information context:

Theorem 24 (Basic Theorem on General Closures).
Let $\mathbb{K}_{\mathfrak{A}}^{inf}(P, \mathcal{L}_G, \mathcal{L}_{\mathfrak{A}}) := (P, \mathcal{D} \times \mathcal{M}, \Delta_{\mathfrak{A}})$ with $((g, \mathfrak{c}), (\mathfrak{D}, M)) :\Leftrightarrow (If\ \mathfrak{c} \to \mathfrak{D}\ then\ g \in M)$. Then the extents of the concepts in $\mathfrak{B}(\mathbb{K}_{\mathfrak{A}}^{inf}(P, \mathcal{L}_G, \mathcal{L}_{\mathfrak{A}}))$ are exactly the general closures with respect to $(\exists_{\mathfrak{A}})$.

Remark 25. At a first look, the information context here seems to be more complicated as the attributes are not pairs of elements as in the concrete approaches but pairs of sets. The reason for this is of course, that the concrete examples provide possibilities to simplify this part. If using object implications as defined in (1) we can describe the $\bigcap$-irreducible closures of $\mathcal{C}(\mathcal{L}_G) = \text{Ext}(G, M, I)$ by the (irreducible) attributes, i.e. for each $N \in \mathcal{M}$ exists an attribute $m \in M$ such that $N = m^I$. If we do not use object implications, the $\bigcap$-irreducible elements are of the form $G \setminus \{h\}$ and allow a simpler representation.

On the (proto-)concept side we had in all cases that for any set of (proto-)concepts $\mathfrak{B}$ we have $\mathcal{C}(\mathfrak{B}) = \mathcal{C}(\nabla \mathfrak{B})$ with $\nabla \in \{\sqcap, \barwedge, \wedge\}$ respectively. Therefore we can replace the closure $\mathfrak{D} \in \mathcal{D}$ by the (proto-)concept $\nabla \mathfrak{D}$, again resulting in a simplification of the representation. Finally, Lemma 8 also suggests to replace the set $\mathcal{D}_\nabla$ of describing protoconcepts by G in the case $(\sqcap)$ or by M otherwise.

6 Conclusion

The solution to the problem of protoconceptual contents respecting object implications presented in Section 3 finalizes the approach started in [HK04]. The influence of background knowledge on the resulting structure of contents has been exemplified in Section 4.

The generalization of all cases as shown in Section 5 allows to be even more flexible. For instance, it might be desirable to restrict the additional implications on objects and (proto-)concepts in (1), (2) and $(2)_\nabla$ to non-empty premises, as they are supposed to correspond to material implications (see [Wi02, Wi03]). Using the results from the generalized approach we can see how this decision influences the structure of contents.

While this approach is rather general with regard to the kind of object and (proto-)concept implications, it does not allow implications between different contexts, for instance "if $((g_1, g_2), \mathfrak{c}) \in Y$ then $((g_1), \mathfrak{d}) \in Y$ for $\mathfrak{c} \in \mathfrak{B}(\mathbb{K}_2)$ and $\mathfrak{d} \in \mathfrak{B}(\mathbb{K}_1)$". Of course, then we cannot restrict ourselves to the investigation of contents of only one context but have to consider the whole power context family. This seems to be an interesting area for further developing the notion of (proto-)conceptual content of (proto-)concept graphs.

Acknowledgements

The authors would like to thank the anonymous reviewers for their detailed and constructive comments. The first author was visitor at the Centro de Ciências Matemáticas at the University of Madeira (Portugal) and appreciated the kind hospitality and support while working on this article.

References

[DK03]   F. Dau, J. Klinger: From Formal Concept Analysis to Contextual Logic. FB4-Preprint, TU Darmstadt 2003.

[GW99]   B. Ganter, R. Wille: Formal Concept Analysis: Mathematical Foundations. Springer Verlag, Berlin–New York 1999.

[HK04]   J. Hereth Correia, J. Klinger: Protoconcept Graphs: The Lattice of Conceptual Contents. In: P. Eklund (Ed.): Concept Lattices, Springer, Berlin Heidelberg New York 2004, 14–27.

[PW99]   S. Prediger, R. Wille: The Lattice of Concept Graphs of a Relationally Scaled Context. In: W. Tepfenhart, W. Cyre (Eds.): Conceptual Structures: Standards and Practices, Springer Verlag, Berlin – New York 1999, 401-414.

[Wi97]   R. Wille: Conceptual Graphs and Formal Concept Analysis. In: D. Lukose, H. Delugach, M. Keeler, L. Searle, J. Sowa (Eds.): Conceptual Structures: Fulfilling Peirce's Dream. Springer, Berlin - Heidelberg - New York 1997, 290 - 303.

[Wi00]   R. Wille: Boolean Concept Logic. In: B. Ganter, G.W. Mineau (Eds.): Conceptual Structures: Logical, Linguistic, and Computational Issues, Springer Verlag, Berlin–New York 2000, 317-331.

[Wi00b] R. Wille: Contextual Logic Summary. In: G. Stumme (Ed.): Working with Conceptual Structures. Contributions to ICCS 2000. Shaker, Aachen 2000, 256-276.

[Wi02]   R. Wille: Existential Concept Graphs of Power Context Families. In: U. Priss, D. Corbett, G. Angelova (Eds.): Conceptual Structures: Integration and Interfaces, Springer Verlag, Berlin–New York 2002, 382-396.

[Wi03]   R. Wille: Conceptual Content as Information - Basics for Contextual Judgment Logic. In: A. de Moor, W. Lex, B. Ganter (Eds.): Conceptual Structures for Knowledge Creation and Communication. Springer Verlag, Berlin–New York 2003, 1-15.

Planarity of Lattices

An Approach Based on Attribute Additivity

Christian Zschalig

Institut für Algebra, TU Dresden, Germany
zschalig@math.tu-dresden.de

Abstract. Popular lattice drawing algorithms do not take planarity into account and find plane diagrams mainly heuristically. We present a characterization of planar lattices based on a theorem of Dushnik and Miller [4] and the "left"-relation introduced by Kelly and Rival [6]. In particular, our work is helpful for drawing plane attribute additive diagrams.

1 Motivation

A lattice is *planar* if it admits a diagram with no edge crossings. There exist algorithms for constructing such plane diagrams (see [3] for an overview), but these do not use the lattice structure and treat the problem as a graph drawing task. Our aim is to automatically construct plane diagrams of planar lattices. Additionally we want to draw them *attribute additively* [7] since this convention provides nice visualizations of lattices.

2 Introduction

Throughout the paper we assume finiteness. For easier notation we use the symbols $\leq$ and $<$ both for lattice order relations and the usual order on $\mathbb{R}$.

2.1 Diagrams of Lattices

A lattice $(\mathfrak{V}, \leq)$ is often represented by a diagram. We draw a small circle for each lattice element and a line for each pair v, w of lattice elements in neighbour relation (i.e. $v < w$ and there is no element z fulfilling $v < z < w$). Lattice diagrams are drawn upward. We define, according to [6]:

Definition 1. *Let $\underline{\mathfrak{V}} = (\mathfrak{V}, \leq)$ be a lattice with the neighbour relation $\prec$. A* diagram *(or* representation *[6]) pos($\underline{\mathfrak{V}}$) of $\underline{\mathfrak{V}}$ is the image of a mapping*

$$\text{pos} : \mathfrak{V} \cup \prec \mapsto \mathbb{R}^2 \cup \mathcal{P}(\mathbb{R}^2)$$

meeting the following conditions for all $v, w, z \in \mathfrak{V}$.

B. Ganter and R. Godin (Eds.): ICFCA 2005, LNAI 3403, pp. 391–402, 2005.
© Springer-Verlag Berlin Heidelberg 2005

1. pos $|_{\mathfrak{V}} : v \mapsto \mathrm{pos}(v) = (x(v), y(v)) \in \mathbb{R}^2$ is an injection.
2. Whenever $v < w$ holds then $y(v) < y(w)$.
3. pos $|_{\prec} : vw \mapsto \{(x_{vw}(y), y) \mid y \in [y(v), y(w)]\} \subseteq \mathbb{R}^2$, where x_{vw} is a continuous function with $x_{vw}(y(v)) = x(v)$ and $x_{vw}(y(w)) = x(w)$ for each pair $v \prec w$.
4. If $\mathrm{pos}(v) \in \mathrm{pos}(wz)$ holds then $v = w$ or $v = z$.

The elements of $\mathrm{pos}(\mathfrak{V}) := \{\mathrm{pos}(v) \mid v \in \mathfrak{V}\}$ are called (diagram) points or nodes, the elements of $\mathrm{pos}(\prec) := \{\mathrm{pos}(vw) \mid v, w \in \mathfrak{V}, v \prec w\}$ are called diagram edges.

Line diagrams are more common, here the diagram edges are just straight line segments.

Definition 2. A *line diagram* (also called *embedding* [6], *Hasse diagram* [1] or simply diagram [2]) of a lattice $\underline{\mathfrak{V}}$ is a diagram (as previously defined), where

$$\mathrm{pos}(vw) = \{t \cdot \mathrm{pos}(v) + (1 - t) \cdot \mathrm{pos}(w) \mid t \in [0, 1]\}.$$

holds for all elements $v \prec w$.

A lattice is planar if it possesses a plane line diagram, i.e. if no diagram edges intersect [6].

2.2   Lattices and Planarity

In this subsection we give some lattice properties characterizing whether a lattice is planar or not.

Definition 3. [4] The (order) dimension $\dim(\underline{P})$ of an ordered set $\underline{P}$ is the smallest cardinal number m such that $\leq$ is the intersection of m linear orders.

Definition 4. [4] A conjugate order L_c on an ordered set $\underline{P} = (P, \leq)$ is a relation meeting the following conditions ($\|$ denotes the incomparability relation in $\underline{P}$).

1. L_c is a strict order
2. $L_c \cup L_c^{-1} = \|$.

Theorem 1. [4] Let $\underline{P} = (P, \leq)$ be an ordered set. Then the following are equivalent:

1. $D(\underline{P}) \leq 2$
2. There exists a conjugate order L_c on $\underline{P}$.

Sketch of proof: Together with L_c, also $R_c := L_c^{-1}$ is a conjugate order. It is easy to show that $L_c \cup \leq$ and $R_c \cup \leq$ are linear orders, the intersection of which is $\leq$. On the other hand, if $K \supseteq \leq$ is a linear order on P then we can show that $K \setminus \leq$ is a conjugate order.

There is an exercise in [2] (p.32, ex. 7c) stating that a finite lattice $\mathfrak{V}$ is planar if and only if there exists a conjugate order on $\mathfrak{V}$. The "$\Rightarrow$" part was proved in [6], see Corollary 1. For the "$\Leftarrow$" part we did not find a published proof, therefore we shall give one in Theorem 3. The above-mentioned result was used to get the well-known characterization.

Theorem 2. *[1] A finite lattice is planar if and only if its order dimension is at most two.*

2.3   Diagrams and Planarity

This section discusses properties of plane diagrams (or, as stated in [6] the "geometry of planar lattices"). In particular, a relation on diagram nodes is introduced which indicates a node to be left (or dually right) of another. In case of plane diagrams this relation can be understood as a conjugate order.

Definition 5. *[6] Let $\mathfrak{V} = (\mathfrak{V}, \leq)$ be a lattice and $\prec$ its neighbour relation. A maximal chain C is a sequence $0_{\mathfrak{V}} = z_0 \prec z_1 \prec \ldots \prec z_n = 1_{\mathfrak{V}}$ of lattice elements z_i. In a diagram $\mathrm{pos}(\mathfrak{V})$ we define the function $x_C : [y(0_{\mathfrak{V}}), y(1_{\mathfrak{V}})] \mapsto \mathbb{R}$ corresponding to C by $x_C(y) = x_{z_i z_{i+1}}(y)$ if $y \in [y(z_i), y(z_{i+1})]$ holds for all $0 \leq i \leq n-1$. Additionally we define $\mathrm{pos}(C) = \{(x_C(y), y) \mid y \in [y(0_{\mathfrak{V}}), y(1_{\mathfrak{V}})]\}$.*

This means that the function $\mathrm{pos}(C)$ is just the join of the sets $\mathrm{pos}(z_i z_{i+1})$ as defined in Definition 1. The function x_C is continuous.

Definition 6. *[6] Let $\mathfrak{V} = (\mathfrak{V}, \leq)$ be a lattice and $\mathrm{pos}(\mathfrak{V})$ a plane diagram of it. Let $\lambda^* \subseteq \mathfrak{V}) \times \mathfrak{V}$ be a relation defined by*

$$v \, \lambda^* \, w : \Longleftrightarrow \quad \exists v^* \in \mathfrak{V} : \; v, w \prec v^* \text{ and}$$
$$x_{vv^*}(m) < x_{wv^*}(m),$$

where $m := \min\{y(z) \mid z \in \mathfrak{V}, z \prec v^*\}$.
Two diagrams are called similar *if their respective λ^* relations are the same.*
The left-relation *$\lambda \subseteq \mathfrak{V} \times \mathfrak{V}$ induced by $\mathrm{pos}(\mathfrak{V})$ is defined by*

$$v \, \lambda \, w : \Longleftrightarrow \quad v \parallel w \text{ and}$$
$$\exists v' \geq v, w' \geq w : \; v', w' \prec (v \vee w) \text{ with } v' \, \lambda^* \, w'.$$

If $v \, \lambda \, w$ holds, we say v is *left* of w. Dually we define $\varrho := \lambda^{-1}$ and say w is *right* of v.

It is shown in [6] that $v \, \lambda \, w$ is equivalent to the existence of a maximal chain $C \ni w$, where $x(v) < x_C(y(v))$ holds. This helps to prove the following.

Proposition 1. *[6] Let $\mathfrak{V}$ be a lattice and $\mathrm{pos}(\mathfrak{V})$ a plane diagram. The relation λ induced by $\mathrm{pos}(\mathfrak{V})$ is a strict order. Additionally $\lambda \cup \varrho = \parallel$ holds.*

Sketch of proof: From $v \, \lambda \, w$ and $w \, \lambda \, v$ we can conclude that there are two maximal chains $C \ni v$ and $D \ni w$ such that $\mathrm{pos}(C)$ and $\mathrm{pos}(D)$ intersect "between" (in terms of the y-coordinate) pos(v) and pos(w). As pos($\mathfrak{V}$) is plane, the intersection point represents a lattice element z fulfilling $v < z < w$ in contradiction to v and w be incomparable. By applying similar arguments we can show λ to be transitive. Hence it is a strict order.

Corollary 1. *Let $\mathfrak{V}$ be a lattice with a plane diagram* pos($\mathfrak{V}$) *and its induced left-relation λ. Then λ is a conjugate order on $\mathfrak{V}$.*

2.4   Attribute Additivity

When drawing lattices, most people tend to intuitively use (at least partially) the convention of attribute additivity. This method is particularly useful for distributive (or "nearly distributive") lattices, as the resulting diagrams look like they are drawn on an n-dimensional grid [8]. An example is given in Figure 1.

Definition 7. *Let $\underline{\mathfrak{B}}(G, M, I)$ be a concept lattice. A line diagram*

$$\text{pos}(\underline{\mathfrak{B}}(G, M, I))$$

is attribute additive if there is a map vec $: M \mapsto \mathbb{R}^2$, *such that the equation*

$$\text{pos}(A, B) = \sum_{m \in B} \text{vec}(m)$$

holds for all concepts $(A, B) \in \mathfrak{B}(G, M, I)$.

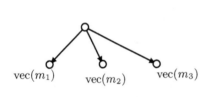

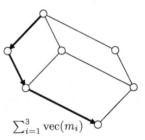

$\sum_{i=1}^{3} \text{vec}(m_i)$

Fig. 1. In the left picture each attribute is mapped to a vector in $\mathbb{R}^2$. This defines (together with the covering relation of the lattice) already the coordinates of other concepts, which are just the sum of all vectors of the attributes in the intent

3   The Left-Relation on Lattices

The characterization of plane diagrams by a left-relation λ is surprisingly intuitive. Comparable elements are understood to be situated below or above each other and the incomparable ones left or right. If the diagram is plane then the left-relation is a conjugate order. We will show in Theorem 3 that the converse holds as well. How can we find conjugate orders on a lattice $(\mathfrak{V}, \leq)$? One way mentioned in [1] is to observe, whether $(\mathfrak{V}, \|)$ is a comparability graph. Another one can be derived from Definition 4 by looking for all conjugate relations[1] L on $\mathfrak{V}$ and picking those, which are strict orders. Let v_- denote the set of all lower neighbours of an element $v \in \mathfrak{V}$. According to Definition 6 there are at most $\prod_{v \in \mathfrak{V}} |v_-|!$ non similar diagrams. This is an upper bound for the number of conjugate relations, which can be realized in a diagram $\text{pos}(\mathfrak{V})$, i.e. for the induced left-relation the equation $\lambda = L$ holds. By our attribute additive approach we can reduce the number of nonsimilar plane diagrams to $|M|!$, where M is the set of $\wedge$-irreducibles or alternatively in a concept lattice the set of attributes of a reduced context[2].

Definition 8. *Let $\mathfrak{V}$ be a finite lattice and $M = M(\mathfrak{V})$ be the set of its $\wedge$-irreducible elements. A strict order $L_a \subseteq M \times M$ is called a* sorting relation *if the following condition holds for all elements $m, n \in M$:*

$$m^* = n^* \iff m\,L_a\,n \text{ or } n\,L_a\,m.$$

The sorting relation just gives a relationship of $\wedge$-irreducibles with common upper neighbour. We extend it to the set of all pairs of incomparable elements.

Definition 9. *Let $\mathfrak{V}$ be a finite lattice with a given sorting relation L_a. For arbitrary lattice elements v and w, we define*

$$M(v, w) = \{(v', w') \subseteq M \times M \mid v \leq v', w \leq w', v \| w', w \| v'\}.$$

We define the relation $L \subseteq \mathfrak{V} \times \mathfrak{V}$ as follows:

$$v\,L\,w : \iff \begin{cases} v\,L_a\,w, & v, w \in M, v^* = w^* \\ \exists (m, n) \in M(v, w) : m\,L\,n, & else \end{cases}$$

L is called left-relation *and $R := L^{-1}$ is called* right-relation *on the lattice $\mathfrak{V}$.*

Consider the picture on the right for an example of calculating the left-relation on the depicted lattice for a given sorting relation. Notice that we are interested just in the underlying lattice, not in the particular diagram used. We assume $m_1\,L_a\,m_2$, i.e. $m_1\,L\,m_2$. Consider now the pair (m_3, v_1). We observe $(m_3, m_2) \in M(m_3, v_1)$ and $(m_1, m_2) \in M(m_3, m_2)$ and conclude $m_3\,L\,v_1$.

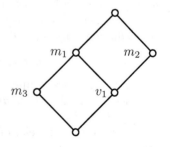

Remark 1. *Let $(\mathfrak{V}, \leq)$ be a finite lattice and M the set of its $\wedge$-irreducible elements. The following properties hold for all $v, w \in \mathfrak{V}$ and $m, n \in M$:*

1. $(m, n) \in M(v, w) \implies m \parallel n$
2. $M(v, w) = \emptyset \iff v \leq w$ or $w \leq v,$
3. $M(m, n) = \{(m, n)\} \implies m^* = n^*,$
4. $(m, n) \in M(v, w) \implies (m, n) \geq (v, w)$ $(: \iff m \geq v$ and $n \geq w)$.
5. $(m, n) \in M(v, w) \iff (n, m) \in M(w, v)$

We give as a first result that the above defined relation acts just on the incomparable pairs of elements.

Lemma 1. *For every left-relation from Definition 9, the identity $L \cup R = \parallel$ holds.*

Proof. Comparable elements obviously never are in left-relation since they are not in sorting relation and $M(v, w) = \emptyset$ holds.

Now let m and n be incomparable and $\wedge$-irreducible. Suppose that neither $m \, L \, n$ nor $n \, L \, m$ holds and that (m, n) is maximal with this property. Then the elements are not in sorting relation and with Remark 1 we know that there exists an element $(m, n) < (m', n') \in M(m, n)$. Since (m, n) is maximal we conclude $(m', n') \in L \cup R$ and with Definition 9 $(m, n) \in L \cup R$ which contradicts our assumption.

Let v and w be arbitrary incomparable lattice elements. There exists a pair of $\wedge$-irreducibles $(m_1, m_2) \in M(v, w)$. Since m_1 and m_2 are in left-relation, we conclude with Definition 9 that $(v, w) \in L \cup R$.

Now we want to show that every conjugate order is a left-relation. First we need the following:

Lemma 2. *Let $\underline{\mathfrak{V}}$ be a finite lattice and L_c be a conjugate partial order on $\underline{\mathfrak{V}}$. Then the following holds for all lattice elements v_1, v_2, w_1, w_2:*

$$v_1 \, L_c \, w_1 \text{ and } (v_1, w_1) \in M(v_2, w_2) \implies v_2 \, L_c \, w_2$$

Proof. Since $M(v_2, w_2) \neq \emptyset$, we know $v_2 \parallel w_2$ and with Definition 4 we conclude either $v_2 \, L_c \, w_2$ or $w_2 \, L_c \, v_2$. We assume $w_2 \, L_c \, v_2$. Since $(v_1, w_1) \in M(v_2, w_2)$ we notice that v_1 and w_2 are incomparable. We conclude that either $v_1 \, L_c \, w_2$ and $v_1 \, L_c \, v_2$, since L_c is transitive, or $w_2 \, L_c \, v_1$ and $v_2 \, L_c \, v_1$. Both cases lead to a contradiction, since v_1 and v_2 are comparable.

Lemma 3. *A conjugate order L_c on a finite lattice $\underline{\mathfrak{V}}$ is a left-relation on $\underline{\mathfrak{V}}$.*

Proof. Let L_c be a conjugate order on $\mathfrak{V}$. For every two incomparable $\wedge$-irreducibles $m, n \in M$ either $m \, L_c \, n$ or $n \, L_c \, m$ holds. Hence there exists a sorting relation $L_a \subseteq L_c$.

Let L be the left-relation generated by L_a. Assume $L \neq L_c$ then we find a maximal pair of elements (v, w) with $v \, L \, w$ and $w \, L_c \, v$. On the one hand we find, by applying Definition 9, a pair $(m, n) \in M(v, w)$ with $m \, L \, n$. On the other hand we know, by applying Lemma 2, that $n \, L_c \, m$. This contradicts our assumption of the maximality of (v, w) since $(m, n) > (v, w)$.

We can subsume the results of this section to the following:

Proposition 2. *Let L be a relation on a finite lattice $\mathfrak{V}$. Then the following are equivalent:*

1. *L is a conjugate order.*
2. *L is a left-relation and a strict order.*

Proof. "1. $\Rightarrow$ 2." follows from Lemma 3.
"2. $\Rightarrow$ 1." follows from Lemma 1 and Definition 4.

Proposition 2 provides a possibility to calculate all conjugate orders: Compute the left-relations from all possible sorting relations (at most $|M|!$) and check whether they are strict orders.

4   The Left-Relation on Diagrams

Now we will define a left-relation on diagrams extending Definition 6 to arbitrary (not necessarily plane) diagrams. This will help us to give a connection between conjugate orders and plane diagrams.

Definition 10. *Let $\mathfrak{V}$ be a finite lattice and $\mathrm{pos}(\mathfrak{V})$ a line diagram of it. The sorting relation $\lambda_a \subseteq M \times M$ induced by $\mathrm{pos}(\mathfrak{V})$ is defined as follows* [3]

$$m \, \lambda_a \, n : \Longleftrightarrow m^* = n^* \wedge \varphi(\mathrm{pos}(m\,m^*)) < \varphi(\mathrm{pos}(n\,n^*)).$$

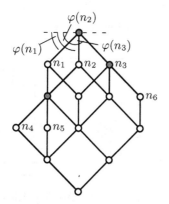

λ_a	m_1	m_2	m_3	m_4	m_5	m_6
m_1		×	×			
m_2			×			
m_3						
m_4					×	
m_5						
m_6						

Fig. 2. A diagram of a lattice with its sorting relation. The symbols n_i label the nodes of the $\wedge$-irreducibles m_i. The grey points represent the set of upper neighbours of $\wedge$-irreducibles

Every sorting relation λ_a in a diagram is a sorting relation on the underlying lattice due to Definition 8 since just the $\wedge$-irreducibles with common upper

---

[3] $\varphi(e)$ denotes the angle of the diagram line e, compare with Figure 2.

neighbour are ordered strict and linear. Notice that two line segments $\mathrm{pos}(v_1 w)$ and $\mathrm{pos}(v_2 w)$ do not have the same angle since this contradicts condition 4 of Definition 1. Conversely every sorting relation L_a in a lattice can be realized in a diagram. See Figure 2 for an example. Now we extend this relation to a left-relation λ induced by the diagram.

Definition 11. *Let $\mathfrak{V}$ be a finite lattice and $\mathrm{pos}(\mathfrak{V})$ a diagram of it. For a maximal chain C,*

$$F_l(C) := \{(x,y) \in \mathbb{R}^2 \mid y \in [y(0_{\mathfrak{V}}), y(1_{\mathfrak{V}})], x < x_C(y)\}$$

is the area left of $\mathrm{pos}(C)$ and dually $F_r(p)$ the area right of $\mathrm{pos}(C)$. We define the left- and the right-relation λ and ϱ induced by $\mathrm{pos}(\mathfrak{V})$ such that

$$v \,\lambda\, w \;:\Longleftrightarrow\; (\exists C \ni w : \mathrm{pos}(v) \in F_l(C)) \wedge (v \parallel w)$$
$$v \,\varrho\, w \;:\Longleftrightarrow\; (\exists C \ni w : \mathrm{pos}(v) \in F_r(C)) \wedge (v \parallel w) \quad\cdot$$

holds for all elements $v, w \in \mathfrak{V}$.

The function $x_C(y)$ is continuous for every maximal chain, since its composing diagram edges are. The equation $\lambda \cup \varrho = \parallel$ holds[4], as for each pair (v, w) of incomparable lattice elements we find a maximal chain $C \ni w$ and $\mathrm{pos}(v)$ is not situated on $\mathrm{pos}(C)$ due to condition 4 of Definition 1. In case of plane diagrams this definition is similar to Definition 6 (see [6], Prop. 1.6). In a plane line diagram the restriction of λ to pairs of nodes of $\wedge$-irreducibles with common upper neighbour is equal to λ_a, i.e. for all $m, n \in M$ holds the equivalence

$$m^* = n^* \text{ and } m \,\lambda\, n \;\Longleftrightarrow\; m \,\lambda_a\, n.$$

In Figure 3 we provide two examples for left-relations induced by diagrams.

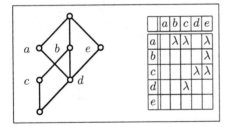
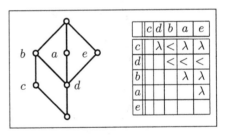

	a	b	c	d	e
a		λ	λ		λ
b					λ
c				λ	λ
d		λ			
e					

	c	d	b	a	e
c		λ	<	λ	λ
d			<	<	<
b				λ	λ
a					λ
e					

Fig. 3. Two diagrams of the same lattice with their left-relations. The left one is not plane and its left-relation not antisymmetric. The right one is plane and its left-relation is a strict order. Notice that $\lambda \cup <$ is a linear order

After these preparations we come to the first result of our work. When having a planar lattice with a conjugate order L_c, it gives us the possibility to actually draw a plane diagram with the left-relation $\lambda = L$.

---

[4] note that here ϱ and λ^{-1} are not necessarily equal.

Theorem 3. *Let $\mathfrak{V}$ be a finite lattice. The following statements are equivalent.*

1. *There exists a plane diagram $\mathrm{pos}(\mathfrak{V})$ with the induced left-relation L.*
2. *L is a left-relation on $\mathfrak{V}$ and a strict order.*

Proof. "1. $\Rightarrow$ 2.": follows from Corollary 1 and Proposition 2.

"2. $\Rightarrow$ 1.": We define two relations $L_< := L \cup <$ and $R_< := R \cup <$. It is easy to show (cf. proof of Theorem 1) that they are linear orders. Let the maps l and r be embeddings of $(\mathfrak{V}, L_<)$ and $(\mathfrak{V}, R_<)$ into $(\mathbb{R}, <)$ and $(\mathbb{R}, >)$ respectively.

Let pos be a map assigning to each $v \in \mathfrak{V}$ the point $(r(v), l(v))$ and to each pair of neighbouring elements a straight line segment connecting them. By the definitions of l and r we realize that pos meets the conditions 1, 2 and 3 of Definition 1. We show now that no line segments cross which makes the image of pos be a plane line diagram of $\mathfrak{V}$.

We assume that the diagram edges corresponding to the elements $v_1 \prec v_3$ and $v_2 \prec v_4$ cross. Let (x_i, y_i) be the coordinates of the node v_i and (x_5, y_5) be the coordinates of the intersection. Since r is order inversing we conclude $x_3, x_4 < x_5 < x_1, x_2$. Furthermore l is order preserving, i.e. $y_1, y_2 < y_5 < y_3, y_4$. It follows $v_2 < v_3$ and $v_1 < v_4$ and therefore $v_1 \parallel v_2$ and $v_3 \parallel v_4$. That means that v_3 and v_4 do not have an infimum in contradiction to $\mathfrak{V}$ being a lattice.

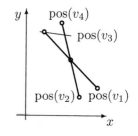

Let λ be the induced left-relation by $\mathrm{pos}(\mathfrak{V})$. We finally show that $\lambda = L$ holds. Let $m\, L_a\, n$ hold for $m, n \in M$. Due to the definitions of r and l the inequalities $x(m^*) < x(m) < x(n)$ and $y(m) < y(n) < y(m^*)$ hold. For the angles $\varphi_m := \varphi(\mathrm{pos}(m\, m^*))$ and $\varphi_n := \varphi(\mathrm{pos}(n\, n^*))$ we get

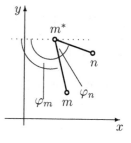

$$\tan \varphi_m = \frac{y(m) - y(m^*)}{x(m) - x(m^*)} \quad \text{and} \quad \tan \varphi_n = \frac{y(n) - y(m^*)}{x(n) - x(m^*)}.$$

We conclude $y(m) - y(m^*) < y(n) - y(m^*) < 0$ and $0 < x(m) - x(m^*) < x(n) - x(m^*)$, hence $\tan \varphi_m < \tan \varphi_n$ and the angles are in the interval $(\pi/2, \pi)$.

In this domain the function arctan is monotonous, so we conclude $\varphi_m < \varphi_n$, i.e. $m\, \lambda_a\, n$. That means, the sorting relations of the lattice and its diagram are the same. With Proposition 2 we conclude $\lambda = L$ since the diagram is plane and therefore λ is a conjugate partial order.

Corollary 2. *A finite lattice $\mathfrak{V}$ is planar if and only if there exists a left-relation on $\mathfrak{V}$ which is a strict order.*

5   The Left-Relation and Attribute Additivity

In the proof of Theorem 3 we gave a possibility to actually draw plane diagrams of a planar lattice. Unfortunately these diagrams are not attribute additive, which was our initial interest. By specifying the mappings l and r defined in the proof of Theorem 3 we can reach this main goal. We consider reduced contexts only.

Theorem 4. *Every finite planar concept lattice possesses a plane attribute additive diagram. Furthermore, is L a conjugate order on a finite concept lattice $\mathfrak{B}(G, M, I)$ there exists a plane attribute additive diagram with the left-relation $\lambda = L$.*

Proof. We define l and r in a recursive way. First we set

$$l(1_{\mathfrak{B}(G,M,I)}) = r(1_{\mathfrak{B}(G,M,I)}) = 0. \tag{1}$$

For the point $\mathrm{pos}(\mu m) = (r(\mu m), l(\mu m))$ of an attribute concept μm we set

$$l(\mu m) = l(C, D) - 1 \iff \mu m \prec_{L_<} (C, D), \tag{2}$$
$$r(\mu m) = r(C, D) + 1 \iff \mu m \prec_{R_<} (C, D), \tag{3}$$

where $\prec_{L_<}$ and $\prec_{R_<}$ are the neighbour relations of $L_<$ and $R_<$. Let $\mathrm{vec}_l(m) = l(\mu m^*) - l(\mu m)$ and $\mathrm{vec}_r(m) = l(\mu m^*) - l(\mu m)$. For all other concepts (A, B) we set

$$\mathrm{pos}(A, B) = (r(A, B), l(A, B)) = \sum_{m \in B} (\mathrm{vec}_r(m), \mathrm{vec}_l(m)). \tag{4}$$

We notice that we can apply a diagram point to each concept, in particular is $< \subseteq L_<, R_<$, i.e. for an arbitrary concept (A, B) the coordinates $\mathrm{pos}(\mu m)$ of the attributes m in its intent can be calculated before $\mathrm{pos}(A, B)$. The equation (4) assures the resulting diagram to be attribute additive. We have to show that l and r are embeddings. Since we consider linear orders only it is enough to show that

$$(A, B) \prec_{L_<} (C, D) \implies l(A, B) < l(C, D) \text{ and} \tag{5}$$
$$(A, B) \prec_{R_<} (C, D) \implies r(A, B) > r(C, D) \tag{6}$$

holds for all concepts $(A, B), (C, D) \in \mathfrak{B}(G, M, I)$. These implications are satisfied for (A, B) being an attribute concept by the equations (2) and (3). Let (A, B) be a concept which is no attribute concept (i.e. not $\wedge$-irreducible) and $(A, B) \prec_{L_<} (C, D)$.

1. If $(A, B) < (C, D)$ holds then $D \subseteq B$, i.e.

$$l(A, B) = \sum_{m \in B} \mathrm{vec}_l(m) < \sum_{m \in D} \mathrm{vec}_l(m) = l(C, D)$$

since $\mathrm{vec}_l(m) < 0$ holds for all attributes m.

2. Let $(A, B)L(C, D)$ hold. Since (A, B) has at least two upper neighbours $(A_1, B_1)L(A_2, B_2)$ we conclude $(A, B) \prec_{L_<} (C, D)L_<(A_1, B_1)L_<(A_2, B_2)$. Since L is transitive we conclude $(C, D) < (A_1, B_1)$ (and not $(C, D)L(A_1, B_1)$) and $(C, D) < (A_2, B_2)$. This is a contradiction, since in this case (A, B) and (C, D) have no supremum.

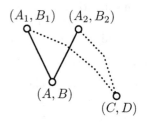

Hence the implications (5) and (by applying similar arguments) (6) hold. With the proof of Theorem 3 we conclude that the constructed diagram is plane and possesses the left-relation $\lambda = L$.

Finally we want to give an example on how to create a plane attribute additive diagram out of a conjugate order. Consider the lattice on the right (do not be confused to see a diagram, we are just interested in the underlying lattice). The nodes labeled with $a, \ldots, f$ are attribute concepts. In Figure 4 you can see the appropriate $L_<$ and $R_<$ relations. The relations l and r are, due to the described construction, calculated "from right to left" in the table.

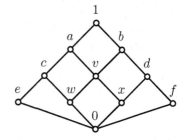

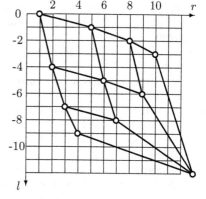

$L_<$	0	e	w	c	x	v	a	f	d	b	1
l	-12	-9	-8	-7	-6	-5	-4	-3	-2	-1	0
vec_l		-2		-3			-4	-1	-1	-1	
$R_<$	0	f	x	d	w	v	b	e	c	a	1
r	12	9	8	7	6	5	4	3	2	1	0
vec_r		2		3			4	1	1	1	

Fig. 4. A plane attribute additive diagram constructed out of a conjugate order L as described in the proof of Theorem 4. The maps vec_l and vec_r actually have attributes m as arguments and no attribute concepts μm, however we identified both for easier notation

6   Results and Further Work

By introducing the left-relation we have shown that the relationship of $\wedge$-irreducibles indeed provides an instrument to characterize whether the lattice under consideration is planar. Unfortunately an efficient algorithm for finding those relations which are strict orders, is not developed yet. Even if we find an polynomial algorithm, it is hard to make it quicker than existing planarity algorithms. This is due to the fact that a planar lattice has at most $\binom{|M|+1}{2}+1$ elements. However, the quickness is not a very important factor, when the considered lattices are small. Diagrams are considered to lose their readability, when exceeding a size of 50 nodes [7].

We gave a possibility to actually construct a plane attribute additive diagram by construction of the maps l and r in the proof of Theorem 4. If the resulting diagrams are not nicely drawn it is possible to employ adequate *force directed placement* methods [3, 5] to improve the drawing without changing λ (see [9]).

It would be interesting to find all plane diagrams of a lattice up to homeomorphisms. It seems that such a classification is very much related to similarity (see Definition 6).

Finally we would like to characterize planar concept lattices by their underlying contexts. So called Ferrers-Relations have been observed already [7], but no quick algorithms have been developed yet.

References

1. K. A. Baker, P. Fishburn, F. S. Roberts: *Partial Orders of Dimension 2.* Networks, 2, 11-28, 1971.
2. G. Birkhoff: *Lattice Theory.* Amer. Math. Soc., Third Edition, 1967.
3. G. DiBattista, P. Eades, R. Tamassia, I. G. Tollis, *Graph Drawing.* Prentice Hall, 1999.
4. B. Dushnik, E.W. Miller: *Partially Ordered Sets.* Amer. J. Math. 63, 1941, pp. 600-610.
5. P. Eades: *A Heuristic for Graph Drawing.* Congressus Numerantium 42, pp. 149-160, 1984.
6. D. Kelly, I. Rival: *Planar Lattices.* Can. J. Math. Vol. 27, No. 3, pp. 636-665, 1975.
7. B. Ganter, R. Wille: *Formal Concept Analysis.* Springer, 1999.
8. M. Skorsky: *Endliche Verbände - Diagramme und Eigenschaften.* PhD thesis, TH Darmstadt, 1992.
9. C. Zschalig: *Ein Force Directed Placement Algorithmus zum Zeichnen von Liniendiagrammen von Verbänden.* Diploma Thesis, TU Dresden, 2002.

Bialgebraic Contexts for Distributive Lattices - Revisited

Jürg Schmid

Inst. of Math., Univ. of Bern, Bern, Switzerland
`juerg.schmid@math-stat.unibe.ch`

Dedicated to the memory of my friend Ivan Rival

Abstract. In [8], Vogt used so-called *bialgebraic* contexts to represent the lattice $Sub(L)$ of all sublattices of a finite distributive lattice L as the substructure lattice of an appropriately defined finite (universal) algebra, based on Rival's description (see [4] and [5]) by means of deleting suitable intervals from L. We show how to extend Vogt's context in order to obtain a conceptually simpler description of $Sub_{01}(L)$ - the lattice of all 0-1-preserving sublattices of L - by means of quasiorders and an associated total binary operation on $J(L)^2$, the set of all pairs of non-zero join-irreducibles of L. Our approach is based on Birkhoff- resp. Priestley-duality, a standard reference is [1].

1 Introduction

In the fall of 1974, when we both were visiting at Caltech, Ivan Rival initiated me to the structure theory of distributive lattices by explaining one of his then favourite results: Given a finite distributive lattice L, a nonempty subset $S \subseteq L$ is a sublattice of L iff S can be obtained from L by deleting some collection of intervals $[j, m]$ where j is join-irreducible or 0 and m is meet-irreducible or 1 (see [4] and [5]). Obviously, this description allows us to compute systematically all sublattices of L and thus to determine the substructure lattice of L. As rather small examples already show, the structure of this lattice may be surprisingly complex, and quite a number of papers have been devoted to its study. A good source is [3] and its bibliography.

Rival's characterization of sublattices may be phrased as follows: Membership of an element $x \in L$ in some sublattice of L is equivalent to non-membership of x in some collection of special intervals in L. Writing $\mathcal{D}(L)$ for the collection of all *nonempty* intervals of the type considered, let $\mathbb{K}_L$ be the formal context $(L, \mathcal{D}(L), \notin)$. It follows at once that the extents of $\mathbb{K}_L$ are just the sublattices of L, while the intents are those subsets $B \subseteq \mathcal{D}(L)$ closed under the closure operator given by $\sigma(B) := \{[j, m] \in \mathcal{D}(L); [j, m] \subseteq \bigcup B\}$. Consequently, the substructure lattice of L is dually isomorphic to the lattice of all closed sets of the closure system $(\mathcal{D}(L), \sigma)$. It is the purpose of this note to replace this

B. Ganter and R. Godin (Eds.): ICFCA 2005, LNAI 3403, pp. 403–407, 2005.

description by the lattice of all subalgebras of a (universal) algebra naturally connected with L.

Unless otherwise stated, L will always denote a finite distributive lattice with bounds 0 and 1 (freely confounded with its carrier set), operations $\wedge$ and $\vee$ and order relation $\leq$. We denote by $J(L)$ the set of all join-irreducible elements of L (excluding 0) and by $M(L)$ the set of all meet-irreducibles (excluding 1). For $x \in L$ we write $\downarrow x$ for $\{y \in L, \ y \leq x\}$, and similarly $\uparrow x$ for $\{y \in L, \ y \geq x\}$. Without loss of generality but some gain in the smoothness of presentation, we will only consider 0-1-sublattices of L in section 3: Writing $Sub(L)$ resp. $Sub_{01}(L)$ for the lattice of all sublattices resp. of all 0-1-sublattices of L, we have $Sub(L) \cong Sub_{01}(0' \oplus L \oplus 1')$ with $0' < x < 1'$ for all $x \in L$. A *quasiorder* Q on a set X is a reflexive and transitive binary relation X; given such Q, a subset of $D \subseteq X$ is called a *Q-down-set* iff $x \in X$, $d \in D$ and $(x, d) \in Q$ jointly imply that $x \in D$ (so, in particular, $\downarrow x$ is a $\leq$-down-set in L).

2   Bialgebraic Contexts

In order to represent the subalgebra lattice of a given algebra by that of another (hopefully simpler) algebra, Vogt introduced so-called bialgebraic contexts, implicitly in [7] and explicitly in [8], and used them to describe - in [8] - the lattice $Sub(L)$ of general sublattices of L. We briefly summarize the idea and main results of [8]:

A context $\mathbb{K} = (G, M, I)$ is called *bialgebraic* provided the object set G as well as the attribute set M are carriers of algebras (G, F_G) resp. (M, F_M) such that the extents of $\mathbb{K}$ coincide with the (carriers of) the subalgebras of (G, F_G) and the intents of $\mathbb{K}$ with the (carriers of) the subalgebras of (M, F_M). Note that there is no formal condition linking the types of F_G resp. F_M (although this may be desirable in a concrete application), and that the fundamental operations in of F_G resp. F_M are allowed to be partial.

Looking at the formal context $\mathbb{K}_L$ defined in section 1, we see that it is algebraic on the object half while for the attribute half we have the explicitly given closure operator σ describing the intents. What is missing in order to make $\mathbb{K}_L$ bialgebraic is thus a bunch of (possibly partial) *operations* defined on $\mathcal{D}(L)$ such that the σ-closed subsets of $\mathbb{K}_L$ are exactly those closed under these operations. Finding such operations - in fact, a single partial binary operation will do in the end - takes the major part of [8].

Since the intents of $\mathbb{K}_L$ are exactly the subsets of $\mathcal{D}(L)$ respecting all attribute implications of $\mathbb{K}_L$, Vogt considers the latter as multioperations on $\mathcal{D}(L)$ and extracts from them the sought partial operation(s). The highly nontrivial key step consists in proving that each implication in the canonical minimum Duquenne-Guiges base for (the implications of) $\mathbb{K}_L$ may be simplified in such a way that the resulting set of implications has premises of cardinality at most two and all conclusions are singletons. These conclusions are then considered as the images of their premises under a single partial operation $* : \mathcal{D}(L) \times \mathcal{D}(L) \longrightarrow \mathcal{D}(L)$, defined by cases, and the intents of $\mathbb{K}_L$ coincide with the $*$-closed subsets of $\mathcal{D}(L)$.

3   A Simple Bialgebraic Context for $Sub_{01}(L)$

One of the distinctive features of finite distributive L lattices is the following relationship between $J(L)$ and $M(L)$: Given $j \in J(L)$, there is a unique maximal element $\bar{j}$ in $\{x \in L;\; x \not\geq j\}$ and $\bar{j} \in M(L)$, and analogously, given $m \in M(L)$, there is a unique minimal element $\bar{m}$ in $\{x \in L;\; x \not\leq m\}$ and $\bar{m} \in J(L)$. We call j and $\bar{j}$ resp. m and $\bar{m}$ conjugates. It follows L is the disjoint union of $\uparrow j$ and $\downarrow \bar{j}$ resp. of $\downarrow m$ and $\uparrow \bar{m}$, for any $j \in J(L)$ resp. $m \in M(L)$ (this is why we insisted on $0 \notin J(L)$ and $1 \notin M(L)$).

Using conjugate elements, comparabilities and non-comparabilities between join-irreducibles and meet-irreducibles may be translated into such involving only one type of irreducibles: It immediately follows from the definitions above that, e.g., $j \not\leq m$ iff $\bar{m} \leq j$ and $j \leq m$ iff $\bar{m} \not\leq j$. Hence, given any $j \in J(L)$ and $m \in M(L)$, the interval $[j, m]$ is empty iff $\bar{m} \leq j$ resp. nonempty iff $\bar{m} \not\leq j$. Finally, for any $x \in L$, we have $x \notin [j, m]$ iff $(x \not\geq j$ or $x \not\leq m)$ iff $(x \geq j$ implies $x \not\leq m)$ iff $(x \geq j$ implies $x \geq \bar{m})$.

This leads us to set up a new formal context $\mathbb{K}'_L$ as follows: Let $G := L$, $M := J(L)^2$ and define incidence I between $x \in G$ and $(j_1, j_2) \in M$ by $(x\, I\, (j_1, j_2)$ iff $x \geq j_2 \Rightarrow x \geq j_1$ (read as "x respects (j_1, j_2)").

How does $\mathbb{K}'_L$ compare with $\mathbb{K}_L$? Introducing conjugates, replace (j_1, j_2) by the pair $(j_2, \bar{j}_1)$ and consider the latter to represent the interval $[j_2, \bar{j}_1]$. Now for any $x \in L$, we have $x \notin [j_2, \bar{j}_1]$ iff $(x \geq j_2$ implies $x \not\leq \bar{j}_1)$ iff $(x \geq j_2$ implies $x \geq j_1)$ iff x respects (j_1, j_2). However, attributes of $\mathbb{K}_L$ were restricted to the nonempty intervals of the type under consideration; for these, incidence in $\mathbb{K}'_L$ thus coincides with that in $\mathbb{K}_L$. But $[j_2, \bar{j}_1] = \emptyset$ iff $j_1 \leq j_2$, in which case every $x \in L$ will respect (j_1, j_2). Summing up, the cross table of $\mathbb{K}'_L$ may be obtained from that of $\mathbb{K}_L$ by adding a full column for each pair (j_1, j_2) with $j_1 \leq j_2$, representing a particular instance of the empty interval (making $\mathbb{K}'_L$ a very unclarified context). But it follows at once that the concept lattices of $\mathbb{K}'_L$ and $\mathbb{K}_L$ are canonically isomorphic, and so the extents of $\mathbb{K}'_L$, ordered by set inclusion, still form a copy of the lattice $Sub_{01}(L)$. It remains to identify the intents of $\mathbb{K}'_L$.

Claim 1. *The intents of $\mathbb{K}'_L$ are the quasiorders on $J(L)$ extending (the restriction to $J(L)$ of) $\leq$.*

Proof: Consider $A \subseteq L$ and $x \in A$. Assume x respects (j_1, j_2) and (j_2, j_3). Hence $x \geq j_3$ implies $x \geq j_2$ implies $x \geq j_1$, so x respects (j_1, j_3) and A' is transitive. Certainly x respects (j_1, j_2) whenever $j_1 \leq j_2$, so A' contains $\leq|_{J(L)}$.

We have to show that every quasiorder Q on $J(L)$ extending $\leq$ is an intent. Writing $\hat{x}$ for $\{j \in J(L);\; j \leq x\}$, we have $L \cong (\{\hat{x};\; x \in L\}, \subseteq)$ by (finite) Birkhoff duality. Now x respects (j_1, j_2) iff $x \geq j_2 \Rightarrow x \geq j_1$ iff $j_2 \in \hat{x} \Rightarrow j_1 \in \hat{x}$.

Hence, x respects *all* $(j_1, j_2) \in Q$ iff $(j_1, j_2) \in Q$ and $j_2 \in \hat{x}$ jointly imply that $j_1 \in \hat{x}$; in other words, iff $\hat{x}$ is a Q-down-set (and thus a $\leq$-down-set, fortiori). Hence, $Q' \cong (\{Q\text{-down-sets in } J(L)\}, \subseteq)$.

Assume now that $(j_1, j_2) \in Q''$, that is, every Q-down-set respects (j_1, j_2). In particular, the Q-down-set $\downarrow_Q j_2 = \{j \in J(L); \ ((j, j_2) \in Q\}$ respects (j_1, j_2). As $j_2 \in \downarrow_Q j_2$, this implies $j_1 \in \downarrow_Q j_2$, that is, $(j_1, j_2) \in Q$. Hence $Q'' \subseteq Q$ and thus $Q = Q''$ as desired.

$\square$

Write $Q_\leq(L)$ for the *lattice of all quasiorders* (under set inclusion) on $J(L)$ extending $\leq|_{J(L)}$. Then we have

Corollary 2. $Sub_{01}(L)$ *is dually isomorphic to* $Q_\leq(L)$.

This description of $Q_\leq(L)$ is contained, in a general form covering arbitrary distributive lattices, in [6].

It is now straightforward to convert $\mathbb{K}'_L$ into a bialgebraic algebraic context. Define $\star : J(L)^2 \times J(L)^2 \longrightarrow J(L)^2$ by

$$(j_1, j_2) \star (j'_1, j'_2) = \begin{cases} (j_1, j'_2) & if \quad j_2 = j'_1 \\ (j_1, j_2) & if \quad j_2 \neq j'_1 \end{cases}$$

and let $c_{jj'}$ be the nullary operation on $J(L)^2$ taking constant value (j, j'). It is immediate that a subset $B \subseteq J(L)^2$ is a quasiorder on $J(L)$ iff B is closed under $\star$ and c_{jj} for all $j \in J(L)$, and a quasiorder extending $\leq|_{J(L)}$ iff B is closed under $\star$ and $c_{jj'}$ for all $j, j' \in J(L)$ with $j \leq j'$. So our result may be phrased as follows:

Proposition 3. *The triple* $((L; \wedge, \vee, 0, 1), (J(L)^2; \star, (c_{jj'})_{j \leq j'}), \text{"respects"})$ *is a bialgebraic context whose concept lattice is isomorphic to* $Sub_{01}(L)$ *and dually isomorphic to* $Q_\leq(L)$.

Proposition 3 fills a gap in [6] where the connection between Vogt's work and a couple of other representations of $Sub_{01}(L)$ was left open. Also, it seems likely that combining the vast resources of algorithms available for concept analysis with those developed for transitive closure, one might obtain an efficient algorithm for actually computing $Sub_{01}(L)$ resp. $Sub(L)$.

It is interesting to compare Vogt's $*$ and the operation $\star$ above in hindsight: Replacing intervals of type $[j, m]$ by pairs $(\overline{m}, j) \in J(L)^2$, Vogt's $*$ translates into a partial operation on the set $\mathrm{Inc}_{J(L)}$ of all *incomparable* pairs in $J(L)^2$. A subset $B \subseteq \mathrm{Inc}_{J(L)}$ is then obviously closed under $*$ iff it has the form $B = \mathrm{Inc}_{J(L)} \cap Q$ for some quasiorder $\leq \subseteq Q \subseteq J(L)^2$. So $*$ must simulate the effects of transitive closure on $\mathrm{Inc}_{J(L)} \cup \leq|_{J(L)}$ restricted to $\mathrm{Inc}_{J(L)}$ without using $\leq|_{J(L)}$ – which explains the somewhat involved definition of $*$ in [8].

4   Beyond Finiteness

In an *infinite* distributive lattice (still with 0 and 1) prime ideals take over the rôle of join-irreducible elements which no longer need exist. Let X_L be the set of all prime ideals in L. A topology τ_L on X_L is defined by taking all sets of type $\{P \in X_L;\ x \in P\}$ and of type $\{P \in X_L;\ x \notin P\}$ for $x, y \in L$ as an (open) subbase. Finally, order X_L by set inclusion $\subseteq$. The triple $(X_L, \tau_L, \subseteq)$ is called the *Priestley* space of L and the main theorem of *Priestley duality* says that L is isomorphic to the lattice of all clopen $\subseteq$-down-sets.

Accordingly, a context $\mathbb{K}'_L$ is defined by $G := L$ and $M := X_L^2$ with incidence I between $x \in G$ and $(P_1, P_2) \in M$ given by $(x\,I\,(P_1, P_2)$ iff $x \notin P_2 \Rightarrow x \notin P_1$ (read as "x *respects* (P_1, P_2)"). Note that this definition includes the finite case: If L is finite, its prime ideals are of the form $\downarrow m$ for $m \in M(L)$, or equivalently, $L \backslash \uparrow j$ for $j \in J(L)$. But $x \notin L \backslash \uparrow j$ is the same as $x \geq j$, so incidence reduces to that considered in section 3.

It is not hard to see that the extents of $\mathbb{K}'_L$ still are the 0-1-sublattices of L and that the intents are quasiorders on X_L extending $\subseteq$. However, not *every* quasiorder on X_L containing $\subseteq$ is an intent. Indeed, let Q be such a quasiorder. It is shown in [6–Prop. 3.4] that Q represents a 0-1-sublattice of L iff Q is "τ_L-separated" (or equivalently, iff τ_L is "totally quasiorder-disconnected"), meaning that whenever $(P_1, P_2) \notin Q$ there exist a τ_L-clopen Q-down-set in the Priestley space $(X_L, \tau_L, \subseteq)$ such that $P_2 \in Q$ but $P_1 \notin Q$. So what is missing in order to make $\mathbb{K}'_L$ a bialgebraic context also in this case is a bunch of continuous operations on X_L such that, given $S \subseteq X_L^2$ containing $\subseteq$, the closure of S under these operations is just the smallest τ_L-separated quasiorder on X_L.

References

1. Davey, B. A., and Priestley, H. A., *Introduction to lattices and order* (2nd ed.), Cambridge Univ. Press, Cambridge, 2002
2. Ganter, B., and Wille, R., *Formal concept analysis*, Springer, New York, 1999
3. Lengvárszky, Z., and McNulty, G., *Covering in the lattice of subuniverses of a finite distributive lattice*, J. Austral. Math. Soc. (Series A) **65** (1998), 333 – 353
4. Rival, I., *Maximal sublattices of finite distributive lattices*, Proc. Amer. Math. Soc. **37** (1973), 417 – 420
5. Rival, I., *Maximal sublattices of finite distributive lattices, II*, Proc. Amer. Math. Soc. **44** (1974), 263 – 268
6. Schmid, J., *Quasiorders and Sublattices of Distributive Lattices*, Order **19**, (2002), 11 – 34
7. Vogt, F., *Subgroup lattices of finite Abelian groups: structure and cardinality*, In: K. A. Baker and R. Wille (eds.), *Lattice theory and its applications*, Heldermann-Verlag, Berlin, 1995, 241 – 259
8. Vogt, F., *Bialgebraic contexts for finite distributive lattices*, Algebra Universalis **35** (1996), 151 – 165

Which Concept Lattices Are Pseudocomplemented?

Bernhard Ganter and Léonard Kwuida[*]

Institut für Algebra,
TU Dresden,
D-01062 Dresden, Germany

Abstract We give a contextual characterization of pseudocomplementation by means of the arrow relations.

Keywords: lattices, pseudocomplement, closure operator, Formal Concept Analysis, arrow-relation, complete homomorphism.

AMS Subject Classification: 06D15

1 Introduction

A lattice L with 0 is *pseudocomplemented* if for each $x \in L$ there is an element $x^* \in L$ (called the *pseudocomplement*[1] of x) such that

$$x \wedge y = 0 \iff y \leq x^*.$$

In this case, $x \mapsto x^*$ defines a unary operation on L, called *pseudocomplementation*, which is automatically antitone and square extensive, i.e., which satisfies

$$x \leq y \Rightarrow y^* \leq x^* \qquad \text{and} \qquad x \leq x^{**}$$

for all $x, y \in L$. These two properties together imply the *join de Morgan law*

$$(x \vee y)^* = x^* \wedge y^*.$$

In fact, we get

Proposition 1. *If L is pseudocomplemented, then*

$$\left(\bigvee X \right)^* = \bigwedge \{ x^* \mid x \in X \},$$

whenever $\bigvee X$ exists in L.

Proof. From $x \in X$ we infer $x \leq \bigvee X$ and thus $(\bigvee X)^* \leq x^*$. Therefore $(\bigvee X)^*$ is a lower bound of $\{ x^* \mid x \in X \}$. Conversely let y be a lower bound of $\{ x^* \mid x \in X \}$, i.e., $y \leq x^*$ for all $x \in X$. Then $y^* \geq x^{**} \geq x$ for all $x \in X$ and thus $y^* \geq \bigvee X$. From this we get $y \leq y^{**} \leq (\bigvee X)^*$. This proves that $(\bigvee X)^*$ is the greatest lower bound of $\{ x^* \mid x \in X \}$. $\square$

---

[*] Supported in part by the Gesellschaft von Freunden und Förderern der TU Dresden.

[1] The pseudocomplement of x (if it exists) is its greatest semicomplement.

B. Ganter and R. Godin (Eds.): ICFCA 2005, LNAI 3403, pp. 408–416, 2005.

This proposition offers an easy way to prove pseudocomplementedness: it suffices to exhibit a join-dense set J of elements with pseudocomplement. The pseudocomplement of any other element x is then obtained as the meet of the pseudocomplements of those elements in J which are below x.

For complete lattices there is a simple and rather obvious condition for having a pseudocomplement:

Proposition 2. *Let x be an element of a complete lattice L. Then*

$$x^* \text{ exists} \quad \Longleftrightarrow \quad x \wedge \bigvee \{y \in L \mid x \wedge y = 0\} = 0.$$

If x^* exists then

$$x^* = \bigvee \{y \in L \mid x \wedge y = 0\}.$$

Pseudocomplemented lattices, also known as *p-algebras*, have been widely investigated. One of the general sources is the survey by Katriňák [Ka80], but also the books by Balbes and Dwinger [BD74] and Grätzer [Gr71] for the distributive case. Varieties of distributive *p*-algebras have been described by Lee [Lee70]. The free algebras in these varieties can be used for modeling inconsistent information in databases, see Schmid [Sc88], Sofronie-Stokkermans [So98]. Generalized notions of negation have been studied in [Kw04].

2 Pseudocomplemented Closure Systems

For a concept lattice, being pseudocomplemented is naturally expressed in terms of the closure system of extents. We therefore formulate our observations in the language of closure systems. Let $\mathcal{E}$ be a closure system on a set G, and let

$$A \mapsto A''$$

be the corresponding closure operator. For simplicity we assume $\emptyset'' = \emptyset$ and $g'' = h'' \Rightarrow g = h$; these are merely technical conditions. The closure system $\mathcal{E}$ is called pseudocomplemented if each closed set has a pseudocomplement.

Proposition 3. *A closed set $A \in \mathcal{E}$ has a pseudocomplement A^* if and only if*

$$A \cap \{g \in G \mid A \cap g'' = \emptyset\}'' = \emptyset.$$

In this case,

$$A^* = \{g \in G \mid A \cap g'' = \emptyset\}.$$

If $\{g \in G \mid A \cap g'' = \emptyset\}$ is closed, then it is the pseudocomplement of A.

Proof. The first claim is the same as in Proposition 2 with A in the rôle of x, except that we have simplified the right hand side:

$$\bigvee \{B \in \mathcal{E} \mid A \cap B = \emptyset\} = \left(\bigcup \{B \in \mathcal{E} \mid A \cap B = \emptyset\} \right)''$$
$$= \{g \in G \mid A \cap g'' = \emptyset\}'',$$

the latter equality being true since g is contained in some closed set B with $A \cap B = \emptyset$ iff $A \cap g'' = \emptyset$.

For the second claim, Proposition 2 yields

$$A^* = \{g \in G \mid A \cap g'' = \emptyset\}''.$$

But if $A \cap \{g \in G \mid A \cap g'' = \emptyset\}'' = \emptyset$, then

$$\{g \in G \mid A \cap g'' = \emptyset\}'' = \{g \in G \mid A \cap g'' = \emptyset\}.$$

The third claim is now immediate. $\qquad \square$

It is not necessary to verify the condition of Proposition 3 for every closed set. Let us call a subset $\mathcal{C} \subseteq \mathcal{E}$ *co-initial*, if every nonempty closed set in $\mathcal{E}$ contains a nonempty set from $\mathcal{C}$. The set $\{g'' \mid g \in G\}$ of one-generated closed sets is always co-initial. A closure system is called *atomic*, if the set of its one-element closures (*"atoms"*) is co-initial. The atoms must be part of any co-initial set.

Proposition 4. *If $\mathcal{C}$ is co-initial in $\mathcal{E}$ and every $C \in \mathcal{C}$ has a pseudocomplement, then $\mathcal{E}$ is pseudocomplemented.*

Proof. Suppose $A \in \mathcal{E}$ has no pseudocomplement. Then, by Proposition 3,

$$X := A \cap \{g \in G \mid A \cap g'' = \emptyset\}'' \neq \emptyset.$$

Since $\mathcal{C}$ is co-initial, we find some non-empty $C \in \mathcal{C}$ with $C \subseteq X$. $C \subseteq A$ implies that

$$\{g \in G \mid A \cap g'' = \emptyset\}'' \subseteq \{g \in G \mid C \cap g'' = \emptyset\}'',$$

and therefore

$$C \cap \{g \in G \mid C \cap g'' = \emptyset\}'' = C \neq \emptyset,$$

which means that C also has no pseudocomplement. $\qquad \square$

Finite closure systems are atomic. Proposition 4 is then another version of a result by Chameni Nembua and Monjardet which states that "finite lattices are pseudocomplemented if and only if all its atoms have pseudocomplements" [CM93].

A variant of Proposition 3 is the following:

Proposition 5. *If $\mathcal{C}$ is co-initial in $\mathcal{E}$ then $\mathcal{E}$ is pseudocomplemented if for each $C \in \mathcal{C}$ the set*

$$\{g \in G \mid C \cap g'' = \emptyset\}$$

is closed.

A set $T \subseteq G$ is a *transversal* of the closure system $\mathcal{E}$ iff every nonempty closed set contains some element of T:

$$T \text{ transversal} \quad :\Longleftrightarrow \quad T \cap E \neq \emptyset \text{ for all } E \in \mathcal{E} \setminus \{\emptyset\}.$$

If T is a transversal then

$$\mathcal{C} := \{t'' \mid t \in T\}$$

is co-initial. The set G is always a transversal. If the closure system is atomic, then $G_{\min} := \{g \in G \mid g'' = \{g\}\}$ is a transversal, and is contained in every transversal of $\mathcal{E}$. The elements of $G_{\min}$ will also be referred to as atoms.

Now let $\mathcal{E}$ be a closure system on G with a fixed transversal T. For subsets $A \subseteq G$ we define

$$s(A) := \{t \in T \mid \exists_{g \in A} \, t \in g''\}.$$

Note that $s(\{g\}) = g'' \cap T$ and $s(A'') = A'' \cap T$.

Proposition 6. *The operator $[\cdot]$ defined on subsets of G by*

$$[A] := \{g \in G \mid s(\{g\}) \subseteq s(A)\}$$

is a closure operator.

Proof. It is obvious from the definition that the operator is monotone and extensive. To prove idempotency, we must show that $s([A]) = s(A)$. If $t \in s([A])$, then there is some $h \in [A]$ such that $t \in s(\{h\})$. But h is in $[A]$ iff $s(\{h\}) \subseteq s(A)$. Therefore $t \in s(A)$. □

Proposition 7. $A \in \mathcal{E}$ *has a pseudocomplement iff* $[T \setminus A]$ *is in* $\mathcal{E}$.

Proof. We know from Proposition 3 that A^* exists iff $\{g \mid g'' \cap A = \varnothing\}$ is closed. We find that

$$
\begin{aligned}
g'' \cap A = \varnothing &\iff g'' \cap A \cap T = \varnothing \\
&\iff s(g) \cap A = \varnothing \\
&\iff s(g) \subseteq T \setminus A \\
&\iff g \in [T \setminus A].
\end{aligned}
$$

□

Combining this with Proposition 4 we get

Theorem 1. *The closure system $\mathcal{E}$ with transversal T is pseudocomplemented iff all sets*

$$[T \setminus t''], \quad t \in T,$$

are closed in $\mathcal{E}$.

In the atomic case we can say more about the closure operator $[\cdot]$. A silent assumption is that in the atomic case the closure operator $[\cdot]$ is always defined with respect to the transversal $G_{\min}$, unless explicitly stated otherwise.

Proposition 8. *If $\mathcal{E}$ is atomic and if for each $a \in G_{\min}$ the set $[G_{\min} \setminus \{a\}]$ is closed in $\mathcal{E}$, then each set $[A]$ is closed in $\mathcal{E}$.*

Proof. Note that $[G_{\min} \setminus \{a\}] \cap G_{\min} = G_{\min} \setminus \{a\}$ for all $a \in G_{\min}$. We will prove that

$$[A] = \bigcap_{a \in G_{\min} \setminus A} [G_{\min} \setminus \{a\}].$$

We denote the set of atoms in $[A]$ by $A_{\min}$, thus

$$A_{\min} := [A] \cap G_{\min}.$$

It follows from the definition of $[\cdot]$ that $[A] = [A_{\min}]$. We therefore have to show that

$$[A_{\min}] = \bigcap_{a \in G_{\min} \setminus A_{\min}} [G_{\min} \setminus \{a\}].$$

It is clear that $[A_{\min}]$ is included in $\bigcap_{a \in G_{\min} \setminus A_{\min}} [G_{\min} \setminus \{a\}]$. To prove equality, consider an element $h \in G$ which is not in $[A_{\min}]$:

$$h \notin [A_{\min}] \Rightarrow s(\{h\}) \nsubseteq s(A_{\min}) = A_{\min}$$
$$\Rightarrow \exists_{a \in G_{\min} \setminus A_{\min}}\ a \in s(\{h\})$$
$$\Rightarrow \exists_{a \in G_{\min} \setminus A_{\min}}\ h \notin [G_{\min} \setminus \{a\}]$$
$$\Rightarrow h \notin \bigcap_{a \in G_{\min} \setminus A_{\min}} [G_{\min} \setminus \{a\}].$$

$\square$

Together with Theorem 1 this yields

Theorem 2. *An atomic closure system $\mathcal{E}$ is pseudocomplemented if and only if each set $[A]$ is closed in $\mathcal{E}$.*

We close the section with a simple observation:

Proposition 9. *If $\mathcal{E}$ is atomic and $A \in \mathcal{E}$ is closed then*

$$A \subseteq [A \cap G_{\min}].$$

Proof. If $g \in A$ then $s(\{g\}) \subseteq A \cap G_{\min}$, therefore $g \in [A \cap G_{\min}]$. $\square$

3   Atomic Concept Lattices

We now apply our findings to concept lattices (see [GW99] for basic notions). To keep things simple, we restrict to the case that the lattice $\mathfrak{B}(G, M, I)$ under consideration is atomic (which means that the closure system of extents is atomic). Later we shall also assume that $\mathfrak{B}(G, M, I)$ is doubly founded, in order to have the arrow-relations at hand. These conditions include all finite lattices. Moreover we suppose w.l.o.g. that the formal context (G, M, I) is clarified with $M' = \emptyset$. As a corollary to Theorem 2 we get

Corollary 1. *An atomic concept lattice is pseudocomplemented if and only if for each $A \subset G$ the subset $[A]$ is an extent.*

We already know that we can restrict to sets of the form $[G_{\min} \setminus \{a\}]$. When are those extents?

Proposition 10. *If $[G_{\min} \setminus \{a\}]$ is an extent and $a \in G_{\min}$, then there is an attribute $m \in M$ such that $m' = [G_{\min} \setminus \{a\}]$.*

Proof. Note that $a \notin [G_{\min} \setminus \{a\}]$. If $[G_{\min} \setminus \{a\}]$ is an extent, then

$$[G_{\min} \setminus \{a\}] = \bigcap_{[G_{\min} \setminus \{a\}] \subseteq m'} m',$$

and at least one of these attributes cannot be incident with a. Let m_a be such an attribute. By Proposition 9,

$$m_a' \subseteq [m_a' \cap G_{\min}] \subseteq [G_{\min} \setminus \{a\}],$$

thus

$$m_a' = [G_{\min} \setminus \{a\}].$$

$\square$

Proposition 11. *An atomic concept lattice is pseudocomplemented if and only if for each atom a there is some attribute m_a such that*

- *m_a is incident with all atoms except a.*
- *$a \notin n'$ implies $n' \subseteq m_a'$.*

Proof. We know from Proposition 10 that in the pseudocomplemented case there is an attribute m_a with $m_a' = [G_{\min} \setminus \{a\}]$. Let n be an attribute with $a \notin n'$, then $h \in n'$ implies

$$s(\{h\}) \subseteq n' \cap G_{\min} \subseteq G_{\min} \setminus \{a\}.$$

Therefore $n' \subseteq m_a'$.

Conversely let m_a be an attribute satisfying the two conditions of the proposition, and let $h \notin m_a'$. Each attribute n of h satisfies $n' \not\subseteq m_a'$ (because $h \in n'$) and thus $a \in n'$ (because of the second condition). Therefore $a \in h''$ and thus $h \notin [G_{\min} \setminus \{a\}]$. Consequently $[G_{\min} \setminus \{a\}] \subseteq m_a'$. Equality follows from Proposition 9, and the existence of a pseudocomplementation from Proposition 8. $\square$

Theorem 3. *Let $\mathfrak{B}(G, M, I)$ be atomic and doubly founded. Then $\mathfrak{B}(G, M, I)$ is pseudocomplemented if and only if the following condition holds for all $g \in G$:*

If $g \swarrow n$ for all $n \notin g'$ and $g \nearrow m$ then

- *if $h \swarrow m$ then $g' = h'$, and*
- *if $g \nearrow n$ then $n' = m'$.*

Proof. Let us first rephrase the condition: $g \swarrow n$ for all $n \notin g'$ is equivalent to g being an atom. If the context is clarified, as we may assume w.o.l.g., then $g' = h'$ is the same as $g = h$ and $m' = n'$ is the same as $m = n$. Then the condition simplifies to the following:

> *For each atom g there is a unique attribute m with $g \nearrow m$, and g is the only object with $g \swarrow m$.*

Suppose $\underline{\mathfrak{B}}(G, M, I)$ is pseudocomplemented. By Proposition 11 there is for each atom a an attribute m_a with $m_a' = [G_{\min} \setminus \{a\}]$. It is clear that $a \nearrow m_a$ holds, and since $a \notin n'$ implies $n' \subseteq m_a'$ there cannot be another attribute n with $a \nearrow n$. Suppose $h \swarrow m_a$ for some $h \neq a$. Then $h \notin m_a' = [G_{\min} \setminus \{a\}]$, and thus $a \in s(\{h\})$, which implies $h' \subseteq a'$ and thereby contradicts $h \swarrow m_a$.

For the other direction we presuppose the condition of the theorem. Let a be an atom and let m be the unique attribute with $a \nearrow m$. Then the two conditions of Proposition 11 are satisfied, which can be seen as follows: If g is any atom with $g \notin m'$ then $g \swarrow m$ is implied and the condition forces $g = a$. Thus m is incident with all atoms except a. If n is an attribute with $a \notin n'$, then there must be some attribute $\tilde{n}$ with $n' \subseteq \tilde{n}'$ and $a \nearrow \tilde{n}$. According to the condition, this attribute must be m, and consequently $n' \subseteq m'$. $\square$

The arrow configuration described in Theorem 3 is displayed in Figure 1. The subcontext $(G_{\min}, \{m_a \mid a \in G_{\min}\})$ is a contranominal context, with exactly one double arrow in each row and column and crosses elsewhere. The rows of the atoms $G_{\min}$ contain no empty cells (arrowless non-incidences) and no upward arrows except for the double arrows mentioned. The columns corresponding to the attributes $\{m_a \mid a \in G_{\min}\}$ have no other downward arrows.

It is evident from Figure 1 that the subcontext $(G_{\min}, \{m_a \mid a \in G_{\min}\})$ is arrow closed and therefore compatible. This leads to our final theorem:

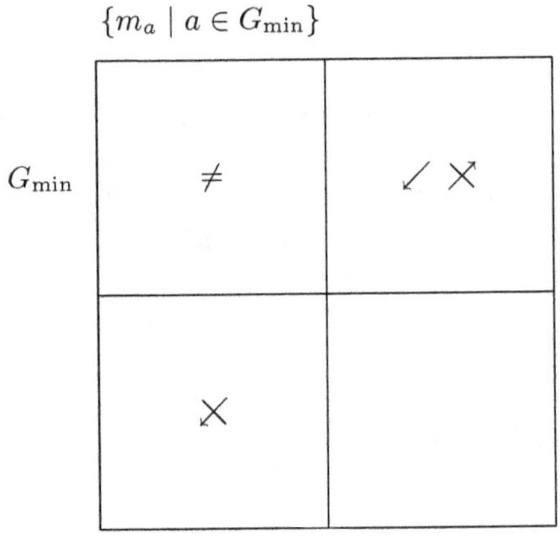

Fig. 1. Arrow configuration in the context of an atomic pseudocomplemented concept lattice

Theorem 4. *An atomic doubly founded complete lattice L is pseudo-complemented if and only if there is a complete homomorphism of L onto a Boolean lattice, mapping atoms to atoms.*

Proof. What the theorem expresses is that the configuration displayed in Figure 1 is characteristic for such lattices. It is apparent from this figure that the condition must be fulfilled in the pseudocomplemented case.

For the converse, suppose that L admits a complete homomorphism φ onto $(\mathfrak{P}(G_{\min}), \subseteq)$, mapping atoms to atoms. Necessarily then 0 is the only element mapped to 0. Let a be an atom and let b denote the largest element of the congruence class $\varphi^{-1}(\neg\varphi(a))$. Then

$$x \leq b \iff \varphi(x) \leq \varphi(b)$$
$$\iff \varphi(x) \leq \neg\varphi(a)$$
$$\iff \varphi(x) \wedge \varphi(a) = 0$$
$$\iff x \wedge a = 0.$$

Therefore b is the pseudocomplement of a. Every atom has a pseudocomplement, and it thus follows from Proposition 4 that L is pseudocomplemented. $\square$

Our result is related to a celebrated theorem by Glivenko (1929, see [CG00]):

Let L be a p-algebra. The operation $x \mapsto x^{**}$ is a closure on L. The set

$$S(L) := \{x \in L \mid x = x^{**}\}$$

of closed elements is called the **skeleton** of L. The operation $\oplus$ defined by $x \oplus y := (x^* \wedge y^*)^*$ turns $\underline{S}(L) := (S(L), \wedge, \oplus, {}^*, 0, 1)$ into a Boolean algebra (this is due to O. Frink (1962)). The Glivenko mapping is given by

$$g : L \to S(L)$$
$$x \mapsto x^{**}.$$

If a is an atom of L, then $g(a)$ is an atom of $\underline{S}(L)$ since $g(a)$ is the smallest closed element above a. Thus g maps atoms onto atoms. Moreover

$$g(x \vee y) = (x \vee y)^{**} = (x^* \wedge y^*)^* = (g(x)^* \wedge g(y)^*)^* = g(x) \oplus g(y)$$

Note that

$$g(x \wedge y) = (x \wedge y)^{**} = x^{**} \wedge y^{**} = g(x) \wedge g(y) \qquad \text{(Glivenko, 1929 see [CG00])}$$

The Glivenko mapping realizes a homomorphism of Theorem 4. The corresponding Boolean algebra is isomorphic to the skeleton of the pseudocomplemented lattice. Theorem 4 adds a converse to Glivenko's result.

4   Conclusion

A simple arrow configuration (see Figure 1) characterizes pseudo-complementedness of doubly founded concept lattices.

It leads to a characterization of atomic doubly founded pseudocomplemented lattices.

References

[BD74]   R. Balbes & P. Dwinger. *Distributive lattices.* University of Missouri Press. (1974).

[CG00]   I. Chajda & K. Glazek. *A basic course on general algebra.* Zielona Góra: Technical University Press. (2000).

[CM93]   C. Chameni Nembua & B. Monjardet. *Finite pseudocomplemented lattices and "permutoedre".* Discrete Math. **111**, No.1-3, 105-112 (1993).

[GW99]   B. Ganter & R. Wille. *Formal Concept Analysis – Mathematical Foundations.* Springer (1999).

[Gr71]   G. Grätzer.   *Lattice Theory. First concepts and distributive lattices.* W. H. Freeman and Company (1971).

[Ka80]   T. Katrinak. *P-algebras. Contributions to lattice theory, Szeged/Hung. 1980,* Colloq. Math. Soc. Janos Bolyai **33**, 549-573 (1983).

[Kw04]   L. Kwuida. *Dicomplemented lattices. A contextual generalization of Boolean algebras.* Dissertation, TU Dresden (2004).

[Lee70]   K. B. Lee. *Equational classes of distributive pseudo-complemented lattices.* Can. J. Math. **22**, 881-891 (1970).

[Sc88]   J. Schmid. *Lee classes and sentences for pseudo-complemented semilattices.* Algebra Univers. **25**, No.2, 223-232 (1988).

[So98]   V. Sofronie-Stokkermans. *Representation theorems and automated theorem proving in certain classes of non-classical logics.* Proceedings of the ECAI-98 workshop on many-valued Logic for AI applications.

Glossary

- A *formal context* (G, M, I) consists of two sets G, M, and a binary relation $I \subseteq G \times M$ between these two sets.

  The elements of G are usually called the *objects* of the formal context (G, M, I), while those of M are the *attributes*.

  We write $g \, I \, m$ or $(g, m) \in I$ to express that g and m are in relation I.

  $g \, I \, m$ may be read as "the object g *has* the attibute m".
- If $A \subseteq G$ is a set of objects, then

$$A' := \{m \in M \mid g \, I \, m \text{ for all } g \in A\}$$

  is the set of attributes that are common to all objects in A. Dually, if $B \subseteq M$ is a set of attributes, then

$$B' := \{g \in G \mid g \, I \, m \text{ for all } m \in B\}$$

  is the set of those objects in G that have all the attributes from B.
- The two operators

$$A \mapsto A' \qquad \text{and} \qquad B \mapsto B'$$

  form a *Galois connection*. Their compositions

$$A \mapsto A'' \qquad \text{and} \qquad B \mapsto B''$$

  form *closure operators* on G and M, respectively.
- A *formal concept* of a formal context (G, M, I) is a pair (A, B) of sets $A \subseteq G$, $B \subseteq M$, with $A' = B$ and $A = B'$.

  The set A is called the *extent* of the formal concept (A, B). The set B is its *intent*.
- Formal concepts are naturally ordered by the *subconcept-superconcept relation* defined as follows:

$$(A_1, B_1) \leq (A_2, B_2) : \Longleftrightarrow A_1 \subseteq A_2 \quad (\Longleftrightarrow B_2 \subseteq B_1).$$

- The set $\mathfrak{B}(G, M, I)$ of all formal concepts of a formal context (G, M, I) with this order is called the *concept lattice* of (G, M, I).

  It is indeed a complete lattice in the sense of mathematical order theory. Some authors use the name *Galois lattice of the relation I* instead of "concept lattice of (G, M, I)".

Author Index

Lecture Notes in Artificial Intelligence (LNAI)

Vol. 3155: P. Funk, P.A. González Calero (Eds.), Advances in Case-Based Reasoning. XIII, 822 pages. 2004.

Vol. 3139: F. Iida, R. Pfeifer, L. Steels, Y. Kuniyoshi (Eds.), Embodied Artificial Intelligence. IX, 331 pages. 2004.

Vol. 3131: V. Torra, Y. Narukawa (Eds.), Modeling Decisions for Artificial Intelligence. XI, 327 pages. 2004.

Vol. 3127: K.E. Wolff, H.D. Pfeiffer, H.S. Delugach (Eds.), Conceptual Structures at Work. XI, 403 pages. 2004.

Vol. 3123: A. Belz, R. Evans, P. Piwek (Eds.), Natural Language Generation. X, 219 pages. 2004.

Vol. 3120: J. Shawe-Taylor, Y. Singer (Eds.), Learning Theory. X, 648 pages. 2004.

Vol. 3097: D. Basin, M. Rusinowitch (Eds.), Automated Reasoning. XII, 493 pages. 2004.

Vol. 3071: A. Omicini, P. Petta, J. Pitt (Eds.), Engineering Societies in the Agents World. XIII, 409 pages. 2004.

Vol. 3070: L. Rutkowski, J. Siekmann, R. Tadeusiewicz, L.A. Zadeh (Eds.), Artificial Intelligence and Soft Computing - ICAISC 2004. XXV, 1208 pages. 2004.

Vol. 3068: E. André, L. Dybkjær, W. Minker, P. Heisterkamp (Eds.), Affective Dialogue Systems. XII, 324 pages. 2004.

Vol. 3067: M. Dastani, J. Dix, A. El Fallah-Seghrouchni (Eds.), Programming Multi-Agent Systems. X, 221 pages. 2004.

Vol. 3066: S. Tsumoto, R. Słowiński, J. Komorowski, J.W. Grzymała-Busse (Eds.), Rough Sets and Current Trends in Computing. XX, 853 pages. 2004.

Vol. 3065: A. Lomuscio, D. Nute (Eds.), Deontic Logic in Computer Science. X, 275 pages. 2004.

Vol. 3060: A.Y. Tawfik, S.D. Goodwin (Eds.), Advances in Artificial Intelligence. XIII, 582 pages. 2004.

Vol. 3056: H. Dai, R. Srikant, C. Zhang (Eds.), Advances in Knowledge Discovery and Data Mining. XIX, 713 pages. 2004.

Vol. 3055: H. Christiansen, M.-S. Hacid, T. Andreasen, H.L. Larsen (Eds.), Flexible Query Answering Systems. X, 500 pages. 2004.

Vol. 3048: P. Faratin, D.C. Parkes, J.A. Rodríguez-Aguilar, W.E. Walsh (Eds.), Agent-Mediated Electronic Commerce V. XI, 155 pages. 2004.

Vol. 3040: R. Conejo, M. Urretavizcaya, J.-L. Pérez-de-la-Cruz (Eds.), Current Topics in Artificial Intelligence. XIV, 689 pages. 2004.

Vol. 3035: M.A. Wimmer (Ed.), Knowledge Management in Electronic Government. XII, 326 pages. 2004.

Vol. 3034: J. Favela, E. Menasalvas, E. Chávez (Eds.), Advances in Web Intelligence. XIII, 227 pages. 2004.

Vol. 3030: P. Giorgini, B. Henderson-Sellers, M. Winikoff (Eds.), Agent-Oriented Information Systems. XIV, 207 pages. 2004.

Vol. 3029: B. Orchard, C. Yang, M. Ali (Eds.), Innovations in Applied Artificial Intelligence. XXI, 1272 pages. 2004.

Vol. 3025: G.A. Vouros, T. Panayiotopoulos (Eds.), Methods and Applications of Artificial Intelligence. XV, 546 pages. 2004.

Vol. 3020: D. Polani, B. Browning, A. Bonarini, K. Yoshida (Eds.), RoboCup 2003: Robot Soccer World Cup VII. XVI, 767 pages. 2004.

Vol. 3012: K. Kurumatani, S.-H. Chen, A. Ohuchi (Eds.), Multi-Agents for Mass User Support. X, 217 pages. 2004.

Vol. 3010: K.R. Apt, F. Fages, F. Rossi, P. Szeredi, J. Váncza (Eds.), Recent Advances in Constraints. VIII, 285 pages. 2004.

Vol. 2990: J. Leite, A. Omicini, L. Sterling, P. Torroni (Eds.), Declarative Agent Languages and Technologies. XII, 281 pages. 2004.

Vol. 2980: A. Blackwell, K. Marriott, A. Shimojima (Eds.), Diagrammatic Representation and Inference. XV, 448 pages. 2004.

Vol. 2977: G. Di Marzo Serugendo, A. Karageorgos, O.F. Rana, F. Zambonelli (Eds.), Engineering Self-Organising Systems. X, 299 pages. 2004.

Vol. 2972: R. Monroy, G. Arroyo-Figueroa, L.E. Sucar, H. Sossa (Eds.), MICAI 2004: Advances in Artificial Intelligence. XVII, 923 pages. 2004.

Vol. 2969: M. Nickles, M. Rovatsos, G. Weiss (Eds.), Agents and Computational Autonomy. X, 275 pages. 2004.

Vol. 2961: P. Eklund (Ed.), Concept Lattices. IX, 411 pages. 2004.

Vol. 2953: K. Konrad, Model Generation for Natural Language Interpretation and Analysis. XIII, 166 pages. 2004.

Vol. 2934: G. Lindemann, D. Moldt, M. Paolucci (Eds.), Regulated Agent-Based Social Systems. X, 301 pages. 2004.

Vol. 2930: F. Winkler (Ed.), Automated Deduction in Geometry. VII, 231 pages. 2004.

Vol. 2926: L. van Elst, V. Dignum, A. Abecker (Eds.), Agent-Mediated Knowledge Management. XI, 428 pages. 2004.

Vol. 2923: V. Lifschitz, I. Niemelä (Eds.), Logic Programming and Nonmonotonic Reasoning. IX, 365 pages. 2003.

Vol. 2915: A. Camurri, G. Volpe (Eds.), Gesture-Based Communication in Human-Computer Interaction. XIII, 558 pages. 2004.

Vol. 2913: T.M. Pinkston, V.K. Prasanna (Eds.), High Performance Computing - HiPC 2003. XX, 512 pages. 2003.

Vol. 2903: T.D. Gedeon, L.C.C. Fung (Eds.), AI 2003: Advances in Artificial Intelligence. XVI, 1075 pages. 2003.

Vol. 2902: F.M. Pires, S.P. Abreu (Eds.), Progress in Artificial Intelligence. XV, 504 pages. 2003.

Vol. 2892: F. Dau, The Logic System of Concept Graphs with Negation. XI, 213 pages. 2003.

Vol. 2891: J. Lee, M. Barley (Eds.), Intelligent Agents and Multi-Agent Systems. X, 215 pages. 2003.

Vol. 2882: D. Veit, Matchmaking in Electronic Markets. XV, 180 pages. 2003.

Vol. 2872: G. Moro, C. Sartori, M.P. Singh (Eds.), Agents and Peer-to-Peer Computing. XII, 205 pages. 2004.

Vol. 2871: N. Zhong, Z.W. Raś, S. Tsumoto, E. Suzuki (Eds.), Foundations of Intelligent Systems. XV, 697 pages. 2003.